USMG   NATO ASI

LES HOUCHES

Session XXXVI

1981

COMPORTEMENT CHAOTIQUE
DES SYSTEMES DETERMINISTES

CHAOTIC BEHAVIOUR
OF DETERMINISTIC SYSTEMS

CONFÉRENCIERS

M.V. BERRY
A. CHENCINER
M. HÉNON
M. HERMAN
A. KATOK
O.E. LANFORD III
M. MISIUREWICZ
S. NEWHOUSE

J.P. Eckmann, U. Frisch, J.P. Gollub, D.D. Joseph, A. Libchaber,
R.M. May, S.A. Orszag, Y. Pomeau, D. Ruelle

Université de Grenoble.
École d'été de Physique
Théorique,
Publications,
Sep/E
PHYS

USMG

NATO ASI

LES HOUCHES

SESSION XXXVI

29 Juin–31 Juillet 1981

COMPORTEMENT CHAOTIQUE
des systèmes déterministes

CHAOTIC BEHAVIOUR
of deterministic systems

édité par

GERARD IOOSS, ROBERT H.G. HELLEMAN, RAYMOND STORA

1983

NORTH-HOLLAND PUBLISHING COMPANY

AMSTERDAM · NEW YORK · OXFORD · TOKYO

69880049
PHYS

©NORTH-HOLLAND PUBLISHING COMPANY, 1983

All rights reserved. No part of this publication may be reproduced, stored in a retrieval system or transmitted, in any form or by any means, electronic, mechanical, photocopying, recording or otherwise, without the prior permission of the copyright owner.

ISBN 0 444 86542 x

Published by:
NORTH-HOLLAND PUBLISHING COMPANY
AMSTERDAM·NEW YORK·OXFORD

Sole distributors for the USA and Canada:
ELSEVIER SCIENCE PUBLISHING COMPANY, INC.
52 VANDERBILT AVENUE
NEW YORK, N.Y. 10017

Library of Congress Cataloging in Publication Data
Main entry under title:

Comportement chaotique des systèmes déterministes.

English and French.
At head of title: USMG, NATO ASI.
"Les Houches, Ecole d'été de physique théorique ...,
session XXXVI" — P. v.
Includes bibliographies.
1. Chaotic behavior in systems—Congresses.
2. Dynamics—Congresses. 3. Nonlinear theories—
Congresses. I. Iooss, Gérard. II. Helleman, Robert H.G.
III. Stora, Raymond. IV. NATO Advanced Study Institute.
V. Ecole d'été de physique théorique (France).
VI. Title: Chaotic behaviour of deterministic systems.
QA843.C65 1983 531'.11 83-11434
ISBN 0-444-86542-X

Printed in The Netherlands

QA843
C65
1983
PHYS

LES HOUCHES
ÉCOLE D'ÉTÉ DE PHYSIQUE THÉORIQUE

ORGANISME D'INTÉRÊT COMMUN DE L'UNIVERSITÉ
SCIENTIFIQUE ET MÉDICALE DE GRENOBLE ET DE
L'INSTITUT NATIONAL POLYTECHNIQUE DE
GRENOBLE

AIDÉ PAR LE COMMISSARIAT A L'ÉNERGIE
ATOMIQUE

Membres du conseil: J.J. Payan, président, D. Bloch, vice-président,
P. Averbuch, R. Balian, M.-Th. Béal-Monod, C. DeWitt, J. Ducuing,
J. Gavoret, S. Haroche, M. Jacob, M. Jean, R. Pauthenet, Y. Rocard,
M. Soutif, J. Teillac, D. Thoulouze, G. Weill

Directeur: Raymond STORA, Division Théorique, CERN, CH 1211
Genève 23

SESSION XXXVI

INSTITUT D'ÉTUDES AVANCÉES DE L'OTAN
NATO ADVANCED STUDY INSTITUTE
29 Juin–31 Juillet 1981

Directeurs scientifiques de la session: Gerard Iooss, Institut de Mathémati-
ques et Sciences Physiques, Université de Nice, Parc Valrose, F06034
Nice Cédex
Robert H.G. Helleman, Department of Theoretical Physics, Twente
University of Technology, P.O. Box 217, 7500 AE Enschede, The
Netherlands (et The La Jolla Institute, P.O. Box 1434, La Jolla, CA
92038, U.S.A.)

SESSIONS PRÉCÉDENTES

* Sessions ayant reçu l'appui du Comité Scientifique de l'OTAN.

PARTICIPANTS

Baesens, Claude, Université Libre de Bruxelles, Service de Chimie-Physique II, C.P. 231 Campus Plaine, Boulevard du Triomphe, 1050 Bruxelles, Belgium.

Benettin, Giancarlo, Istituto di Fisica dell'Università, Via Marzolo 8, 35100 Padova, Italy.

Campanino, Massimo, Istituto Matematico "G. Castelnuovo", Università degli studi di Roma, Piazzale A. Moro, Roma, Italy.

Casasayas Mas, Josefina, Universitat de Barcelona, Facultat de Matemàtiques, Departament de Teoria de Funcions, Placa Universitat, Barcelona, Spain.

Collet, Pierre, Physique Théorique, Ecole Polytechnique, 91128 Palaiseau Cedex, France.

Crutchfield, James, Physics Board of Studies, University of California, Santa Cruz, C.A. 95064, U.S.A.

Curry, James, Mathematics Department, Box 426, University of Colorado, Boulder, Colorado 80309, U.S.A.

Douady, Raphael, Ecole Normale Supérieure, 45, rue d'Ulm, 75005 Paris, France.

Doveil, Fabrice, Laboratoire PMI, Ecole Polytechnique, 91128 Palaiseau Cedex, France.

Froeschlé, Claude, Observatoire de Nice, B.P. 252, 06007 Nice Cedex, France.

Gambaudo, Jean-Marc, Faculté des Sciences de Nice, Parc Valrose, 06034 Nice Cedex, France.

Goroff, Daniel, Department of Mathematics, Fine Hall Box 37, Princeton University, Princeton, N.J. 08544, U.S.A.

Hannay, Jonathan, H.H. Wills Physics Laboratory, Tyndall Avenue, Bristol BS8 ITL, England.

Heldstab, Juerg, Institut f. Theoretische Physik, Klingelbergstrasse 82, CH-4056 Basel, Switzerland.

Jowett, John, Division ISR, CERN, CH-1211 Geneve 23, Switzerland.

King, Peter, TCM Group, Cavendish Laboratory, Madingley Road, Cambridge CB3 OHE, England.

Langford, William F., Department of Mathematics and Statistics, University of Guelph, Guelph, Ontario, Canada.

Le Calvez, Patrice, Ecole Normale Supérieure, 45, rue d'Ulm, 75005 Paris, France.

Legras, Bernard, Laboratoire de Météorologie Dynamique, Ecole Normale Supérieure, 24, rue Lhomond, 75231 Paris Cedex 05, France.

De La Llave, Rafael, Department of Mathematics, Fine Hall, Princeton, N.J. 08544, U.S.A.

McCreadie, Geoffrey, Department of Physics, University of Warwick, Coventry CV4 7AL, England.

MacKay, Robert, Plasma Physics Lab., P.O. Box 451, Princeton University, Princeton, N.J. 08544, U.S.A.

Mayer-Kress, Gottfried, 1. Inst. Theor. Physik, Universität Stuttgart, Pfaffenwaldring 57/4, D-7000 Stuttgart 80, R.F.A.

Nadzieja, Tadeusz, Uniwersytet Wrocławski, Instytut Matematyczny, Wrocław, pl. Grunwaldzki 2/4, Poland.

Newman, Richard, Department of applied Mathematics, Queen Mary College, Mile End Road, London E14NS, England.

Nierwetberg, Johannes, Fakultät Physik, Universität Regensburg, Universitätsstr., D-8400 Regensburg, R.F.A.

Noronha Da Costa, Ana Maria, CFMC, Av. Prof. Gama Pinto, 2, 1699 Lisboa-codex, Portugal.

Ozorio de Almeida, Alfredo, Instituto de Fisica "Gleb Wataghin", Universidade Estadual de Campinas, Campinas, 13100, SP, Brazil.

Patera, Anthony, Math. Dept., M.I.T., 77 Mass. Ave, Cambridge, M.A. 02139, U.S.A.

Pelce, Pierre, Université de Provence, Centre de St. Jérôme, Rue Henri Poincaré, 13397 Marseille Cedex 13, and: E.N.S. rue d'Ulm, 75005 Paris, France.

Pignataro, Thea, Physics Department, Jadwin Hall, Princeton University, P.O. Box 708 Princeton, N.J. 08544, U.S.A.

Politi, Antonio, Istituto Nazionale di Ottica, Largo E. Fermi 6, 50125 Firenze, Italy.

Ramaswamy, Ramakrishna, Chemical Physics Group, T.I.F.R., Homi Bhabha Road, Colaba, Bombay 400 005, India.

Rapp, Paul, Dept. Physiology and Biochemistry, Medical College of Pennsylvania, 3300 Henry Ave., Philadelphia, P.A. 19129, U.S.A.

Robnik, Marko, Institut F. Astrophysik, Universität Bonn, auf dem Hügel 71, D-5300 BONN 1, F.R.G.

Santini, Paolo, Istituto di Fisica "G. Marconi", Piazzale A. Moro, 00185 Roma, Italy.

Shraiman, Boris, Jefferson Lab., Department of Physics, Harvard University, Cambridge, M.A. 02138, U.S.A.

Sivaramakrishnan, Anand, Astronomy Department, University of Texas, Austin, TX 78712, U.S.A.

Siegberg, Hans, University of Bremen, Fachbereich Mathematik, Forschungsschwerpunkt "Dynamische Systeme", 2800 Bremen, B.R. Deutschland.

Sommeria, Joël, Institut de Mécanique de Grenoble, B.P. 53, 38041 Grenoble Cedex, France.

Strelcyn, Jean-Marie, Departement de Mathématiques, Centre Scientifique et Polytechnique de l'Univ. Paris-Nord, Av. J.-B. Clément, 93430 Villetaneuse, France.

Szewc, Bolesław, SGGW-Warsaw Agricultural University, Inst. of Appl. Math. and Statyst., Nowoursynowska 166, 02-766 Warsaw, Poland.

Tresser, Charles, Equipe de Mécanique Statistisque, Parc Valrose, 06034 Nice Cedex France.

Ushiki, Shigehiro, Department of Mathematics, Faculty of Science, Kyoto University, 606 Kyoto, Japan.

Van Zeijts, Johannes B.J., Twente Univ. of Technology, P.O. Box 217, 7500 AE Enschede, The Netherlands.

Voros, André, DPhT-CEA, CEN Saclay, 91191 Gif-sur-Yvette Cedex, France.

Wayne, Clarence, Department of Physics, Harvard University, Cambridge, M.A. 02138, U.S.A.

Wisdom, Jack, 170-25 Caltech, Pasadena, C.A. 91125, U.S.A.

LECTURERS

Berry, Michael, H.H. Wills Physics Laboratory, Tyndall Avenue, Bristol BS8 1TL, England.

Chenciner, Alain, Département de Mathématiques, Université Paris VII, 2 place Jussieu, 75251 Paris cedex 05, France.

Eckmann, Jean-Pierre, Département de Physique Théorique, Université de Genève, 2 quai Ernest Ansermet, CH-1211 Genève 4, Switzerland.

Frisch, Uriel, Observatoire de Nice, B.P. 252, 06007 Nice cedex, France.

Gollub, Jerry P., Haverford College, Physics Department, Haverford, P.A. 19041, U.S.A.

Hénon, Michel, Observatoire de Nice, B.P. 252, 06007 Nice cedex, France.

Herman, Michael, Centre de Mathématiques, Ecole Polytechnique, 91128 Palaiseau cedex, France.

Joseph, Daniel D., Department of Aerospace engineering and Mechanics, Akerman Hall, University of Minnesota, Minneapolis, MN 55455, U.S.A.

Katok, Anatole, Department of Mathematics, University of Maryland, College Park, M.D. 20742, U.S.A.

Lanford III, Oscar E., Department of Mathematics, University of California, Berkeley, C.A. 94720, U.S.A.

Libchaber, Albert, Ecole Normale Supérieure, Laboratoire de Physique, 24, rue Lhomond, 75005 Paris, France.

May, Robert M., Department of Biology, Princeton University, Princeton, N.J. 08540, U.S.A.

Misiurewicz, Michał, Institute of Mathematics, Warsaw University, PKiN IX p., 00–901 Warszawa, Poland.

Newhouse, Sheldon, Mathematic department, University of North Carolina, Chapel Hill, N.C. 27514, U.S.A.

Orszag, Steven A., Dept. Mathematics, M.I.T., Cambridge, MA 02139, U.S.A.

Pomeau, Yves, DPhT-CEA, Orme des Merisiers, 91191 Gif-sur-Yvette
cedex, France.

Ruelle, David, I.H.E.S., 91440 Bures-sur-Yvette, France.

PRÉFACE

Le titre de cet ouvrage: *"Comportement chaotique des systèmes déterministes"*, peut sembler contradictoire à ceux qui ne connaissent de la Mécanique que les cours usuels. La dernière décade a vu des progrès aussi nombreux que passionnants en Dynamique non-linéaire. Il est apparu d'après ces nouveaux résultats que la plupart des systèmes mécaniques déterministes, avec plus de un degré de liberté, présentent une transition d'un comportement bien régulier (périodique ou quasi-périodique par exemple comme le montrent les ouvrages classiques de mécanique) vers un comportement chaotique. Ce dernier pourrait être appelé comportement "ergodique" s'il intervient dans un système conservatif, ou comportement "turbulent" s'il intervient dans un système dissipatif. En fait on peut démontrer que certaines propriétés de nombreuses orbites de tels systèmes déterministes sont *aléatoires*! La transition d'un comportement régulier vers un comportement chaotique devient apparente lorsqu'on fait croître un certain paramètre μ, e.g. l'Energie dans un système conservatif, ou un paramètre comme le nombre de Reynolds dans un système dissipatif. Bien que quelques articles de revue sur le comportement chaotique aient paru dans la littérature de Physique Théorique, nous pensons que ces notes constituent le premier ouvrage imprimé, qui contienne simultanément un survol des résultats abstraits et des applications, et qui soit susceptible de servir de base à des cours destinés à des étudiants de bon niveau. La plupart des participants à cette Ecole d'Eté des Houches étaient en fait étudiants de Physique Théorique et Physique-Mathématique au niveau du PhD.

Il est prématuré de présenter un historique (fût-il bref) de ce domaine neuf. En revanche, l'histoire générale de la Mécanique ou Dynamique Non-linéaire, est trop longue et trop riche pour être décrite ici alors qu'elle est bien traitée dans plusieurs livres (voir par exemple, pour une version moderne, l'Introduction de "Foundation of Mechanics", 2nd Ed., par R. Abraham et J. Marsden).

Nous nous contenterons donc dans ce qui suit de donner une description simplifiée de la classe des phénomènes étudiés dans le présent ouvrage.

Le comportement chaotique est inhérent à la plupart des systèmes mécaniques puisque beaucoup d'orbites ont une *Dépendance très Sensible par rapport aux conditions initiales.* Cela signifie qu'une modification infinitésimale des conditions initiales peut causer rapidement de grandes modifications de l'orbite qui en est issue. Ceci implique qu'indépendamment de la connaissance du comportement passé, nous ne pouvons pas prédire beaucoup du comportement à venir de telles orbites, à cause des erreurs expérimentales lors de ces observations. Un exemple familier est le billard incliné à plots (sans flippers et avec de vrais plots) où l'on a les conditions initiales de la boule métallique entre les mains, littéralement. Cependant, quel que soit le soin pris pour frapper la boule, nous ne pouvons pas prédire quels plots seront heurtés et où la boule tombera finalement. Néanmoins, en répétant l'"expérience" nous trouvons que la répartition de l'endroit où la boule tombe suit une loi de probabilité bien déterminée. C'est donc en termes de probabilités et statistiques que le résultat de cette expérience de Mécanique déterministe doit être décrit.

Un comportement chaotique semblable peut intervenir même à partir d'équations du mouvement plus régulières que celles de l'exemple précédent (coins, bords des plots,...): les équations différentielles analytiques sans aucune singularité engendrent le même type de phénomènes. Dans les cours ci-après de tels systèmes aussi bien dissipatifs que conservatifs sont discutés.

Dans les deux cas les applications potentielles sont abondantes. Pour certaines, on souhaite comprendre le comportement chaotique afin de l'*éviter*, e.g. pour certaines applications d'ingénierie mécanique et électrique, de mécanique céleste (orbites de satellites), de dynamique des populations, d'anneaux de stockage en physique des hautes énergies, d'hydrodynamique, de physique des plasmas (fusion), de biophysique, etc.... Pour d'autres applications on souhaite au contraire *obtenir* un comportement chaotique, e.g. pour certaines applications de mécanique statistique classique (et quantique), d'hydrodynamique (turbulence), de cinétique chimique, etc.... Des problèmes aussi nombreux que différents, existent dans l'un ou l'autre de ces domaines de la Dynamique non-linéaire. Il est ainsi quelque peu surprenant, mais tout à fait fascinant, de voir qu'une approche commune des problèmes chaotiques de tous ces domaines, ergodiques et turbulents, est en train de se développer. Il est d'ailleurs encore plus surprenant que plusieurs des phénomènes chaotiques découverts dans des modèles à un petit nombre de variables, dont il est question tout au long de ce livre, soient en accord avec certains résultats expérimentaux en turbulence, cf. les cours de A. Libchaber et J.P. Gollub, et celui de Y. Pomeau.

Les outils mathématiques de base utilisés dans l'étude du comportement chaotique sont présentés par O.E. Lanford lors du premier cours. Le comportement chaotique dans les systèmes conservatifs est étudié principalement dans les cours de M. Hénon, et M.V. Berry (traitement quantique).

Il est effleuré lors des cours de A. Chenciner et S. Newhouse. Pour la classe la plus simple de systèmes dissipatifs: "Les Applications de l'Intervalle", il y a une théorie très développée, traitée ici dans les cours de J.P. Eckmann et M. Misiurewicz, et utilisée dans les cours de R.M. May et Y. Pomeau. Les systèmes dissipatifs plus compliqués avec un grand nombre de degrés de liberté peuvent souvent se réduire à un petit nombre, grâce à la "Théorie des Bifurcations". Ce sujet est traité dans les cours de A. Chenciner et D.D. Joseph. Les bases mathématiques pour l'étude du comportement chaotique dans les équations différentielles et aux différences sont traitées en détail dans le cours de S. Newhouse. Enfin, les cours sur la vraie et bien concrète turbulence par S.A. Orszag, U. Frisch, A. Libchaber et J.P. Gollub, nous montrent l'importance et le nombre des problèmes qui restent à résoudre sur la base des principes fondamentaux développés ci-dessus.

Il est néanmoins passionnant de détecter des progrès sur ces anciens problèmes de comportement chaotique et de confronter de façon satisfaisante la théorie avec l'expérience, comme indiqué ci-dessus.

Le programme initial de cette session comprenait d'importantes contributions par D.V. Anosov, V.I. Arnold, Ya G. Sinaï, d'une part, J. Moser, d'autre part. Celles-ci ne sont malheureusement pas incluses dans ce volume, leurs auteurs n'ayant pu se libérer comme prévu. Nous déplorons également l'absence des notes du cours de A. Katok ("Hyperbolicity in smooth dynamical systems") et de celui de M. Herman ("Introduction to area-preserving diffeomorphisms of 2-manifolds).

48 étudiants appartenant à 15 nations et 16 conférenciers venant de 5 pays ont participé à cette session de l'Ecole de Physique Théorique des Houches, la 36e depuis la création de l'école.

Remerciements

La réalisation de la session XXXVI de l'Ecole des Houches et la publication de ce volume sont le résultat de nombreuses contributions:
  – le soutien financier de l'Université Scientifique et Médicale de Grenoble, et les subventions de la Division des Affaires Scientifiques de l'OTAN, qui a inclus cette session dans son programme d'Instituts d'Etudes Avancées, et du Commissariat à l'Energie Atomique;
  – l'orientation et le soutien effectif du Conseil de l'Ecole;
  – le soin apporté par Marie-Claude Pophillat dans la préparation et la frappe des manuscripts;
  – la coopération de tous les participants qui ont contribué de façon inestimable à la richesse du programme scientifique, par leur aide active au cours de l'élaboration des notes de cours et les compléments sous forme de séminaires;
  – l'aide d'Henri et Nicole Coiffier, d'Anny Battendier, et de toute l'équipe qui a rendu la vie de tous les jours aussi confortable que possible;
  – Thea Pignataro pour son aimable contribution au musée photographique de la session et Gottfried Mayer-Kress pour son graphisme dynamique.
  Nous tenons finalement à remercier les conférenciers pour avoir donné leur temps à la préparation et à la rédaction des cours et pour avoir fait preuve d'une inépuisable disponibilité tout au long de la session.

G. Iooss
R. Helleman
R. Stora

PREFACE

The title of this volume, *"Chaotic Behavior of Deterministic Systems"*, might seem like a contradiction-in-terms to some people who know mechanics from a standard graduate course. Over the past ten years however, many exciting developments have taken place in Nonlinear Dynamics and in Mechanics in general. From these new results it has become apparent that most deterministic mechanical systems (with more than one degree of freedom) exhibit a transition from the nice regular (quasi-)periodic behavior we all know from our graduate mechanics textbooks, to Chaotic Behavior. The latter might be called "ergodic" behavior when occurring in a conservative system or "turbulent" behavior when occurring in a dissipative system. Actually some properties of many orbits of such (deterministic) systems can be proven to be *random*! The transition from regular to chaotic behavior becomes clearly visible as we increase some parameter μ, e.g. the Energy in a conservative system or a parameter like the Reynolds Number in a dissipative system. While some review articles on chaotic behavior have appeared in the theoretical physics literature, we expect these Proceedings to be the first printed book, providing an overview of abstract results as well as applications, which can also serve as a textbook for serious students. Most participants in this Les Houches Summer School were in fact students of theoretical and mathematical physics at, or near, the PhD level. In view of the above it would be premature to present here the (brief) history of this new field. The general history of Mechanics or Nonlinear Dynamics, on the other hand, is too long and rich to describe here and is well treated in many texts (for a modern version see, e.g., the Introduction to "Foundations of Mechanics", 2nd Ed., by R. Abraham and J. Marsden).

Chaotic behavior is inherent in most mechanical systems since many orbits have a *"Sensitive Dependence on their Initial Conditions"*. It means that (arbitrarily) small changes in the initial conditions can rapidly cause very large changes in the ensuing orbit. This implies that no matter how much of the past behavior we experimentally observe, we cannot predict much of the future behavior of such orbits, due to the experimental error in that observation. One familiar exemple is the pinball machine (i.e. an old-fashioned one, without flippers and with real pins) where we have the initial conditions of the metal ball in our own hands, literally. Yet, no matter how careful we are in shooting away the ball, we cannot predict which pins will be hit or where the ball will land at the bottom. Nevertheless repeating this "experiment" we find that the balls spread out over the bottom according to a certain probability distribution. So, at the moment the best way to describe the outcome of this deterministic mechanics experiment is in terms of probability and statistics. Similar chaotic behavior can arise from even nicer equations of motion, i.e. from analytic differential equations without any of the above singularities (at the edge of the pins and balls). In these Proceedings many such dissipative as well as conservative systems are discussed. In both cases important potential applications abound. In some cases one would like to understand chaotic behavior in order to *avoid* it, e.g. for certain applications of mechanical and electrical engineering, celestial mechanics (satellite orbits), population dynamics, the storage rings of high energy physics, hydrodynamics, plasma physics (fusion), biophysics, etc. In other cases one would like to *obtain* chaotic behavior, e.g. for certain applications of classical (and quantum) statistical mechanics, hydrodynamics (turbulence), chemical kinetics, etc. Many diverse and notorious problems of Nonlinear Dynamics exist in one or another of those fields. Hence it is somewhat surprising, but utterly fascinating, to see that a common approach is being developed to the chaotic, ergodic and turbulent problems of all those fields. It is even more surprising that several of the chaotic phenomena discovered in model equations, with few variables, discussed at length in these proceedings, already agree with some experimental turbulence results, of the lectures by A. Libchaber and J.P. Gollub and by Y. Pomeau.

The basic mathematical tools used in the study of chaotic behavior are introduced by O.E. Lanford in the opening lecture. Chaotic behavior in conservative systems is mainly discussed in the lectures of M. Hénon, and M.V. Berry (quantum treatment), and is touched upon in the lectures of A. Chenciner and S. Newhouse. For the simplest class of dissipative systems, "Mappings of the Interval", there exists a well developed theory which is treated here in the lectures of J.P. Eckmann and M. Misiurewicz and is employed in the lectures of R.M. May and Y. Pomeau. More complicated dissipative systems with many degrees of freedom can often be reduced thanks to "Bifurcation theory". This subject is treated in the lectures of A. Chenciner and D.D. Joseph. The mathematical basis for chaotic behavior in difference- and differential equations is treated in detail in the lectures of S. Newhouse. Finally, the lectures on full fledged real turbulence, by S.A. Orszag, U. Frisch, A. Libchaber and J.P. Gollub show us how many outstanding problems still remain to be explained from first principles. Nevertheless it is exciting to detect progress on these old problems of chaotic behavior and see some agreement with experiment, as noted above.

The important contributions to the modern theory of dynamical systems by D.V. Anosov, V.I. Arnold, Ya G. Sinaï on the one hand, J. Moser on the other hand, which had been planned and expected during the preparation of the session, are unfortunately absent from these proceedings due to unexpected difficulties. We are also sorry that the notes by A. Katok ("Hyperbolicity in smooth dynamical systems") and by M. Herman ("Introduction to area-preserving diffeomorphisms of 2-manifolds), could not be included in this volume.

In this session of the Les Houches Summer School, the 36th since its creation, 48 students from 15 nations, and 16 lecturers from 5 nations participated.

Acknowledgements

The XXXVI Session of the Les Houches Summer School and the publication of this volume of lecture notes would not have been possible without:
  — the financial support from the Université Scientifique et Médicale de Grenoble, the NATO Scientific Affairs Division (who included this session in its Advanced Study Institute Programme) and the Commissariat à l'Energie Atomique;
  — the guidance of the school board;
  — the careful preparation and typing of the manuscripts by Marie-Claude Pophillat;
  — the cooperation of all the participants who contributed in an invaluable manner to the whole scientific programme, including substantial help in the preparation of many of the lecture notes and providing complements

to the courses in the form of carefully chosen seminars;

– the help of Henri and Nicole Coiffier, Anny Battendier and the whole team, in making everyday's life as comfortable as possible;

– Thea Pignataro for kindly looking after the session's photographic museum and Gottfried Mayer-Kress for his dynamical graphism.

Finally we wish to thank the lecturers for giving so much of their time in preparing the lectures, and writing them up, as well as making themselves available for discussion throughout their stay in Les Houches.

Gérard Iooss
Robert Helleman
Raymond Stora

CONTENTS

Course 8. *Nonlinear problems in ecology and resource management, by R.M. May*   *513*

CHAOS IS IN

COURSE 1

INTRODUCTION TO THE MATHEMATICAL THEORY OF DYNAMICAL SYSTEMS

Oscar E. LANFORD III*[†]

*Department of Mathematics,
University of California, Berkeley, CA 94720, U.S.A.*

*Work supported in part by the National Science Foundation (MCS78-06718).
[†] Present address: IHES, 91440 Bures-sur-Yvette, France.

*G. Iooss, R.H.G. Helleman and R. Stora, eds.
Les Houches, Session XXXVI, 1981 − Comportement Chaotique des Systèmes Déterministes/
Chaotic Behaviour of Deterministic Systems*
© *North-Holland Publishing Company, 1983*

Contents

0. Introduction

These notes are intended as a brief survey of a few topics in the theory of differentiable dynamical systems and as preparation for subsequent lectures in this volume. All the topics treated are standard and the discussion is either standard or distinctly old-fashioned. More complete treatments may be found, for example, in Shub (1978), Bowen (1978), Bowen (1975) and Smale (1967).

I have tried to keep the level of discussion fairly informal, emphasizing central ideas rather than technical details. Accordingly, I have generally not tried to give the most refined versions of results, nor have I worried too much about making the hypotheses as weak as possible. Notably, I have generally not specified how smooth mappings need to be; except when the contrary is specifically stated, it should generally be assumed that any smooth mapping is infinitely differentiable. I have also confined my attention to iterates of a single mapping, ignoring special features of flows.

In general, T will denote a smooth mapping acting on a state space M. Except in a very few places, it does no harm at all to take M to be an open set in Euclidean space $\mathbb{R}^m$ and all formulas are written as if this were the case, in spite of the occasional inconsistencies this engenders. The reader familiar with the definition of smooth manifold should have no difficulty at all in translating my concrete and co-ordinate dependent statements into co-ordinate free form.

I am very grateful to Mr Rafael de la LLave for his invaluable assistance in preparing this text.

[See p. 51 for a note added in proof.]

1. Invariant manifolds

We will be concerned, in this section, with the behavior of orbits of a differentiable mapping in the neighborhood of a fixed point. The analysis can be applied to periodic points as well by considering the pth iterate of the mapping, where p is the period.

Given a differentiable mapping and a point z, the *derivative* $DT(z)$ means simply the Jacobian matrix of partial derivatives of the components of T at z. If δz is small, then,

$$T(z + \delta z) = T(z) + DT(z)\delta z + \text{higher order terms in } \delta z.$$

A linear operator L on a finite dimensional vector space E is said to be *hyperbolic* if its spectrum does not intersect the unit circle, i.e. if it has no eigenvalue of modulus one. A fixed point z_0 for T is said to be *hyperbolic* if the derivative $DT(z_0)$ is hyperbolic.

The following proposition is elementary linear algebra:

Proposition (1.1). Let L be a hyperbolic linear operator on a finite-dimensional real vector space E. We can then write

$$E = E_s \oplus E_u$$

where

$LE_s \subset E_s$ and the spectrum of $L\big|_{E_s}$ is inside the unit circle.

$LE_u \subset E_u$ and the spectrum of $L\big|_{E_u}$ is outside the unit circle.

E_s and E_u may be characterized as:

$$E_s = \{\xi \in E : L^n \xi \to 0 \quad \text{as } n \to +\infty\}$$

$$E_u = \{\xi \in E : L^{-n} \xi \to 0 \quad \text{as } n \to +\infty\} \text{ (assuming } L \text{ to be invertible)}.$$

The idea is simply that, for example, E_s is spanned by the eigenvectors (and generalized eigenvectors if L is not diagonalizable) with eigenvalues whose moduli are less than one. Note, however, that although the eigenvalues and eigenvectors may be complex, E_s and E_u are real subspaces.

The subspace E_s is called the *contracting* or *stable* subspace for L, E_u the *expanding* or *unstable* subspace. Observe that, contrary to naive intuition, E_u is not the space of all vectors which grow under iteration of L; any vector outside E_s grows asymptotically in an exponential way. It is for this reason that E_u has to be characterized in a round-about way as the space of vectors which decay under the iteration of L^{-1}.

Now let z_0 be a hyperbolic fixed point for T. For z near z_0, the action of T is given approximately by the linearization of T at z_0:

$$T(z_0 + \delta z) \approx z_0 + DT(z_0)\delta z.$$

To what extent are properties of iterates of $DT(z_0)$ reflected in properties of iterates of T itself? One important property which goes over is the existence of invariant *stable* and *unstable* manifolds which are nonlinear analogues of the contracting and expanding subspaces for $DT(z_0)$. Roughly, the *stable manifold* W^s for T at z_0 is a smooth surface with dimension equal to that of E_s, passing through z_0, tangent there to E_s, mapped into itself by T, and composed of points x such that $T^n z \to z_0$ as $n \to +\infty$. The *unstable manifold* W^u can be described economically as the stable manifold for T^{-1}.

Let us now make this more explicit. We note first that the problem of constructing a stable manifold splits into two parts, constructing a small piece of stable manifold in the vicinity of z_0 (local stable manifold), and then building up from it the whole stable manifold. We start with the local problem.

1.1. Preliminaries

Make a change of coordinates putting z_0 at the origin, and identifying the state space near z_0 with the unit ball in $E_s \oplus E_u$ so that

$$T(x, y) = [\Lambda_s x + t_s(x, y), \Lambda_u y + t_u(x, y)]$$

$x =$ component in E_s and $y =$ component in E_u. Here: the spectra of Λ_s and Λ_u^{-1} are inside the unit circle and t_s and t_u vanish together with their first derivatives at $(0,0)$.

Choose a norm on $E_s \oplus E_u$ so that

$$\|\Lambda_s\| \le \lambda < 1; \quad \|\Lambda_u^{-1}\| \le \lambda < 1.$$

A *local stable manifold* will be a surface $W^s_{(0)}$ defined as the graph of a mapping w_s from E_s to E_u

$$W^s_{(0)} = \{(x, w_s(x)) : x \text{ small}\}.$$

This surface will be required to pass through the fixed point $(0,0)$:

$$w_s(0) = 0$$

and to be a tangent there to E_s:

$$D_x w_s(x)\big|_{x=0} = 0$$

Theorem (1.2) (Local Stable Manifold Theorem). For sufficiently small positive δ, there is exactly one continuously differentiable function w_s defined on $\{x \in E_s : \|x\| < \delta\}$, taking values in E_u, such that

$$w_s(0) = 0,$$

$$D_x w_s(0) = 0,$$

$W_{(0)}^s = \{(x, w_s(x)) \text{ with } \|x\| < \delta\}$ is mapped into itself by T.

We have furthermore

Let $U_\delta = \{z = (x, y) : \|x\| < \delta, \|y\| < \delta\}$. If δ is small enough, then any point such that $T^n z \in U_\delta$ for $n = 0, 1, 2\ldots$ must be in $W_{(0)}^s$. (Any point in U_s but not in $W_{(0)}^s$ is eventually pushed out in the expanding direction.)

If $z \in W_{(0)}^s$, then $T^n z \to 0$ ($= z_0$).

In particular: any orbit $T^n z$ converging to z_0 must lie in $W_{(0)}^s$, for all sufficiently large n.

We now define the full stable manifold as

$$W^s = \{z : T^n z \to z_0 \text{ as } n \to \infty\}.$$

By the above, this is the same as

$$\{z : T^{n_0} z \in W_{(0)}^s \text{ for some } n_0\}$$

i.e.

$$W^s = \bigcup_{n_0 \geq 0} T^{-n_0} W_{(0)}^s.$$

In general, the global W^s is very complicated, doubles back on itself infinitely often, etc. (It cannot, however, cross itself.) The preceding formula shows that it is made up of infinitely many pieces each of which is a smooth surface with dimension equal to that of E_s.

1.2. Sketch of proof of local stable manifold theorem

We follow, loosely, Hartman (1964), Chapter IX. Notation can be simplified substantially if we employ a device which can be called magnification: working on a small neighborhood of the fixed point is equivalent to working on a neighborhood of fixed size and assuming that the nonlinear terms in T are small.

Write $T(x, y) = [\Lambda_s x + t_s(x, y), \Lambda_u y + t_u(x, y)]$ and write $x = \delta x'$, $y = \delta y'$.

In terms of the new coordinates, T becomes

$$T'(x', y') = \left(\Lambda_s x' + \frac{1}{\delta} t_s(\delta x', \delta y'), \Lambda_u y' + \frac{1}{\delta} t_u(\delta x', \delta y') \right)$$

so

$$DT' = DT(\delta x', \delta y').$$

Since the derivatives of nonlinear terms vanish at the origin and are continuous, we can, by taking δ small, arrange to make these derivatives as small as we like on

$$\{(x', y') : \|x'\| \le 1 ; \|y'\| \le 1\}.$$

Thus, dropping the primes, for any fixed positive θ, we can assume that

$$\|D_x t_s(x, y)\| \le \theta \qquad \text{for } \|x\| \le 1; \ \|y\| \le 1.$$
$${}_{y\ u}$$

We can treat θ as an "adjustable parameter" in the course of the argument. Ultimately, adjusting θ just means adjusting δ, i.e. determining the size of the neighborhood of the fixed point in which the construction works.

Step 1. The functional equation. Ask, semi-formally, what condition the function w must satisfy if its graph is to be mapped into itself by T. Given (x, y), write temporarily $(\bar{x}, \bar{y})$ for $T(x, y)$. The invariance condition is just:

$$y = w(x) \qquad \text{implies } \bar{y} = w(\bar{x}).$$

(We drop the subscript "s" on w to simplify the formulas.)
  Using the particular form of T in our co-ordinates:

$$\bar{y} = \Lambda_u y + t_u(x, y); \qquad \bar{x} = \Lambda_s x + t_s(x, y)$$

the equation $\bar{y} = w(\bar{x})$ becomes

$$\Lambda_u y + t_u(x, y) = w[\Lambda_s x + t_s(x, y)]$$

or

$$y = \Lambda_u^{-1} \{ w[\Lambda_s x + t_s(x, y)] - t_u(x, y) \}.$$

Using $y = w(x)$, we see that the condition for invariance reduces to

$$w(x) = \Lambda_u^{-1} \{ w[\Lambda_s x + t_s(x, w(x))] - t_u(x, w(x)) \}.$$

Hence, if we denote the right-hand side of this equation by $(\mathscr{F}w)(x)$, the condition for invariance is simply:

$$w = \mathscr{F}w.$$

Step 2. $\mathscr{F}$ is well defined. From

$$\|D_x t_s{}_u (x, y)\| \le \theta$$

it follows that

$$\|t_s{}_u (x, y)\| \le 2\theta$$

for $\|x\| \le 1$; $\|y\| \le 1$. Recall also that

$$\|A_s\| \le \lambda < 1; \quad \|A_u^{-1}\| \le \lambda < 1.$$

Then if $\lambda + 2\theta \le 1$ we have

$$\|A_s x + t_s(x, y)\| \le \lambda + 2\theta \le 1$$

for $\|x\| \le 1$, $\|y\| \le 1$. Hence, if $w(x)$ is defined and has norm ≤ 1 for $\|x\| \le 1$, $\mathscr{F}(x)$ is also defined for $\|x\| \le 1$.

Step 3. Propagated bounds. Define $\|w\| = \sup \{\|w(x)\| : \|x\| \le 1\}$. We want to show that $\|w\| \le 1$ implies $\|\mathscr{F}w\| \le 1$. But

$$\|\mathscr{F}w(x)\| \le \lambda [1 + 2\theta],$$

so we need only require

$$\lambda(1 + 2\theta) \le 1.$$

Step 4. Propagated bounds on Dw. We have

$$D\mathscr{F}w(x) = A_u^{-1} \{ Dw(-)[A_s + (D_x t_s)(x, w(x)) + (D_y t_s)(x, w(x))D_x w]$$
$$- D_x t_u(x, w(x)) - (D_y t_u)(x, w(x))D_x w(x)\}.$$

We assume $\|D_x w\| \le 1$ for $\|x\| \le 1$ and we want the same to be true for $D_x \mathscr{F}w$. The right hand side may be estimated as

$$\lambda \{1 \cdot (\lambda + 2\theta) + \theta + \theta \cdot 1\} = \lambda(\lambda + 4\theta).$$

We therefore get the desired propagated bound on Dw if we require

$$\lambda(\lambda + 4\theta) \le 1.$$

Step 5. Contractivity (The main step). Show that, if $\|w_1\| \le 1$ and $\|w_2\| \le 1$ then

$$\|\mathcal{F}w_1 - \mathcal{F}w_2\| \le \kappa\|w_1 - w_2\|, \qquad \kappa < 1.$$

Things are actually a tiny bit more complicated. We need to know that

$$\|Dw_1\|_2 \le 1$$

as well. If δw denotes an infinitesimal change in w:

$$\delta\mathcal{F}w(x) = \Lambda_u^{-1}\{\delta w[\Lambda_s x + t_s(x, W(x))]$$
$$+ Dw \cdot D_y t_s \delta w(x) - D_y t_u \cdot \delta w(x)\},$$
$$\|\delta\mathcal{F}\| \le \lambda(1 + 2\theta)\|\delta w\|,$$

so we can take $\kappa = \lambda(1 + 2\theta)$ which we can make less than one by making θ small.

Step 6. Existence of a fixed point. The contraction mapping principle almost – but not quite – applies. Let $w_0(x) \equiv 0$; $w_{n+1} = \mathcal{F}w_n$. Then

$$\|w_{n+1} - w_n\| \le \kappa^n\|w_1 - w_0\|.$$

Since $\kappa < 1$, $\Sigma_n\|w_{n+1} - w_n\| < \infty$ which implies that

$$\lim_{n \to \infty} w_n = w$$

exists.

Moreover,

$$w = \lim_{n \to \infty} w_{n+1} = \lim_{n \to \infty} \mathcal{F}w_n = \mathcal{F}w$$

(using again contractivity). As w is a uniform limit of functions each of which has derivative of norm no larger than one

$$\|w(x_1) - w(x_2)\| \le \|x_1 - x_2\|$$

(i.e. w is Lipschitz continuous with Lipschitz constant one). At this point, it is easy to check that the formal calculation leading to the functional equation $\mathcal{F}w = w$ can be applied to show that the graph of is mapped into itself by T.

Step 7. Continuous differentiability of $w(x)$. Prove that the Dw_n are equicontinuous and apply the Arzelà–Ascoli Theorem. We omit the details.

Step 8. Existence of higher derivatives. Assuming T is r times continuously differentiable, it is straightforward to obtain a bound on the $\|D^r w_n\|$ which is uniform in n simply by repeated differentiation of the definition of $\mathscr{F}w$. It follows immediately that w is $r-1$ times continuously differentiable (with a Lipschitz continuous $(r-1)^{st}$ *derivative*). *Hence, if T is infinitely differentiable, so is w.* It can, in fact, be shown that w is r times continuously differentiable, assuming the same about T, by proving that the $D^r w_n$ are then equicontinuous (not just uniformly bounded). If T is analytic then, by making a contraction argument in a space of complex-analytic w's, one proves the existence of an analytic stable manifold.

This proves the existence of a smooth invariant manifold tangent to E_s at the fixed point. To prove uniqueness, and to show that every point z in $U = \{(x, y): \|x\| \le 1, \|y\| \le 1\}$ which does not lie on the graph of w is eventually driven out of U argue as follows: consider two points z_1 and z_2 in U and write:

$$\bar{z}_i = (\bar{x}_i, \bar{y}_i) \qquad \text{for} \qquad Tz_i.$$

Show that if $\|x_1 - x_2\| \le \|y_1 - y_2\|$ (the separation between z_1 and z_2 is *predominantly vertical*), then

$$\|\bar{x}_1 - \bar{x}_2\| \le \|\bar{y}_1 - \bar{y}_2\|:$$

predominantly vertical separations are preserved by T.

$$\|\bar{y}_1 - \bar{y}_2\| \ge \sigma \|y_1 - y_2\|$$

for some fixed $\sigma > 1$: pairs of orbits with predominantly vertical separation diverge exponentially as long as they remain in U.

Thus, for a given x, there is at most one y such that $T^n(x, y) \in U$ for all n. We have already exhibited one such y, namely $w(x)$. This proves both uniqueness and that points of U not on the graph of w are eventually pushed out of U by repeated application of T.

The existence part of the above argument can be extended quite a bit.

(A) Pick some ϱ between zero and one such that no eigenvalue of $DT(z_0)$ has modulus ϱ. The above argument can be repeated essentially word for word with E_s replaced by the spectral subspace corresponding to the part of the spectrum inside $\{\zeta:|\zeta|<\varrho\}$ and E_u replaced by the spectral subspace corresponding to the part of the spectrum *outside* that disk. (The only change necessary is to define $\|w\|=\sup\{\|w(x)\|/\|x\|\}$ rather than as $\sup\{\|w(x)\|\}$.) The resulting invariant manifold will be a submanifold of the stable manifold; it can be characterized as the set of z's such that the distance from $T^n z$ to the fixed point z_0 decreases (asymptotically) faster than ϱ^n.

(B) A more interesting extension is to allow for the possibility of eigenvalues with modulus 1. Let:

E_{cs} denote the spectral subspace associated with the part of the spectrum inside and on the unit circle (cs stands for "centerstable").

E_u the spectral subspace associated with the part of the spectrum outside the unit circle.

The formal calculation leading up to the definition of $\mathscr{F}$ goes through as before. However, the careful definition of $\mathscr{F}w$ which would require that

$$\|\Lambda_{cs}x+t_{cs}(x,y)\|\le1 \qquad \text{if } \|x\|\le1 \text{ and } \|y\|\le1$$

fails. The way around this is to "cut off" T by multiplying its non-linear parts t_{cs} and t_u by a smooth function which is one in the neighbourhood of zero and zero if $\|x\|>1$ or $\|y\|>1$. Then

$$\|t_{cs}\|, \|t_u\|, \|D_x t_{cs}\|_{y\ u}$$

are small *globally* (not just for $\|x\|\le1$ and $\|y\|\le1$) and the contraction argument goes just as before to construct a unique globally invariant manifold for the cut-off T, which is a *locally* invariant manifold tangent to E_{cs} for the original T. Such a manifold is called a *center-stable* manifold for T at z_0.

(Formally: by a *locally invariant manifold* for T at z_0 we mean a manifold W passing through z_0 such that, for z sufficiently near to z_0, $z\in W$ implies $Tz\in W$).

A *center manifold* for T at z_0 will mean a locally invariant manifold tangent to the spectral subspace E_c associated with the part of the spectrum *on* the unit circle. To prove the existence of center manifolds, first construct a center-stable manifold; restrict T to it, and invert. The derivative of the inverse has spectrum outside and on the unit circle, but none inside. Hence, a center-stable manifold for the inverse is in fact a center manifold, and hence, also a center manifold for T.

Remarks.

(1) Proceeding as outlined to construct center-stable and center manifolds appears to cost a fraction of a derivative. To prove that w is r times continuously differentiable it is necessary to assume that $D^r T$ is Hölder continuous. I do not know whether this loss of differentiability is genuine or not.

(2) Unlike stable manifolds, center-stable and center-manifolds are not generally unique. Here is a simple example (from Marsden and Mc Cracken 1976, pp. 44–46). The example is a flow rather than a mapping but to get a mapping one can simply take the time-one map. The flow is two dimensional and given by:

$$\mathrm{d}x/\mathrm{d}t = -x; \qquad \mathrm{d}y/\mathrm{d}t = y^2$$

Any orbit in the lower half plane is asymptotic to the origin and its tangent is asymptotically vertical. The positive y-axis together with *any one* of these orbits in the lower half-plane comprises a center manifold for the fixed point at the origin.

(3) If T is infinitely differentiable, it has C^r center manifolds for each r, but need not have an infinitely differentiable center manifold. Pugh has constructed an example (see Marsden and Mc Cracken 1976). There also exist analytic mappings with no analytic center manifold.

(4) The above construction of (local) stable and center-stable manifold works as well for non-invertible mappings as for invertible ones. Without doing any more work, one gets immediately the existence of unstable, center-unstable, and center manifolds for invertible mappings by applying the above analysis to T^{-1}. It is possible to give an alternative direct construction of these objects which does not require the invertibility of T. See section 2 of Marsden and Mc Cracken (1976) or Hirsch et al. (1977).

2. Horseshoe and solenoid examples

We now take up the subject of recurrence of orbits of a dynamical system. A point x_0 is said to be a wandering or transient point if there is an open set V containing x_0 such that

$$T^n V \cap V = \emptyset \qquad \text{for sufficiently large } n.$$

(Equivalently, one can require $T^n V \cap V = \emptyset$ for $n \geq 1$, but the former definition extends more directly to flows.) The *non-wandering* set $\Omega(T)$ is the complement of the set of wandering points. By definition, the set of wandering points is open so the non-wandering set is closed. Note that, for x_0 to be non-wandering, it is not quite necessary for the orbit of x_0 itself to return to the vicinity of x_0 but only that there are points x arbitrarily near whose orbits return to the vicinity of x_0 for arbitrarily large n. Fixed and periodic points are trivial examples of non-wandering points. A less trivial example is provided by a Hamiltonian system such that $\{(q,p):H(q,p)\leq E\}$ has finite volume for each E. For such a system, *every* point is non wandering. See Theorem 4.1. The main objective of this section is to discuss two very rich examples with complicated non-wandering sets but whose orbit structure can, in a certain sense, be analyzed completely using the techniques of *symbolic dynamics*.

The first example is the Smale horseshoe. The basic idea is to examine a mapping which acts on a rectangle by squeezing, stretching, and bending it back onto itself (see fig. 1) and to determine the structure of the set of orbits remaining in the rectangle for all positive and negative times. In the above, we have only described the action of the mapping on the rectangle itself, and the rectangle is not mapped into itself. To fix ideas, it is convenient (and customary) to extend the mapping outside the rectangle as follows: Adjoin caps on the two ends of the rectangle to make a stadium shaped region U (see fig. 2) which will be mapped into itself.

Both left and right hand caps are mapped into the left-hand cap; the left-hand cap is to contain an attracting fixed point which pulls in everything in that cap. Thus, any orbit which eventually gets into either the left or right cap is attracted to that fixed point and so is non-recurrent.

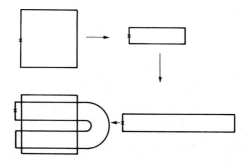

Fig. 1.

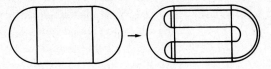

Fig. 2.

Let R denote the central rectangle. The set of points in R which are mapped back into R by the transformation T consists of two roughly vertical rectangles whose images are the two roughly horizontal rectangles in $TR \cap R$. We will call the two vertical rectangles Δ_0 and Δ_1 respectively (see fig. 3). To permit a complete and transparent analysis of the mapping T, we will construct it in a very special and restrictive way; it is true – but entirely non-trivial – that the qualitative outcome would be the same with considerably more flexibility in the construction. The respect in which the construction is special is that the mapping T will be assumed to be affine on each of Δ_0 and Δ_1 or, even more specifically

$$T\begin{pmatrix} x \\ y \end{pmatrix} = \begin{pmatrix} \alpha x + \gamma_{0,x} \\ \beta y + \gamma_{0,y} \end{pmatrix} \qquad \text{on } \Delta_0,$$

$$= \begin{pmatrix} -\alpha x + \gamma_{1,x} \\ -\beta y + \gamma_{1,y} \end{pmatrix} \qquad \text{on } \Delta_1,$$

with $\alpha > 1$ and $0 < \beta < 1$. (For some purposes, it is interesting to allow the α and β on Δ_1 to be different from what they are on Δ_0 but we will not pursue this possibility.) With these assumptions, T expands uniformly in the horizontal direction, and contracts uniformly in the vertical direction on $\Delta_0 \cup \Delta_1$. We also, for definiteness, take the rectangle R to be in the unit square.

The question we want to take up is: what is the structure of the set of orbits remaining in R for all times, positive and negative? Let $(z_n, -\infty < n < \infty)$ be such an orbit. In order that $z_{n+1} = Tz_n$ be in R, it is

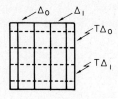

Fig. 3.

necessary that z_n be in either Δ_0 or Δ_1. Since Δ_0 and Δ_1 do not intersect, z_n cannot be in both; there is a uniquely determined index i_n, either zero or one, such that $z_n \in \Delta_{i_n}$. Doing this for all n, we associate with the orbit (z_n) a sequence (i_n) of zeros and ones, called the *symbolic description* of the orbit; we can think of it as a function of z_0 and write it as $[i_n(z_0)]$ or $i(z_0)$. Evidently, just from the definition:

$$i_n(Tz_0) = i_{n+1}(z_0)$$

or

$$i(Tz_0) = \sigma i(z_0),$$

where σ is the *shift transformation*

$$(\sigma i)_n = i_{n+1}$$

which moves the sequence one place to the *left*. The key to analyzing the recurrence of orbits for the Smale horseshoe is to make detailed use of the correspondence between orbits remaining in R and symbol sequences. To do this effectively, we need to know more about the correspondence, namely:

Is the orbit uniquely determined by its symbol sequence?

Does *every* symbol sequence correspond to some orbit?

In the case of the horseshoe, unlike most cases to which symbolic dynamics is applied, the answer to these questions is unqualifiedly affirmative. The reasoning is as follows:

given a finite sequence $i_0, \ldots, i_{n-1}$ of zeros and ones, we let $\Delta_{i_0}, \ldots, i_{n-1}$ denote $\{z_0 : T^j z_0 \in \Delta_{i_j} \text{ for } j = 0, 1, \ldots, n-1\}$ i.e. $\Delta_{i_0, \ldots, i_{n-1}}$ is the set of initial points whose symbolic orbits begin with $i_0, \ldots, i_{n-1}$. Using the fact that T expands horizontal separations by a factor of α, it is easy to see by induction that

$$\Delta_{i_0, \ldots, i_{n-1}}$$

is a vertical rectangle of width $(1/\alpha)^n$ and height 1. (Remember that we are taking the central rectangle R to be the unit square.)

Similarly, given $(i_{-n}, i_{-n+1}, \ldots, i_{-1})$, the set of z_0's such that $T_j z_0 \in \Delta_{i_j}$ for $-n \leq j \leq -1$ is simply

$$T^n \Delta_{i_{-n}, \ldots, i_{-1}}$$

which is again easily seen to be a horizontal rectangle of width 1 and height

β^n. These two rectangles intersect in a small rectangle i.e. – the set of z_0's such that $T^j z_0 \in \Delta_{ij}$ for $-n \le j \le n-1$ is a rectangle of height β^n and width α^{-n}. Here, $(i_{-n}, \ldots, i_{n-1})$ is *any* sequence of zeros and ones of length 2_n (see fig. 4).

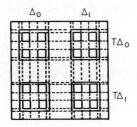

Fig. 4.

Now pick any infinite sequence (i_j) of zeros and ones. The sets

$$G_n = \{z_0 : T^j z_0 \in \Delta_{i_j} \text{ for } -n \le j \le n-1\}$$

are a decreasing sequence of non-empty closed rectangles of diameter going to zero. They have, therefore, exactly one point in common. Call that point $\pi(i) = z_0$. Then,

$$T^j z_0 \in \Delta_{i_j} \text{ for all } j.$$

No other initial point gives an orbit with this property.

In other words: the correspondence between symbol sequences and initial points for orbits remaining in R for all time is bijective. (The correspondence is also continuous if the space of sequence is topologized appropriately, but we will not pursue this point.) We will denote this set of initial points by Λ. The above picture makes it apparent that Λ is a product of two one-dimensional Cantor sets. It may also be worth noting that the x co-ordinate of a point in Λ is determined by $(i_0, i_1, i_2, \ldots)$, i.e. by the present and future of its symbolic orbit whereas the y co-ordinate is determined by $(\ldots, i_{-2}, i_{-1})$, i.e. by its past symbolic orbit.

The correspondence between orbits and symbol sequences makes it possible to infer the existence of orbits with various kinds of recurrence from the existence of sequences with analogous properties. For example, an orbit whose symbol sequence is periodic must also be periodic; from this is follows readily that Λ contains infinitely many periodic orbits for T. It is easy to see, in fact, that periodic orbits are dense in Λ: Let $z_0 \in \Lambda$; let

i be its symbol sequence; let $\boldsymbol{i}$ be the periodic sequence of period $2n$ (or smaller) agreeing with i from $-n$ through $n-1$ and let $\hat{z}_0$ be the point in Λ whose symbol sequence is $\boldsymbol{i}$. Then, since the symbol sequence for $\hat{z}_0$ agrees with that for z_0 from $-n$ through $n-1$, $\hat{z}_0$ and z_0 both sit in the same rectangle of size α^{-n} by β^n. Thus, by making n large, we can make $\hat{z}_0$ as close as we like to z_0. Since every periodic orbit is non-wandering and since the non-wandering set is closed, it follows that every point of Λ is non-wandering. It is quite easy to see that the only non-wandering point for T outside of Λ is the fixed point in the left-hand cap.

Much more can be done. For example:

Λ contains a dense orbit. Let $\boldsymbol{i}$ be any sequence such that *every* finite sequence of 0's and 1's appears somewhere as a subsequence of $(i_1, i_2, \dots)$ and let z_0 be the corresponding point of Λ. Then, by the argument of the preceding paragraph the future orbit of z_0 (i.e., $\{T^n z_0 : n \geq 0\}$) is dense in Λ.

Given any f between zero and one inclusive, the set of orbits which spend asymptotically a fraction f of their time in Δ_0 is dense in Λ_0.

Let i_j be defined for positive j by:

$$
\begin{aligned}
i_j &= 0 \text{ for } & 1 \leq j \leq 10 \\
&= 1 \text{ for } & 11 \leq j \leq 100 \\
&= 0 \text{ for } & 101 \leq j \leq 1000
\end{aligned}
$$
...

(the i_j's for $j < 0$ can be anything), and let z_0 be the corresponding point of Λ. Then, the asymptotic fraction of the time that the orbit of z_0 spends in Δ_0 does not converge. By modifying i_j in a finite number of places, we can even construct a dense set of z_0's with this property.

As already noted, the mapping T from which we constructed Λ is extremely special. It turns out, however, that the qualitative properties of Λ are not destroyed by a small perturbation on T. In fact, T is *structurally stable*, which in this case just means that any small perturbation on T can be undone by a continuous change of variables.

Horseshoes appear in many mappings arising in practice. A very general mechanism producing them is the transverse crossing of stable and unstable manifolds of a hyperbolic fixed or periodic point. Phenomena associated with these crossings will be discussed by S. Newhouse in Course 6.

Despite its many nice features, the horseshoe has the major disadvantage of *not* being an attractor – the set of initial points whose orbits converge to the Cantor set Λ is a set of Lebesgue measure zero in the ambient space. We turn now to another example, the solenoid, which is an attractor. It has the further advantage that a simple formula can be written for the mapping which produces it. Nevertheless, we first describe it geometrically. The mapping in question sends the solid three-dimensional torus into itself, schematically as shown in fig. 5. (Actually, there is no need to cut and re-join the torus; the mapping can be visualized instead as stretching out the torus; then folding it over and putting it back inside itself.)

To describe this mapping analytically, represent the solid torus as the product of a circle with a disk

$$\{(z, w) : |z| = 1; \ |w| \leq 1; \ z, w \in \mathbb{C}\},$$

z is essentially an angular co-ordinate giving longitudinal position; w gives transverse position. In terms of these coordinates, we can simply take:

$$T : (z, w) \mapsto (z^2, \tfrac{1}{2}z + \tfrac{1}{4}w).$$

The $z \mapsto z^2$ part maps the longitudinal circle onto itself in a 2-to-1 way, transverse disks are shrunk by a factor of $\tfrac{1}{4}$ and shifted by an amount depending on longitudinal position.

The image of the solid torus under T^n is then a long thin tube, of transverse radius $(\tfrac{1}{4})^n$, winding around the torus 2^n times. Taking the intersection of this tube with the transverse plane $z = 1$ gives a set of 2^n circles, each of radius $(\tfrac{1}{4})^n$ (see fig. 6). Then:

$$\bigcap_{n=0}^{\infty} T^n \, (\text{torus})$$

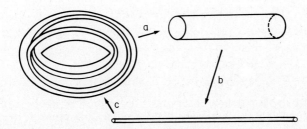

Fig. (5). (a) Cut the torus transversally and straighten it out. (b) Squeeze and stretch the resulting cylinder. (c) Bend the thin cylinder into a double loop; join the ends together and put it back into the solid torus.

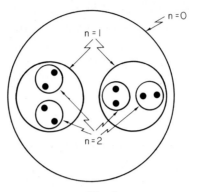

Fig. 6.

is a bundle of lines (with no thickness) wrapping around the torus. This limiting structure is called the *solenoid*. Its intersection with the transverse plane $\{z = 1\}$ is a Cantor set. Since the Cantor set has uncountably many points, whereas each line making up the solenoid has only countably many intersections with $\{z = 1\}$, the solenoid is not a single line but an uncountable union of lines. We will refer to a whole line as a wire; cutting the solenoid along the $\{z = 1\}$ splits up each wire into countably many *loops**. Any successive 2^n loops on a wire thread through each of the 2^n circles in $T^n(\text{torus}) \cap \{z = 1\}$ and from this it follows that each wire is dense in the solenoid. It is an informative exercise to figure out when two points on the transverse Cantor set are on the same wire.

We turn now to the introduction of symbolic dynamics for orbits on the solenoid Λ. The idea, roughly, is to let

$$\Delta_0 = \{(e^{i\theta}, w) : 0 \leq \theta \leq \pi; \ w \text{ arbitrary}\}$$
$$\Delta_1 = \{(e^{i\theta}, w) : \pi \leq \theta \leq 2\pi; \ w \text{ arbitrary}\}$$

and to associate with an orbit $T^n x_0$ a sequence of zeros and ones by

$$\begin{aligned} i_n &= 0 &&\text{if } T^n x_0 \in \Delta_0 \\ i_n &= 1 &&\text{if } T^n x_0 \in \Delta_1. \end{aligned}$$

In contrast to the horseshoe case, this prescription is no longer unambiguous: if the z component of $T^n x_0$ is ± 1, we may choose either $i_n = 0$ or $i_n = 1$. Let us explore the relation between symbol sequences and orbits,

---

* Note: Contrary to conventional usage, these loops are not closed.

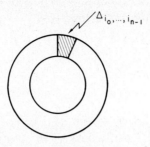

Fig. 7.

ignoring for the moment this ambiguity. For any finite sequence $i_0,\dots,i_{n-1}$ of zeros and ones, we let $\Delta_{i_0,\dots,i_{n-1}}$ denote the closure of the set of x_0's in the solid torus such that $T^j x_0$ is in the interior of Δ_{i_j}, for $j = 0, 1, 2, \dots, n-1$. Then, $\Delta_{i_0,\dots,i_{n-1}}$ is a "sector" of angle $2\pi/2^n$; it is represented in fig. 7 as viewed from above. Similarly, given any sequence $(i_{-n},\dots,i_{-1})$

$$T^n \Delta_{i_{-n}\cdots i_{-1}}$$

is the closure of the set of x_0's in the solid torus such that $T^j x_0$ is in the interior of Δ_{i_j} for $j = -n, -n+1, \dots, -1$; this set is a tube of radius $(\tfrac{1}{4})^n$ going once around the solid torus (see fig. 8). Just as in the case of the horseshoe, we can now fix a two-sided infinite sequence $(\dots, i_{-2}, i_{-1}, i_0, i_1, \dots)$ and form the decreasing sequence of sets

$$G_n = \Delta_{i_0\dots i_{n-1}} \cap T^n \Delta_{i_{-n}\dots i_{-1}}.$$

Each G_n is closed and nonempty; the diameter of G_n goes to zero as n goes to infinity. We let $\pi(i)$, or x_0, denote the single point belonging to all of

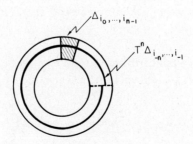

Fig. 8.

the G_n's. It is nearly immediate that:

$$T^j x_0 \in \Delta_{i_j} \text{ for all } j$$

and that:

$$T\pi(i) = \pi(\sigma i).$$

It is also easy to see that every point of Λ can be written as $\pi(i)$ for some i. Thus, the mapping π, from sequences to points of Λ, is unambiguously defined and has the formal properties that would be expected of the inverse of the "mapping" from points to sequences described above, if that mapping were defined. This, incidentally, is a characteristic feature of symbolic dynamics arguments: the mapping from sequences to orbits is likely to be less troublesome than the mapping from orbits to sequences.

We want to use the correspondence between symbol sequences and orbits to analyze recurrence of orbits on the solenoid, as we did for the horseshoe. We therefore need to know to what extent the projection is not one-to-one. It is clear that, in order for x_0 to correspond to more than one sequence, it is necessary that $T^n x_0$ be on the surface $\{z = 1\}$ for some n (and hence for all subsequent n's); otherwise, the sequence corresponding to x_0 could simply and unambiguously be read off by looking at whether each of the various $T^n x_0$'s is in Δ_0 or Δ_1. There are now two cases to consider:

(1) $T^n x_0$ is on the surface $\{z = 1\}$ for all n (positive or negative). Since T acts contractively on the surface $\{z = 1\}$, this forces x_0 to be the unique fixed point for T, which indeed corresponds both to the sequence $(\ldots 0, 0, 0, \ldots)$ and to the sequence $(\ldots 1, 1, 1, \ldots)$ (but no other).

(2) There is a first n for which $T^{n+1} x_0$ is on the surface $\{z = 1\}$; then $T^n x_0$ is on the surface $\{z = -1\}$ and $T^j x_0$ is in the interior of either Δ_0 or Δ_1 for all $j < n$, so i_j is unambiguously determined for $j < n$. It is then possible to take i_n equal to either 0 or 1; but, if i_n is put equal to 0 then necessarily $i_{n+1} = i_{n+2} = \ldots = 1$ whereas if i_n is put equal to 1 then necessarily $i_{n+1} = i_{n+2} = \ldots = 0$.

Thus:

No orbit corresponds to more than two sequences;

The only pairs of sequences which give the same orbit are those of the form

$$(\ldots i_{n-1}, 1, 0, 0, \ldots) \qquad \text{and } (\ldots i_{n-1}, 0, 1, 1, \ldots)$$

and the limiting case

$$(\ldots 0,0,0,\ldots) \qquad \text{and} \quad (\ldots 1,1,1,\ldots).$$

It follows in particular that the projection π is one-to-one from the set of periodic sequences (excluding the two constant sequences) to points of Λ. Thus, exactly as for the horseshoe, periodic points are dense in Λ, so every point of Λ is non-wandering. Similarly, one proves the existence of a dense orbit, and of various kinds of reasonable and unreasonable recurrence, just as for the horseshoe.

Note also: From the analysis given above, it follows that the future symbolic orbit $(i_0, i_1 \ldots)$ determines the longitudinal position of x_0; the past symbolic orbit determines which loop of the solenoid x_0 lies on. In fact, if we write:

$$x_0 = (e^{2\pi i \phi_0}, w_0), \ 0 \le \phi \le 1$$

then:

$$\phi_0 = \frac{i_0}{2} + \frac{i_1}{4} + \frac{i_2}{8} + \ldots,$$

i.e. $i_0, i_1, i_2, \ldots$ is the binary representation of ϕ_0. The nonuniqueness of the sequence corresponding to certain x_0's is just a more complicated version of the non-uniqueness of the binary representation of dyadic rationals.

3. Hyperbolic sets and Axiom A

The horseshoe and solenoid share a property reminiscent of hyperbolicity of a fixed point: there are well-defined directions in which separations between nearby orbits decay, and a complementary set of directions in which separations "grow" in the sense that they decay under iteration of T^{-1}. One of the most important developments in the theory of differentiable dynamical systems has been the systematic study of such generalized hyperbolicity properties. This investigation was initiated by Anosov for conservative systems and by Smale for more general dynamical systems.

Let T be a differentiable mapping on $\mathbb{R}^m$ or a region in $\mathbb{R}^m$ (for definiteness). Let Λ be a compact set mapped onto itself by T, and assume that T is invertible on a neighborhood of Λ. We write DT_x for the $m \times m$ matrix $(\partial T_i / \partial x_j)$ of partial derivatives of T at x.

We say that Λ is a *hyperbolic set* if, for each x in Λ, we have a splitting

$\mathbb{R}^m = E_x^s \oplus E_x^u$ such that

(a) $DT_x\, E_x^s = E_{T_x}^s$

   $DT_x\, E_x^u = E_{T_x}^u$.

(b) There are constants $C > 0, \lambda$ with $0 < \lambda < 1$, such that

   if $\zeta \in E_x^s$, $\|DT_x^n\, \zeta\| \le C\lambda^n \cdot \|\zeta\|$

   if $\zeta \in E_x^u$, $\|DT_x^{-n}\, \zeta\| \le C\lambda^n \cdot \|\zeta\|$.

(c) E_x^s and E_x^u vary continuously with x.

Remarks.

(1) Condition (b) says that orbits whose initial points are displaced infinitesimally from x in directions belonging to E_x^s converge exponentially back to the orbit of x as $n \to +\infty$, and that orbits whose initial points are displaced infinitesimally from x in directions belonging to E_x^u converge exponentially back as $n \to -\infty$. It is essential that the decay rate is uniform in x.

(2) The splitting is only assumed to be defined on Λ, not throughout the ambient space or even on a neighborhood of Λ.

(3) The definition implies in particular that every periodic points in Λ is hyberbolic.

(4) The meaning of continuity in (c) may be defined in various equivalent way. One elementary and concrete definition is the following. Any vector ζ can be written uniquely

$$\zeta = \zeta_x^s + \zeta_x^u \qquad \text{with } \zeta_x^s \in E_x^s \text{ and } \zeta_x^u \in E_x^u,$$

continuity of E_x^s and E_x^u in x just means that ζ_x^s varies continuously with x for each ζ.

(5) The presence of the constant C in (b) is a technical nuisance; it says that the contractivity need only become effective after a number of iterations. Fortunately, C can be made equal to 1 by changing the norm (and perhaps making λ closer to 1). The necessary change in norm may, however, vary with x, i.e. instead of working with a single (Euclidean) norm on $\mathbb{R}^m$ one needs to use a Riemannian metric. A metric in which C can be put equal to one is called an *adapted* metric. We will always assume that we are using an adapted metric. Although not necessary, this simplifies the formulation of many results. The existence of adapted

metrics follows from a simple argument of Mather which is given in Hirsch and Pugh (1970).

(6) In the more general case where T acts not on Euclidean space but on a manifold M (which we can think of as imbedded in a larger-dimensional Euclidean space) then what should be required is the existence of a splitting of the indicated type of the *tangent space* to the manifold M at each point of Λ.

In terms of the notion of hyperbolic set, we can formulate *Smale's Axiom A*: we say that T satisfies Axiom A if

— the non-wandering set $\Omega(T)$ is compact and hyperbolic,

and

— periodic points are dense in $\Omega(T)$.

Anosov condition. A mapping of a compact manifold M to itself is said to be an *Anosov* mapping if M itself is hyperbolic.

Associated with a hyperbolic fixed point there are stable and unstable manifolds which are non-infinitesimal versions of the stable and unstable eigenspaces. Something similar is true for general hyperbolic sets. In formulating there results we will exploit the simplifications made possible by using an adapted metric.

Given, in addition, a positive number ε, we define also

$$W^s(x) = \{y \in \Lambda : d(T^n x, T^n y) \to 0 \qquad \text{as } n \to +\infty\},$$

$$W^u(x) = \{y \in \Lambda : d(T^{-n} x, T^{-n} y) \to 0 \qquad \text{as } n \to +\infty\}.$$

Given in addition, a positive number ε, we define also

$$W_\varepsilon^s(x) = \{y \in \Lambda : d(T^n x, T^n y) \le \varepsilon \qquad n \ge 0\},$$

$$W_\varepsilon^u(x) = \{y \in \Lambda : d(T^{-n} x, T^{-n} y) \le \varepsilon \qquad n \ge 0\}.$$

We have already seen that, if x is a hyperbolic fixed point and if ε is small, then

$W_\varepsilon^s(x)$ is a smooth surface tangent at x to $E^s(x)$,

$W_\varepsilon^s(x)$ is contained in $W^s(x)$,

$$W^s(x) = \bigcup_{n=0}^{\infty} T^{-n} \, W_\varepsilon^s(x).$$

The following theorem extends these results to general points of a hyperbolic set.

Theorem (3.1). (Invariant Manifolds Theorem for Hyperbolic Sets). For sufficiently small $\varepsilon > 0$, and for every $x \in \Lambda$,

(1) $W_\varepsilon^s(x)$ and $W_\varepsilon^u(x)$ are submanifolds, as smooth as T is, passing through x, and tangent there to E_x^s and E_x respectively.

(2)   If $y \in W_\varepsilon^s(x)$, $d(T^n x, T^n y) \leq \lambda^n d(x, y)$ $(n \geq 0)$.

   If $y \in W_\varepsilon^u(x)$, $d(T^{-n}x, T^{-n}y) \leq \lambda^n d(x, y)$ $(n \geq 0)$.

(3) $T W_\varepsilon^s(x) \subset W_\varepsilon^s(Tx)$,

   $T^{-1} W_\varepsilon^u(x) \subset W_\varepsilon^u(T^{-1}x)$.

(4) $W^s(x) = \bigcup_{n=0}^{\infty} T^{-n} W_\varepsilon^s(T^n x)$,

   $W^u(x) = \bigcup_{n=0}^{\infty} T^n W_\varepsilon^u(T^{-n} x)$.

(5) $W_\varepsilon^s(x)$ and $W_\varepsilon^u(x)$ vary continuously with x.

Remark. It may be necessary to take the λ appearing in this theorem slightly larger than the λ appearing in the definition of hyperbolicity.

Example. The solenoid. Then:

   $W_\varepsilon^s(x)$ is a small piece of transverse disk, centered at x, of radius ε.
   $W_\varepsilon^u(x)$ is a small segment of the wire through x, of length about 2ε,
      centered at x.
   $W^u(x)$ is the full wire through x.

Note for future reference two facts about this example:
(1) $W^u(x)$ (but not $W_\varepsilon^s(x)$) *is actually contained in* Λ.
(2) For each x, $W^u(x)$ is dense in Λ.

All we will say about the proof of the invariant manifold theorem for hyperbolic sets is that, although more complicated to express, it can be given in a way which closely parallels the proof for a hyperbolic fixed point. Roughly, what is done is to apply the argument for the fixed-point case at every point of Λ simultaneously. For the details, see the original paper of Hirsch and Pugh (1970) or the monograph of Hirsch et al. (1977).

Here is one immediate consequence of the invariant manifolds theorem. At any point x, E_x^s and E_x^u are transverse to each other, so $W_\varepsilon^s(x)$ and $W_\varepsilon^u(x)$ cross transversally at x. For sufficiently small ε, therefore, $W_\varepsilon^s(x)$ and $W_\varepsilon^u(x)$ have only the point x in common. It is not hard to see, using the compactness of Λ, that a single ε can be found which works for all x. Now suppose we have another point y such that $d(T^n x, T^n y) \le \varepsilon$ for all n, positive and negative. The validity of this condition for positive n says that $y \in W_\varepsilon^s(x)$ and its validity for negative n says that $W_\varepsilon^u(x)$. But the only point belonging both to $W_\varepsilon^s(x)$ and $W_\varepsilon^u(x)$ is x itself, so y must equal x. In other words, we have proved:

Proposition (3.2). There exists $\varepsilon > 0$ such that, if $x \in \Lambda$ and $y \in M$, and $y \neq x$, then

$$d(T^n x, T^n y) > \varepsilon \text{ for at least one } n.$$

The formulation of Axiom A is adapted to analysis of the global structure of mappings. The analysis proceeds by showing that, if T satisfies Axiom A, then the non-wandering set can be decomposed into a finite number of irreducible pieces called *basic sets*. Among the basic sets are *attractors*, which are the objects governing the long-time behavior of "typical" orbits. We will not describe the decomposition theory but turn directly to the subject of Axiom A attractors.

We will not, however, give a general definition of the term *attractor*. The question of how this term is best defined is still a subject of active investigation; see, for example, the recent article of Ruelle (1981). Nevertheless, an important element in the definition is likely to be something like the following: A set X is said to be an *attracting set* for T, if there exists an open set U containing X such that

$$TU \subset U$$

and

$$\bigcap_{n=1}^{\infty} T^n U = X,$$

i.e. the forward images of U under T shrink down to X. The *basin of attraction B* for an attracting set is the set of all points x whose forward orbits $(T^n x)_{n \geq 0}$ converge to X; we have

$$B = \bigcup_{n=1}^{\infty} T^{-n} U$$

so B is open.

We now define: an *Axiom A attractor* means a compact hyperbolic set Λ such that

            − Λ is an attracting set
            − there is an x in Λ such that its forward orbit $(T^n x)_{n \geq 0}$ is dense
              in Λ (Λ is *topologically transitive*).

Remark. A hyperbolic attracting set contains the unstable manifold of each of its points; conversely, a hyperbolic set containing the unstable manifold of each of its points is attracting.

Proof. Let Λ be a hyperbolic attracting set, U a contracting neighborhood of Λ (i.e. an open set such that $TU \subset U$ and $\bigcap_{n=1}^{\infty} T^n U = \Lambda$). Let $x \in \Lambda$ and let $y \in W^u(x)$; we want to show that $y \in \Lambda$. Since $d(T^{-n}y, T^{-n}x)$ goes to zero as $n \to \infty$, $T^{-n}y \in U$ for all sufficiently large n. Thus $y \in T^n U$ for all sufficiently large n (and hence for all n as $TU \subset U$). But $\bigcap_n T^n U = \Lambda$, so $y \in \Lambda$.

In the other direction we give only a heuristic argument: suppose Λ is a hyperbolic set which contains the unstable manifold of each of its points. What we will argue is that, given $x \in \Lambda$, the union of the $W_\varepsilon^s(y)$'s for y near x in Λ fill up a neighborhood of x. Let $m_s = \dim (E_x^s)$; $m_u = \dim (E_x^u)$, so $m_s + m_u = m$. Also write W for $W_\varepsilon^u(x)$. By assumption, $W \subset \Lambda$. For each y in W, $W_\varepsilon^s(y)$ is a surface of dimension m_s transverse to W as y runs over the m_u-dimensional surface W, the $W_\varepsilon^s(y)$'s must sweep out a region of dimension $m_s + m_u = m$, i.e., must sweep out a neighbourhood of x. This argument also indicates that every orbit which converges to a hyperbolic attractor must in fact converge to some particular orbit on the attractor.

Perhaps, the most important analytical tool for the detailed study of the structure of mappings satisfying Axiom A is the notion of pseudo-orbit. Let T be any mapping; δ any positive number. A sequence

$$\ldots x_{-2}, \; x_{-1}, \; x_0, \; x_1, \; x_2, \ldots$$

is said to be a δ-pseudo-orbit if

$$d(Tx_n, x_{n+1}) \le \delta \qquad \text{for all } n$$

i.e., if x_{n+1} is obtained from x_n by first applying T and then making a jump of magnitude no larger than δ. A true orbit

$$\ldots \hat{x}_{-2}, \ \hat{x}_{-1}, \ \hat{x}_0, \ \hat{x}_1, \ \hat{x}_2, \ldots (T\hat{x}_n = \hat{x}_{n+1})$$

is said to ε-*shadow* the pseudo-orbit if

$$d(\hat{x}_n, x_n) \le \varepsilon \qquad \text{for all } n.$$

The definitions are extended to finite sequences $(x_n, \ldots, x_m)$ in the obvious way.

Theorem (3.3) (Shadow Lemma, Anosov–Bowen Theorem). Let Λ be a hyperbolic attracting set for T. For any $\varepsilon > 0$ there is a $\delta > 0$ such that every δ-pseudo-orbit which remains within a distance δ of Λ is ε-shadowed by a true orbit on Λ.

This result is a somewhat non-standard member of a family of related results. The standard versions are
  – T is taken to be an Anosov mapping, Λ to be the whole space
and
  – Λ is taken to be an Axiom A basic set, (not necessarily an attractor) and the pseudo-orbit is taken to lie on Λ (not just near it).
  Before starting on the formal proof, we consider, somewhat imprecisely, two trivial but enlightening model situations.
  (a) Assume T to be contractive: $d(Tx, Ty) \le \lambda d(x, y)$, $\lambda < 1$. Let $(x_0, x_1, \ldots x_n)$ be a δ-pseudo-orbit, and let

$$\hat{x}_j = T^j x_0 \qquad j = 0, 1, \ldots, n.$$

Then

$$d(\hat{x}_{j+1}, x_{j+1}) = d(T\hat{x}_j, x_{j+1}) \le d(T\hat{x}_j, Tx_j) + d(Tx_j, x_{j+1}) \le \lambda d(\hat{x}_j, x_j) + \delta.$$

From this estimate, it follows by induction that

$$d(\hat{x}_j, x_j) \le \frac{\delta}{1-\lambda} \qquad \text{for all } j$$

so we have shown that our δ-pseudo-orbit is ε-shadowed by a true orbit with $\varepsilon = \delta/(1 - \lambda)$; furthermore, the appropriate true orbit is very easy to identify; its initial point is x_0. The argument given evidently extends without difficulty to pseudo-orbits (x_j) defined for all positive j. To deal with the case of pseudo-orbits defined for all j, positive and negative, we can preceed as follows: pick N and find a true orbit $\hat{x}_j^{(N)}$ which ε-shadows our pseudo-orbit for $-N \le j \le N$. Then

$$d(\hat{x}_0^{(N)}, x_0) \le \varepsilon \qquad \text{for all } N.$$

By compactness, choose a subsequence N_k of N's so that

$$\hat{x}_0 = \lim_{k \to \infty} \hat{x}_0^{(N_k)}$$

exists. Then assuming that T is continuous with a continuous inverse – or at least continuous local inverses – it follows that

$$\lim_{k \to \infty} \hat{x}_j^{(N_k)} = T^j \hat{x}_0 \qquad \text{for all } j.$$

But since

$$d(\hat{x}_j^{(N_k)}, x_j) \le \varepsilon, \qquad \text{for } |j| \le N_k$$

if follows that

$$d(T^j \hat{x}_0, x_j) \le \varepsilon \qquad \text{for all } j,$$

i.e., the orbit of $\hat{x}_0$ ε-shadows the pseudo-orbit (x_j). This device is evidently quite general, and shows that there is no loss of generality in proving the Shadow Lemma only for pseudo orbits indexed by $(0, 1, \ldots, n)$, provided that the relation obtained between ε and δ does not depend on n.

(b) Assume T to be expansive:

$$d(Tx, Ty) \ge \frac{1}{\lambda} d(z, y)$$

for x near y; $\lambda < 1$. The mapping $x \mapsto 2x$ (mod. 1) of the circle to itself is an example. The idea here is that, if T is expansive then T^{-1} should be contractive so that the preceding analysis applies. The non-existence of a global inverse is not really a problem; local inverses suffice.

Exercise. Give the details for the mapping $x \mapsto 2x$ (mod. 1). Let x_0 be any number between 0 and 1; δ some small number, and define x_n for

$-\infty < n < \infty$ by

$$x_{n+1} = 2x_n + \delta \text{ (mod. 1)}.$$

Find $\hat{x}_0$ for the true orbit which shadows this pseudo-orbit.

We now take up the proof of the Shadow Lemma itself. We will proceed in a slightly unsystematic way in that we will give an outline of the construction which is a bit oversimplified and correct it as we go along. As noted above, we need only consider pseudo-orbits (x_j) defined for $j = 0, 1, 2,\ldots n$. We will give ourselves a δ-pseudo-orbit and show that it is ε-shadowed by a true orbit, where ε depends only on δ (and on the structure of Λ) and goes to zero with δ. Recall that we are only assuming that the pseudo-orbit is near Λ but that we want the true orbit which shadows it to be actually *on* Λ.

The construction of the shadowing orbit goes as follows:

– Find $\tilde{x}_0$ on Λ near x_0 such that $x_0 \in W^s_\eta(\tilde{x}_0)$. (Here and in what follows, η will denote some fixed number which, although small, is to be much larger than δ).

– Construct recursively $\tilde{x}_j, j = 1, 2,\ldots,n$ by

$$\tilde{x}_j \in W^u_\eta(T\tilde{x}_{j-1}) \qquad \text{and} \quad x_j \in W^s_\eta(\tilde{x}_j)$$

In other words: $\tilde{x}_j$ is obtained from $\tilde{x}_{j-1}$ by applying T; then moving around in the unstable direction until an orbit asymptotic in the forward direction to that of x_j is found.

– Take $\hat{x}_j = T^{j-n}\tilde{x}_n$.

The proof that this construction works amounts to showing that $\hat{x}_j$ stays near x_j. We do this in two stages:

(1) Show that $\tilde{x}_j$'s exist and that $d(\tilde{x}_j, x_j) \le K_1\delta$ for all j;

(2) Show that $d(\tilde{x}_j, \hat{x}_j) \le K_2\delta$ for all j.

(1) Existence of $\tilde{x}_j$. At this point we need some detailed information about how the stable and unstable manifolds for Λ are organized.

Lemma (3.4). There are positive numbers ζ, η such that, if $x \in \Lambda$ and $d(x, y) < \zeta$, then there is a unique $z \in W^u_\eta(x)$ such that $y \in W^s_\eta(z)$. There is, moreover, a constant B such that

$$d(x, z) \le B\, d(x, y),$$

$$d(y, z) \le B\, d(x, y).$$

The idea is indicated in fig. 9. Roughly, B may be large if the angle between $W_\eta^u(x)$ and $W_\eta^s(z)$ is small. But since stable and unstable manifolds always intersect transversally, and Λ is compact, it is possible to find a B large enough to work for all points of Λ.

The lemma is a by-product of the proof of the invariant manifolds theorem for hyperbolic sets, taking advantage of the fact that Λ is assumed to be an attractor.

Using the lemma, and the fact that $d(x_0, \Lambda) \leq \delta$, it follows that we can find $\tilde{x}_0$ satisfying $d(x_0, \tilde{x}_o) \leq B\delta$. We now look at the recursive construction of the successive $\tilde{x}_j$'s.

Since $\tilde{x}_j$ is supposed to be a point on $W_\eta^u(T\tilde{x}_{j-1})$ such that $x_j \in W_\eta^s(\tilde{x}_j)$. The existence of $\tilde{x}_j$ will follow from the lemma provided

$$d(T\tilde{x}_{j-1}, x_j) \leq \zeta.$$

Now

$$d(T\tilde{x}_{j-1}, x_j) \leq d(T\tilde{x}_{j-1}, Tx_{j-1}) + d(Tx_{j-1}, x_j).$$

Using the fact that x_{j-1} is on the stable manifold of $\tilde{x}_{j-1}$, we get

$$d(T\tilde{x}_{j-1}, Tx_{j-1}) \leq \lambda d(\tilde{x}_{j-1}, x_{j-1})$$

and by definition of pseudo-orbit, we have

$$d(Tx_{j-1}, x_j) \leq \delta.$$

As long as

$$d(T\tilde{x}_{j-1}, x_j) \leq \zeta,$$

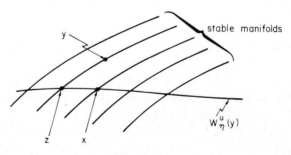

Fig. 9.

we can thus apply the estimates of the lemma to conclude:

$$d(\tilde{x}_j, x_j) \le B\ d(T\tilde{x}_{j-1}, x_j)$$

$$\le B\lambda\ d(\tilde{x}_{j-1}, x_{j-1}) + B\delta.$$

If $B\lambda < 1$, we thus get a propagated bound

$$d(\tilde{x}_j, x_j) \le \frac{B\delta}{1 - B\lambda}$$

and by considering only small δ's we can arrange to make

$$d(T\tilde{x}_{j-1}, x_j) \le (B\lambda)d(\tilde{x}_{j-1}, x_{j-1}) + B\delta \le \zeta$$

so the estimates hold together and we get $d(\tilde{x}_j, x_j) \le K_1\delta$. Unfortunately, there is no reason why $B\lambda$ should be less than 1, so we need to modify our strategy. The modification consists in considering T^M rather than T. We make three remarks:

(i) Replacing T by T^M replaces λ by λ^M; by making M large, we can make λ^M as small as we like.

(ii) B depends only on the geometry of the stable and unstable manifolds and on the metric used. The stable and unstable manifolds for T^M are the same as they are for T, so replacing T by T^M does not change B.

(iii) If the Shadow Lemma is true for T^M, it is true for T (with a different relation between ε and δ).

In view of these remarks, there is no genuine loss of generality in assuming $B\lambda < 1$, so the proof of (1) is complete.

(2) *We want to estimate* $d(\tilde{x}_j, \hat{x}_j)$ *where:*

$\tilde{x}_j$ is constructed in (1).

$$\hat{x}_j = T^{j-n}\tilde{x}_n.$$

We write

$$d(\tilde{x}_j, \hat{x}_j) \le d(\tilde{x}_j, T^{-1}\tilde{x}_{j+1}) + d(T^{-1}\tilde{x}_{j+1}, T^{-2}\tilde{x}_{j+2})$$

$$+ \ldots + d(T^{-(n-j)+1}\tilde{x}_{n-1}, T^{-(n-j)}\tilde{x}_n).$$

The right-hand side is a sum of terms each of which has the form

$$d(T^{-l}T\tilde{x}_{m-1}, T^{-l}\tilde{x}_m),$$

and there is one term for each $l = 1, 2 \ldots n - j$. But, by construction, $\tilde{x}_m$ is on the unstable manifold of $T\tilde{x}_{m-1}$, so

$$d(T^{-l}T\tilde{x}_{m-1}, T^{-l}\tilde{x}_m) \le \lambda^l d(T\tilde{x}_{m-1}, \tilde{x}_m)$$

$$\le \lambda^l [d(T\tilde{x}_{m-1}, Tx_{m-1}) + d(Tx_{m-1}, x_m) + d(x_m, \tilde{x}_m)]$$

$$\le \lambda^l \cdot K_3 \cdot \delta \quad \text{for some constant } K_3$$

Summing over l we get

$$d(\tilde{x}_j, \hat{x}_j) \le \frac{K_3 \lambda}{1-\lambda} \delta$$

which is the desired estimate.

The Shadow Lemma provides a powerful tool from proving the existence of many kinds of orbits. As an illustration, we will use it to show that periodic points are dense in any Axiom A attractor. By our definition, an Axiom A attractor contains a dense orbit $(T^n x_0)_{n \ge 0}$. Pick a point $x \in \Lambda$, and a very small positive ε. Take ε in particular small enough so that

$$d(T^n u, T^n v) \le \varepsilon \quad \text{for all } n \text{ implies } u = v.$$

Then take δ small enough so that any δ-pseudo-orbit on Λ is $\varepsilon/2$-shadowed by a true orbit. Then, choose n_1 so that

$$d(T^{n_1} x_0, x) \le \varepsilon$$

and finally choose $p > 0$ so that $d(T^{n_1+p} x_0, T^{n_1} x_0) \le \delta$. Make a periodic δ-pseudo-orbit with period p by

$$y_j = T^j(T^{n_1} x_0) \qquad 0 \le j \le p-1$$

and repeating periodically. Then $Ty_{j-1} = y_j$ except when j is a multiple of p; in the latter case

$$d(Ty_{j-1}, y_j) = d(T^{n_1+p} x_0, T^{n_1} x_0) \le \delta$$

so this does define a δ-pseudo-orbit. Let $(T^n \hat{x}_0)$ be an exact orbit which $\varepsilon/2$-shadows this pseudo-orbit. By the periodicity of the pseudo-orbit

$$d(T^{n+p} \hat{x}_0, T^n \hat{x}_0) \le \varepsilon$$

for all x. But this implies $T^p \hat{x}_0 = \hat{x}_0$ is a periodic point. Furthermore,

$$d(x, \hat{x}_0) \le d(x, T^{n_1} x_0) + d(T^{n_1} x_0, \hat{x}_0) \le \varepsilon + \frac{\varepsilon}{2} = \frac{3}{2}\varepsilon;$$

since ε can be taken arbitrarily small, the result follows.

The Shadow Lemma is often regarded as evidence for the reliability of numerical computation of chaotic behavior. The reasoning goes as follows. Suppose we use a computer to find approximate orbits for a mapping T which admits an Axiom A attractor, and suppose furthermore that we start from a point very near the attractor. What the computer produces is a δ-pseudo-orbit with a small δ, and the Shadow Lemma can then be applied to show that there is a true orbit on the attractor which stays near to the computed approximate orbit. So far, the reasoning is impeccable. We shall see later, however, that there is a well-defined notion of a *typical* orbit on a Axiom A attractor, and the real question is: Is the approximate orbit found by the computer approximately typical? In practice, the answer seems to be usually affirmative, but a simple example shows that caution is required. The example is the mapping

$$x \mapsto 2x \ (\text{mod. } 1) \text{ on } [0, 1)$$

for which, as we have seen, the Shadow Lemma is valid. Most computers doing binary arithmetic will actually evaluate this function exactly, but the resulting orbit will be highly non-typical – it will converge to zero after relatively few steps. Although it is fair to regard this example as contrived, it does show that there is no guarantee even in technically very favorable cases, that observed behavior on the computer will reflect the typical behavior of the underlying dynamical system.

4. Ergodic theory

In this section, we develop some ideas about the statistical theory of dynamical systems. It will be our point of view that the statistical element will be introduced through taking time averages along individual orbits rather than either through a random choice of initial condition or through the influence of external noise on the system.

The first point to be make is that the existence of time averages along orbits is *not* automatic for general systems (although it is automatic for Hamiltonian systems). To start with, it is not difficult to find a bounded sequence of numbers for which no limiting average exists. For example, we can take

$$a_n = 0 \qquad 0 \le n \le 10,$$
$$= 1 \qquad 11 \le n \le 100,$$
$$= 0 \qquad 101 \le n \le 1000,$$

etc.; then

$$\frac{1}{N} \sum_{j=0}^{N-1} a_j$$

oscillates as $N \to \infty$ and does not approach a limit. This example may seem quite remote from iterates of a mapping but a simple construction due to Mather shows that something similar can indeed be realized. The example is a differential equation in the plane, with two stationary solutions A and B and invariant curves connecting them (see fig. 10). Solution curves either inside or outside the resulting circle spiral respectively outward and inward towards the circle. When they get near to the circle, they must almost stop each time they arrive at A or B; the velocity fields near A and B are arranged so that orbits move essentially parallel to the circle there and so that the time it takes the orbit to get past is approximately inversely proportional to its distance from the circle. Between A and B, orbits move with reasonable speed, but in passing from A to B, or from B to A, an orbit moves roughly ten times closer to the circle. Thus, asymptotically, an orbit spends most of its time either stuck at A or stuck at B, and each sticking time is roughly ten times as long as the preceding one. Thus, the fraction of time that an orbit spends near A does not approach a limit as the length of time it is observed increases indefinitely.

In spite of the existence of examples like this one, in practice it always seems to happen that time averages along orbits exist. A general theory accounting for this fact does not exist although certain particular cases − such as that of an orbit converging to an Axiom A attractor − are well understood.

We will next develop some general machinery for analyzing the situation when time averages do exist. To begin with, we will say that an orbit

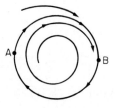

Fig. 10.

$(T^n x_o)_{n=0, 1, \ldots}$ is *statistically regular* if the limiting time average

$$\lim_{N \to \infty} \frac{1}{N} \sum_{n=0}^{N-1} f(T^n x_o)$$

exists for all continuous functions f. To avoid complicating the formulation of results, we will assume in this section that all orbits of the transformation T are bounded. It then follows from the Riesz Representation Theorem that there is a probability measure μ_{x_o} such that

$$\lim_{N \to \infty} \frac{1}{N} \sum_{n=0}^{N-1} f(T^n x_o) = \int f \, d\mu_{x_o}$$

for all continuous functions f. We will refer to μ_{x_o} as the *asymptotic distribution* of the orbit. Loosely, we can interpret $\mu_{x_o}(A)$ as the fraction of time that the orbit spends in A, but this interpretation cannot be taken too literally – it frequently happens that $\mu_{x_o}(A) \neq 0$ even when the orbit never actually enters A. We can give it an alternative interpretation with a somewhat more probabilistic flavor by saying that, if an n is chosen "at random" in $0, 1, 2, \ldots$ then $\mu_{x_o}(A)$ is the probability of finding $T^n x_o$ in A.

Now if x_o is any point whose orbit is statistically regular, and if f is any continuous function, then the time average of $f \circ T$ along the orbit of x_o is the same as that of f itself

$$\lim_{N \to \infty} \frac{1}{N} \sum_{n=0}^{N-1} f \circ T(T^n x_o) = \lim_{N \to \infty} \frac{1}{N} \sum_{n=1}^{N} f(T^n x_o)$$

$$= \lim_{N \to \infty} \frac{1}{N} \sum_{n=0}^{N-1} f(T^n x_o) + \lim_{N \to \infty} \frac{1}{N} [f(T^N x_o) - f(x_o)]$$

$$= \lim_{N \to \infty} \frac{1}{N} \sum_{n=0}^{N-1} f(T^n x_o).$$

In terms of the asymptotic distribution

$$\int f \, d\mu_{x_o} = \int f \circ T \, d\mu_{x_o},$$

i.e. the asymptotic distribution μ_{x_o} is *invariant* under the action of T. (A measure μ is said to be *invariant* under a measurable mapping if $\mu(T^{-1}A) = \mu(A)$ for all measurable sets A. Standard arguments show that, in the case at hand, this is equivalent to

$$\int f \circ T \, d\mu = \int f \, d\mu \qquad \text{for all continuous } f.)$$

Quite a lot can be said about measure preserving transformations in the abstract, i.e. without regard to questions of differentiability, continuity, etc. We will use the term *abstract dynamical system* to refer to a pair consisting of a measure and a measure preserving transformation. We are mostly interested in *probability* measures (i.e. measures assigning measure one to the whole space), and it is convenient to include in the definition of abstract dynamical system the requirement that the measure in question be a probability measure. Asymptotic distributions are, of course, probability measures. On the other hand, we definitely do not want to assume, in general, that T is invertible. It may be worth noting a very trivial example of an invariant measure. If x_o is a periodic point for T with period p, then the measure which assigns measure $1/p$ to each point of the orbit of x_o and measure zero to the rest of the space is manifestly an invariant measure. A striking early result of the theory of invariant measures is:

Theorem (4.1) Poincaré Recurrence Theorem). Let (T,μ) be an abstract dynamical system, A a measurable set, and let $A_0 = \{x \in A : T^n x$ eventually leaves A permanently$\}$. Then $\mu(A_0) = 0$.

Proof. Let $D_n = \{x : T^n x \in A; \ T^j x \notin A$ for $j > n\}$. (Here, we need only consider $n \geq 0$ and T need not be invertible.) Evidently, $D_n \cap D_m = \emptyset$ if $n \neq m$. Also, x is in D_n if and only if $T^n x$ is in D_0, i.e.

$$D_n = T^{-n} D_0,$$

so

$$\mu(D_n) = \mu(D_0) \qquad \text{for all } n.$$

Hence

$$\mu(D_0 \cup \ldots \cup D_{n-1}) = n\mu(D_0) \leq \mu(X) = 1.$$

Since n is arbitrary, $\mu(D_0) = 0$ so

$$\mu\left(\bigcup_{n=0}^{\infty} D_n\right) = 0.$$

But

$$A_0 = A \cap \left(\bigcup_{n=0}^{\infty} D_n\right)$$

so $\mu(A_0) = 0$ as desired.

From this result it follows easily that, if T is, for example, a continuous mapping on $\mathbb{R}^m$, then any T-invariant probability measure must assign measure zero to the complement of the non-wandering set for T. If, in particular, we have a Hamiltonian system such that

$$\{(q;p): H(q,p) \leq E\}$$

has finite volume for each E, we can apply Liouville's Theorem and the Poincare Recurrence Theorem, to conclude that the set of wandering points must have phase space volume zero. Since the set of wandering points is open, it must be empty, i.e. every point is non-wandering.

Theorem (4.2) (Birkhoff Pointwise Ergodic Theorem). Let (T, μ) be an abstract dynamical system f a measurable function such that

$$\int |f| \, d\mu < \infty.$$

Then

$$\bar{f}(x) = \lim_{N \to \infty} \frac{1}{N} \sum_{n=0}^{N-1} f(T^n x)$$

exists for almost all x and

$$\int \bar{f} d\mu = \int f \, d\mu$$

Furthermore,

$$\frac{1}{N} \sum_{n=0}^{N-1} f(T^n x)$$

converges to $\bar{f}$ in the L^1 norm, i.e.

$$\lim_{N \to \infty} \int d\mu \left| \bar{f}(x) = \frac{1}{N} \sum_{n=0}^{N-1} f(T^n x) \right| = 0.$$

For a proof, see, for example, Billingsley (1965).

Although this theorem seems to ensure that time averages exist, it is important not to read too much into it. It can be read simply as saying that the set of x's for which the average over n of $f(T^n x)$ does not exist is a set of measure zero for every invariant probability measure. Depending on the situation, we may or may not be willing to regard a set of points assigned measure zero by every finite invariant measure as "negligible". It is certainly reasonable to do so for Hamiltonian systems satisfying the sort of finite-volume conditions formulated above. On the other hand, if Λ is an attracting set and B its basin of attraction, then every point of $B \setminus \Lambda$ is wandering so any finite invariant measure assigns measure zero to $B \setminus \Lambda$. Thus, the Birkhoff Theorem says nothing about time averages along orbits starting off an attracting set and converging to it. Since most orbits for dissipative systems seem to fall into this category, the Birkhoff Theorem is not very helpful in proving the existence of time averages for dissipative systems.

An abstract dynamical system (T, μ) is said to be *ergodic* if the only measurable sets A such that $T^{-1}A = A$ are either sets of measure zero or complements of sets of measure zero; in other words, if the space cannot be split in a measure-theoretically non-trivial way into two invariant pieces. Ergodicity is a property of the pair (T, μ); it makes no sense to say that a mapping is ergodic without specifying an invariant measure (For a Hamiltonian system, it is usually understood that the measure is to be the microcanonical ensemble on an energy surface). The definition is easily seen to imply that an invariant measurable function – i.e., one satisfying $f \circ T = f$ – must be equal almost everywhere to a constant. (Proof: For any α, $\{f \leq \alpha\}$ is an invariant measurable set; hence, is either a null set or the complement of a null set. Then f is equal almost everywhere to sup $\{\alpha:$ $\{f \leq \alpha\}$ is a null set$\}$.) By confronting this fact with the Pointwise Ergodic Theorem, we get a remarkably strong consequence of ergodicity.

Proposition (4.3). Let (T, μ) be an ergodic dynamical system, f a measurable function with $\int |f| \, d\mu < \infty$. Then

$$\lim_{N \to \infty} \frac{1}{N} \sum_{n=0}^{N-1} f(T^n x) = \int f \, d\mu$$

for almost all x.

Proof. The limiting time average is an invariant function; hence, by ergodicity, must be equal to a constant almost everywhere. The constant equals its own integral, which by the Pointwise Ergodic Theorem is equal to the integral of f. (There is one fussy detail: the limiting time average need not – and generally will not – exist everywhere. To deal with this, apply the above argument to a function equal to the limiting time average where the latter exists and equal to zero elsewhere.)

This result is frequently formulated as: for ergodic systems, time averages equal ensemble averages. In general, there will be a set of x's where the limiting time averages either do not exist or have values different from the integral of f, and this pathological set will depend on f. In the case of greatest interest to us, where T is a continuous mapping on a region of Euclidean space or on a manifold, and where the measure μ has compact support, it is easy to show that the union of these pathological sets for all continuous functions is again a set of measure zero. (To do this, show that if the time average of f_k starting from x_0 equals the integral of f_k, for each of a sequence of f_k's dense in the supremum norm, then the same is true for every continuous f.)

Proposition (4.4). Let T be as in the preceding sections and let μ be an ergodic T-invariant measure with compact support. Then all x's except those in a set of μ-measure zero are statistically regular with asymptotic distribution μ.

For a Hamiltonian system ergodic with respect to the microcanonical ensemble, this says that almost every orbit asymptotically distributes itself uniformly over the energy surface.

The preceding proposition looks paradoxical if we apply it to two distinct ergodic measures μ_1 and μ_2; it says that, on the one hand, almost every orbit is asymptotically distributed according to μ_1, and, on the other hand, almost every orbit is asymptotically distributed according to μ_2. There is of course no contradiction: μ_1 and μ_2 give very different meanings to "almost every". There is in fact a general measure-theoretic result as follows:

Proposition (4.5). Let μ_1 and μ_2 be two distinct ergodic probability measures for a measurable transformation T. There is then a measurable set E such that

$$\mu_1(E) = 1; \qquad \mu_2(E) = 0$$

(i.e., μ_1 and μ_2 are concentrated on E and the complement of E respectively).

In practice, it is not usually feasible to prove that a concrete mapping is ergodic directly from the definition. The following criterion is more likely to be verifiable.

Criterion. If (T, μ) is ergodic, then

$$\lim_{N \to \infty} \frac{1}{N} \sum_{n=0}^{N-1} \int d\mu \, f \circ T^n \cdot g = \left(\int f \, d\mu \right) \cdot \left(\int g \, d\mu \right)$$

for all square-integrable $f \cdot g$. Conversely, if

$$\lim_{N \to \infty} \frac{1}{N} \sum_{j=0}^{N-1} \int d\mu \, f \circ T^n \cdot g = \left(\int f \, d\mu \right) \cdot \left(\int g \, d\mu \right)$$

for a set of f's and g's whose linear combination are dense in the space of square integrable functions, then (T, μ) is ergodic.

The direct statement follows easily from the Pointwise Ergodic Theorem and the Dominated Convergence Theorem for f and g bounded; the proof for general f, g follows from the bounded case by a straightforward approximation argument. To prove the converse statement, apply the same approximation argument to prove the limiting relation for all square-integrable f and g; if, then, A is an invariant measurable set, we put f and g equal to the characteristic function of A to obtain

$$\mu(A) = \mu(A)^2$$

so

$$\mu(A) = 0 \qquad \text{or} \quad \mu(A) = 1.$$

In particular, (T, μ) is ergodic if and only if

$$\lim_{N \to \infty} \frac{1}{N} \sum_{n=0}^{N-1} \mu(T^{-n} A \cap B) = \mu(A) \cdot \mu(B) \quad \text{for all } A, B,$$

(T, μ) is said to be *mixing* if

$$\lim_{n \to \infty} \mu(T^{-n} A \cap B) = \mu(A) \mu(B)$$

for all measurable sets A, B. It is thus immediate that the mixing property implies ergodicity. Exactly as with ergodicity, it is equivalent to the validity of

$$\lim_{n \to \infty} \int f \circ T^n \cdot g \, d\mu = \int f \, d\mu \cdot \int g \, d\mu$$

for any set of f's and g's whose linear combinations are dense in the space of square integrable functions.

We now discuss a few very simple examples.

(1) Irrational rotation. Let T denote the mapping $x \mapsto x + \alpha$ (mod. 1) on $[0,1]$, where α is irrational. Lebesgue measure dx is invariant under T. The abstract dynamical system (T, dx) is ergodic but not mixing: Ergodicity is easily proved by verifying that

$$\lim_{N \to \infty} \frac{1}{N} \sum_{n=0}^{N-1} \int dx \, f \circ T^n \cdot g = \int f \, dx \cdot \int g \, dx$$

for f, g of the form $e^{2\pi i k x}$; mixing disproved by considering

$$\int dx \, f \circ T^n \cdot g$$

for

$$f = e^{2\pi i x}; \qquad g = e^{-2\pi i x}.$$

(2) $x \mapsto 2x$ (mod. 1) on $[0,1]$. This mapping is not invertible and in fact is two-to-one. Nevertheless, it preserves Lebesgue measure: if A is any measurable set in $[0,1]$ then $T^{-1}A = (\frac{1}{2}A) \cup (\frac{1}{2}A + \frac{1}{2})$. The two pieces are disjoint: each has measure half that of A; so the total measure of $T^{-1}A$ is exactly the same as the measure of A.

This time, the abstract dynamical system (T, dx) is mixing. This can be seen in an elementary way by showing that

$$\lim_{n \to \infty} \int f \circ T^n \cdot g \, dx = \int f \, dx \cdot \int g \, dx$$

for any integrable f and any g which is a linear combination of characteristic function of integrals with dyadic endpoints. A moment's reflection shows that it will suffice to show that

$$\int_{(j-1)/2^n}^{j/2^n} f \circ T^n \, dx = \frac{1}{2^n} \int_0^1 f \, dx$$

for all j, n. But T^n maps $[(j-1)/2^n, j/2^n)$ onto $[0,1]$ linearly (with slope 2^n) so the desired equality follows by a simple change of variables.

(3) $x \mapsto 1 - 2x^2$ on $[-1, 1]$. Like the preceding example, this mapping is two-to-one. It does not preserve Lebesgue measure. It turns out, however, that there is a very special trick which makes its properties transparent; the

trick is to introduce a new variable θ related to x by

$$x = -\cos\theta \qquad 0 \le \theta \le \pi$$

i.e.

$$\theta = \arccos(-x).$$

Then

$$1 - 2x^2 = 1 - 2\cos^2\theta = -(2\cos^2\theta - 1) = -\cos(2\theta)$$

so, in terms of the θ variable, the mapping takes the form

$$\theta \mapsto \arccos(\cos(2\theta))$$

or

$$\theta \mapsto 2\theta \qquad 0 \le \theta \le \pi/2,$$
$$\mapsto 2\pi - 2\theta \qquad \pi/2 \le \theta \le \pi.$$

This mapping does preserve the probability measure $d\theta/\pi$ and, for exactly the same reason as in the preceding example, the resulting abstract dynamical system is mixing. We translate this statement back to the x variable.

$$\frac{d\theta}{\pi} = \frac{1}{\pi}\frac{dx}{(1-x^2)^{1/2}}.$$

Thus: the mapping $x \mapsto 1 - 2x^2$ leaves the measure $\pi^{-1}dx(1-x^2)^{-1/2}$ invariant, and the resulting abstract dynamical system is mixing. Since mixing implies ergodicity, we conclude in particular: there is a set X_0 in $[-1,1]$, of Lebesgue measure zero, such that any $x \notin X_0$ gives an orbit under $x \mapsto 1 - 2x^2$ which is statistically regular with asymptotic distribution $dx/\pi(1-x^2)^{1/2}$. This strikingly specific result can be "verified experimentally" by iterating the mapping on a computer starting from a randomly chosen initial point and collecting statistics about the distribution of the resulting orbit. Such experiments do indeed produce the predicted result.

It needs to be noted, however, that although the set X_0 has Lebesgue measure zero, it may appear quite large in other respects. For example: if $T(x)$ denotes $1 - 2x^2$, then it is easy to see that the graph of $T^n(x)$ runs from -1 to 1 and back 2^{n-1} times, from which it follows that $T^n(x)$ has at least 2^n fixed points. (By looking instead at the θ variable, one sees that it has exactly 2^n, and that they are all repelling.) Each of these fixed points lies on a periodic orbit for T which, although statistically regular, has a discrete asymptotic distribution. It is also possible, using simple symbolic-dynamics ideas to prove the existence of plentiful orbits which are statistically regular, whose asymptotic distributions are not discrete, but whose asymptotic distributions are not equal to $dx/\pi(1-x^2)^{1/2}$. (It follows from the ergodicity of $dx/\pi(1-x^2)^{1/2}$ that these non-standard asymptotic distributions cannot be absolutely continuous measures.) Similarly, there are many orbits which are not statistically regular. All these orbits must lie in X_0.

5. The Bowen–Ruelle ergodic theorem

The purpose of this section is to introduce a theorem which establishes a very satisfactory statistical theory for Axiom A attractors and to prove this theorem in the special case of the solenoid.

Theorem (5.1). Let Λ be an Axiom A attractor, B its basin of attraction. There is a subset X_0 of B, of Lebesgue measure zero, and an invarious probability measure μ on Λ such that

Every $x_0 \in B \setminus X_0$ is statistically regular with asymptotic distribution μ.

In other words, almost all orbits converging to an Axiom A attractor give rise to the same time averages. Moreover, the expression "almost all" is to be understood here in the intuitively natural sense of volume in the state space rather than in the sense of some invariant measure. Its seems very reasonable to regard the common asymptotic distribution μ of almost all orbits as the "equilibrium ensemble" for the attractor.

This theorem is due to Ruelle (1976). An improved version, and extension to flows, was given in Bowen and Ruelle (1975). For a systematic exposition, see Bowen (1975).

We will not attempt to outline the proof of this theorem, which relies on analytical tools – notably Markov partitions – which we have not developed. Instead, we will outline a simple and direct proof for the solenoid.

Recall that the solenoid is constructed from a mapping T of the solid torus $\{(z, w):|z|=1, \ |w| \leq 1\}$ into itself; T is given by the simple formula

$$T(z, w) = (z^2, \tfrac{1}{2}z + \tfrac{1}{4}w).$$

As a preliminary remark, we note that if x_1, x_2 are two points of the solid torus with the same z co-ordinate – i.e., belonging to the same transverse disk – then

$$d(T^n x_1, T^n x_2) \to 0 \qquad \text{as } n \to \infty.$$

Since statistical regularity is defined in terms of time averages of *continuous* functions, it follows that x_1 is statistically regular if and only if x_2 is and that their asymptotic distributions are then the same. In short: Asymptotic properties of $T^n x$ depend only on the z co-ordinate of x.

Now write $z = e^{2\pi i \theta}$ with $0 \leq \theta \leq 1$. The θ coordinate of Tx does not depend on w and is given simply by 2θ (mod. 1). For present purposes, write $t(\theta)$ for 2θ (mod. 1). We have seen that t leaves $d\theta$ invariant and that the abstract dynamical system $(t, d\theta)$ is ergodic. Hence: for almost all θ_0, the orbit $[t^n(\theta_0)]_{n=0,1,2,\dots}$ is statistically regular with asymptotic distribution $d\theta$. Let

$$\Theta_0 = \{\theta_0 \in [0,1): \ t^n(\theta_0) \text{ is } not \text{ statistically regular with asymptotic} \\ \text{distribution } d\theta\}$$

and let

$$X_0 = \{(e^{2\pi i \theta_0}, w): \theta_0 \in \Theta_0\}.$$

Since Θ_0 is a set of linear Lebesgue measure zero, X_0 is a set of three-dimensional Lebesgue measure zero. We are going to show that

– Every point of the solid torus which is *not* in X_0 is statistically regular
– These points all have the same asymptotic distribution.

To do this we pick an x_1 outside X_0 and a continuous function and show that

$$\lim_{N \to \infty} \frac{1}{N} \sum_{n=0}^{N-1} f(T^n x_1)$$

has a value independent of x_1. By our choice of X_0, this is immediately true if $f(z, w)$ depends only on z. In that case:

$$\lim_{N \to \infty} \frac{1}{N} \sum_{n=0}^{N-1} f(T^n x_1) = \int_0^1 d\Theta \, f(e^{2\pi i \theta}, w).$$

The general idea is now the following: take any continuous f and some reasonably large m. Then $f(T^m x)$ is almost independent of w (since changing w with θ fixed only moves $T^m x$ around in a disk of radius $(\frac{1}{4})^m$), so the preceding argument *almost* implies the existence of the average over n of $f(T^{n+m} x_1)$. But this average is the same as the average over n of $f(T^n x_1)$.

Here is one way to make this precise. Given a continuous function and given m, we define

$$f_m^+(z, w) = \max_{w'} f[T^m(z, w')],$$

$$f_m^-(z, w) = \min_{w'} f[T^m(z, w')].$$

Although we have written f_m^+ and f_m^- as functions of z and w, they evidently depend only on z. By a preceding remark, we can make

$$\max_{z, w} \{f_m^+(z, w) - f_m^-(z, w)\}$$

as small as we like by making m large. Now

$$\int_0^1 f_m^+(e^{2\pi i\theta}, w)\,d\theta = \lim_{N \to \infty} \frac{1}{N} \sum_{n=0}^{N-1} f_m^+(T^n x_1)$$

$$\geq \lim_{N \to \infty} \sup \frac{1}{N} \sum_{n=0}^{N-1} f(T^{n+m} x_1) = \lim_{N \to \infty} \sup \frac{1}{N} \sum_{n=0}^{N-1} f(T^n x_1)$$

$$= \lim_{N \to \infty} \inf \frac{1}{N} \sum_{n=0}^{N-1} f_m^-(T^n x_1) = \int_0^1 f_m^-(e^{2\pi i\theta}, w)\,d\theta.$$

Simplifying by removing some intermediate terms:

$$\int_0^1 f_m^+(e^{2\pi i\theta}, w)\,d\theta \geq \lim_{N \to \infty} \sup \frac{1}{N} \sum_{n=0}^{N-1} f(T^n x) \geq \lim_{N \to \infty} \inf \frac{1}{N} \sum_{n=0}^{N-1} f(T^n x_1)$$

$$\geq \int_0^1 f_m^-(e^{2\pi i\theta}, w)\,d\theta.$$

We now let $m \to \infty$; the two quantities at the ends of this inequality, which do not depend on x_1, approach each other, and so we get that

$$\lim_{N \to \infty} \frac{1}{N} \sum_{n=0}^{N-1} f(T^n x_1)$$

exists. We also see that it equals, for example,

$$\lim_{m \to \infty} \int_0^1 f_m^+(e^{2\pi i\theta}, w)\,d\theta.$$

To see how to compute the standard time average as an ensemble average, we can proceed as follows: pick any w_0, and form

$$f_m(\theta) = f[T^m(e^{2\pi i\theta}, w_0)]$$

then

$$\int_0^1 f_m^-(e^{2\pi i\theta}, w)\,d\theta \le \int_0^1 f_m(\theta)\,d\theta \le \int_0^1 f_m^+(e^{2\pi i\theta}, w)\,dw.$$

Hence,

$$\lim_{N \to \infty} \frac{1}{N} \sum_{n=0}^{N-1} f(T^n x_1) = \lim_{m \to \infty} \int_0^1 f_m(\theta)\,d\theta$$

for all $x_1 \notin X_0$ and all continuous f. Now $\int_0^1 f_m(\theta)\,d\theta$ is an average of f over a curve which winds 2^m times around the torus and is contained in the image of T^m. We can view that average as obtained by
— averaging f with respect to θ over each of the turns around the torus.
then
— forming the unweighted average of the resulting 2^n numbers.
It is now easy to see how to pass to the limit $m \to \infty$. The resulting prescription for computing $\int f d\mu$ can be formulated as follows:
— Cut the attractor along the plane $\{z = 1\}$. It falls apart into infinitely many loops, which are labelled by points of the transverse Cantor set or, equivalently, by sequences $(\dots i_{-2}, i_{-1})$ of zeros and ones.
— Integrate with respect to around each loop. This gives a numerical function on the transverse Cantor set.
— Average over the Cantor set with respect to the measure assigning "equal weight to each point" i.e., which makes the discrete co-ordinates $i_{-1}, i_{-2}, \dots$ mutually independent and assigns probability $\frac{1}{2}$ to each of the values zero and one.

Roughly, we can say that the equilibrium ensemble assigns the same probability to each loop and is uniformly distributed with respect to angles over the loops.

The simple and explicit form we have obtained for the equilibrium ensemble depends on the simple form of the mapping producing the solenoid. If the mapping is subjected to a small perturbation, the formulas cease to hold but the general scheme of the argument persists. Both the averages over loops and those over the transverse Cantor set acquire weights. It turns out that computing these weights reduces to finding the thermodynamic equilibrium state for a one-dimensional lattice spin system (spin $\frac{1}{2}$) with a translation-invariant rapidly decreasing many body interaction. For full details, see the references cited above.

Note added, March 1983: On reviewing these notes nearly two years after writing them, I find naturally enough many things I would now do differently. One of these is serious enough so the reader should be warned about it: it is the proof of the existence of shadowing orbits (Theorem 3.3). I have been entirely converted to the more abstract but much more powerful approach based on reducing the construction of a shadowing orbit to a fixed point problem for a nonlinear operator with hyperbolic derivative. This approach yields more comprehensive results than Theorem 3.3, with simpler proofs. A systematic discussion of shadowing from this point of view can be found in Shub (1978).

References

P. Billingsley, Ergodic Theory and Information (Wiley, New York, 1965).

R. Bowen, Equilibrium states and the ergodic theory of Anosov diffeomorphisms, Lectures Notes in Math., vol. 470 (Springer, Berlin, 1975).

R. Bowen, On Axiom A Diffeomorphisms, Regional Conference Series no. 35 (1978).

R. Bowen and D. Ruelle, The ergodic theory of Axiom A flows, Inv. Math. 29 (1975) 81–202.

P. Hartman, Ordinary Differential Equations (Wiley, New York, 1964).

M.W. Hirsch and C. Pugh, Stable manifolds and hyperbolic sets in: Global Analysis, AMS Proc. Symp. Pure Math., Vol. 14, (1970) M. 133–163.

M.W. Hirsch, C. Pugh and M. Shub, Invariant Manifolds, Lecture Notes in Math., Vol. 583 (Springer, Berlin, 1977).

J. Marsden and M. Mc Cracken, The Hopf Bifurcation and its Applications, Appl. Math., Vol. 19 (Springer, Berlin, 1976).

D. Ruelle, A measure associated with Axiom A attractors, Am. J. Math. 98 (1976) 619–654.

D. Ruelle, Small random perturbations of dynamical systems and the definition of attractors, Commun. Math. Phys. 82 (1981) 137–151.

M. Shub, Stabilité globale des systèmes dynamiques, Astérisque 56 (1978).

S. Smale, Differentiable dynamical systems, Bull. AMS 73, (1967) 747.

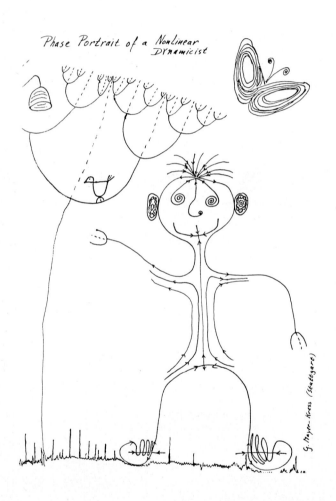

Phase Portrait of a NonLinear Dynamicist

G. Mayer-Kress (Stuttgart)

COURSE 2

NUMERICAL EXPLORATION OF HAMILTONIAN SYSTEMS

Michel HÉNON

C.N.R.S., Observatoire de Nice,
B.P. 252, 06007 Nice Cedex, France

G. Iooss, R.H.G. Helleman and R. Stora, eds.
Les Houches, Session XXXVI, 1981 — Comportement Chaotique des Systèmes Déterministes/
Chaotic Behaviour of Deterministic Systems
© *North-Holland Publishing Company, 1983*

Contents

1. Generalities

The central subject of this course will be the use of numerical experiments for exploring the fascinating properties of dynamical systems. Actually only Hamiltonian systems will be considered, for three reasons: first, space is limited; second, this is the case which I know best (most astronomical problems fall into this category); third, the case of non-Hamiltonian or dissipative systems will be described in detail in other courses, in this volume.

At the same time, since this course is almost at the beginning of the volume, I have tried to organize it in such a way that it can serve as a first introduction to dynamical systems. No previous knowledge of the subject will be assumed, and the general level is rather elementary; this course is intended for newcomers, not for specialists.

Numerical experiments are just what their name implies: experiments. In describing and evaluating them, one should enter the state of mind of the experimental physicist, rather than that of the mathematician. Numerical experiments cannot be used to prove theorems; but, from the physicist's point of view, they do often provide convincing evidence for the existence of a phenomenon. We will therefore follow an informal, descriptive and non-rigorous approach. Briefly stated, our aim will be to *understand* the fundamental properties of dynamical systems rather than to *prove* them. Mathematical results will be mentioned only when they play an essential role in the description of the events (e.g. the KAM theorem, section 6.5, or the theory of Lyapunov characteristic exponents, section 6.7).

Finally, this is a course and not a review paper. By now there exists a large collection of numerical studies on Hamiltonian systems. I shall not attempt to review this collection; instead, only a few illustrative examples will be described. The selection of these examples does not imply that they are in any way superior to others; it reflects only my degree of familiarity with them. An extensive list of references on dynamical systems can be found in the recent review paper by Helleman (1980).

1.1. Dynamical systems

From the point of view of the physicist, a dynamical system can be defined as any physical system such that

(i) its state at a given time is completely defined by the values of N variables $x_1, \dots, x_N$;

(ii) its evolution is given by a system of N ordinary differential equations:

$$
\begin{aligned}
\mathrm{d}x_1/\mathrm{d}t &= f_1(x_1, \dots, x_N), \\
&\;\;\vdots \\
\mathrm{d}x_N/\mathrm{d}t &= f_N(x_1, \dots, x_N).
\end{aligned}
\tag{1}
$$

The N dependent variables can represent arbitrary physical quantities: positions, velocities, angles, temperatures, pressures, concentrations, etc. Many real problems can be represented in this way, not only in physics but also in other sciences, notably biology and chemistry.

By defining a vector X of components $x_1 \dots x_N$ and a vector F with components $f_1 \dots f_N$, we can write the differential system more simply as

$$
\mathrm{d}X/\mathrm{d}t = F(X).
\tag{2}
$$

N is the *order* of the dynamical system.

The fact that all equations are assumed to be of first order is not a restriction as it might seem at first view: a differential system can always be brought into this form by introducing additional variables.

The system (1) is *autonomous*, which means that t does not appear on the right-hand side. Again this is not really a restriction: a non-autonomous system can be transformed into an autonomous system by the introduction of additional variables.

1.2. Phase space

A very useful representation of a dynamical system is the *phase space*. This is simply an N-dimensional space with $x_1, \dots, x_N$ as coordinates. The state of the system at a given time is represented by a point in phase space. This point moves with time; its velocity vector is F. This velocity is known from eq. (1); in fact, given the system (1), one can immediately draw the whole velocity field in phase space (fig. 1). The representative point describes a curve, called *trajectory* or *orbit*, which is tangential to the vector field at

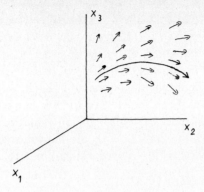

Fig. 1.

every point. Thus, simply by plotting the velocity field in phase space and without doing any integrations, one can already obtain some idea of the shape of the solutions. Note that this is possible only because all differential equations in eq. (1) are of the first order and the system is autonomous.

Through a given point of phase space, there passes in general one trajectory and only one. Physically: if the state of the system is known at a given time, then its future evolution is determined. Its past is also determined, because eq. (1) can be integrated backwards in time.

As a consequence, two trajectories started at two different points at a given time can never come to coincide at some later time.

1.3. Integrals

An integral of a differential system is a function $I(x_1,\ldots,x_N)$ whose value is constant on any given trajectory. In other words, it identically satisfies the equation

$$\frac{\partial I}{\partial x_1} f_1 + \ldots + \frac{\partial I}{\partial x_N} f_N = 0. \tag{3}$$

Consider in phase space the subspace defined by $I(X) = C$, where C is a constant. It is an $(N-1)$-dimensional manifold. If now we give to C all possible values, we obtain a one-parameter family of such manifolds, which fills phase space completely. A given trajectory is constrained to stay on a particular manifold. Therefore, if we know an integral I, we can effectively reduce the order of the system by one unit. To do this, we choose a definite value for C; we solve the equation $I(X) = C$ for x_n, for instance; we substitute this expression for x_n in the first $N-1$ equations of eq. (1), and we drop the last equation. A system of order $N-1$ results.

If we are interested in a particular trajectory, then the value of C is determined by the initial conditions. If, on the other hand, we want to study all solutions of the system, then C must be considered as a parameter, and the $(N-1)$th-order system must be studied for all values of this parameter.

More generally, if p integrals $I_1, \dots, I_p$ are known, then each trajectory is constrained to stay on a $(N-p)$-dimensional manifold, defined by

$$I_1(X) = C_1, \dots, I_p(X) = C_p, \tag{4}$$

and the order of the system can be lowered to $N-p$.

Integrals are frequently derived from physical considerations (conservation laws). Apart from that, unfortunately, no systematic procedure is known for finding the integrals of a given system.

1.4. Various cases

Given a dynamical system in the form (1), two cases can arise (fig. 2):

(1) The general solution can be written explicitly. It is in the form $X(a_1, \dots, a_N, t)$, where the a_i are the integration constants. An example is the gravitational two-body problem. This case is often encountered in textbooks, much more rarely in practice! The problem can then be considered as completely solved, since any desired information can usually be easily obtained from the explicit solution; for instance, obtaining the trajectory

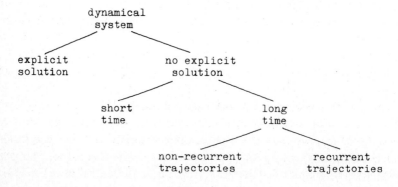

Fig. 2.

which corresponds to given initial conditions amounts to solving a system of N ordinary equations for the a_i. Therefore this case will be left aside.

(2) No explicit general solution is known. An example is the gravitational three-body problem. We can again distinguish two cases:

(2.1) The solution is desired only for a finite length of time, of the same order as the natural time scale of the system. Example: computation of the positions of the planets for the next hundred years; computation of the orbit of an interplanetary probe. Straightforward numerical integration then gives the desired answers.

(2.2) One is interested in the solution for a time very long compared to the time scale of the system. Examples: problem of the long-term stability of the solar system: construction of a particle accelerator or storage ring, where the particles may undergo up to 10^{11} revolutions. Essentially, what one wants to know in such cases is the asymptotic behaviour of solutions for $t \to \infty$.

We must finally distinguish between two kinds of trajectories:

(2.2.1) *Non-recurrent* trajectories are those which, roughly speaking, do not come back to regions which they have already visited; the simplest example is a trajectory which goes to infinity. In that case, not much can be said in general: since the trajectory is constantly entering new territory in phase space, anything can happen, depending on the shape of the velocity field which is found there. Thus no general theory is possible. One example is the hyperbolic case of the two-body problem, where the two bodies escape to infinity.

(2.2.2) *Recurrent* trajectories, on the contrary, are those which come back again and again, indefinitely, to any part of phase space which they have once visited. This is the case which gives rise to the most interesting and complex behaviour. The fact that the same regions of phase space are visited again and again means that the trajectory cannot be completely arbitrary; certain regularities must be present. This gives a hold to the theory, and allows one to find properties which are true for large classes of dynamical systems, not just for one. This last case, which also happens to include the most interesting real dynamical systems, is the one which we will consider from now on (and also in most other courses in this volume).

It should be noted that a given dynamical system can exhibit both recurrent and non-recurrent trajectories; witness again the two-body problem. Attention will then be confined to that part of phase space which corresponds to recurrent trajectories. (Non-recurrent and recurrent trajectories are also called, respectively, *wandering* and *non-wandering* trajectories.)

1.5. Approaches

We have another logical tree (fig. 3) for the possible approaches to the study of dynamical systems. We distinguish two cases:

(1) The interest lies with one particular instance of a dynamical system, usually representing a specific concrete problem. The aim is then to gain as much information as possible on that specific system. In particular, the equations should be made as realistic as possible, even if a more complicated system results.

(2) Conversely, one may be interested in dynamical systems in general, independently of applications. This approach should be again subdivided, depending on the method used:

(2.1) One may try to prove rigorously certain properties of dynamical systems, usually under some set of assumptions. This is the mathematical approach; it will be described in several courses in this volume.

(2.2) The other possible approach, which will be the subject of the present course, belongs to what I like to call *experimental mathematics*. This is actually an approach of wide applicability throughout mathematical problems. In the case of dynamical systems, it consists in choosing a particular system, and in studying its properties through numerical computation, usually but not necessarily with a computer. Superficially, this might look similar to case (1) above; but in fact the motivations are completely different, and these two approaches are really very far apart (as shown by fig. 3). One wants now to throw light on the general properties of dynamical systems. Accordingly, the criteria for the choice of the basic differential

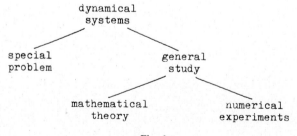

Fig. 3.

equations are completely different. One has, in fact, almost complete freedom in the choice of these equations; but this freedom should be used responsibly. The equations should be as simple as possible, in order to facilitate the study; but not simple to the point that the problem becomes trivial. Also, the equations should be selected so as to exhibit in the best possible way the particular properties which one wants to study.

I would also like to give the following essential advice to those of you who might be contemplating numerical studies for the first time: never run blindly into numerical experiments, but plan them carefully, just as you would for laboratory experiments. The ease with which a model computer can be programmed and run has its dangerous side, and one can easily produce huge stacks of worthless computer output! You should first have in mind a well-defined question, precisely formulated, about dynamical systems. Then you should give careful thought to how this question can best be attacked by numerical experiments; and then only should you engage into actual computations, always keeping in mind that these computations are not an end in themselves, and that the aim is not to compute thousands of trajectories in record time but to arrive at a better understanding of the behaviour of dynamical systems in general.

In a similar spirit, before embarking on a detailed description of numerical results we will review in sections 2 to 4 a number of basic facts and definitions.

2. Surfaces of section, mappings

2.1. Definition and properties

Since we are interested in the asymptotic behaviour of a trajectory, it is not really necessary to follow this trajectory in great detail; it might be sufficient to "sample" it from time to time. This is the idea behind the method of the surface of section. For the exposition, it will be convenient to consider first the case $N = 3$, so that phase space can be pictured as the familiar three-dimensional space. We select in this phase space a *surface of section* Σ which in the present case will be an ordinary two-dimensional surface; and we consider the successive intersections Y_0, Y_1, Y_2,... of the trajectory with Σ (fig. 4). Since we assume the trajectory to be recurrent, there will be an infinite sequence of such points. The sequence can also be extended towards the past: ... Y_{-3}, Y_{-2}, Y_{-1}, Y_0,... .

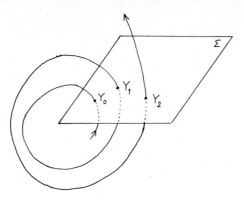

Fig. 4.

This sequence has the fundamental property that if one of its points Y_i is given, then the next point Y_{i+1} can be deduced. This is quite obvious: all one has to do is to follow the trajectory from Y_i, by integrating the differential equations, until it intersects Σ again: this new intersection is Y_{i+1}. We have thus a mapping G of Σ on itself which is called the *Poincaré map*. Thus:

$$Y_{i+1} = G(Y_i). \tag{5}$$

More generally, one has

$$Y_{i+j} = G^j(Y_i). \tag{6}$$

Since the trajectory can be integrated towards both directions of time, the mapping is invertible: G^{-1} exists, and

$$Y_{i-1} = G^{-1}(Y_i). \tag{7}$$

More generally, eq. (6) holds for any j, positive or negative.

The generalization to an arbitrary order N is obvious: we select, in the N-dimensional phase space, an $(N-1)$-dimensional subspace Σ, which should now be properly called a *space of section*, but is in practice frequently referred to as a "surface of section". We then consider the successive intersections... Y_{-1}, Y_0, Y_1, Y_2,... of the trajectory with Σ. We shall call $M = N - 1$ the dimension of the space of section. We introduce in this space a system of coordinates $y_1,\ldots,y_M$, and we represent the coordinates of the point Y_i by $y_{i,1},\ldots,y_{i,M}$. Then the mapping G can be written as

$$y_{i+1,1} = g_1(y_{i,1},\dots,y_{i,M}),$$
$$\vdots \tag{8}$$
$$y_{i+1,M} = g_M(y_{i,1},\dots,y_{i,M}).$$

When considering the mapping itself, rather than the sequence of points, it will often be convenient to adopt a simpler notation: two consecutive points Y_i and Y_{i+1} are represented by Y and Y', and eq. (8) becomes

$$y'_1 = g_1(y_1,\dots,y_M),$$
$$\vdots \tag{9}$$
$$y'_M = g_M(y_1,\dots,y_M).$$

Now comes the fundamental step: we decide to consider only the sequence of points Y_i from now on, and to forget about the rest of the trajectory, i.e. the detail of what happens between two intersections. The sequence is considered as being defined directly by the mapping G, and not any more by the system of differential equations, which will also be forgotten. The justifications for doing this are as follows:

(1) Experience shows that the essential properties of the differential system are reflected in equivalent properties of the mapping. A close correspondence exists between these two objects. Example: a simple periodic orbit of the differential system, closing back upon itself after one revolution, corresponds to a fixed point of the mapping G; the periodic orbit is stable if and only if the fixed point is stable; and so on.

(2) The new problem is much simpler: we have only the ordinary equations (9) to consider instead of the differential equations (1). Also the dimension of the relevant space has been reduced by one unit: a dynamical system of order N gives rise to a mapping in an $(N-1)$-dimensional space. Therefore the study is easier, both for theory and for numerical experiments.

(3) The essential properties of the system, related to the long-term behaviour, are more clearly seen because the irrelevant details of the short-term evolution have been eliminated. (Conversely, the surface-of-section would be of no use for a short-term problem, such as the trajectory of an interplanetary probe!)

(4) The graphical representation of the results is easier. For instance for $N=3$, the surface of section has two dimensions and is much more easily represented than the three-dimensional phase space.

(5) Computing time at first is not improved by the introduction of a surface of section. If our problem is originally defined by a set of differential equations, and we introduce the surface of section afterwards, then in general it is not possible to obtain explicit equations for the mapping G; the only way to compute the image of a point Y_i is to go back to the differential equations and to integrate the trajectory until the next intersection is reached.

It is here, however, that the "philosophy" of numerical experiments comes into play in a crucial way. Remember that we are not interested in a particular dynamical system, but rather in the general properties of dynamical systems. Now the study of dynamical systems can be reduced to the study of iterated mappings, through the introduction of a surface of section. We may as well, then, attack this latter subject directly. Instead of defining the mapping G implicitly through the differential equations, we shall define it by giving explicitly the functions $g_1,\ldots,g_M$ in eq. (9).

This reduces the computing time drastically (typically by a factor of the order of 1000), since all we have to do now is to evaluate expressions instead of integrating differential equations. The accuracy is also much better; only round-off errors are left.

The chosen mapping G should be invertible, in order to correspond to the case of a dynamical system. This seems to be the only condition. Numerous experiments have shown that essentially the same properties are found in mappings defined implicitly from dynamical systems and in mappings defined explicitly from given equations.

When studying a mapping, we shall sometimes refer to the infinite sequence of points Y_i as a "trajectory", or "orbit", since this is now the equivalent of the original trajectory in phase space.

2.2. Fixed points

A fixed point is a point Y^* which satisfies

$$Y^* = G(Y^*). \tag{10}$$

This is a system of M ordinary equations for M unknowns; thus, finding fixed points poses no problem in principle.

Fixed points play a very important role in mappings because they force a definite structure of the trajectories in their neighbourhood. To study this, we write

$$Y = Y^* + U \qquad (11)$$

where U is taken to be small; substituting in eq. (5) and developing in series, we obtain

$$U_{i+1} = (\partial G/\partial Y)_{Y=Y^*} U_i + O(U_i^2). \qquad (12)$$

$\partial G/\partial Y$ is a $M \times M$ matrix, called the *Jacobian* of the mapping. If we neglect the last term in eq. (12), we have a linear mapping for U, which can be analysed in the standard way. I shall consider here only the general case where the matrix has M distinct eigenvalues. What happens when we iterate the mapping G depends on the nature of these eigenvalues. We consider first a real eigenvalue λ, and an associated real eigenvector V. If we choose as the initial displacement from the fixed point $U_0 = V$, then the successive values of U are

$$U_1 = \lambda V, \ldots, U_j = \lambda^j V, \ldots. \qquad (13)$$

Thus, the sequence of points Y_j lies on a straight line passing through the fixed point. The arrangement of the points depends on the value of λ; there are essentially four distinct cases, represented on fig. 5 (not counting the particular cases $\lambda = 0$, $\lambda = 1$, and $\lambda = -1$). Successive points move away from the fixed point Y^* if $|\lambda| > 1$ (cases a and d), and tend towards Y^* if $|\lambda| < 1$ (cases b and c). They stay on the same side if $\lambda > 0$ (cases a and b), or jump regularly from one side to the other if $\lambda < 0$ (cases c and d).

Next we consider the case of a pair of complex conjugate eigenvalues;

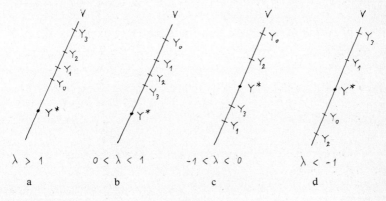

Fig. 5.

it will be convenient to write them as $\varrho e^{\pm i\varphi}$, with ϱ and φ real. They are associated with a pair of complex conjugate eigenvectors $V_1 \pm V_2 i$, with V_1 and V_2 real. We are interested only in real points Y_j, so we consider an initial displacement $U_0 = aV_1 + bV_2$, with a and b real. Then it is easily computed that

$$U_j = \varrho^j [(aV_1 + bV_2) \cos j\varphi + (bV_1 - aV_2) \sin j\varphi]. \tag{14}$$

We consider first the case $\varrho = 1$, i.e. the two eigenvalues have modulus 1. The first term inside the bracket of eq. (14) represents an oscillatory motion along the real direction $aV_1 + bV_2$. The second term represents a similar oscillation, but 90° out of phase, along another real direction $bV_1 - aV_2$. The sum of the two terms therefore represents an elliptical motion: the sequence of points Y_j lies on an ellipse in the (V_1, V_2) plane (fig. 6a).

Now if $\varrho > 1$, this elliptical motion is combined with an expansion due to the factor ϱ^j, and we have an outwards spiralling motion (fig. 6b). Similarly, if $\varrho < 1$, we have an inwards spiralling motion (fig. 6c).

In all cases (λ real or complex), the sequence of points Y_j moves away from the fixed point Y^* if $|\lambda| > 1$, and tends toward that point if $|\lambda| < 1$.

Finally, the trajectory starting from an arbitrary initial displacement can be obtained, as usual, by decomposing U_0 into its components along the eigenvectors, finding the evolution of each component from the above rules, and summing again (since we are using here a linear approximation of the mapping).

2.3. Cycles

A *cycle of period n*, or *n-cycle*, is a finite sequence of points $Y_0, \ldots, Y_{n-1}$,

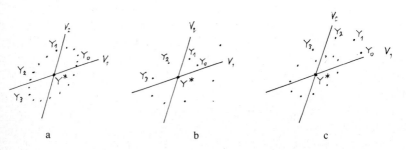

Fig. 6.

such that

$$Y_1 = G(Y_0), \dots, Y_{n-1} = G(Y_{n-2}), \qquad Y_0 = G(Y_{n-1}). \tag{15}$$

In other words, it is a periodic trajectory, which comes back to the same point after n iterations of the mapping. Every point Y_j of the cycle satisfies

$$Y_j = G^n(Y_j) \tag{16}$$

and is therefore a fixed point of the mapping G^n.

A fixed point of G is simply a cycle of period 1. Thus, there is no fundamental difference between fixed points and cycles. All that has been said in section 2.2 is also applicable to cycles, simply by considering G^n instead of G. In particular, a cycle is characterized by M eigenvalues, which determine the properties of neighbouring trajectories (it can be shown easily that these eigenvalues are the same for all points of the cycle).

2.4. Invariant manifolds

We come back to the vicinity of a fixed point, considered in the linear approximation. We consider the set of eigenvectors corresponding to eigenvalues with modulus less than 1. They define a *contracting subspace*, or *stable subspace*, V^s. If the initial displacement U_0 of a trajectory is in V^s, then the sequence of points Y_j tends towards the fixed point Y^* for $j \to +\infty$. Similarly, the eigenvectors corresponding to eigenvalues with a modulus greater than 1 define an *expanding subspace*, or *unstable subspace*, V^u. If the initial displacement U_0 is in V^u, the sequence of points Y_j tends towards the fixed point Y^* for $j \to -\infty$.

If now we move away from the fixed point, dropping the linear approximation and considering again the exact mapping G, we can extend these concepts as follows: we define the *stable invariant manifold* W^s of a fixed point Y as the set of points Y_0 such that the trajectory passing through Y_0 tends to Y^* for $j \to +\infty$; and similarly, the *unstable invariant manifold* W^u as the set of points Y_0 such that the trajectory passing through Y_0 tends to Y^* for $j \to -\infty$.

In the vicinity of Y^* and in the linear approximation, W^s and W^u usually reduce to V^s and V^u.

These invariant manifolds also play an important role in determining the general properties of the trajectories. To see why, we consider the case

$M = 2$, and a fixed point with two real eigenvalues, one being larger than 1 and the other less than 1 in modulus. V^s and V^u both are straight lines (fig. 7); W^s and W^u are curves, tangent to V^s and V^u at Y^*. These curves can be easily computed in practice: for instance W^u is determined by taking points on V^u and close to Y^*, and computing the trajectories which originate from these points. W^s is determined in the same way by taking initial points on V^s and iterating backwards, i.e. using G^{-1} instead of G.

Consider now a point Y_0 near W^s. The sequence of points emanating from Y_0 will first move towards Y^* and W^s, and then away from Y^* along a branch of W^u (fig. 7). If we displace slightly the initial point into Y_0' on the other side of the curve W^s, the other branch of W^u is followed. Thus, the future of the trajectory is completely different when the initial point is on one side of W^s or the other. Similarly, the points lying on one side or the other of W^u have completely different past trajectories. For this reason, these curves are sometimes called *separatrices*.

Stable and unstable invariant manifolds are defined in a similar way for a cycle of arbitrary period n. W^s now consists of n pieces, emanating from the n points of the cycle, and images of each other under G; the same is true of W^u.

The *phase portrait* of a dynamical system is the figure obtained by representing the most important fixed points and cycles and their stable and unstable invariant manifolds.

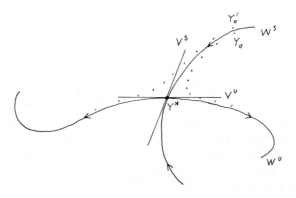

Fig. 7.

3. Hamiltonian systems

3.1. Definition and properties

Hamiltonian systems are a particular case of dynamical systems, but a very important one since many applications fall within their province. Also their properties are very different from those of the general, non-Hamiltonian case.

A Hamiltonian system is characterized, first, by an even number of dimensions:

$$N = 2n. \tag{17}$$

The number n is traditionally called the *number of degrees of freedom*; it should not be confused with N, the dimension of phase space. The $2n$ variables are also traditionally called

$$q_1, \dots, q_n, \qquad p_1, \dots, p_n. \tag{18}$$

The system is completely defined by a single function of the $2n$ variables (instead of N functions in the general case), called the *Hamiltonian*:

$$H(q_1, \dots, p_n) \tag{19}$$

and the basic differential equations for the variables are

$$\frac{dq_i}{dt} = \frac{\partial H}{\partial p_i}, \qquad \frac{dp_i}{dt} = -\frac{\partial H}{\partial q_i} \qquad (i = 1, \dots, n). \tag{20}$$

q_i and p_i are called *conjugate variables*.

One integral is immediately available: the Hamiltonian itself. It can be verified from eq. (20) that H is constant on a trajectory. Therefore the order of the system can be reduced to $2n - 1$ (see section 1.3).

If now we introduce a space of section, the problem is reduced to the study of a mapping in a space with $2n - 2$ dimensions.

A good way to do these reductions is to first use the integral H to eliminate one variable, for instance p_i, and then to define the space of section by the equation $q_i = 0$, where q_i is the variable conjugate to p_i. In this way, one pair of conjugate variables is eliminated, and the $n - 1$ pairs left serve as natural coordinates in the space of section. The mapping G obtained in this fashion can be shown to be *symplectic*; this means that its Jacobian $\partial G / \partial Y$ satisfies identically the relation

$$\left(\widetilde{\frac{\partial G}{\partial Y}}\right)J\left(\frac{\partial G}{\partial Y}\right)=J, \tag{21a}$$

where the tilde represents the transposed matrix, and J is defined by

$$J=\begin{pmatrix} 0 & I \\ -I & 0 \end{pmatrix}, \tag{21b}$$

where I is the $(n-1)\times(n-1)$ unit matrix. A particular consequence of eq. (21a) is

$$|\partial G/\partial Y|=1, \tag{21c}$$

which indicates that the mapping G is *volume-preserving*.

3.2. *Periodic orbits and their stability*

A periodic orbit of the Hamiltonian system corresponds to a fixed point or a cycle in the space of section; we consider the case of a fixed point Y^* for simplicity. This fixed point is characterized by $M=2n-2$ eigenvalues. It can be shown that, as another consequence of the symplectic property, these eigenvalues are not entirely arbitrary but can be grouped in pairs, the product of the two eigenvalues in each pair being equal to 1. Thus, if λ is an eigenvalue, then λ^{-1} is also an eigenvalue.

We shall explore the consequences of this property for the important case of two degrees of freedom. For $n=2$, we have $N=2n=4$, and $M=2n-2=2$. The space of section is a two-dimensional surface of section. A fixed point has a single pair of eigenvalues, inverse of each other; therefore the eigenvalue equation

$$\left|\frac{\partial G}{\partial Y}-\lambda I\right|=0 \tag{22}$$

must be of the form

$$\lambda^2-2a\lambda+1=0. \tag{23}$$

a is a real number, called the *stability index*. The value of this single number characterizes completely the properties of the neighbouring trajectories. The two eigenvalues are

$$\lambda=a\pm(a^2-1)^{1/2} \tag{24}$$

and therefore three cases can be distinguished:

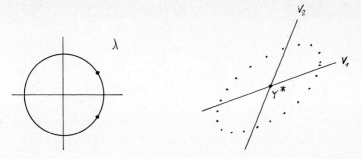

Fig. 8.

(1) $-1 < a < 1$: the two eigenvalues are complex conjugate, and lie on the unit circle (fig. 8, left). In the linear approximation, the sequence of points for a neighbouring trajectory lies on an ellipse (fig. 8, right; cf. fig. 6a). The fixed point is then called *elliptic*. It is also said to be *linearly stable*; this means that in the linear approximation, a trajectory starting near the fixed point will remain near the fixed point. It is important to remark here that linear stability does not imply that the fixed point is stable when the exact mapping is considered; this is a much more difficult problem, to which we shall return in section 6.5.

(2) $a > 1$: the two eigenvalues are real and positive, one being less than 1 and the other greater than 1. Points of a neighbouring trajectory lie on a branch of hyperbola (fig. 9); V and V' are the eigenvectors. The fixed point is called *hyperbolic*, or *linearly unstable*, since nearby trajectories run away from it.

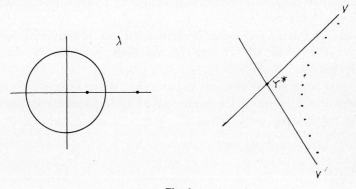

Fig. 9.

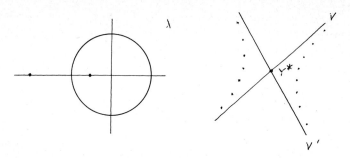

Fig. 10.

(3) $a < -1$: this case is very similar to the previous one. The two eigen-values are real and negative. Points of a neighbouring trajectory now lie on both branches of a hyperbola (fig. 10). The fixed point is again called hyperbolic, or linearly unstable.

We shall not consider here the marginal cases $a = 1$ and $a = -1$, which are more complicated.

A similar analysis can be made for an n-cycle, simply by considering G^n instead of G, and the n-cycle will also be called elliptic or hyperbolic accor-ding to the case.

4. Integrable and ergodic systems

4.1. Integrable systems

One can try to simplify a Hamiltonian system by an appropriate change of variables. If the new variables $Q_1, \ldots, Q_n$, $P_1, \ldots, P_n$ are such that the equations of motion can again be derived from a Hamiltonian function $H(Q_1, \ldots, P_n)$, then the change of variables is called a *canonical transformation*.

The new form will be simpler, in particular, if one or more of the new variables do not appear in the expression of H. Suppose for instance that H does not depend on Q_n. Then

$$dP_n/dt = -\partial H/\partial Q_n = 0, \tag{25}$$

and therefore

$$P_n(t) = P_n(0) = \text{constant}. \tag{26}$$

This constant value can be considered as a parameter. For a given value of that parameter, the Hamiltonian depends now only on $2n-2$ variables, forming $n-1$ pairs of conjugate variables. The number of degrees of freedom has been decreased by one unit, and the order of the system by two units.

The ideal case is when one can find a canonical transformation such that the new Hamiltonian does not depend on any Q_i, i.e. has the form $H(P_1,\dots,P_n)$. This is called a *normal form* of the system. The new variables are often called *action–angle variables*, the P_i being the "actions" and the Q_i the "angles". We have then

$$P_i(t) = \text{constant} = C_i \qquad (i=1 \text{ to } n), \tag{27}$$

and:

$$dQ_i/dt = \partial H/\partial P_i. \tag{28}$$

The right-hand member of this equation is a function of the P_i, i.e. of the C_i; we shall write this function as $\omega_i(C_1,\dots,C_n)$. The ω_i are called *frequencies*. If we also write for the initial values $Q_i(0) = D_i$, eq. (28) is immediately integrated into

$$Q_i(t) = \omega_i t + D_i. \tag{29}$$

We have thus the general solution in explicit form, given by eqs. (27) and (29); the C_i and the D_i are the $2n$ integration constants. For this reason, a Hamiltonian system which can be reduced to a normal form is called an *integrable system*.

The actions $P_1,\dots,P_n$ are integrals of the system, since their value is constant on any trajectory. Conversely, if n integrals are known in a Hamiltonian system, it is in general possible to devise a canonical transformation such that the new P_i are equal to these integrals. Then the new Hamiltonian will depend on $P_1,\dots,P_n$ only, i.e. will be in normal form, and the general solution can be written down. Thus, a Hamiltonian system can be completely solved if only $n = N/2$ integrals are known; this is in contrast to the general case of a dynamical system, in which it is necessary to know N integrals, and is again due to the particular structure of Hamiltonian systems.

One word of warning: it is often the case that one must first divide phase space into a number of regions, and then use a different canonical transformation, leading to different action–angle variables, in each region. For instance, for the simple pendulum, three regions of phase space must be distinguished, corresponding respectively to oscillation, direct rotation, and retrograde rotation.

4.2. *Example*

We consider as an example a system of n masses $m_1,\ldots,m_n$ suspended from a beam by springs of strengths $k_1,\ldots,k_n$ (fig. 11). If z_i is the altitude of mass m_i, the equations of motion are

$$m_i \ddot{z}_i = - k_i z_i. \tag{30}$$

This system is put into Hamiltonian form by taking as variables

$$q_i = z_i, \qquad p_i = m_i \dot{z}_i, \tag{31}$$

and as Hamiltonian function

$$H = \tfrac{1}{2} \sum_{i=1}^{n} \left(k_i q_i^2 + \frac{p_i^2}{m_i} \right).$$

This is not a normal form. We effect now the change of variables

$$q_i = \mu^{-1} (2P_i)^{1/2} \sin Q_i, \qquad p_i = \mu (2P_i)^{1/2} \cos Q_i, \tag{32}$$

with

$$\mu = (k_i m_i)^{1/4}. \tag{33}$$

It is easily verified that the equations of motion in the new variables derive from the new Hamiltonian

$$H = \sum_{i=1}^{n} \omega_i P_i, \tag{34}$$

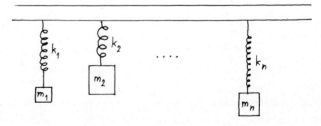

Fig. 11.

with

$$\omega_i = (k_i/m_i)^{1/2}. \tag{35}$$

This is a normal form; the system is therefore integrable. The general solution is

$$P_i = C_i, \qquad Q_i = \omega_i t + D_i. \tag{36}$$

An intuitive interpretation of this formalism is as follows: if we consider separately the motion of oscillator i and represent it by plotting velocity against position with proper scalings, namely using μq_i and $\mu^{-1} p_i$ as coordinates, then all trajectories are circles, described with uniform velocity, as shown by eqs. (32) and (36) (fig. 12). For a given trajectory, the radius of the circle is $(2P_i)^{1/2}$, and the angle, measured in the counter-clockwise direction from the vertical axis, is Q_i. Thus, the change of variables, eq. (32), amounts simply to replacing the cartesian variables by polar variables which are more natural for this problem.

This is in fact a highly special example of an integrable system, because the ω_i are constants. Now imagine that by some unspecified mechanism, each strength k_i is made to depend on the amplitudes of the n oscillations; specifically, we assume that each k_i is some function $k_i(P_1, \ldots, P_n)$. We have then an example of the general case of an integrable system. All the above equations are still valid; but the frequencies ω_i now differ from one trajectory to another.

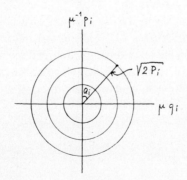

Fig. 12.

4.3. Tori and trajectories

In an integrable system, a trajectory in the $2n$-dimensional phase space is restricted to an n-dimensional subspace defined by $P_1 = C_1, \dots, P_n = C_n$. What is the shape of this subspace?

We remark in the above example that each Q_i is a *cyclical variable*: if Q_i is increased by 2π, the system returns to the same state. In other words, Q_i is defined modulo 2π. This is generally true for recurrent systems: if one of the Q_i were not cyclical, then as shown by eq. (29) the system would constantly enter new regions of phase space. So each Q_i must be cyclical, and by a simple scaling its period can be made equal to 2π.

Consider first the case $n = 1$. A trajectory is represented by a segment of length 2π on the Q_1 axis (fig. 13a). The representative point moves uniformly towards the right, with a velocity ω_1 (assumed to be positive), and when it reaches the point $Q_1 = 2\pi$, it jumps back to $Q_1 = 0$. This discontinuity can be avoided by representing the trajectory as a circle, with Q_1 as angular coordinate of the representative point (fig. 13b).

We consider next the case $n = 2$. In the (Q_1, Q_2) plane, motion takes place inside a square with sides of length 2π (fig. 14a). The point moves with a constant velocity, whose components are (ω_1, ω_2); when it hits a side of the square, it reappears on the opposite side. Here again, these discontinuities can be avoided by "sewing" first the left side of the square to the right side, and then the lower side to the upper side. The surface thus obtained is an ordinary *torus* (fig. 14b). Q_1 and Q_2 are now the two angular coordinates which specify the position of a point on the torus.

More generally, for n degrees of freedom, the motion can be pictured as taking place either inside an n-dimensional hypercube, in the cartesian space $(Q_1, \dots, Q_n)$, with discontinuities at the sides; or on an n-torus, with the Q_i as angular coordinates, and without discontinuity in the motion. Both representations are useful.

a   b

Fig. 13.

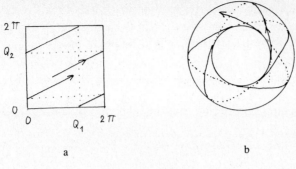

Fig. 14.

We inquire now into the shape of the trajectory itself on the n-torus. For $n = 1$, the trajectory simply coincides with its supporting 1-torus (circle). For $n = 2$, things are not so simple. If the ratio ω_1/ω_2 is rational, then after a while the trajectory closes back upon itself: it is periodic. (physicists often speak of a *resonance* in this case.) If ω_1/ω_2 is irrational (the general case), the trajectory never closes; and it can be shown to fill densely the square (fig. 14a) or the 2-torus (fig. 14b). This is called a *quasi-periodic trajectory*.

More generally, for n degrees of freedom, if the ω_i are mutually incommensurable, i.e. if there exists no commensurability relation

$$k_1\omega_1 + \ldots + k_n\omega_n = 0 \tag{37}$$

(with the k_i integers and not all zero), then the trajectory is dense on the n-torus; this is the general case. If there exist p independent commensurability relations, then the trajectory occupies only a $(n - p)$-dimensional subspace on the n-torus.

4.4. Surface of section

How do these properties appear when a space of section is taken?

It will be convenient to define the space of section simply by: $Q_n = 0$ (mod. 2π). Then the intersection of the n-torus which supports a trajectory with the space of section is simply an $(n - 1)$-torus, described by the coordinates $Q_1, \ldots, Q_{n-1}$. Thus, the sequence of points representing a trajectory in the space of section lies on an $(n - 1)$-torus.

The uniform variation of Q_n with time is given by eq. (29). We call Y_0

the point of a sequence corresponding to $Q_n = 0$. Then the next point Y_1 of the sequence corresponds to $Q_n = 2\pi$ (we assume $\omega_n > 0$). More generally, point Y_k corresponds to $Q_n = 2k\pi$, and to a time

$$t = \frac{2k\pi}{\omega_n} - \frac{D_n}{\omega_n}. \tag{38}$$

Using again eq. (29), we find that the other coordinates of Y_k are given by

$$Q_{k,i} = \left(D_i - \frac{\omega_i}{\omega_n} D_n \right) + 2\pi \frac{\omega_i}{\omega_n} k. \tag{39}$$

The first term is a constant. The second term shows that the successive values $Q_{k,i}$ of the coordinate Q_i are equally spaced. If we represent Q_i on a circle, as usual (fig. 15), the successive points are separated by equal angles. Thus, for Q_i considered in isolation, the mapping consists in a simple rotation, of angle $2\pi\omega_i/\omega_n$ (in radians).

We introduce a quantity

$$v_i = \omega_i/\omega_n, \tag{40}$$

called the *rotation number* for coordinate Q_i; it will play a very important role. v_i can be simply interpreted as the angle between successive points in fig. 15, measured in revolutions instead of in radians.

If ω_i and ω_n are not commensurable, the rotation number v_i is irrational, and the sequence of points covers the circle densely in fig. 15. More generally, if there is no commensurability relation between the ω_i, then the sequence of points is dense on the $(n-1)$-torus. If there are p commensurability relations, the points occupy only an $(n-p-1)$-dimensional subspace.

In order to illustrate clearly what the rotation number means, we consider again the case $n = 2$ and the representation of the trajectory in the

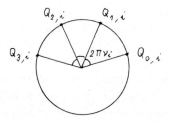

Fig. 15.

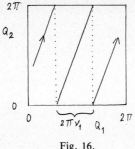

Fig. 16.

(Q_1, Q_2) plane (fig. 16). The surface of section is defined by $Q_2 = 0$, and corresponds therefore to the lower side of the square. The rotation number v_1 is then essentially (apart from a factor 2π) the distance between two consecutive points of intersection of the trajectory with the lower side.

4.5. Over-integrable systems

A trajectory of an integrable system lies on an n-torus. This n-torus is characterized by the n constant values $C_1, \ldots, C_n$ of the coordinates P_i [see eq. (27)]. The n frequencies $\omega_1, \ldots, \omega_n$ are in general functions of the C_i; they change continuously, and independently, when one or more of the C_i are varied. Therefore the n-tori for which there exists a commensurability relation between the ω_i form a set of measure zero. (For $n = 2$, for instance, they correspond to rational values of ω_1/ω_2.) Thus, in general, the trajectory fills the n-torus densely. One consequence of this is that no other integrals exist apart from the C_i.

However, there exist degenerate systems, called *over-integrable*, for which one or more commensurability relations are identically satisfied, i.e. hold true whatever the values of the C_i. We have then additional integrals (which are the commensurability relations themselves), and every trajectory is restricted to a subspace with less than n dimensions.

As an illustration, we consider the motion of a particle in the (x, y) plane, under the effect of a central field. This is a Hamiltonian system with 2 degrees of freedom; it is integrable since we know two integrals, the total energy and the angular momentum. The two frequencies ω_1 and ω_2 can be interpreted as the frequency of radial oscillation and the mean angular velocity around the origin. For an arbitrary force law, there is no other integral; the frequencies ω_1 and ω_2 are not related. Trajectories are quasi-

M. Hénon

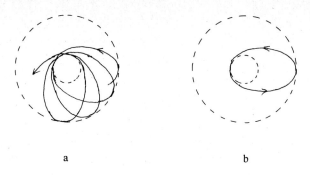

a   b

Fig. 17.

periodic and fill a 2-torus in the 4-dimensional phase space. The projection of a trajectory on the (x, y) plane is a rosette orbit, which fills an annulus (fig. 17a).

But in the particular case of an inverse-square law of force, the system becomes over-integrable. There is an additional integral (the direction of the pericentre). The frequencies ω_1 and ω_2 are always equal. Therefore every trajectory is a simple closed curve (fig. 17b).

4.6. Relevance of integrable systems

Given a Hamiltonian system, is it possible in general to reduce it to a normal form? The answer is no. In other words, Hamiltonian systems are generally non-integrable.

The study of integrable systems is nevertheless of great interest; first, because some interesting real systems happen to be integrable; and second, because a good knowledge of the integrable case will be of great help in understanding the more general non-integrable case.

4.7. Ergodic systems

In a Hamiltonian system, a given trajectory is restricted in phase space to a subspace $H = $ constant, since H is an integral. (This subspace is often called an *energy surface*, since in many cases H can be physically interpreted as the total energy of the system.)

Loosely speaking, the system will be called *ergodic* if any trajectory (except for a set of measure 0) fills densely its energy surface.

Correspondingly, every sequence of points in the space of section (except again for a set of measure 0) fills densely the space of section.

Ergodic systems constitute a special case among Hamiltonian systems (Markus and Meyer 1974). Precisely because of this, however, it is possible to push the theory much farther than in the general case, and to understand the behaviour of trajectories quite well. This is the subject of *ergodic theory*. What has been said in the previous section for integrable systems applies also here: knowledge of the properties of the limiting ergodic case helps to understand the general case. Lack of space, however, prevents me from entering into a detailed description of ergodic systems, and I must refer you to the literature.

5. One degree of freedom

This case is simple and will be quickly disposed of; but here again, it will be useful to review it before moving on to less trivial cases.

A Hamiltonian system with one degree of freedom has always one integral, H; therefore *it is always integrable*. Phase space has only two dimensions (q_1, p_1) and can be easily represented. The subspaces $H = $ constant are curves; therefore a trajectory essentially coincides with a curve $H = $ constant. This means that the system is *ergodic*. (It may happen, however, that $H = $ constant corresponds to several disconnected curves; in that case, a trajectory will lie on one of the curves only.)

The space of section has a dimension $2n - 2 = 0$; it reduces to a point, or a finite sequence of points forming a cycle. This shows that the case of one degree of freedom is in a sense trivial.

As an example, we consider the motion of a point on the x axis under the effect of a potential $V(x)$:

$$\ddot{x} = - dV/dx. \tag{41}$$

This is brought to Hamiltonian form by taking

$$q_1 = x, \ p_1 = \dot{x}, \ H = \tfrac{1}{2}p_1^2 + V(q_1). \tag{42}$$

If, for instance, V has the shape indicated by fig. 18a, then from eq. (42) one finds that the curves $H = $ constant are as represented on fig. 18b. All curves are closed trajectories, corresponding to an oscillatory motion of the particle in the potential well.

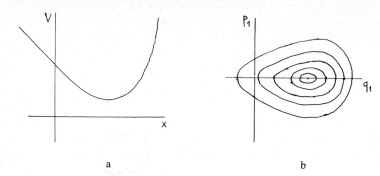

Fig. 18.

6. Two degrees of freedom

Here we come to the really interesting and non-trivial cases. For a Hamiltonian system with two degrees of freedom, in general, only one integral is known: the Hamiltonian $H(q_1, q_2, p_1, p_2)$. Therefore, the system is generally non-integrable. One cannot solve it analytically and write down an explicit general solution. However, the problem can be attacked by numerical computations. I shall now describe in detail a number of examples.

6.1. First example: triangular potential

We consider the motion of a point in a plane two-dimensional potential $V(x, y)$:

$$\ddot{x} = -\partial V/\partial x, \qquad \ddot{y} = -\partial V/\partial y. \tag{43}$$

If we picture x and y as horizontal coordinates in ordinary space and V as vertical coordinate, then $V(x, y)$ represents a surface, and eqs. (43) describe approximately the motion of a marble rolling on this surface. The Hamiltonian form of the problem is

$$q_1 = x, \quad q_2 = y, \quad p_1 = \dot{x}, \quad p_2 = \dot{y}, \tag{44}$$

$$H = \tfrac{1}{2}(p_1^2 + p_2^2) + V(q_1, q_2). \tag{45}$$

According to our rules, V should be taken as simple as possible; a polynomial in x and y seems a good choice. If this polynomial if of second order only, the equations of motion (43) are linear, and the problem is trivially integrable. Therefore, we should use a polynomial of degree 3 at least. It will be seen that degree 3 is in fact sufficient to produce a nontrivial problem, with apparently all the characteristics of the most general case. We select the following form for V (Hénon and Heiles 1964):

$$V(x, y) = \tfrac{1}{2}(x^2 + y^2 + 2x^2y - \tfrac{2}{3}y^3). \tag{46}$$

The reason for these seemingly bizarre coefficients is that V thus defined has a ternary symmetry, and is thus "simplest" in some sense among thirdorder polynomials. This becomes apparent if we use polar coordinates defined by $x = \varrho \cos \theta$, $y = \varrho \sin \theta$; we have then

$$V = \tfrac{1}{2}\varrho^2 + \tfrac{1}{3}\varrho^3 \sin 3\theta. \tag{47}$$

This potential can be pictured by drawing the equipotential lines $V =$ constant in the (x, y) plane (fig. 19). Near the origin, second-order terms of eq. (46) dominate and the equipotentials are approximately circular; this corresponds to small values of V. As we move away from the origin, the curves are distorted; the ternary symmetry can be observed. Finally, for $V = 1/6$, the equipotential is an equilateral triangle.

Actually the whole plane (x, y) is filled with equipotentials; fig. (19) represents only that part of the plane occupied by closed equipotential

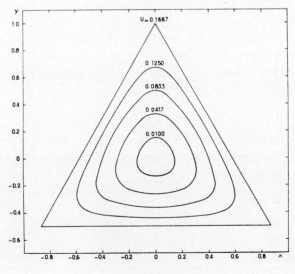

Fig. 19.

curves, corresponding to the range of values $0 \leq V \leq 1/6$. The reasons for this limitation will shortly become apparent.

The system has one integral: the Hamiltonian itself. Its constant value for a given trajectory will be represented by E; it is the total energy per unit mass of the particle. From eq. (45) we have:

$$V(x, y) \leq E; \tag{48}$$

therefore the trajectory is constrained to remain in the part of the (x, y) plane where this inequality is satisfied. Thus, if the trajectory is started inside the equipotential line $V(x, y) = E$ in fig. 19, it must lie entirely within that line.

The velocity is also constrained; from eq. (45) we have

$$\tfrac{1}{2}(p_1^2 + p_2^2) \leq E, \tag{49}$$

since V is positive in the region represented by fig. 19. So the trajectory must lie inside a finite volume of phase space. It follows then from the classical recurrence theorem of Poincaré that the trajectory is necessarily recurrent. It is because of this nice property that we shall restrict our attention to values of the energy in the range $0 \leq E \leq 1/6$, and to the inside of the triangle of fig. 19.

For $V > 1/6$, the equipotential lines open, and the trajectories can no more be guaranteed to be recurrent; in fact, most trajectories "escape" to infinity in the (x, y) plane.

No integral other than H is known; we must therefore resort to numerical integration. As usual, we first lower the order of the system by choosing a definite value for E, and using it to eliminate the coordinate p_1. Next, we define a surface of section by

$$q_1 = o. \tag{50}$$

The coordinates in the surface of section will be q_2 and p_2. We can verify that these two numbers define a starting point for a trajectory: q_1 is given by eq. (50), and p_1 by eq. (45):

$$p_1 = \pm (2E - p_2^2 - q_2^2 + \tfrac{2}{3}q_2^3)^{1/2}. \tag{51}$$

The $\pm$ sign, however, is annoying because it means that the correspondence is not one-to-one: to one point of the surface of section correspond two possible trajectories. We can eliminate this ambiguity simply by redefining the surface of section as:

$$q_1 = 0 \qquad \text{and} \quad p_1 \geq 0. \tag{52}$$

In other words, we consider only intersections with $q_1 = 0$ in the positive direction (q_1 increasing), i.e. every other intersection.

We come now to the numerical results. We take first a comparatively small value of the energy: $E = 1/12$. Figure 20 shows a typical sequence of points; successive points have been numbered. They seem to lie on a curve; in order to show this, a tentative curve has been actually drawn through them on fig. 20. In other words, the results suggest that there exists an *invariant curve*, i.e. a curve invariant under the mapping G. (It is important, however, to realize that only the points are results of the computation in fig. 20. The curve is an *interpretation*, added later by the "observer".) Moreover, successive points appear to rotate regularly around the curve. This is exactly what we would expect for an integrable system (see section 4.4). In fact we can even obtain an estimate for the rotation number v from fig. 20: when we go from 1 to 9 we do not quite complete a revolution, therefore $v < 1/8$; but 10 lies after 1, therefore $v > 1/9$; after two more revolutions, we reach 27 which lies slightly before 1, therefore $v < 3/26$, and so on; the value of v can be progressively refined as more points are computed.

Figure 21 shows, for the same energy $E = 1/12$, an overall view of the surface of section. A number of trajectories are represented (in particular the trajectory of fig. 20 can be seen again). In each case, the sequence of

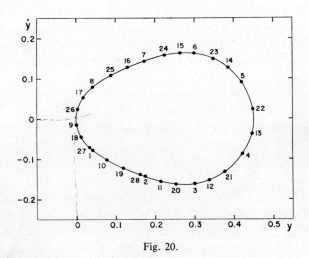

Fig. 20.

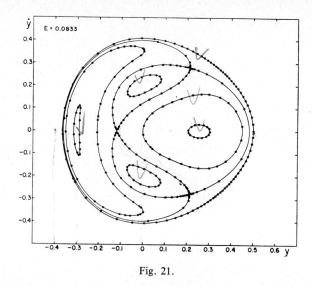

Fig. 21.

points appears to lie on a well-defined invariant curve. The outer curve in fig. 21 marks the boundary of the accessible region in the (q_2, p_2) surface of section; this is defined by the condition that p_1 given by eq. (51) must be real, i.e.

$$p_2^2 + q_2^2 - \tfrac{2}{3} q_2^3 < 2E. \tag{53}$$

This accessible region seems to be filled with a one-parameter family of invariant curves. Thus, fig. 21 strongly suggests that the system is integrable.

Incidentally, fig. 21 indicates the existence of four elliptic fixed points, inside the four little oval curves, and also of three hyperbolic points, where curves appear to cross. For the latter points, the tangents to the curves correspond to the directions of the eigenvectors, while the curves themselves coincide with the stable and unstable invariant manifolds.

However, if we do the same computations for a somewhat higher energy, $E = 1/8$, we get a surprise (fig. 22). For some initial conditions, we still find that the sequence of points lies on a curve; but in other cases, it seems to fill a two-dimensional region. All the isolated points of fig. 22 correspond to one and the same trajectory; and clearly it is not possible to draw a simple curve through these points. This becomes more and more apparent when more points are computed. Besides, if one observes the order in which the points appear while the figure is being plotted, he finds that they

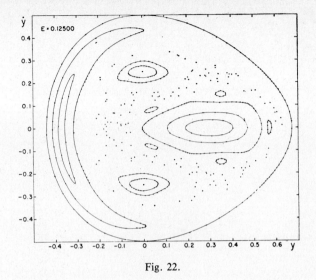

Fig. 22.

seem to jump in a more or less random fashion from one part of the figure to another; this contrasts with the regular motion of the points on a curve. This is called a *chaotic trajectory*, and the region which it occupies is called a *chaotic region* (a number of other adjectives are also in use: irregular, wild, erratic, stochastic, semi-ergodic, aperiodic, turbulent, strange,...).

The curves which have survived on fig. 22 are organized in nested sets; and in the center of each set one finds that there is an elliptic fixed point.

These findings destroy completely our previous notion that the system could be integrable, and show that it has in fact a much more complex behaviour. This is confirmed by trying a still higher energy: $E = 1/6$ (fig. 23). Here again, all isolated points are intersections of the surface of section by a single trajectory; these points fill almost completely the accessible region.

One of the striking aspects of these results is that the picture changes completely for a moderate variation of the energy. This can be shown in a more quantitative way. For each value of the energy, one can measure the relative area which is covered by curves. (A criterion for distinguishing easily between quasi-periodic and chaotic trajectories will be presented in section 6.7.) Figure 24 shows how this relative area evolves with E. (The points represent actual measures; the curve is an interpretation.) For low energies, practically the whole area is covered by curves; as the energy is

M. Hénon

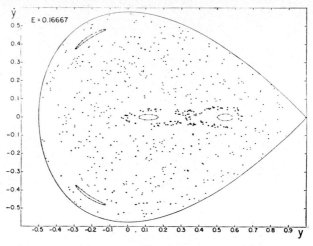

Fig. 23.

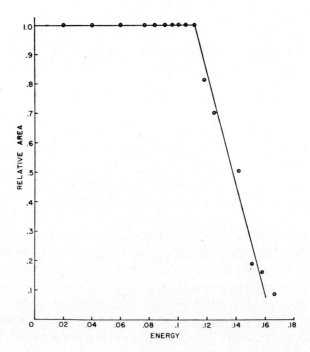

Fig. 24.

increased, quite abruptly the proportion begins to decrease below 1, and it rapidly drops to low values. (Note, however, that the sharp corner shown on fig. 24 is probably slightly rounded in reality.) The curve cannot be continued beyond $E = 1/6$, because the equipotential lines are then open and the accessible area becomes infinite.

6.2. Second example: restricted problem

A famous old problem in astronomy is the *N-body problem*: find the motion of N point masses under the effect of their mutual inverse-square attraction. This problem is of great interest both for applications and from a mathematical point of view.

For $N = 2$, the problem is integrable, and even over-integrable; one finds the classical elliptic, parabolic and hyperbolic solutions. So the next case, $N = 3$, has been attacked; but it has proved to be more difficult by many orders of magnitude. As directly formulated, it is a Hamiltonian system with 9 degrees of freedom; by using all the known integrals, this can be lowered to 4 degrees of freedom, but not further.

Instead of attacking the 3-body problem in all its generality, one should therefore begin by considering particular cases. One such case, which has received much attention, is the *restricted problem* of three bodies, defined by the following three restrictions:

(1) The third body M_3 has zero mass. Therefore it does not influence bodies M_1 and M_2, whose motions are given as a solution of a 2-body problem.

(2) A particular motion is chosen for M_1 and M_2: they describe circular orbits around their common center of mass.

(3) The motion of M_3 takes place in the orbital plane of M_1 and M_2.

The restricted problem is of interest both because it represents a useful first approximation to many real problems, and because it is the simplest unsolved case of the N-body problem.

The problem is reduced to a non-dimensional form as follows: (i) one takes the sum of the masses of M_1 and M_2 as unit of mass; these two masses are then called $1-\mu$ and μ respectively; (ii) one takes as unit of length the constant distance between M_1 and M_2; therefore the radii of their orbits are respectively μ and $1 - \mu$; (iii) one chooses the unit of time in such a way that the gravitational constant $G = 1$. It follows from these choices that the angular velocity of M_1 and M_2 is also equal to 1. It is convenient to use a system of axes (x, y) which rotates with M_1 and M_2, with the x axis pointing towards M_2. In this system, M_1 has fixed coordinates $(-\mu, 0)$ and M_2 has fixed coordinates $(1 - \mu, 0)$. M_3 moves in the (x, y) plane, and its equations of motion are easily shown to be

$$\ddot{x} = 2\dot{y} + x - (1 - \mu)\frac{x + \mu}{r_1^3} - \mu\frac{x - 1 + \mu}{r_2^3},$$

$$\ddot{y} = -2\dot{x} + y - (1 - \mu)\frac{y}{r_1^3} - \mu\frac{y}{r_2^3}, \tag{54}$$

with

$$r_1 = [(x + \mu)^2 + y^2]^{1/2}, \qquad r_2 = [(x - 1 + \mu)^2 + y^2]^{1/2}. \tag{55}$$

This is brought into the form of a Hamiltonian system with two degrees of freedom by

$$q_1 = x, \quad q_2 = y, \quad p_1 = \dot{x} - y, \quad p_2 = \dot{y} + x, \tag{56}$$

$$H = \tfrac{1}{2}(p_1^2 + p_2^2) + p_1 q_2 - p_2 q_1 - \frac{1 - \mu}{r_1} - \frac{\mu}{r_2}. \tag{57}$$

Therefore there exists an integral; it is customary to define it as $C = -2H$, and to call it the *Jacobi integral*. In the original variables, it is given by

$$C = x^2 + y^2 + \frac{2(1 - \mu)}{r_1} + \frac{2\mu}{r_2} - \dot{x}^2 - \dot{y}^2. \tag{58}$$

No other integral is known. We follow the usual procedure: we choose a particular value of C, and we define a surface of section by

$$y = 0, \qquad \dot{y} > 0. \tag{59}$$

The coordinates in the surface of section will be x and $\dot{x}$.

We will see the results for the case $\mu = 1/2$, i.e. equal masses for M_1 and M_2. Fig. 25 (Hénon 1966a) represents the surface of section for $C = 4.5$. As before, points linked by a curve correspond to the same trajectory. The dashed lines represent the boundaries of the accessible region. The system seems to be integrable: the whole accessible region appears to be covered with curves. The successive points have been numbered in one of the sequences; and fig. 26 shows the corresponding orbit in the (x, y) physical plane, with the points of intersection with the surface of section, eq. (59),

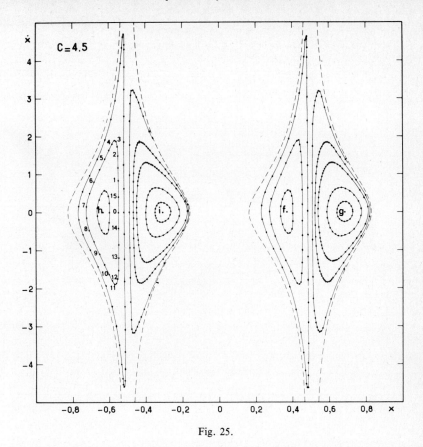

Fig. 25.

again represented. This orbit shows a great regularity, which is typical of quasi-periodic orbits in general. M_3 stays comparatively close to M_1, so that in a first approximation, the effect of M_2 can be neglected and the orbit is simply a two-body elliptical motion around M_1, seen in rotating axes. However, the symmetry around the origin is not perfect because of the perturbation from M_2.

We note also on fig. 26 that at a given point of the (x, y) plane, only two directions appear to be possible for the velocity. This again is typical of quasi-periodic orbits, and corresponds to the fact that the trajectory in phase space is a 2-dimensional torus.

Finally, we note on fig. 25 the existence of four elliptic fixed points f, g, h, i. They correspond to periodic orbits around M_1 or M_2, in the direct

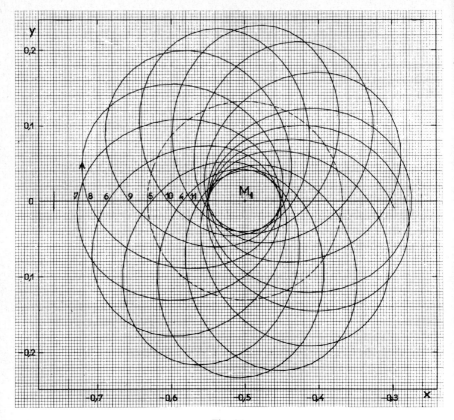

Fig. 26.

or the retrograde direction. For instance, the orbit corresponding to point *h* is represented on fig. 26 as a dashed line.

We consider now a somewhat lower value: $C = 4$. (This corresponds to a greater H). Figure 27 shows the surface of section. The curves do not fill the accessible region any more; all isolated points correspond to a single trajectory, which clearly has a chaotic character. For a still lower value, $C = 3.5$ (fig. 28), the chaotic region increases in extent. A number of successive points have been numbered on the trajectory which occupies that region, in an attempt to show how the points jump quasi-randomly from one place to another. A similar chaotic trajectory (for $C = 3$) is represented in the (x, y) plane on fig. 29, which should be contrasted with fig. 26. This orbit has a very disordered character. The third body M_3 describes a few

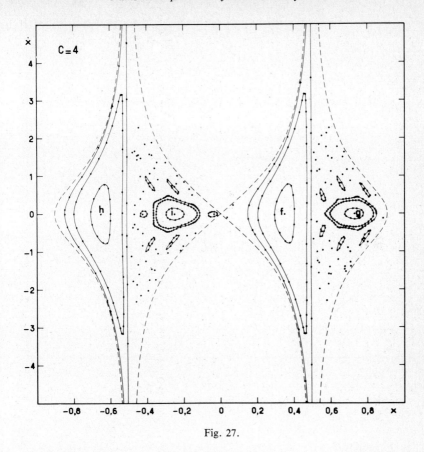

Fig. 27.

loops around M_1, then jumps over to the vicinity of M_2, then back to M_1, in a highly irregular fashion. We note also that at a given point, the velocity can apparently have any direction. This figure is typical of chaotic orbits in general.

6.3. Third example: quadratic mapping

The two previous examples dealt with dynamical systems, defined by systems of differential equations. Now we shall use the idea of defining directly a mapping in the surface of a section (see section 2.1).

We take as the surface of a section a plane with coordinates (x, y). The mapping is then defined by two equations

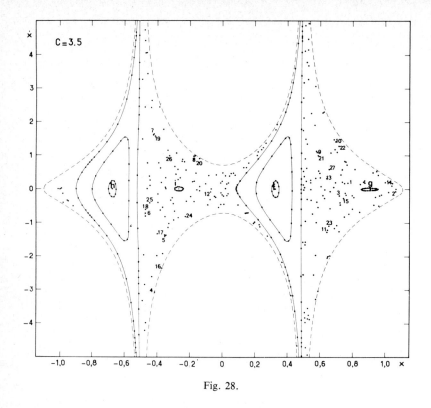

Fig. 28.

$$x' = g_1(x, y),$$
$$y' = g_2(x, y). \tag{60}$$

As seen in the previous examples, invariant curves are generally closed curves surrounding an elliptic fixed point. Thus, elliptic fixed points play a central role in the general organization of the figure; typically, each elliptic fixed point appears to be surrounded by a set of nested closed curves, which extend outwards to some distance; beyond this, a chaotic region begins. We shall therefore assume that the origin is an elliptic fixed point in our mapping.

Here again, one of the simplest ways to define the mapping is to use polynomials for g_1 and g_2. We consider first the case of first-order polynomials:

$$x' = ax + by,$$
$$y' = cx + dy, \tag{61}$$

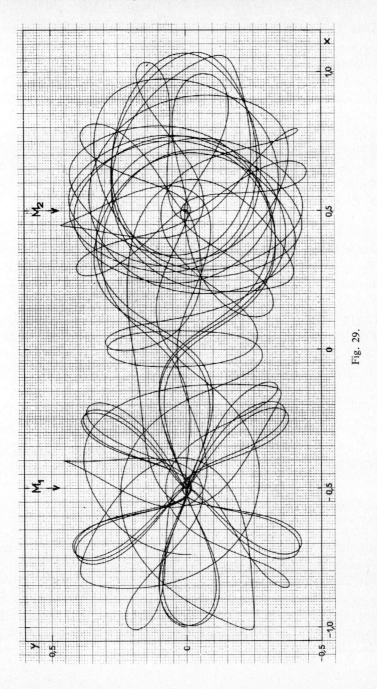

Fig. 29.

with a, b, c, d constants. In this case, the linear approximation of the mapping near the fixed point coincides with the exact mapping; the results of the linear analysis (section 2.2) are applicable to the whole plane. Since the origin is by hypothesis an elliptic fixed point, it has two complex conjugate eigenvalues of modulus 1, which we write: $e^{\pm i\alpha}$. The corresponding complex eigenvectors are of the form $V_1 \pm V_2 i$, with V_1 and V_2 real. We make a linear change of coordinates, taking V_1 and $-V_2$ as a new basis. The new coordinates will be renamed x and y, so that the mapping is still of the form of eq. (61) (with new values for a, b, c, d). The two eigenvectors are now respectively $(1, -i)$ and $(1, i)$. By substitution in eq. (61), writing that the associated eigenvalues are $e^{\pm i\alpha}$, one easily finds that

$$a = \cos\alpha, \quad b = -\sin\alpha, \quad c = \sin\alpha, \quad d = \cos\alpha. \tag{62}$$

Thus, in the new coordinates, the mapping is a simple rotation of angle α.

So we must go to at least second order to obtain a non-trivial mapping. The results will show that second order is in fact sufficient. We take, as above, V_1 and $-V_2$ for a new basis of coordinates, so that the linear terms reduce to the form of eq. (62). The most general second-order mapping with an elliptic fixed point at the origin can thus be brought into the form

$$
\begin{aligned}
x' &= x\cos\alpha - y\sin\alpha + a_{20}x^2 + a_{11}xy + a_{02}y^2, \\
y' &= x\sin\alpha + y\cos\alpha + b_{20}x^2 + b_{11}xy + b_{02}y^2.
\end{aligned}
\tag{63}
$$

However, we want our mapping to be area-preserving since it is intended to represent a Hamiltonian system (section 3.1). The determinant of the Jacobian of eq. (63) must therefore be identically equal to 1:

$$
\begin{vmatrix}
\cos\alpha + 2a_{20}x + a_{11}y & -\sin\alpha + a_{11}x + 2a_{02}y \\
\sin\alpha + 2b_{20}x + b_{11}y & \cos\alpha + b_{11}x + 2b_{02}y
\end{vmatrix} \equiv 0.
\tag{64}
$$

On developing this, we obtain a number of conditions to be satisfied by the coefficients. This reduces the number of independent parameters from 7 to 3 (see Hénon 1969 for details).

We can also take advantage of the fact that it is still possible to make some linear changes of coordinates without affecting the form of the linear terms in eq. (63), namely rotations around the origin and changes of scale. This can be used to eliminate two more parameters, and to reduce the mapping to a standard form:

$$G: \begin{cases} x' = x\cos\alpha - y\sin\alpha + x^2\sin\alpha, \\ y' = x\sin\alpha + y\cos\alpha - x^2\cos\alpha, \end{cases} \qquad (65)$$

in which only the parameter α is left. This last parameter cannot be eliminated, since it is related to the eigenvalues of the fixed point, which are invariant under any change of coordinates; it is an essential parameter of the mapping.

The mapping should also be invertible (section 2.1). This is the case for G defined by eq. (65); the inverse mapping is easily written down:

$$G^{-1}: \begin{cases} x = x'\cos\alpha + y'\sin\alpha, \\ y = -x'\sin\alpha + y'\cos\alpha + (x'\cos\alpha + y'\sin\alpha)^2. \end{cases} \qquad (66)$$

We note that G^{-1} is itself of second order. It is not in the standard form, eq. (65), but could be brought to it by an appropriate linear change of coordinates.

We try now to find some fixed points and cycles of G. An n-cycle will correspond to a solution of the system of $2n$ equations

$$\left. \begin{array}{l} x_{i+1} = x_i\cos\alpha - (y_i - x_i^2)\sin\alpha \\ y_{i+1} = x_i\sin\alpha + (y_i - x_i^2)\cos\alpha \end{array} \right\} \quad (i = 0 \text{ to } n-1), \qquad (67)$$

with $x_n = x_0$, $y_n = y_0$. The y_i are conveniently eliminated with the help of eq. (66a), leaving a system of n equations only:

$$x_i^2 \sin\alpha + 2x_i \cos\alpha - x_{i-1} - x_{i+1} = 0 \quad (i = 0 \text{ to } n-1) \qquad (68)$$

with $x_n = x_0$. This is a system of n second-order equations, which can be explicitly solved up to $n = 4$. The results are as follows: there are two 1-cycles or fixed points, one being elliptic (the origin) and the other hyperbolic; there is no 2-cycle; there are two 3-cycles if $\cos\alpha < 1 - 2^{1/2}$; and there are two 4-cycles if $\cos\alpha < 0$. These cycles play a prominent role in the general structure of the mapping.

Now we show some numerical results (Hénon 1969). Figure 30 represents a number of trajectories (sequences of points) for $\cos\alpha = 0.4$. The picture is qualitatively similar to what we found previously for surfaces of section of dynamical systems: a regular structure in a neighbourhood of the elliptic fixed point at the origin, and farther away a chaotic region. No curves have been drawn here; only the successive points of the sequences have been plotted by the computer. Owing to the much greater speed of computation, it is possible to compute a large number of points in each sequence; and in many places the plotted points are so dense that they give the illusion of a continuous curve.

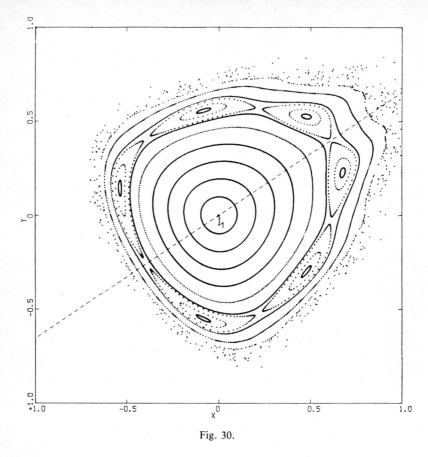

Fig. 30.

Near the origin, the "curves" are almost circular, because linear terms dominate in eq. (65). As we move outwards, the curves become distorted. (The chain of little closed curves, or "islands", will be discussed in section 6.4.) Finally we reach a chaotic region, where apparently there are no more curves and the points fill a two-dimensional region. This is not truly the case, however, because of a drawback of this model: the accessible region is the whole plane, therefore infinite in extent, and there is no guarantee that orbits will be recurrent. In fact the outermost trajectory, after about 600 iterations (corresponding to the isolated points seen on the outside of fig. 30), "escapes" to infinity. It can easily be seen that once points are sufficiently far from the origin, they never come back, but move ever farther: at large distances, terms of second order dominate in eq. (65), and from the first equation we have

$$x_{i+1} \sin\alpha \simeq (x_i \sin\alpha)^2. \tag{69}$$

The quantity $x_i \sin\alpha$ is approximately squared at each iteration, and thus diverges to infinity extremely quickly. This, however, is a non-essential phenomenon. Mappings can be devised for which the accessible region is finite (see for instance below, fig. 65).

Many other mappings have been studied numerically, and in most cases one finds the same general picture: part of the plane is filled with invariant curves, and the rest is occupied by chaotic trajectories. The proportion of these two regions varies as parameters are varied. It seems that one could describe the general situation as follows: each elliptic fixed point is surrounded by a "continent", of variable size, made of nested invariant curves, and these "continents" are immersed in a "chaotic sea".

6.4. Islands

But a closer look will reveal that the real situation is far more complex than this. First we go back for a while to integrable systems (section 4.4). If: (i) the system has been reduced to normal form; (ii) a value of P_2 is chosen; (iii) a surface of section is defined by $Q_2 = 0$; and (iv) P_1 and Q_1 are taken as polar coordinates in the surface of section, then each sequence of points lies on a circle (fig. 15). We have thus a one-parameter family of invariant curves, formed by concentric circles around the origin; the parameter is P_1. On a given circle, successive points are separated by a constant angle, equal to the rotation number v_1 (in revolutions). This number is given by $v_1 = \omega_1/\omega_2$ [eq. (40)]; and the frequencies ω_1 and ω_2 are functions of P_1 and P_2. P_2 is fixed for the surface of section which we are considering; therefore, v_1 is a function of P_1. In general this function does not reduce to a constant; the rotation number is not the same on different circles, and the mapping can be described as a *differential rotation*, or *twist mapping*.

Circles which have an irrational rotation number v_1 are covered densely by the points of a sequence; no point of the circle is ever visited twice. On circles with a rational rotation number $v_1 = i/j$ (i and j being relatively prime integers), the situation is completely different: every sequence comes back to the initial point after j iterations, and then repeats itself periodically. Every point of the circle belongs to a j-cycle; there are thus an infinity of such cycles on the circle.

(Incidentally, this shows that an integrable system is a very particular and degenerate case. For an arbitrary mapping G, the fixed points of G^j are solutions of a system of 2 equations for 2 unknowns; we expect therefore a finite number of solutions, or at the most a countably infinite number. For instance, in the quadratic mapping of section 4.3, a finite number of cycles was found for $j = 1, 2, 3, 4$. For an integrable system, on the other hand, we have an uncountable number of solutions, forming a one-parameter family.)

We assume now that the integrable system is slightly perturbed (for instance by adding small terms in the equations) so that it becomes non-integrable. What happens to all the j-cycles which formed the curve with $v_1 = i/j$? Because of what has just been said, we expect that all but a finite number of them will disappear. This is indeed what happens. Typically, one finds that only two cycles survive; one is of elliptic type and the other of hyperbolic type, and the points of both cycles alternate. (In some cases, more than two cycles survive; we shall not consider this possibility here.) Consider now a point of the elliptic cycle. It is an elliptic fixed point of G^j. Therefore, everything that has been said so far about the neighbourhood of an elliptic fixed point applies also to this point, provided that we consider G^j instead of G, i.e. that we take only every jth point of the sequence. In particular, this point will be surrounded by a set of nested invariant curves; this feature is often called an *island*. If now we apply G once, this island is mapped into a similar structure surrounding the next point of the elliptic cycle. We obtain thus a *chain of islands* (fig. 31; the elliptic cycle is represented by dots and the hyperbolic cycle by crosses).

Successive points of a sequence have here a composite motion. At each application of the mapping G, the point jumps from one island to another; specifically, to the ith next island. After j iterations, it comes back to the same island, but not to the same point. If now we consider only every jth point of the sequence, we obtain a subsequence of points which all belong to the same island, and which rotate regularly around one of its curves.

Figure 30 shows a chain of 6 large islands, corresponding to a rotation number $1/6$. As another example, fig. 32 shows the quadratic mapping for $\cos\alpha = -0.95$; two chains are visible, corresponding respectively to $v_1 = 3/7$ and $v_1 = 5/12$.

In the integrable system, the rotation number v_1 generally varies continuously as a function of the radius. Therefore it goes through an infinity of rational values. We come thus to a shattering but unescapable conclu-

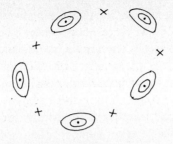

Fig. 31.

sion: in the perturbed system, there must exist in reality not just one or two, but an infinity of concentric chains of islands, present everywhere!

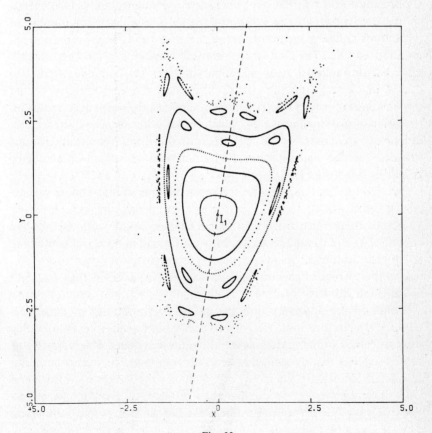

Fig. 32.

The picture is much more complex than it appeared at first view.

One immediately asks: why are not all these islands observed on the figures which we have seen? The answer is that the radial width of the islands decreases rapidly as the number j of islands increases; therefore, in a numerical exploration where initial conditions are chosen more or less at random, only chains corresponding to small j have some chance of being detected. However, if one searches systematically for chains of islands (by locating places where the rotation number takes rational values), one can find many of them, until eventually the limit of resolution of the computer is reached.

This is not yet the end of the story. As already said, the features of the mapping as a whole should appear again in each island, if one considers G^j instead of G. In particular, each little curve in an island is characterized by a *second-order rotation number*, equal to the fractional number of revolutions made by the point on the curve, in the average, after one application of G^j. (The first-order rotation number is i/j.) This second-order rotation number again varies continuously as one moves across the curves which make up the island. Therefore, it passes infinitely often through rational values. Every time this happens, the little curve breaks into a chain of still smaller islands, called *second-order islands*. If the second-order rotation number is i'/j', then there are j' second-order islands in one first-order island, and therefore the total number of second-order islands visited by a trajectory is jj'.

This process continues: again the same features should appear in each second-order island, if we isolate it by considering the mapping $G^{jj'}$. Therefore there must exist third-order islands; and so on. Islands are organized in an infinite hierarchy, with each island at one level containing an infinite number of islands of the next level. (Incidentally, the total picture in the surface of section, such as seen for instance on fig. 30, might be called an island of order 0.)

We thus have a glimpse into the incredible complexity of dynamical systems, even in the simple case of two degrees of freedom, as soon as they cease to belong to the special class of integrable systems. The hierarchy of islands suggests that dynamical systems are *inexhaustible*, in the sense that new details constantly emerge as one examines the system on a finer and finer scale. This might be compared to the case of geographical maps, where new details also constantly emerge when we look at maps drawn on a larger and larger scale. It also suggests that it will never be possible to obtain the "general solution" of the problem, i.e. an explicit formula giving the position in phase space as a function of initial position and time (or the position in the surface of section as a function of initial position and number of iterations).

To illustrate the hierarchy of islands, we look first at fig. 33, which represents the quadratic mapping for $\cos\alpha = 0.24$. There is a chain of 5 conspicuous islands, surrounding the 5 points of an elliptic cycle. The positions of the intermediate 5 points of the hyperbolic cycle are also well marked; these are the points where apparently two branches of "curve" intersect. Now we enlarge the vicinity of the rightmost hyperbolic point (with approximate coordinates $x = 0.57$, $y = 0.16$): we obtain fig. 34. The three

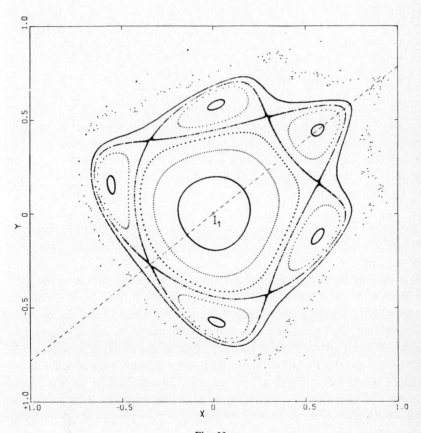

Fig. 33.

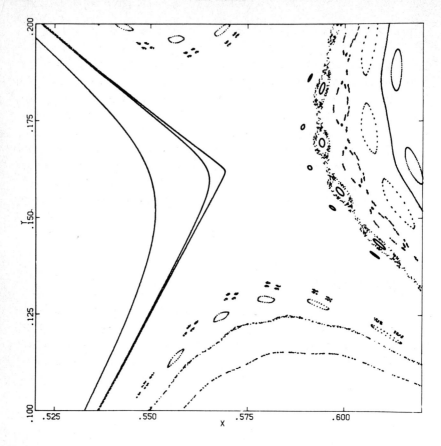

Fig. 34.

well-defined curves on the left are parts of invariant curves around the origin (or "islands of order 0"). Similarly, the portion of curve near the upper right corner is a part of an invariant curve which surrounds the chain of 5 islands, and therefore can also be labelled as an island of order 0. Next, the two curves (not completely filled with points) near the bottom of fig. 34 belong to one of the 5 first-order islands. Immediately above these two curves, a chain of second-order islands can be observed; and above this, a trajectory consisting of third-order islands. (Parts of these last two trajectories can also be seen at the top of the picture, which corresponds to the next first-order island.)

6.5. KAM theorem

This fundamental theorem plays a central role in the theory of Hamiltonian dynamical systems, and explains many of the observed features. We shall therefore digress momentarily from the description of numerical results and give an elementary description of the theorem. The initials stand for Kolmogorov–Arnold–Moser; the method of proof was suggested by Kolmogorov (1954), and proofs were constructed by Arnold (1963) and Moser (1962). These two proofs differ in some details; I shall follow here Moser's formulation.

As in the previous section, we consider first a mapping of the plane corresponding to an integrable system, and given in polar coordinates P, Q (we drop the subscript 1 for convenience):

$$P' = P,$$
$$Q' = Q + 2\pi v(P). \tag{70}$$

For technical reasons, the mapping is supposed to be defined only inside an annulus: $a \le P \le b$, with a and b positive constants. Inside this interval, the rotation number v is assumed to be a monotonically increasing function of P. This is a simple twist mapping. Now we perturb it by adding two small terms:

$$P' = P + f_1(P, Q),$$
$$Q' = Q + 2\pi v(P) + f_2(P, Q). \tag{71}$$

f_1 and f_2 must be periodic functions of Q, with period 2π. They are assumed to be small, and also differentiable a sufficient number of times. Moreover, eq. (71) must still be area-preserving. The interesting question is: do the invariant curves of eq. (70) still exist in eq. (71), with slightly perturbed shapes? Thus formulated, however, the question is too vague and cannot be answered by yes or no; we must be more specific. It turns out that the essential property which characterizes an invariant curve is its rotation number, and that the correct way to ask the question is as follows: given a particular invariant curve of eq. (70), with a rotation number v, does there exist in eq. (71) an invariant curve with a slightly different shape, and on which successive points are arranged according to the same rotation number v?

Moser's theorem answers partly the question by giving a *sufficient* condition for the existence of the perturbed invariant curve. The condition is that v must be "sufficiently far from all rational values". What this means precisely is that v must satisfy the unequality

$$\left| v - \frac{i}{j} \right| > \varepsilon j^{-5/2} \tag{72}$$

for all rational numbers i/j. ε is a small constant, which depends on the amplitude of the perturbation terms f_1 and f_2.

The condition (72) excludes a finite interval around each rational, and intuitively one would think that it excludes therefore all numbers. Surprisingly, this is not the case; on the contrary, most values of v are not excluded! To show this, we note first that v is defined modulo 1, so that it is sufficient to consider the $[0,1]$ interval. For a given j, the possible values of i are then 0 to $j-1$, and the sum of all intervals excluded by eq. (72) is $2\varepsilon j^{-3/2}$. It remains only to sum over j. In so doing, we overestimate the measure L of the excluded values, because many intervals will be counted more than once. Therefore, L verifies

$$L < \sum_{j=1}^{\infty} 2\varepsilon j^{-3/2} \simeq 5.224\varepsilon. \tag{73}$$

Thus, for a small perturbation, i.e. a small ε, most values of v satisfy eq. (72); in other words, Moser's theorem states that most invariant curves will be preserved under a small perturbation.

Arnold's theorem presents some differences:

(i) it considers a dynamical system, defined by a Hamiltonian, rather than the induced mapping;

(ii) it applies to any number of degrees of freedom, while Moser's theorem corresponds only to the case of two degrees of freedom;

(iii) on the other hand, it assumes that the Hamiltonian is analytic, while Moser's theorem assumes only that the mapping is differentiable a finite number of times.

In essence, however, Arnolds's theorem is quite similar. One considers first an integrable Hamiltonian system, and in this system a particular n-torus, characterized by its n frequencies $\omega_1, \ldots, \omega_n$ (see section 4). Then one adds a small perturbation to the Hamiltonian, and one asks whether an invariant n-torus still exists with these frequencies. Again the answer is yes, if the ω_i are sufficiently far from all commensurability relations; precisely, this means that they must satisfy

$$\left| \sum_{i=1}^{n} k_i \omega_i \right| > c \left| \sum_{i=1}^{n} |k_i| \right|^{-\gamma}, \tag{74}$$

where c and γ are constants, for any set of integers k_i (not all 0). γ is fixed, but c depends on the amplitude of the perturbation.

It should be borne in mind that the KAM theorem gives only sufficient conditions for the existence of invariant curves or invariant tori. It says nothing about what happens in the regions excluded by eqs. (72) or (74). In other words, and contrary to what is sometimes believed, the theorem *does not* state that invariant curves and tori are destroyed in the vicinity of resonances, or if the perturbation becomes large. It does, however, suggest that something like this happens; and the suggestion becomes stronger when one considers also the results from numerical experiments.

For this theorem explains quite well what we have observed numerically. Consider for instance the quadratic mapping studied in section 6.3. Near the origin, linear terms dominate, and quadratic terms represent a small perturbation, whose relative amplitude is proportional to the distance to the origin. Therefore Moser's theorem applies, and it predicts that one should mostly find invariant curves.

The theorem, as initially formulated (Moser 1962), applies only to a very small region around the origin: the relative perturbation, i.e. essentially the distance to the origin, must be less than 10^{-48} (Hénon 1966b). Here the numerical experiments bring more information, by showing, or at least suggesting, that invariant curves are still preserved for much larger perturbations; typically, one finds that the curves begin to break down only when the perturbations reach a size comparable to that of the main terms. In figs. 30, 32, 33, for instance, it can be seen that curves apparently exist up to distances of the order of unity, for which the quadratic terms in eq. (65) are comparable to the linear terms in magnitude.

Next, we note that the breaking of invariant curves into chains of islands happens precisely in the regions which are excluded by the theorem. The complexity of the picture, with its infinite number of chains of islands, corresponds to the complex structure of the set of points excluded by eq. (72). The theorem also appears to explain why the islands decrease rapidly in size as j increases: this is a consequence of the exponent $-5/2$ in eq. (72).

One of the most interesting consequence of the KAM theorem is the rigorous proof of the stability of elliptic fixed points, in the case of a two-dimensional mapping corresponding to a dynamical system with two degrees of freedom. Before this theorem was proved, these points could only be said to be linearly stable, i.e. stable in the linear approximation (section 3.2); but nothing could be said about what happens when the non-linearity is taken into account. From the KAM theorem, it can be deduced that (excluding a few particular cases) closed invariant curves exist in any neighbourhood, however small, of an elliptic fixed point. On the other hand, a trajectory which starts inside such a closed invariant curve must stay inside it at all times, because the inside region is mapped into itself. This is sufficient to prove stability.

(The excluded cases are those where the eigenvalues are $e^{\pm i\alpha}$ with $\alpha = 2\pi p/q$, p and q integers, and $q \le 4$; see Moser 1955, 1958).

More generally, each annular region limited by two concentric invariant curves is itself an invariant region, which maps into itself, and from which trajectories cannot escape.

For more than two degrees of freedom, the situation is rather different; this will be treated in section 7.3.

6.6. Chaotic regions

We turn now our attention to the chaotic regions. Since quasi-periodic trajectories and invariant curves are associated with elliptic fixed points, we might suspect that chaotic regions are associated with the hyperbolic fixed points, which so far have not entered the show. We shall therefore consider these points now. Once more, we start with the integrable case. We cannot use here the normalized action-angle variables, because hyperbolic fixed points generally lie precisely on the boundaries between regions where different sets of action-angle variables must be used (see section 4.1). So we revert to the unnormalized variables (q_1, q_2, p_1, p_2), which are valid for the whole phase space. There are two integrals: the Hamiltonian H, and another integral which we call I_2. For a given trajectory, we have

$$H(q_1, q_2, p_1, p_2) = C_1, \tag{75a}$$

$$I_2(q_1, q_2, p_1, p_2) = C_2, \tag{75b}$$

with C_1, C_2 constants. The introduction of a surface of section corresponds to a third relation

$$\Sigma\,(q_1, q_2, p_1, p_2) = 0. \tag{75c}$$

We take q_1 and p_1 as coordinates in the surface of section, and as usual we select a definite value for C_1. Then we can solve eqs. (75a) and (75c) for q_2 and p_2 and substitute in eq. (75b), obtaining a relation of the form

$$I_2(q_1, p_1) = C_2. \tag{76}$$

This describes a one-parameter family of curves in the surface of section, which are the familiar invariant curves. It will be convenient to imagine a three-dimensional space in which q_1 and p_1 are the horizontal coordinates, while C_2 is the vertical coordinate. Then eq. (76) is the equation of a 3-dimensional surface, and the one-parameter family of curves represents the level lines of that surface. An elliptic fixed point is surrounded by closed invariant curves (section 3.2); therefore it corresponds to a "hill", or to a "hollow" (fig. 35a). In the vicinity of a hyperbolic fixed point, the picture is quite different: invariant curves have a hyperbolic shape (fig. 35b). In particular, the stable and unstable invariant manifolds form one particular level line, passing through the fixed point. This corresponds to a "pass" of the surface. This "pass" is surrounded by two "valleys" and two "mountains". Now suppose that we start from the pass and walk, keeping a constant altitude; we follow then one of the branches of the invariant manifolds, for instance an unstable branch. In the typical situation, what happens then is that we walk around one of the "mountains" and come back to the pass (fig. 35c); we come back exactly to it since we have kept a constant altitude. Moreover, we come back along a branch of the stable invariant manifold. Thus, branches of the stable and unstable manifolds

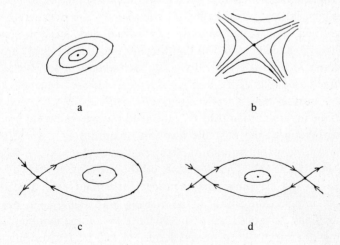

Fig. 35.

typically join each other to form a unique curve, beginning and ending at the hyperbolic fixed point. The same thing happens, of course, on the left of the picture.

We have reasoned in terms of continuous motion for convenience. Actually a trajectory in the surface of section is made of a discrete sequence of points. What happens, then, is that a trajectory which lies on the unstable invariant manifold lies also on the stable invariant manifold; and therefore the points Y_j of the trajectory tend towards the fixed point both for $j \to -\infty$ and for $j \to +\infty$.

More generally, a hyperbolic n-cycle corresponds to n "passes" which are all exactly at the same altitude. The typical situation is then represented by fig. 35d: walking away from one pass at constant altitude, following an unstable invariant manifold, one reaches exactly the next pass, i.e. the next point of the cycle, along a branch of the stable invariant manifold corresponding to that point.

We consider now the non-integrable case. There exists no more a conserved quantity C_2. Therefore, if we start a trajectory on an unstable invariant manifold, there is no reason anymore why it should come back exactly to the hyperbolic fixed point (or reach exactly the next point of the hyperbolic cycle); and in general it does not. If the system is close to an integrable case, the trajectory will come back almost exactly to the fixed point, and the discrepancy may be too small to be seen; this is what happened, for instance, in fig. 21, where the invariant manifold emanating from one hyperbolic point seemed to end exactly into another.

If we trace out simultaneously the stable and unstable invariant manifolds, W^s and W^u, we find that they do not join smoothly anymore, but intersect at an angle (fig. 36a). The intersection point P_0 is called a *homoclinic point*. The existence of this point will have far-reaching consequences. Consider the successive images P_1, P_2,..., of P_0. They all belong to the stable invariant manifold W^s, and tend towards the fixed point. But they also belong to the unstable invariant manifold W^u. Therefore W^u must manage to pass through all these points. It does so by oscillating around W^s (fig. 36b). The direction of the crossing of W^s by W^u is preserved by the mapping, and therefore there must exist a second sequence of intersection points Q_j alternating with the P_j. (In particular cases there may be more than two sequences.) Now the mapping is area-preserving; therefore the successive loops formed by W^u on one side of W^s must have equal areas. But the base of these loops tends to 0 as j tends

Fig. 36.

to ∞; therefore their length must constantly increase. They become more and more thin and elongated; this elongation takes place along the unstable direction of the fixed point (fig. 36b). The story does not end here; in fact it has just begun! The loops must constantly get longer, at an exponential rate; therefore they start creeping along W^u, which they follow closely on an almost parallel course. Soon they are long enough to go around the entire figure, and then to engage into oscillations of their own. Each "first-order loop" thus starts producing second-order loops. These loops in turn grow in length, reach around and start producing third-order loops, and so on. It soon becomes impossible to make a proper sketch of W^u. We again find here a situation of indescribable complexity, with an infinite hierarchy of structures of decreasing size. Not least surprising is the fact that in spite of all these loops, W^u manages to never cross itself! Starting from P_0, we can also go backwards and consider the sequence P_{-1}, $P_{-2}\ldots$. Exactly the same phenomena happen, but involving now the continuation of the stable invariant manifold. Figure 37 is a sketch of the first-order loops of both W^u and W^s. There is thus an infinite set of points of intersection of W^u and W^s, becoming more dense as one approaches the fixed point; all these points are homoclinic points.

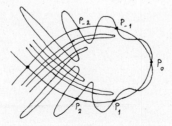

Fig. 37.

We illustrate all this by a computed example from Jenkins and Bartlett (1972). They studied the mapping defined by

$$x' = x + a(y - y^3),$$
$$y' = y - a(x' - x'^3). \tag{77}$$

Fig. 38 shows the initial development of two invariant manifolds, for $a = 1.6$. They belong to a 6-cycle, of which two points h_0 and h_1 appear on the figure, rather than to a fixed point; but the phenomena are in essence the same. Figure 39 shows the further development of one of the manifolds; second-order loops are clearly seen. Only the end part of these loops has been represented, for greater clarity. (Other invariant manifolds,

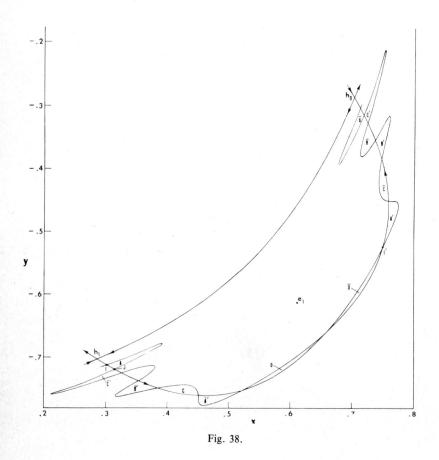

Fig. 38.

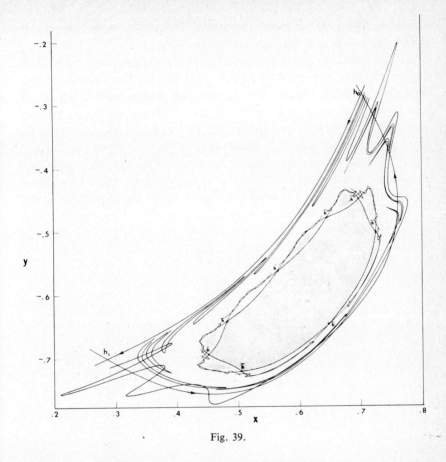

Fig. 39.

related to a different cycle, are also represented in the center of the picture.)

These figures suggest that the invariant manifolds, with their multitude of loops, fill densely a region of finite extend around the hyperbolic fixed point. A trajectory emanating from one point of an invariant manifold will consist of points scattered along the loops, and therefore it seems likely that these points also will fill a two-dimensional region. Another plausible conjecture is that any trajectory started inside that region, even if it does not strictly lie on the invariant manifold, will have similar properties. Moreover, because of the exponential elongation of the loops, successive points of the sequence will present an appearance of randomness. We have thus arrived at least at an understanding of how a two-dimensional chaotic region could be generated; and this seems definitely to be associated with hyperbolic points and their invariant manifolds.

We come back now to the chains of islands considered in section 6.4. It was stated there that for rational values of the rotation number, the invariant curves break down generally into one elliptic and one hyperbolic cycle. The elliptic cycle gives birth to a chain of islands. From what has just been said, we expect the hyperbolic cycle to give birth to a chaotic region. This indeed is what one finds numerically. Figure 40 shows an example; this is the quadratic mapping again, for $\cos\alpha = 0.22$. We see a chain of 5 islands, surrounding the five points of an elliptic cycle, and also a chaotic region, which is thickest around the five points of the hyperbolic cycle.

Such a chaotic region is necessarily limited in extent. Closer to the origin,

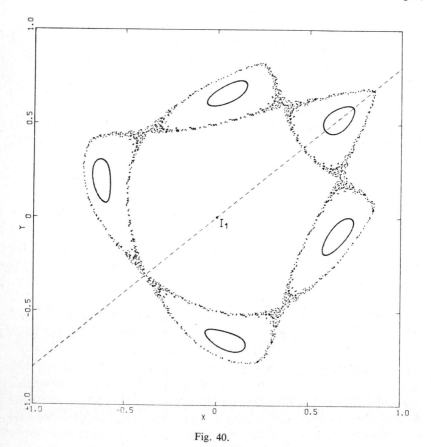

Fig. 40.

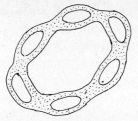

Fig. 41.

invariant curves will reappear when the rotation number is sufficiently far from the rational value i/j. Similarly, invariant curves generally exist on the outer side. Finally, the islands themselves are domains which cannot be invaded by the chaotic region. The region left free between these various boundaries has the shape represented schematically in fig. 41: and observed chaotic orbits indeed have that kind of shape (fig. 40), which suggests that they tend to fill the space which is left to them.

For every rational value of the rotation number, there exists a chain of islands *and* a chaotic region. Therefore there exists an infinity of chaotic regions, present everywhere in the mapping, and intimately mixed with the invariant curves at all levels of the hierarchy. The general picture is thus still more complex than the description which was reached in section 6.4! In practice, however, many of these chaotic regions are very thin and difficult to see.

The outer "chaotic sea" surrounding apparently a "region of curves" which is seen in many experiments, is not fundamentally different in nature from any of the chaotic regions which exist within the curves; it simply happens to be larger and more visible. Thus, our earlier description of the surface of section as consisting of one or more "curve regions" or "regular regions", surrounded by a single large "chaotic region" is now seen to be only a first approximation, which can be useful in some applications, but which does not at all do justice to the true complexity of the picture.

6.7. Separation of orbits

In this section we consider two trajectories initially close to each other, and we ask: how does their distance evolve with time? Do they remain close to each other or do they diverge, and if so, how fast? This question might be initially motivated by an inquiry into the propagation of computer errors, or into the effect of ill-defined initial conditions; but we will find that the question has in fact a much greater significance, and is intimately connected with fundamental properties of the trajectories.

For an integrable system, the answer is easy. The general solution is given by eqs. (27) and (29) in action–angle variables. We consider two trajectories, characterized respectively by initial conditions C_i, D_i and C'_i, D'_i. At time t we have

$$P'_i - P_i = C'_i - C_i, \qquad Q'_i - Q_i = D'_i - D_i + (\omega'_i - \omega_i)t. \tag{78}$$

For large times, the term in t dominates, and the distance between the two points in phase space grows approximately linearly with time. Thus, nearby trajectories diverge linearly in general. This can be intuitively understood as an effect of the differential rotation which takes place for each pair of conjugate coordinates.

We consider now non-integrable systems. In a "regular" region, consisting mostly of quasi-periodic trajectories, numerical experiments show that the divergence is still approximately linear with time. Figure 42 is an example, taken from Casati and Ford (1975); it corresponds to the "unequal-mass Toda lattice", which is a one-dimensional system of two

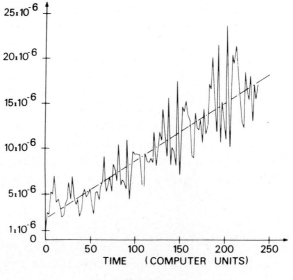

Fig. 42.

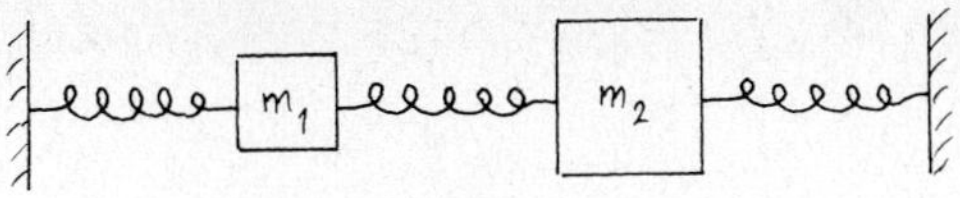

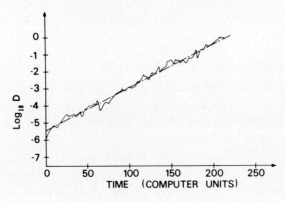

Fig. 43.

mobile masses m_1 and m_2 and two fixed walls, linked by non-linear springs (fig. 43); the Hamiltonian is

$$H = \tfrac{1}{2}\left(\frac{p_1^2}{m_1} + \frac{p_2^2}{m_2}\right) + e^{-q_1} + e^{q_1 - q_2} + e^{q_2}. \tag{79}$$

The quantity D plotted as ordinate on fig. 42 is the ordinary distance in phase space. Initially this distance is $D = 10^{-6}$. The straight dashed line indicates the average linear growth. Short-period fluctuations arise because the q_i and p_i are not normal coordinates; the change of coordinates from Q_i, P_i to q_i, p_i multiplies the distance by a factor which depends on the position on the torus and which varies quasi-periodically. We observe that the linear divergence is a quite mild one: the distance D, although increasing, remains very small.

In a chaotic region, things are very different. Figure 44 shows an example for the same problem, using the same parameter values and the same energy; only the initial positions are different. Note that the scale for D is now logarithmic, while the horizontal scale for time is still linear. The figure thus shows an approximately *exponential* increase of the distance with time.

Many other experiments have shown that this behaviour is typical of

Fig. 44.

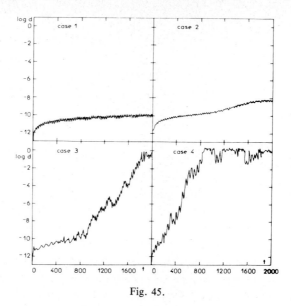

Fig. 45.

chaotic regions. Figure 45 shows an example taken from Froeschlé and Scheidecker (1975) (see section 8.1 for the definition of the problem). Here a logarithmic scale has been used for the distance in all cases, and the initial separation is 10^{-12}. Cases 1 and 2 correspond to initial conditions in the regular region, and exhibit linear separation; cases 3 and 4 correspond to initial conditions in the chaotic region, and show exponential separation. Case 4 shows a phenomenon known as *saturation*: D cannot be larger than the dimensions of the accessible region in phase space, which in the present case are of order unity. When D reaches this value, the two orbits have become completely different; they are as far apart as any two randomly chosen orbits.

Case 3 is a borderline case, which shows that the long-term behaviour may not always be immediately apparent: the distance increases first slowly, in a manner comparable to cases 1 and 2, and then suddenly begins to increase exponentially. Fortunately, however, most cases show a clear-cut behaviour of the kind of cases 1, 2 or 4.

These properties provide in practice a very useful criterion for distinguishing between quasi-periodic and chaotic trajectories; it becomes especially useful for more than two degrees of freedom (see sections 7.1 and 8.1).

One can also consider the separation of the corresponding sequence of points in the surface of section, or in a mapping. The distance between the points of two trajectories are now considered as a function of the number of iterations j, rather than as a function of time t. Essentially the same results are obtained: the separation grows linearly in regions populated by invariant curves, and exponentially in chaotic regions.

Thus, in chaotic regions, the distance between two nearby orbits grows approximately as $e^{\lambda t}$ in phase space, or $e^{\lambda j}$ for a mapping. λ is a constant, which so far characterizes only the particular pair of orbits chosen. How does it change if we change these orbits?

First we fix one orbit, which we write $X^*(t)$, and we consider the second orbit as a perturbation by writing it as $X^*(t) + U(t)$, with U small. This second orbit is then determined by the initial displacement $U(0)$. We experiment therefore with various values of $U(0)$. Numerical results show that, within statistical fluctuations, λ is independent of the choice of $U(0)$. In other words, the number λ seems to characterize the orbit $X^*(t)$: all nearby orbits separate from it as $e^{\lambda t}$.

Next we change the orbit $X^*(t)$ itself, by changing its initial point $X^*(0)$. Numerical experiments then reveal another remarkable result: within a given chaotic region, λ appears to be again a constant! It characterizes therefore not just one orbit, but a whole chaotic region. This can be understood intuitively from the fact that any orbit belonging to the chaotic region tends to go more or less everywhere in that region; λ can thus be interpreted as an "average rate of separation", the average being taken over the entire chaotic region.

On the other hand, for different (non-communicating) chaotic regions, one generally finds different values of λ. Examples of this will be shown below in section 6.8.

Figure 46, from Benettin et al. (1976), shows results for the triangular potential (section 6.1), for the energy $E = 1/8$. What is plotted here as ordinate is not D, but essentially the quantity

$$\chi = \frac{1}{t} \ln \frac{D(t)}{D(0)}; \tag{80}$$

log–log scales are used. If D is proportional to $e^{\lambda t}$, χ will be independent of time, and equal to λ. The three upper curves correspond to three different trajectories in the same chaotic region (the large chaotic region of fig. 22). They appear to converge towards a unique, well-defined value as

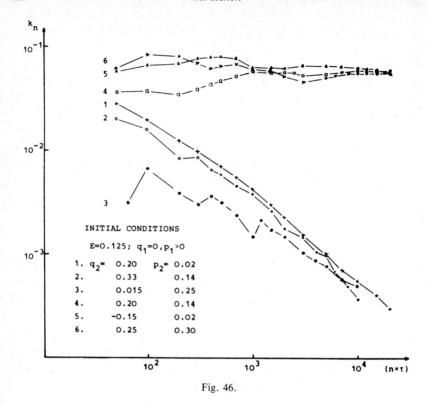

Fig. 46.

time increases; this is the value of λ for the chaotic region. (The convergence corresponds to the fact that λ becomes better defined as a trajectory is computed for a longer time.)

The three lower curves correspond to trajectories in the region of invariant curves. Then D is proportional to t, and χ behaves as $t^{-1}\ln t$, or very nearly t^{-1}.

These results can be explained by a theory which goes back to Lyapunov; I shall only describe it very briefly and refer you to Benettin et al. (1980) for a detailed review. This theory applies to all dynamical systems, and not only to Hamiltonian systems; therefore we revert temporarily to the general case (2).

For a rigorous treatment, it is necessary to go to the limit of infinitely small displacements $U(t)$; but for the exposition it will be convenient to continue to think of $U(t)$ as a finite, small quantity. We consider a given orbit $X^*(t)$ and a given initial displacement $U(0)$. If we substitute

$X(t) = X^*(t) + U(t)$ into the equations of motion (2) and expand in U, we obtain

$$\frac{dU}{dt} = \left(\frac{\partial F}{\partial X}\right)_{X=X^*} U + O(U^2). \tag{81}$$

These equations are known as the *variational equations*. $(\partial F / \partial X)_{X=X^*}$ is an $N \times N$ matrix, whose elements are known functions of time. In the limit of vanishing U, we can drop the last term in eq. (81), and we have a set of linear homogeneous equations, which can in principle be solved; the solution has the form

$$U(t) = M(t)\, U(0). \tag{82}$$

$M(t)$ is the $N \times N$ *transition matrix*.

It can be shown, under fairly general conditions, that for a given $U(0)$ the following limit exists:

$$\lambda = \lim_{t \to \infty} t^{-1} \ln|U(t)|. \tag{83}$$

This is *Osedelec theorem*. An equivalent theorem holds for mappings, with j in place of t. λ is called the *Lyapunov characteristic exponent*, or *Lyapunov characteristic number* (or LCN for short) for the given orbit $X^*(t)$ and the given initial displacement $U(0)$.

This already explains the observed exponential separation between two given orbits. Other properties can also be deduced:

(1) It is immediately seen that if $U(0)$ is multiplied by a constant k, λ does not change. $U(t)$ is multiplied by k, and the additional term $t^{-1} \ln k$ in eq. (83) tends to zero for $t \to \infty$.

(2) It is not difficult to show (see for instance Benettin et al. 1980), that if $X^*(t)$ is kept fixed and $U(0)$ is varied, λ can take at most N different values. This is again a consequence of the linearity. The simplest case is when there are actually N different values $\lambda_1 > \lambda_2 > ... > \lambda_N$. They can be associated with N vectors $U_1(0), U_2(0), ... , U_N(0)$, with the following properties: the N vectors form a basis; and the LCN for the initial displacement $U_i(0)$ is λ_i.

The evolution of an arbitrary initial perturbation $U(0)$ is now easily traced. We write it as

$$U(0) = \sum_{i=1}^{N} c_i U_i(0). \tag{84}$$

Since the equations are linear, we have

$$U(t) = \sum_{i=1}^{N} c_i U_i(t). \tag{85}$$

For $t \to \infty$, the component $c_i U_i(t)$ behaves as $e^{\lambda_i t}$, provided that $c_i \neq 0$. In the long run, therefore, the term with the largest λ_i will dominate. On the other hand, in numerical experiments initial conditions are taken more of less at random, and in general all c_i will be non-zero. Therefore the largest λ_i will in general be λ_1, and $U(t)$ will behave as $e^{\lambda_1 t}$ for $t \to \infty$. This explains why one finds numerically always the same value of λ, irrespective of $U(0)$, and we can now interpret this observed λ as the *maximal Lyapunov characteristic number*.

In fact, even if one tries to start from a $U(0)$ with $c_1 = 0$, numerical errors soon introduce a small non-zero component along U_1, which grows and eventually dominates. This is illustrated by an experiment from Benettin et al. (1980) (fig. 47) for the "Anosov mapping":

$$x' = 2x + y, \qquad y' = x + y \pmod{1} \tag{86}$$

The abscissa represents the number of iterations (here called r). The ordinate is the quantity χ defined by eq. (80) (with r substituted for t). The two LCNs are $\lambda_1 = 0.96242\ldots$, $\lambda_2 = -0.96242\ldots$ The initial displacement is taken along $U_2(0)$. In an exact computation, therefore, χ would converge

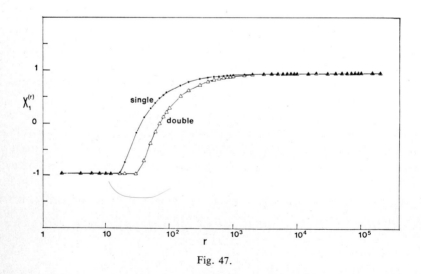

Fig. 47.

towards λ_2 for $r \to \infty$. This is what seems to happen at first. But after a while the component of c_1 created by numerical errors begins to emerge on the figure (around $r = 15$ for a single precision computation, $r = 30$ in double precision), and in the end χ tends towards λ_1.

Because of this, finding numerically the other LCNs is not as easy as finding the largest one λ_1; special tricks are necessary. Intuitively, the idea is as follows: λ_1 was obtained by considering the evolution of a one-dimensional object, the vector $U(t)$. So we consider next the evolution of a small two-dimensional surface; it will be dominated by the two largest LCNs, and for $t \to \infty$ the surface will behave as $\exp[(\lambda_1 + \lambda_2)t]$; from this λ_2 can be extracted. Next one considers a small three-dimensional volume, which will behave as $\exp[(\lambda_1 + \lambda_2 + \lambda_3)t]$, and so on.

Figure 48 shows an example of such computations for a 6-dimensional mapping (Benettin et al. 1980). The different symbols correspond to dif-

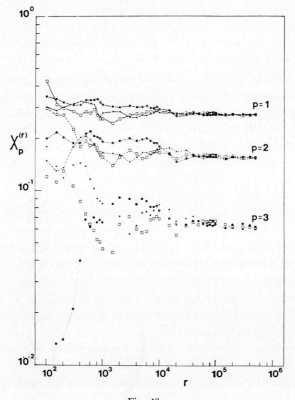

Fig. 48.

ferent orbits. Consider first one particular orbit (for instance the squares). The three curves plotted represent quantities similar to eq. (80), designed to converge towards the three largest LCNs for $r \to \infty$. This convergence is actually observed, and the values of the three LCNs can be read from the figure with good accuracy.

Moreover, fig. 48 shows that these three values are the same for all computed orbits. These orbits were taken in the same chaotic region. This indicates that not only the largest LCN, but in fact the complete set of LCNs characterizes a given chaotic region and is the same for all orbits in that region.

If, for a given orbit $X^*(t)$, there are less than N distinct possible values for the LCNs, the analysis is slightly more complicated but the results are not greatly changed. Some LCNs have then a *multiplicity* greater than 1; this means that they are associated with more than one vector U_i.

So far I have presented the theory as it applies to dynamical systems in general. Now we restrict our attention to Hamiltonian systems. It can be shown in that case that the LCNs must exist in pairs: if λ is a LCN, then $-\lambda$ is also a LCN. (One particular consequence of this is that the sum of all LCNs is zero, a result which follows also from the conservation of volume in phase space.) Therefore only two cases are possible:

(i) Only 0 is a LCN. This case corresponds to quasiperiodic orbits, as is easily seen: for a quasi-periodic orbit, all nearby orbits separate linearly, and λ defined by eq. (83) is always zero.

(ii) There exist positive and negative LCNs. Then λ_1 is greater than 0, and we observe exponential separation; this corresponds to chaotic regions.

It is not possible to have all LCNs negative; this explains why exponential *convergence* of orbits is never observed in Hamiltonian systems.

The exponential separation of orbits in chaotic regions has dramatic consequences. In principle, a dynamical system is deterministic, i.e. from the initial conditions one can deduce future evolution for all time. But in practice, if we are in a chaotic region, the slightest deviation will be amplified exponentially; after some time, saturation is reached, meaning that the computed orbit bears no resemblance any more to the exact orbit which we wished to follow! Deviations will be produced by numerical errors. However, even if we had at our disposal an ideal computer, free of round-off and integration errors, the situation would not be improved, because of two other factors. First, it is impossible to measure the initial state of a physical system with infinite accuracy; and second, the equations used to represent the system are never strictly exact: many approximations are made and many small effects are neglected. From the physicist's point of view, therefore, the system is *not* deterministic. Its evolution cannot be determined for more than a fixed period of time. If λ is large, the system will in fact exhibit a quasi-random behaviour; there will be no apparent relation between its states at two different times, even if the interval between these times is not very large. An extreme example of this can be found in the quasi-random number generators used in computers. These generators can be considered as dynamical systems; the most widely used algorithm consists in a simple one-dimensional map, each new quasi-random number being computed as a function of the previous one (Knuth 1981):

$$x_{i+1} = f(x_i). \tag{87}$$

This is a perfectly deterministic procedure, and there is no computing error. Yet, if the function f is well designed, it will produce a series of numbers which mimic very effectively the properties of true random numbers. In particular, f must be such that the mapping (87) has a very large LCN; a typical value is $\lambda = 10^9$.

The exponential separation also happens if one goes back in time. As a consequence, trajectories in a chaotic region "forget" progressively their initial state; nothing can be deduced, beyond a certain point, about the past history of the system from an examination of its present state.

One might then conclude that there is no point in computing orbits in the chaotic region, and that numerical experiments can tell us nothing about the properties of these regions. Fortunately, the situation is not so bad. It is true that the location of individual points of the trajectory has not much meaning, since the slightest change in the initial state will produce entirely different points. If, however, we consider not individual points but rather the *set* of the points which constitute an orbit, we find that this set has definite statistical properties. More precisely, the points appear to obey a well-defined distribution. In contrast to the rapidly deteriorating accuracy of individual points, this distribution becomes better and better defined as more points are computed. This is illustrated by figs. 49 to 51, for the mapping

$$x' = x + a\sin y, \qquad y' = y + x' \pmod{2\pi}, \tag{88}$$

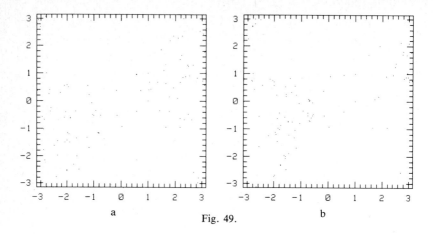

Fig. 49.

(Taylor 1968, Froeschlé 1970b), nowadays called the "standard mapping" (Chirikov 1979, Greene 1979), with $a = 2$. x and y are considered as being defined modulo 2π; therefore the mapping takes place on a square of side 2π. It is convenient to define this square as: $-\pi \leq x < \pi$, $-\pi \leq y < \pi$. Figure 49a shows the result of hundred iterations, starting from $x_0 = 1$, $y_0 = 0$; Figure 49b is the same computation with $x_0 = 1 + 10^{-9}$, $y_0 = 0$. The two sets of points are markedly different: this is the consequence of a fast exponential divergence. Next we compute 10 000 points instead of 100 for the same two trajectories (figs. 50a and b): the two figures are now quite similar, and differ only in small details. They show a large chaotic region, which within

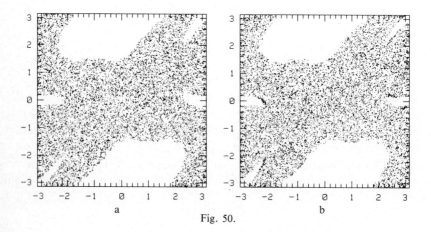

Fig. 50.

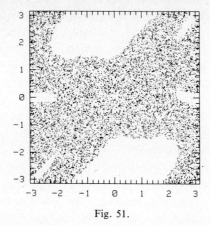

Fig. 51.

statistical fluctuations appears to be uniformly filled with points. (This uniformity is a consequence of the area-preserving property of eq. (88).) The white regions which are not penetrated by the points are regions of invariant curves or "islands".

Figure 51 corresponds to initial conditions $x_0 = 1$, $y_0 = 1$; it indicates that essentially the same distribution is obtained for any trajectory inside the chaotic region.

6.8. Last example: the oval billiard

I shall present one last example of a Hamiltonian system with two degrees of freedom, which illustrates nicely their essential properties. (Note added in proof: see however Hénon and Wisdom.) First we need some generalities about billiards.

The ordinary billiard is a rectangular area (fig. 52a); a ball rolls inside

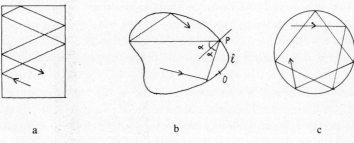

a   b   c

Fig. 52.

it and bounces on the walls. We idealize the problem by assuming that: (i) the ball is a point particle; (ii) there is no friction; (iii) the ball bounces like a ray of light on a mirror, i.e. the incident and reflected paths make equal angles with the normal to the wall, and velocity is preserved. We have then a Hamiltonian system with two degrees of freedom; it can be considered as a particular case of eq. (45), i.e. of the motion of a particle in a given potential, by taking $V=0$ inside the billiard and $V=\infty$ outside. The constancy of the Hamiltonian corresponds to the fact that the velocity of the ball is constant inside the billiard.

In this idealized rectangular billiard, the velocity vector can have only four possible directions for a given trajectory (fig. 52a); this means that there exists a second conserved quantity. (We can take for instance p_1^2 for this quantity.) Therefore the problem is integrable, and not very interesting. We introduce now a *generalized billiard*, in which the wall is an arbitrary closed curve (fig. 52b). In general this corresponds to a non-integrable problem. This problem has been frequently used in the study of dynamical systems, and in that context it is called simply a *billiard*.

A natural choice for the surface of section is the wall itself: the points of a trajectory which belong to the surface of section are those points where it bounces on the wall. As coordinates in the surface of section, we take the curvilinear distance l along the boundary from some origin 0 to the impact point P, and $\sin\alpha$, where α is the oriented angle with the normal (fig. 52b). It can be shown that with these coordinates the mapping is area-preserving. l is defined modulo L, L being the length of the boundary, and $\sin\alpha$ ranges from -1 to $+1$; thus the surface of section is a finite rectangle.

The numerical computation of a trajectory is quite simple. We start from a point on the surface of section. From l and $\sin\alpha$ we determine the emerging path; and we compute the next intersection of this straight line with the boundary curve, which gives the next values of l and $\sin\alpha$. No numerical integration is required: billiards are among those rare dynamical systems for which the equations of the mapping can be obtained in explicit form (at most we will have to solve an ordinary implicit equation). We remark also that it is not necessary to know the magnitude of the velocity of the particle; the picture in the surface of section will be exactly the same for all values of the energy E (the constant value of the Hamiltonian). Thus, a single two-dimensional picture will be sufficient to represent all possible motions on the billiard. (Billiard problems are degenerate in this respect). Because of these simple properties, billiards are very attractive model problems.

One of the simplest boundaries which one might think of is a circle (fig. 52c). In that case, however, α is a constant, and the problem is again integrable.

Benettin and Strelcyn (1978) have introduced the *oval billiard*, defined as follows. We consider four points A, B, C, D, forming a square of side 2, and we join them by four arcs of circle, in such a way that the arcs have a common tangent at their meeting points (fig. 53). The centres a, b, c, d of the four arcs must lie on the symmetry axes of the square; it is easily seen that the billiard itself must be symmetric with respect to these axes, and that it is characterized by a single parameter, which can be taken to be δ, the distance from a to the nearest side AB of the square. For $\delta = 1$, the oval billiard reduces to a circle, and is therefore integrable; the invariant curves are horizontal lines in the $(l, \sin\alpha)$ surface of section. For $\delta = 0$, the two arcs BC and DA become straight lines, while AB and CD become half-circles; this particular case is known as a *stadion*, and stadions have been shown to be ergodic (Bunimovich 1974). The points of a trajectory fill densely the surface of section. Thus, by varying the parameter δ from 0 to 1, we may hope to observe a gradual transition from the extreme case of an integrable system, corresponding to complete order, to the other extreme case of an ergodic system, corresponding to complete disorder.

We start from the ergodic case, $\delta = 0$, for which a single chaotic region fills the rectangular surface of section. Figure 54 represents the surface of section for $\delta = 0.1$. The horizontal coordinate is $\eta = l/L$, ranging from 0 to 1; the origin for l is the point 0 (fig. 53). The vertical coordinate is $\sin\alpha$. The regular lattice arrangement of the points is an artifact, resulting from the use of a printer and the attendant rounding-off of coordinates. Different orbits are identified by different letters or symbols.

A chaotic region still fills most of the rectangle in fig. 54, but two islands

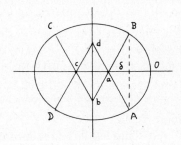

Fig. 53.

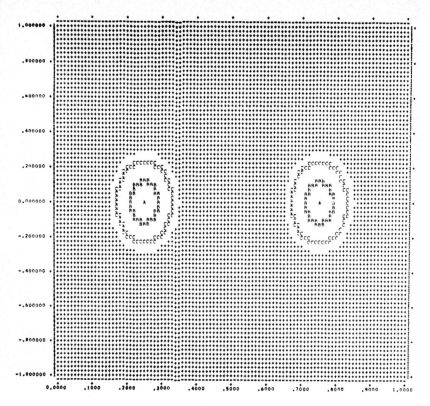

Fig. 54.

have appeared. The two points in the center of the islands, with coordinates (1/4,0) and (3/4,0), form a 2-cycle and correspond to a linearly stable periodic orbit lying on the short axis of the oval. There exists also a periodic orbit on the long axis [points (0,0) and (1/2,0)], but that one is linearly unstable and is lost in the chaotic region.

The figure exhibits a symmetry with respect to $\sin\alpha = 0$, corresponding to a reversal of the direction of motion, and also a symmetry with respect to $\eta = 1/2$, corresponding to a symmetry with respect to the horizontal axis of the oval billiard. From here on we shall therefore show only one-fourth of the surface of section, corresponding to: $0 \le \eta \le 1/2$, $0 \le \sin\alpha \le 1$; this will allow us to see more detail.

For $\delta = 0.3$ (fig. 55), the island has increased in size; also the appearance of other, smaller islands is suggested by the white patches in the chaotic region.

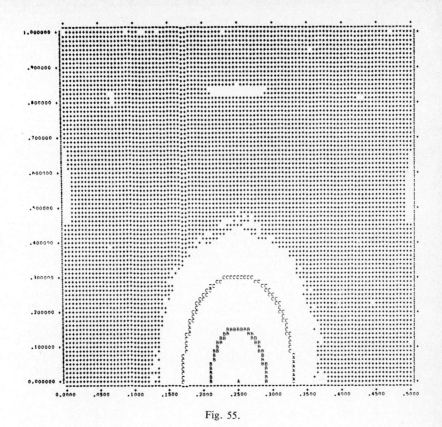

Fig. 55.

For $\delta = 0.6$ (fig. 56), these new islands have much increased, and many more have appeared. The large chaotic region has shrunk; but it is still in one piece.

For $\delta = 0.76$ (fig. 57), we have a new phenomenon: an invariant curve joining the left and the right side has appeared. Therefore there are now two large chaotic regions, which do not communicate; they are represented by different symbols.

For $\delta = 0.85$ (fig. 58), the number of chaotic regions has increased; 6 of them can be identified on the picture. Each boundary between two successive regions corresponds probably to an invariant curve (not represented).

For $\delta = 0.9$ (fig. 59), we observe 8 independent regions. As δ tends towards 1, each of these regions tends to become a horizontal band. The

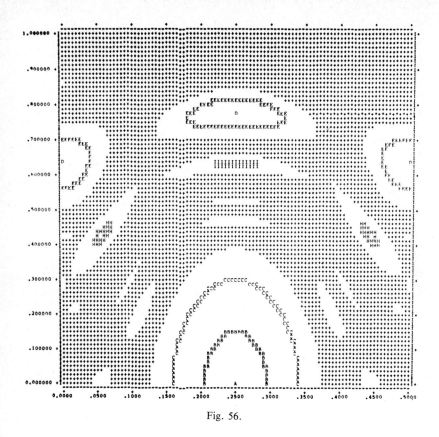

Fig. 56.

separating invariant curves become more and more numerous, and tend towards horizontal lines.

In the limit, for $\delta = 1$, the number of chaotic regions becomes infinite and each of them shrinks to zero thickness, so that the surface of section is filled with horizontal invariant lines.

Benettin and Strelcyn (1978) have also computed the maximal Lyapunov characteristic number, λ_{max}, in the chaotic regions. Figure 60 shows λ_{max} as a function of δ. Up to $\delta = 0.75$ approximately, there is only one large chaotic region, and therefore essentially only one value of λ_{max}. (There must also exist thin chaotic regions inside the islands; but these have very small λ_{max} values, which would be undistinguishable from zero on fig. 60). At $\delta = 0.75$, the chaotic region separates into two distinct regions, and the λ_{max} of these two regions begin to evolve separately: the curve forks

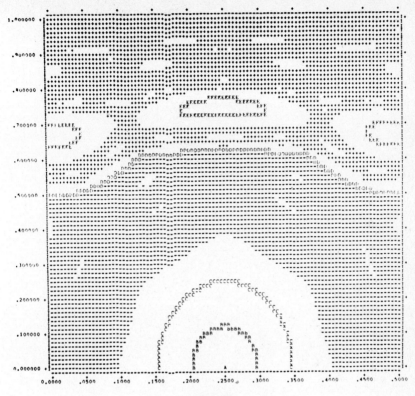

Fig. 57.

on fig. 60 (see enlargement in the upper part). As δ continues to increase, a new branch appears every time a chaotic region is divided. For $\delta = 0.85$, for instance, the 6 values of λ_{max} corresponding to the 6 chaotic regions of fig. 58 are clearly separated.

All values of λ_{max} decrease as δ increases, and appear to be headed for a zero value as $\delta \to 1$ (except for one branch, which corresponds to singular orbits with $\alpha \to \pm\pi/2$ in the limit). This is natural, since for $\delta = 1$ the system is integrable and all LCNs are equal to zero. We observe also that the value of λ_{max} tends to be correlated with the size of the corresponding region; this seems to be true for dynamical systems in general.

(Note added in proof: a recent study (Hénon and Wisdom 1983) has shown that the situation is actually much more complex than fig. 60 suggests.)

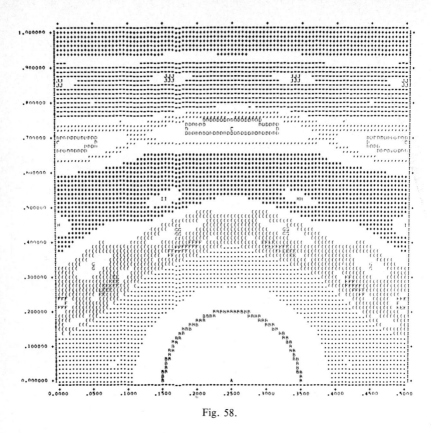

Fig. 58.

7. Three degrees of freedom

7.1. Techniques

This has been much less studied than the case of two degrees of freedom. Phase space has now 6 dimensions; fixing H = constant and taking a section, we obtain a 4-dimensional space of section, which is much more difficult to visualize and to study than the 2-dimensional surface of section of the case of two degrees of freedom.

If the system is integrable, there are, in addition to the Hamiltonian, two other integrals. Therefore the sequence of points should lie on a 2-dimensional subset of the space of section. If the system is non-integrable, but we are in a region where the KAM theorem applies, then

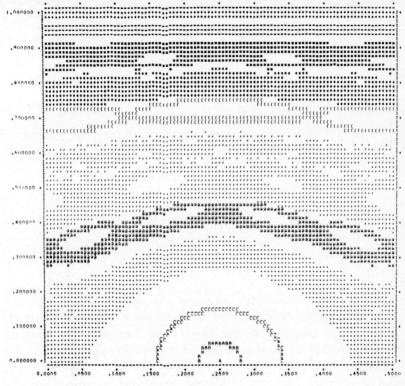

Fig. 59.

we also expect 2-dimensional tori. At the other extreme, for an ergodic system, points should fill the 4-dimensional space of section. Thus, the dimension of the manifold occupied by a sequence of points can range from 2 to 4.

A first technique of study consists in ignoring one coordinate: instead of the 4-dimensional space of section (q_1, q_2, p_1, p_2), we consider only the 3-dimensional space (q_1, q_2, p_1) (for instance). In essence, this is a projection of the space of section on the 3-dimensional space. If the sequence of points lies on a 2-dimensional subset in the space of section, then in projection the points also lie on a 2-dimensional subset of the space (q_1, q_2, p_1). This is easier to test. Two methods have been tried.

(1) One can make stereoscopic projections in order to actually see the arrangement of the points in the 3-dimensional space. This technique has

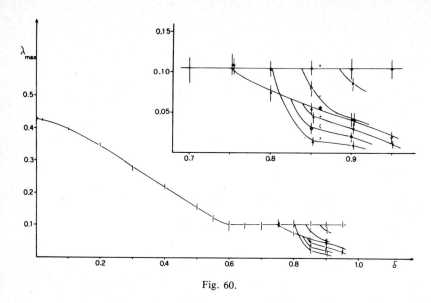

Fig. 60.

been pioneered by Froeschlé (1970a); Figures 61 and 62 show examples for the three-dimensional restricted problem (the motion of M_1 and M_2 is still as described in section 6.2, but M_3 is now free to move in three-dimensional space). Phase space is $(x, y, z, \dot{x}, \dot{y}, \dot{z})$. The space of section is defined by $z = 0$. The three coordinates retained are x, y, $\dot{z}$. In other words, a point of coordinates $(z, y, \dot{z})$ is "plotted" at each intersection of the orbit with the plane $z = 0$. Actually, a pair of stereoscopic projections of this point on the (x, y) plane are plotted. With a stereoscopic viewer, the points of fig. 61 are seen to lie on a smooth surface, which resembles somewhat a cooling tower. This case therefore appears to be a quasi-periodic orbit, with the points lying on a 2-dimensional torus in the space of section, and the orbit itself lying on a 3-dimensional torus in phase space.

For fig. 62, on the other hand, the stereoscope reveals that the points fill more or less a three-dimensional volume. The invariant tori have been destroyed. Beyond this, however, it is not possible to deduce from fig. 62 any information about the arrangement of the points in the true 4-dimensional space of section. The most likely conjecture is that they fill a 4-dimensional region.

This technique has been more recently used by Martinet and Magnenat (1981), in connection with the problem of the motion of a star in a galaxy without symmetry.

(2) We can also try to make a further section through the 3-dimensional space (q_1, q_2, p_1). However, this space is populated by points, not by a continuous trajectory; and the probability of a point falling on an arbitrarily chosen surface is zero. Therefore we must take not a true section, but rather a *slice*, having a small but finite thickness, in order to catch some points. Figures 63 and 64 are examples taken from Froeschlé (1972), for the 4-dimensional mapping

$$x' = x + a_1 \sin(x + y) + b \sin(x + y + z + w),$$

$$y' = x + y,$$

$$z' = z + a_2 \sin(z + w) + b \sin(x + y + z + w), \tag{89}$$

$$w' = z + w.$$

All variables are defined modulo 2π. A trajectory is a set of points in the (x, y, z, w) space. It is first projected on the 3-dimensional (x, y, z) space; and then a series of nine slices are taken, defined by $|z - z_0| < 0.01$ with nine regularly spaced values for z_0. Parameter values are $a_1 = -1.3$, $a_2 = -1$ for both figures; $b = -0.15$ for fig. 63; and $b = 0.5$ for fig. 64. In the first case, the points fall on well-defined curves in each slice; this suggests that they fall on a 2-dimensional surface in (x, y, z) space. In the second case, the situation is rather different: the points appear to fill a 3-dimensional region in (x, y, z) space.

A rather different technique of study consists in determining numerically the Lyapunov characteristic exponents of the trajectories. This allows one to distinguish between quasi-periodic and chaotic orbits (see section 6.7), and to map the regular and chaotic regions in phase space. Other properties can also be deduced (see for instance Gonczi and Froeschlé 1981). However, this technique does not provide any information about the shape of an orbit in phase space, or of the set of points in the space of section; it gives just two numbers for each trajectory.

For systems with two degrees of freedom, the two-dimensional surface of section provided a very direct and striking illustration of the properties of the orbits. Unfortunately, an equivalent representation is lacking in the case of three degrees of freedom.

7.2. Weakly coupled mappings

In the above examples, we jumped straight into the middle of fully

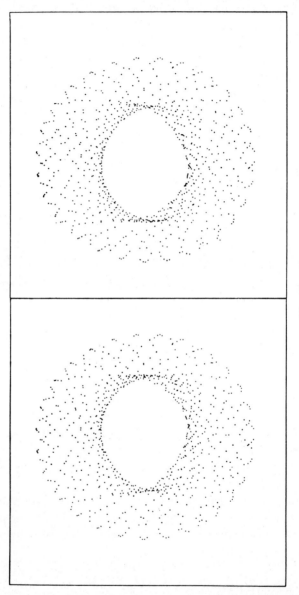

Fig. 61.

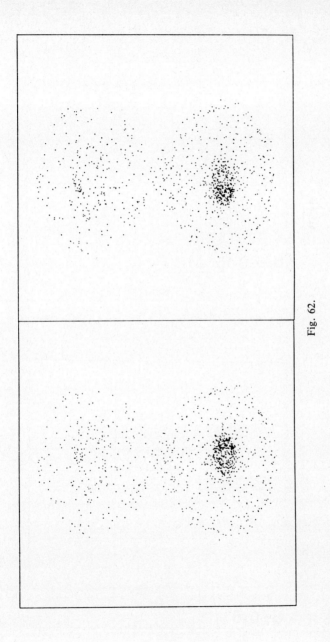

Fig. 62.

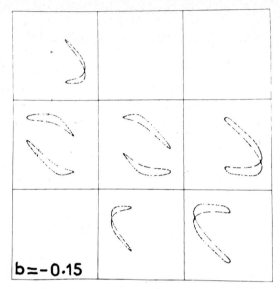

Fig. 63.

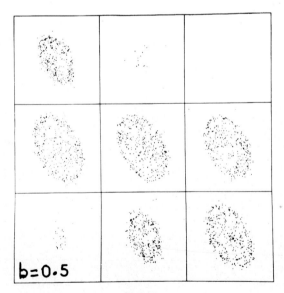

Fig. 64.

developed systems with three degrees of freedom. A more cautious approach consists in introducing the third degree of freedom in a more progressive manner, starting from the case of two degrees of freedom which is well understood. I shall describe a study by Froeschlé (1971). We use again the mapping (89) as an example, taking now $a_1 = a_2 = -1.3$. We observe that for $b = 0$, this 4-dimensional mapping separates into two independent 2-dimensional mappings for (x, y) and (z, w) respectively. The (x, y) mapping is

$$x' = x + a_1 \sin(x + y), \quad y' = x + y, \tag{90}$$

which is the "standard mapping" (88) in a slightly different form. Figure 65 shows the surface of section of this mapping. We see a large region of invariant curves around the origin; some scattered islands; and a chaotic region filling the space between these. The (z, w) mapping is identical to eq. (90), with x and y replaced by z and w.

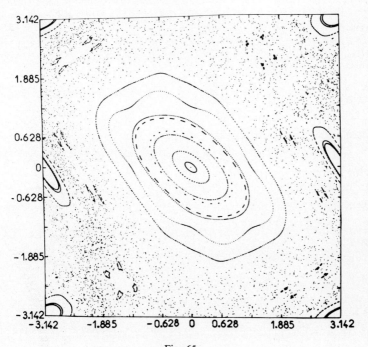

Fig. 65.

For $b \neq 0$, the two mappings become coupled, and we have now a non-separable 4-dimensional mapping, equivalent to a dynamical system with three degrees of freedom. But we shall select a small value of the coupling parameter: $b = 0.01$. We may hope thus to be able to understand the results by reference to the limiting separable case $b = 0$.

First we take the initial points (x, y) and (z, w) inside the region of curves of fig. 65. The KAM theorem leads us to conjecture that the trajectory will then lie on a 2-dimensional torus in the 4-dimensional space of section. To verify this, we use the "slice" technique: first we eliminate w and consider only the (x, y, z) space; then we take a slice in that space, defined by $|z - z_0| < \varepsilon$, and we project on (x, y). Figure 66 shows the result (for details see Froeschlé 1971). The points appear to lie on two well-defined curves, which represent the intersection of the slice with the torus. [The two curves

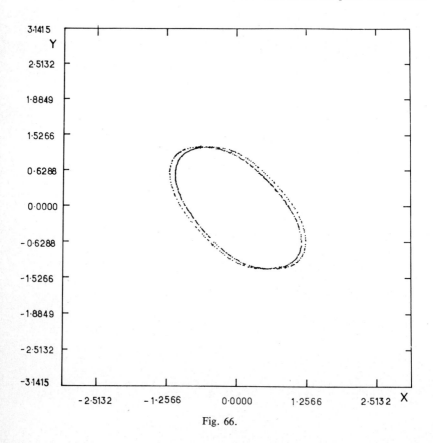

Fig. 66.

can be interpreted as corresponding to the two points of intersection of $z = z_0$ with the invariant curve in (z, w).] So our expectations are confirmed.

Next we take the initial point (x, y) in the region of curves and the inital point (z, w) in the chaotic region. A projection of the trajectory on the (z, w) plane does not show any visible difference with the limiting case $b = 0$: it fills again the chaotic region. The projection on the (x, y) plane is more interesting. Since the behaviour of the z and t coordinates is chaotic, the coupling term in the first equation (89) acts as a small quasi-random perturbation. Therefore, over a short time interval the projections on the (x, y) plane will still lie approximately on an invariant curve; but this curve will slowly move, in quasi-random fashion, inwards and outwards: it will perform a *random walk* among the family of invariant curves. (We forget for the moment that this family is not really continuous.) This is exactly what is observed numerically. Figure 67 represents extracts from a trajectory consisting of 200 000 points. The first frame shows points 10 001 to

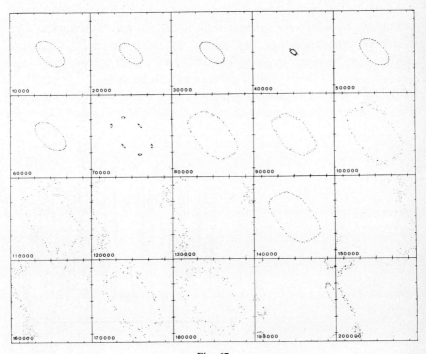

Fig. 67.

10 100, in projection on the (x, y) plane; the second frame shows points 20 001 to 20 100, and so on. In other words, we see a sequence of "snapshots" of the trajectory, the exposure time being equal to 100 iterations. The random walk is clearly exhibited. The diameter of the instantaneous invariant curve goes up and down. At one point $(j = 70\ 000)$, we go through a chain of small islands. Around $j = 100\ 000$, the orbit happens to hit the limit of the region of curves, and penetrates into the chaotic (x, y) region. Later, around $j = 140\ 000$, it briefly comes back to the region of curves; and so on.

This random walk can be illustrated in a different way by defining an instantaneous "dimension of the invariant curve" as

$$D = \langle x^2 - axy - ay^2 \rangle, \tag{91}$$

where $\langle \ldots \rangle$ represents the average over 100 consecutive points. The quadratic expression in eq. (91) is so chosen as to be constant along an invariant curve near the origin, in the linear approximation. Figure 68 represents the evolution of D as a function of the number j of iterations (the numbers along the horizontal axis are actually $j/100$). Up to about $j = 100\ 000$, we observe a random walk inside the region of curves; after $j = 100\ 000$, the trajectory lies mostly in the chaotic region and is characterized by much larger values of D and violent fluctuations. Temporary returns to the region of curves can also be seen.

Finally, fig. 69 is a cumulative picture: each frame shows points of the orbit up to the indicated value of j. This shows that the points tend to fill

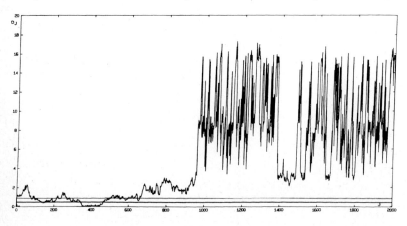

Fig. 68.

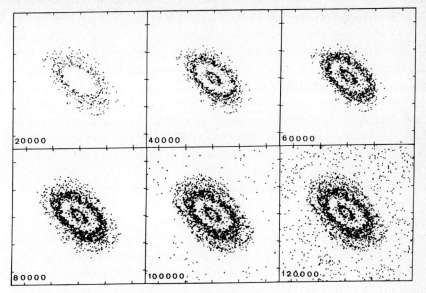

Fig. 69.

progressively the whole square, including both the region of curves and the chaotic region of the original 2-dimensional mapping.

It seems thus quite likely that in the 4-dimensional space of section (x, y, z, w), the points of the trajectory fill a 4-dimensional region. There appears to be no intermediate case in which the points would fill a 3-dimensional region only; as soon as there is chaos for one pair (z, w), and even for a small coupling term b, the other pair (x, y) is "contaminated" and in the long run behaves also chaotically.

The random walk in the region of curves can also be described as a *diffusion*. The variation of D after j iterations should be proportional to $j^{1/2}$, and we can define a "diffusion coefficient" as

$$k = j^{-1} \langle (D_j - D_0)^2 \rangle, \tag{92}$$

where $\langle \dots \rangle$ now represents the expected value in the probabilistic sense. k can be estimated numerically, and one can verify that the random walk really has the characteristic properties of a diffusion process (Froeschlé and Scheidecker 1973).

7.3. Arnold diffusion

This is a very important property, which in principle distinguishes fundamentally the systems with three or more degrees of freedom from those with two degrees of freedom.

For an integrable system with two degrees of freedom, the surface of section exhibits a one-parameter family of invariant curves; each curve is characterized by a rotation number v, and this number varies continuously across the curves. We shall therefore associate each curve with a point of abscissa v on the $[0,1]$ segment.

If a small perturbation is added, making the system non-integrable, then the invariant curves are destroyed and replaced by a chaotic region (and also a chain of islands) in a small interval around each rational value of v (fig. 70). Some overlap and merging of these intervals will occur; but in the end we will still be left with an infinite number of thin chaotic regions, which do not communicate because they are separated by invariant curves. A chaotic trajectory remains imprisoned in one of these regions and can never leave it.

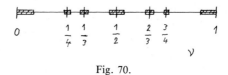

Fig. 70.

We consider now the situation for three degrees of freedom. If the system is integrable, sequences of points in the 4-dimensional space of section lie on invariant 2-dimensional tori. Each of these tori is characterized by two rotation numbers, v_1 and v_2, and can therefore be represented by a point of coordinates (v_1, v_2) in a square of side 1.

Now we add a perturbation. Tori are then destroyed in the vicinity of each commensurability relation:

$$k_1 v_1 + k_2 v_2 + k_3 = 0, \tag{93}$$

where k_1, k_2, k_3 are integers [see eqs. (40) and (74)]. Each relation (93) represents a straight line in the (v_1, v_2) plane; tori will be destroyed and replaced by chaotic trajectories inside *bands* of finite thickness around these lines (fig. 71). This thickness decreases rapidly as the k_i become

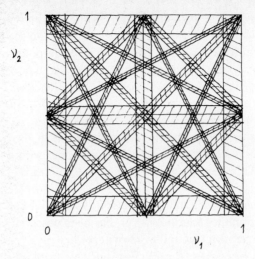

Fig. 71.

larger, and tends to zero when the perturbation tends to zero. We notice on fig. 71 that *the bands communicate*; their union is called the *Arnold web*. Therefore, an orbit started inside the chaotic region of one particular band will be able to wander around and to explore all the other bands. Since the Arnold web is dense on the square, the orbit will be able to go everywhere. Instead of an infinite number of separated chaotic regions, as in the case of two degrees of freedom, we have a single chaotic region, which penetrates everywhere! This phenomenon is known as *Arnold diffusion* (Arnold 1964; for a detailed review see Chirikov 1979).

We can now interpret the results of section 7.2 in a somewhat different light. The mapping, eq. (90), for $a = -1.3$, actually does not correspond to a small perturbation of an integrable system but rather to a large one, and instead of fig. 70 we have a situation which is very schematically sketched on fig. 72: the large shaded interval corresponds to the large chaotic region in fig. 65, while the smaller intervals correspond to the thin chaotic regions inside the region of curves. (We may imagine that many of the intervals of fig. 70 have grown and merged to form a single large interval.)

Fig. 72.

For the 4-dimensional mapping, eq. (89), then, the situation is as sketched on fig. 73. v_1 and v_2 are associated respectively with (x, y) and (z, w). Thus, the random walk shown by figs. 67 to 69 corresponds to a trajectory started in region A, and wandering through the large horizontal band in the upper half of fig. 73. This kind of trajectory has also been observed by Tennyson et al. (1979), who call it "thick layer diffusion". The same authors have computed another trajectory in which the (z, w) point is initially placed inside a thin chaotic region (region B in fig. 73); a random walk is again observed ("thin layer diffusion"), but with a much smaller diffusion rate. This corresponds to motion within a thin horizontal band in fig. 73.

Diffusion along the diagonal band of fig. 73 (region c), corresponding to the commensurability relation $v_1 = v_2$, has been observed by Chirikov et al. (1979), who call it a "coupling resonance".

We comment briefly on the implications of Arnold's diffusion. The phenomenon is always present for non-integrable systems with more than 2 degrees of freedom. Therefore there are in principle two possible kinds of motion:

(i) a quasi-periodic motion on an invariant torus, corresponding to a point which does not belong to any band on fig. 71;

(ii) chaotic motion on Arnold's web, slowly diffusing across the web in a quasi-random way.

However, Arnold's web is dense everywhere; therefore, even if we start with a quasi-periodic motion, the slightest perturbation can send us on the

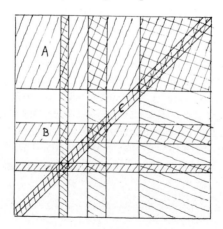

Fig. 73.

web; then the orbit will start diffusing chaotically, and ultimately it will be able to reach every part of phase space! This argument might lead one to think that a system with 3 degrees of freedom or more is practically always chaotic. This would be very bad for many applications.

Fortunately there is a counteracting factor in practice. As one penetrates into the regular region (or equivalently, as the perturbation becomes small), the diffusion rate drops very rapidly, and soon it becomes entirely negligible (see Chirikov 1979). This is confirmed by numerical experiments, which show well-preserved invariant tori in spite of the numerical errors (see for instance fig. 66). In fact, precisely because of this factor, numerical detection of Arnold's diffusion inside the regular region is difficult, and can only be achieved near the border, i.e. close to the chaotic region.

8. Many degrees of freedom

In this section we shall consider the following question: what happens when the number n of degrees of freedom becomes large? This question is of great importance in connection with statistical mechanics, which deals with systems having extremely large values of n. The basic assumption of statistical mechanics is that a system with a large n is ergodic. How are we to reconcile this assumption with the existence of invariant tori predicted by the KAM theorem?

8.1. Parallel sheets

One way to approach this question is to try to find out how the fraction of phase space occupied by invariant tori changes with n. I shall describe first an investigation by Froeschlé and Scheidecker (1975). The model chosen is a self-gravitating system, consisting of n plane infinite sheets in the (x, y, z) physical space. Each sheet is parallel to the (x, y) plane, so that its position is characterized by its altitude z only. The sheets move up and down; they are considered as being not solid, but made up of a large number of point particles, so that they can cross each other freely. This model can be used to study the dynamics of a flat galaxy; it has also been much used in plasma physics. We call σ the mass per unit surface of a sheet; σ is taken to be constant on a sheet, and also has the same value on all sheets. Then, the gravitational field produced at altitude z by a sheet situated at altitude z_0 is (by a simple application of Gauss's theorem):

$$-2\pi G\sigma \,\mathrm{sign}(z-z_0), \tag{94}$$

where G is the gravitation constant, and

$$\mathrm{sign}(x) = \begin{cases} -1 & \text{for } x<0, \\ 0 & \text{for } x=0, \\ +1 & \text{for } x>0. \end{cases} \tag{95}$$

We number the sheets from 1 to n; sheet i has an altitude z_i and a vertical velocity $\dot{z}_i$. Its acceleration, due to the attraction of the other sheets, is

$$\ddot{z}_i = 2\pi G\sigma \sum_{j=1}^{n} \mathrm{sign}(z_j - z_i). \tag{96}$$

We can reduce the problem to Hamiltonian form by taking $q_i = z_i$, $p_i = \sigma\dot{z}_i$, and

$$H = \frac{1}{2\sigma} \sum_{i=1}^{n} p_i^2 + 2\pi G\sigma^2 \sum_{j>i} |q_j - q_i|. \tag{97}$$

The constant value E of the Hamiltonian is the energy per unit (x, y) area of the system.

There exists another integral:

$$I_2 = \sum_{i=1}^{n} p_i, \tag{98}$$

which is the total momentum per unit area. It can be used to reduce the order of the system by one unit; the easiest way to do this is to take the center of mass of the system as origin, so that the positions of $(n-1)$ sheets determine the position of the last. Thus, a system of n sheets has in reality only $(n-1)$ degrees of freedom.

This simple model offers a number of advantages:

(1) The numerical computation is very simple. As long as no sheets cross, each of them feels a constant field as shown by eq. (96), and therefore each sheet has a parabolic motion, which can be written down explicitly: there is no need for numerical integration. The computation is essentially reduced to finding the next time at which two sheets will cross; and this involves nothing more than solving second-order equations. Thus, there are no computing errors except round-off errors, which can easily be kept very small, and the computation is fast.

The space of section can be defined by the condition that two sheets coincide. Then the equations of the mapping can be written explictly. (This is another exceptional case where we have both the differential equations and the mapping equations in explicit form.)

(2) For a given energy, the accessible region in phase space is finite; this is easily seen from the form of eq. (97), plus the relation

$$\sum_{i=1}^{n} q_i = 0. \tag{99}$$

Therefore one can define the fraction of phase space occupied by invariant tori.

(3) The value of the energy can be normalized to a fixed value by a simple change of units. Therefore, for a given number of sheets n, we have, in fact, a unique problem, without any variable parameter, and the fraction of phase space occupied by tori is a definite number, which we call $p(n)$.

For $n = 2$, the system has only 1 degree of freedom and is integrable; therefore $p(2) = 1$. Each sheet has a simple periodic motion, with one period made up of two arcs of parabola (fig. 74).

For $n = 3$, the system has two degrees of freedom. A surface of section (fig. 75) reveals that most trajectories lie on invariant tori; however, a thin chaotic region can also be seen. In order to obtain a quantitative estimate of $p(n)$, Froeschlé and Scheidecker (1975) computed 100 randomly selected orbits; the starting points were taken at random in phase space, with uniform probability. The character of each orbit was determined by observing the separation of nearby orbits. Figure 45, already shown above, shows four typical cases; cases 1 and 2 clearly correspond to quasi-periodic orbits while cases 3 and 4 correspond to chaotic orbits. These last two cases were in fact the only chaotic orbits found in the sample of 100; thus the

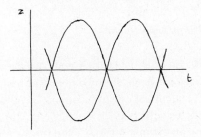

Fig. 74.

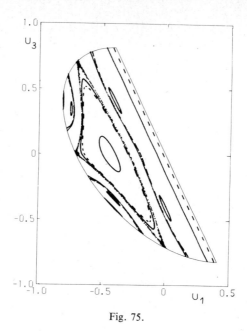

Fig. 75.

result of the experiment is $p(3) \simeq 0.98$. The system is close to integrable.

For $n = 4$, the same technique was used, with again a sample of 100 randomly chosen orbits. Here 88 orbits showed clear exponential separation; 3 showed linear separation; and 9 cases remained undecided. Therefore $p(3)$ should lie somewhere between 0.03 and 0.12.

For $n = 5$, the computations give $p(5) \simeq 0.02$. For $n = 6$, no case of linear separation was found, indicating that $p(6) < 0.01$. For $n > 6$, the same situation appears to prevail. A reasonable conjecture is that tiny "islands" still exist in phase space, but are so small that there is little chance of hitting one with the randomly chosen starting points.

8.2. Fermi–Pasta–Ulam experiment

Before we can describe this classical experiment, we need some generalities. A *one-dimensional lattice* is a system of n identical particles moving along a straight line; $(n + 1)$ identical springs connect neighbour particles, and also the two end particles with two fixed walls (fig. 76). This can be taken as a simple model for an atomic lattice. There exists an equilibrium position in which all springs have the same length. We call x_i the displacement

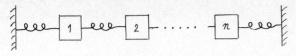

Fig. 76.

of particle i from this equilibrium position. The springs are generally non-linear; the force acting on particle i, from the spring on its right, is some function

$$F(x_{i+1} - x_i). \tag{100}$$

A similar force is produced by the spring on its left; if we assume that the mass of the particle is 1, its motion is given by

$$\ddot{x}_i = F(x_{i+1} - x_i) - F(x_i - x_{i-1}). \tag{101}$$

This equation is valid for all particles, including the end particles 1 and n, provided that we define

$$x_0 = 0, \qquad x_{n+1} = 0. \tag{102}$$

This is a Hamiltonian system with n degrees of freedom. A particular trajectory is described by the n functions $x_i(t)$. It is of interest to consider the spatial Fourier transform, defined by

$$a_k(t) = \sum_{i=1}^{n} x_i(t) \sin \frac{ik\pi}{n+1} \qquad (k = 1 \text{ to } n). \tag{103}$$

A trajectory can just as well be described by the n functions $a_k(t)$; the inverse relations are

$$x_i(t) = \frac{2}{n+1} \sum_{k=1}^{n} a_k(t) \sin \frac{ik\pi}{n+1}. \tag{104}$$

Each term in the sum (104) is called a *mode*; the general solution can thus be described as a superposition of modes. Figure 77 illustrates modes 1 and 2.

If the force law, eq. (100), is taken to be linear:

$$F(y) = y, \tag{105}$$

then it is easily found that the a_k obey the equations

$$\ddot{a}_k = -\omega_k^2 a_k, \tag{106}$$

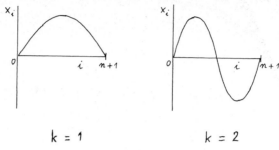

$$k = 1 \qquad\qquad\qquad k = 2$$

Fig. 77.

with

$$\omega_k = 2 \sin \frac{k\pi}{2(n+1)}. \tag{107}$$

Each mode evolves independently of the others: the system is separable. There are n integrals

$$E_k = \tfrac{1}{2}(\omega_k^2 a_k^2 + \dot{a}_k^2), \tag{108}$$

therefore the system is integrable. E_k can be physically interpreted as the energy contained in mode k.

Fermi, Pasta and Ulam (1955) considered the case where the function F has the following form:

$$F(y) = y + \alpha y^2, \tag{109}$$

for a system of 32 particles. The springs are now non-linear, and the energies E_k are no longer constant. According to the basic postulate of statistical mechanics, the system should then be ergodic and wander everywhere on the surface of constant total energy. One particular consequence of this is that the system should reach an *equipartition of energy* between the modes: the average values of the E_k, taken over a long interval of time, should be approximately equal. Also the trajectories are expected to display a chaotic behaviour.

The first experiment was designed to observe this approach to equipartition. Initially, only mode 1 was excited, i.e. only a_1 was different from zero. It was expected that energy would progressively flow from this mode to the others. To everyone's great surprise, however, this did not happen! Figure 78 shows the actual results. The abscissa is time; each curve represents the energy contained in one particular mode, computed from eq.

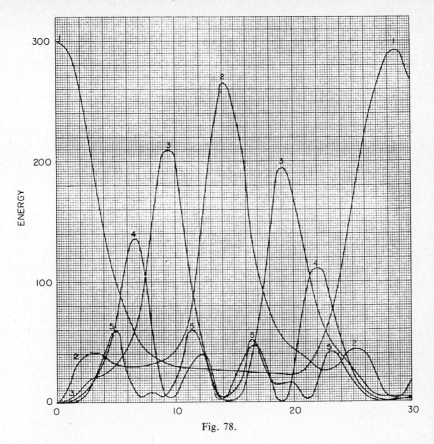

Fig. 78.

(108). Modes 1 to 5 are represented. Some energy is transferred from mode 1 to the other modes; however, there is no tendency towards equipartition. The energy of each mode oscillates around a mean value, which decreases as k increases. The higher modes (not represented) have very little energy. Also the trajectory is not at all chaotic: at $t = 28.6$ the system returns almost exactly to its initial state.

Other forms of $F(y)$ were tried, with the quadratic term in eq. (109) replaced by a cubic term, or with F a broken linear function of y; the results were essentially the same.

These results were very puzzling at the time, because this was before the KAM theorem, and dynamical systems were believed to be generally ergodic. In the light of the KAM theorem, the results are more easily understood: fig. 78 simply represents a quasi-periodic orbit, lying on an invariant torus in phase space. This explains at once the absence of equipartition and the regular appearance of the orbit.

The absence of equipartition is shown in a different way in fig. 79. F is here a broken linear function. The curves show the average kinetic energies of modes 1, 3, 5, 7, the average being taken from the origin to the present time t. These averages are seen to approach definite limits, which are different for the different modes.

Many similar studies followed. Figure 80 shows results obtained by Bocchieri et al. (1970) for a Lennard−Jones force between particles; in normalized units this can be written

$$F(y) = (1 + y)^{-7} - (1 + y)^{-13}. \tag{110}$$

For small amplitudes, the results are similar to those of Fermi, Pasta and Ulam. Only mode 1 is excited initially. Figure 80 represents the mode energy as a function of the mode number; the solid line is an average over

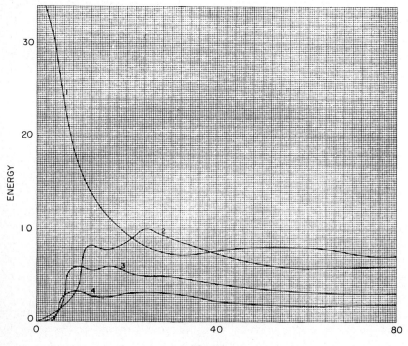

Fig. 79.

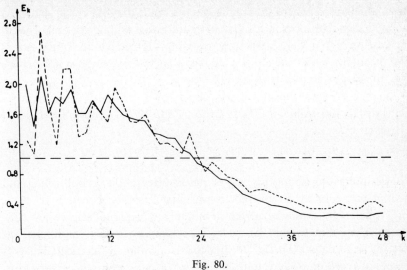

Fig. 80.

the whole evolution, while the dotted line represents the final state. Clearly the system shows no tendency towards equipartition (which would correspond to the horizontal dashed line).

But for higher amplitudes, the results are quite different. Figure 81

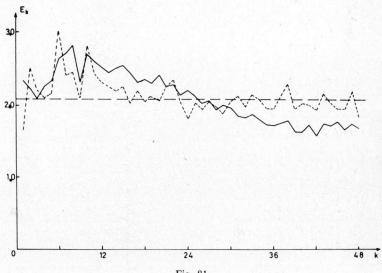

Fig. 81.

shows an example where the total energy is twice that of fig. 80. Now the system is seen to approach equipartition quite well as time goes on. The interpretation is that there exists now probably a large chaotic region, extending throughout most of phase space, so that the system does not behave very differently from a true ergodic system.

One case of a non-linear force is also known for which the system remains strictly integrable: the *Toda lattice* (Toda 1967). In normalized units, F is given by

$$F(y) = 1 - e^{-y}. \tag{111}$$

What is interesting about this case is that its integrability is not trivial, and was not immediately discovered. It was first suspected when Ford et al. (1973) made an extensive series of numerical experiments. (They considered a one-dimensional lattice with periodic boundary conditions, instead of fixed ends as in fig. 76; but this is a minor difference). For two degrees of freedom, they found that the surface of section is covered with invariant curves, even for very high energies (fig. 82; $E = 1$ at left, $E = 256$ at right). Other tests (absence of islands, linear separation of orbits) also suggested strongly that the system is integrable. Integrals were indeed found shortly afterwards (Hénon 1974, and independently Manakov 1974; see also Flaschka 1974).

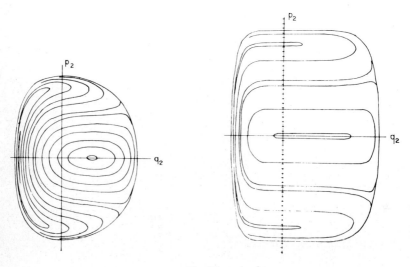

Fig. 82.

On the other hand, an example is known of a system which is ergodic for any number of degrees of freedom. It is a three-dimensional system of hard spheres enclosed in a cubical box, colliding elastically between them and with the walls. Sinai (1970, 1979) was able to prove that this system is ergodic for any number of spheres greater than 1.

Thus there is no clear-cut answer to the question asked at the beginning of section 8. The above examples exhibit widely different behaviours of $p(n)$ (fraction of phase space occupied by invariant tori) as n increases. An explanation for these differences might lie in the concept of *connectance*, defined as the percentage of other particles with which a given particle interacts directly. In the parallel sheet model, each sheet interacts with all other sheets, as shown by eq. (96); the connectance has its maximal value of 1. In Sinai's billiard, any two particles can interact, and the connectance is again 1. By contrast, in a one-dimensional lattice, each particle interacts only with its two neighbours; the connectance is low and equal to $2/n$. Systematic experiments by Froeschlé (1978) suggest that for systems with many degrees of freedom, there exists a critical value of the connectance. Below this value, a significant fraction of phase space is occupied by invariant tori; above this critical value, the fraction becomes negligible.

9. Mappings on a lattice

Some doubt can always remain about the validity of the results of numerical experiments, because of the effect of numerical errors. We can eliminate integration errors by studying an explicitly defined mapping instead of a dynamical system; however, there remains the round-off errors.

It is of interest, therefore, to try to build a mapping in which the round-off errors are themselves eliminated, so that the computation is exact. One way of doing this is to discretize the space of section: we cover it with a fine-meshed lattice of points, and we restrict our attention to the points of the lattice. The mapping will naturally have to be adjusted in such a way that a lattice point is always mapped into another lattice point. The coordinates of each point can then be given as a set of integers; the computations involve only integers and can be made exactly.

I shall describe a study by Rannou (1974). We revert to the case of two degrees of freedom, and we consider first an ordinary mapping, operating on real numbers:

$$G: \begin{cases} x' = x + f(y) \quad \text{(mod. } 2\pi\text{)}, \\ y' = y + g(x') \quad \text{(mod. } 2\pi\text{)}, \end{cases} \tag{112}$$

where f and g are periodic functions (modulo 2π) of period 2π. It can easily be seen that G is area-preserving. The form (112) includes, as a particular case, the standard mapping (88). The surface of section is a square of side 2π. We define a square lattice by dividing each side in m equal parts, with m a large integer, and we introduce new variables a and b:

$$x = \frac{2\pi}{m} a, \qquad y = \frac{2\pi}{m} b. \tag{113}$$

Points of the lattice correspond to integer values of a and b. Substituting in eq. (112), we have

$$G: \begin{cases} a' = a + \dfrac{m}{2\pi} f\left(\dfrac{2\pi}{m} b\right) \quad \text{(mod. } m\text{)}, \\ b' = b + \dfrac{m}{2\pi} g\left(\dfrac{2\pi}{m} a'\right) \quad \text{(mod. } m\text{)}. \end{cases} \tag{114}$$

In general, G does not map a point of the lattice into another point of the lattice. Therefore we modify it slightly by defining

$$G^*: \begin{cases} a' = a + \left[\dfrac{m}{2\pi} f\left(\dfrac{2\pi}{m} b\right) \right] \quad \text{(mod. } m\text{)}, \\ b' = b + \left[\dfrac{m}{2\pi} g\left(\dfrac{2\pi}{m} a'\right) \right] \quad \text{(mod. } m\text{)}, \end{cases} \tag{115}$$

where the brackets represent the operation of rounding to the nearest integer. This can be written as

$$G^*: \begin{cases} a' = a + F(b) \quad \text{(mod. } m\text{)}, \\ b' = b + G(a') \quad \text{(mod. } m\text{)}, \end{cases} \tag{116}$$

where F and G are piecewise constant functions, taking only integer values [fig. 83; the dashed line represents the function inside the brackets in eq. (115a), the solid line is the same term with the brackets added, i.e. $F(b)$].

We restrict now our attention to integer values of a and b, i.e. points of the lattice. The values of the functions F and G can be tabulated once and for all for integer values of their arguments in the interval $[0, m-1]$. The

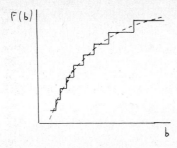

Fig. 83.

computation of the mapping (116) is then extremely simple and fast: it reduces to two additions and two table look-ups! G^* has a simple geometrical interpretation: first each horizontal row of points is displaced cyclically, by an amount $F(b)$ which depends on its vertical coordinate b; then each column of points is displaced cyclically by an amount $G(a)$ which depends on its horizontal coordinate a.

For large m, G and G^* differ only slightly; and numerical experiments show that the resulting trajectories also are only slightly different. Figure 84 represents a set of trajectories for the original mapping G, with

$$f(y) = y, \qquad g(x') = -a \sin x' \tag{117}$$

i.e. again the standard mapping, and $a = 1.3$. Figure 85 represents the modified mapping G^*, with $m = 1800$. The two figures are very similar; in the discretized mapping G^* we observe again the usual features of invariant curves, islands, and chaotic regions. Since G^* is free from numerical errors, this comparison certainly increases one's confidence into the results of numerical studies of mappings.

It could be objected that what we are doing here is not fundamentally different from what the computer does in any case; even when the computer is supposedly operating on real numbers x and y, actually only a finite number of distinct values of x or y can be represented internally, and therefore the computer also operates on a finite lattice of discrete points. There are, however, essential differences. The user has in general no control over the internal rounding-off process of a computer, and this process may well have unwanted properties. For instance, it usually has the effect that the mapping is not strictly one-to-one: because of the rounding, two different points can have the same image. By contrast, the rounding-off which we use to define the mapping G^* is totally under our control; and

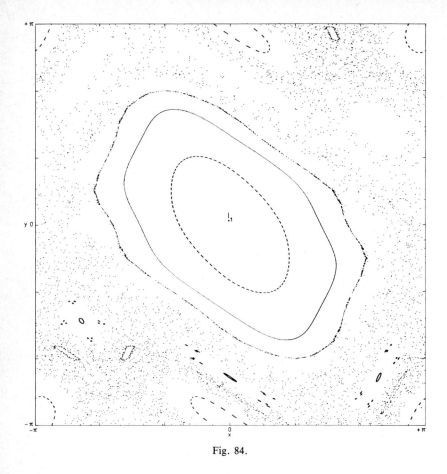

Fig. 84.

in eq. (115) it was chosen in such a way that G^* is strictly one-to-one, or invertible. This is a desirable property since we are studying area-preserving and invertible mappings.

Similarly, the user has almost no control on the size of the round-off error in a computer (all he can do is shift from single to double precision), while in the present process we can vary m at will, with interesting results.

The study of G^* is also interesting in other respects. Since the total number of points is finite, and equal to m^2, it is possible to make an exhaustive study, that is, to find and study all trajectories. In practice this is possible up to values of m of the order of 1000. We remark also that every point belongs to a cycle: since the total number of points is finite,

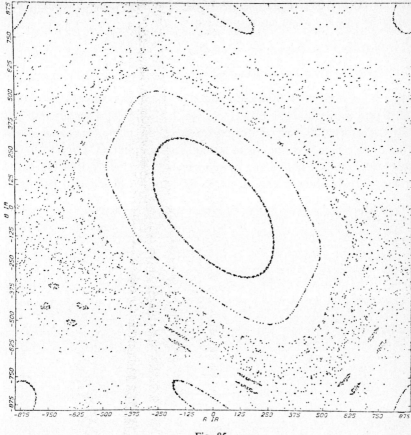

Fig. 85.

a trajectory sooner or later must return to its initial point. (It cannot come back to any other of its points since the mapping is one-to-one.) Thus the mapping consists simply in a collection of cycles. This might seem to be contradictory with the observation of two quite different kinds of trajectories on fig. 85. The answer is that these trajectories differ essentially in the length of their cycles: "invariant curves" are found to correspond to cycles with a period of order m, while "chaotic trajectories" have a period of order m^2. This corresponds to the fact that invariant curves are essentially one-dimensional objects, while chaotic regions are two-dimensional objects. It is of interest, therefore, to make detailed statistics on the lengths of the cycles. For instance, one finds that the longest cycle typically goes through 60 to 70% of the points of the chaotic region. An example is shown by fig. 86, which corresponds to

$$f(y) = y + 1 - \cos y,$$

$$g(x') = -a(\sin x' + 1 - \cos x'),$$

(118)

with $a = 1.3$, and $m = 400$; the total number of points is therefore 160 000. The figure shows a single trajectory, which is a cycle of length 100 384. This trajectory fills very uniformly most of the plane; the white patch around the centre corresponds to a small region of curves. For $a = 10$, this region has vanished, and the chaotic region apparently fills the entire square; fig. 87 shows again a single trajectory, which has a period of

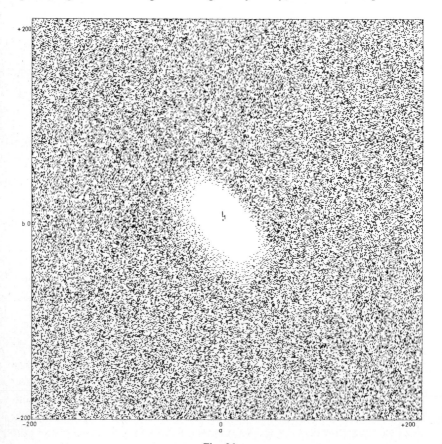

Fig. 86.

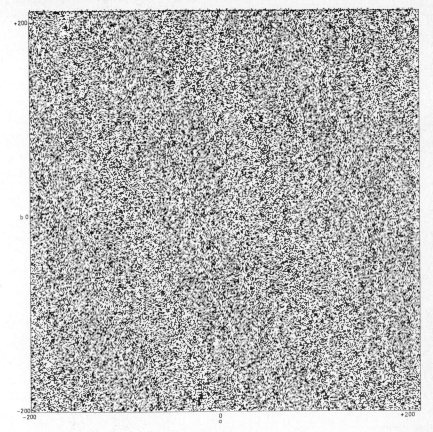

Fig. 87.

104 037.

For this value of a, the mapping appears to behave in a quasi-random way, and it is of interest to compare its properties with those of a truly random mapping. The latter can be defined as follows. We observe that the mapping G^* is one-to-one, and therefore is nothing more than a *permutation* of the m^2 points of the lattice. Conversely, any permutation of these points can be looked at as a one-to-one mapping. We write $m^2 = M$ for brevity. The number of distinct permutations is $M!$. We define then quite simply a *random mapping* as corresponding to a randomly chosen permutation, each permutation being affected with the same probability $1/M!$. (Note that this definition is possible only because we are considering a discrete mapping.) With this definition, a number of properties of the random mapping can be easily derived (see Knuth 1973). For instance, the median value for the period of the longest cycle is found to be approximately $M/e^{1/2} \simeq 0.61\ M$; the probability for a randomly chosen point to belong to a cycle with a given length n is simply $1/M$; the average length of the cycle passing through a given point is $(M+1)/2$; the average number of cycles is approximately $\ln M$, and so on. All these properties are found to be obeyed, within statistical fluctuations, by the mapping of fig. 87; it seems thus that in a deeply chaotic region, the properties of a "deterministic" mapping are actually undistinguishable from those of a truly random mapping.

As a conclusion to this course, I would like to re-quote (after Iooss and Joseph 1980) a precept which is probably valid for science in general (and also for teaching), but seems to me particularly appropriate to the pursuit of numerical experiments:

> "Everything should be made as simple as possible,
> but not simpler"

A. Einstein

Acknowledgements

I thank Claude Froeschlé and Jack Wisdom for reading the manuscript of these lectures and suggesting a number of improvements, and Giancarlo Benettin for kindly communicating figs. 54 to 59 to me.

References

Arnold, V.I. (1963) Usp. Mat. Nauk 18, 13; english translation in Russian Math. Surveys 18, 9.

Arnold, V.I. (1964) Dokl. Akad. Nauk SSSR 156, 9; english translation in Soviet Math. Dokl. 5, 581.

Benettin, G., L. Galgani and J.M. Strelcyn (1976) Phys. Rev. A14, 2338.

Benettin, G. and J.M. Strelcyn (1978) Phys. Rev. A17, 773.

Benettin, G., L. Galgani, A. Giorgilli and J.M. Strelcyn, Meccanica (March 1980) pp. 9 and 21.

Bocchieri, P., A. Scotti, B. Bearzi and A. Loinger (1970) Phys. Rev. A2, 2013.

Bunimovoch, L.A. (1974) Funkt. Anal. Prilog. 8, 73; english translation in Funct. Anal. Appl. 8, 254.

Casati, G. and J. Ford (1975) Phys. Rev. A12, 1702.

Chirikov, B.V. (1979) Phys. Rep. 52, 263.

Chirikov, B.V., J. Ford and F. Vivaldi (1979) in: Nonlinear Dynamics and the Beam–Beam Interaction, AIP Conf. Proc. No. 57, eds., M. Month, J.C. Herrera (American Institute of Physics, New York) p. 323.

Fermi, E., J. Pasta and S. Ulam (1955) Los Alamos Scientific Laboratory Report No. 1940; (1965) Fermi's Collected Works (University of Chicago Press) p. 978.

Flaschka, H. (1974) Phys. Rev. B9, 1924.

Ford, J., S.D. Stoddard and J.S. Turner (1973) Prog. Theor. Phys. 50, 1547.

Froeschlé, C. (1970a) Astron. Astrophys. 4, 115.

Froeschlé, C. (1970b) Astron. Astrophys. 9, 15.

Froeschlé, C. (1971) Astrophys. Space Sci. 14, 110.

Froeschlé, C. (1972) Astron. Astrophys. 16, 172.

Froeschlé, C. (1978) Phys. Rev. A18, 277.

Froeschlé, C. and J.-P. Scheidecker (1973) Astrophys. Space Sci. 25, 373.

Froeschlé, C. and J.-P. Scheidecker (1975) Phys. Rev. A 12, 2137.

Gonczi, R. and C. Froeschlé (1981) Celest. Mech. 25, 271.

Greene, J.M. (1979) J. Math. Phys. 20, 1183.

Helleman, R.H.G. (1980) in: Fundamental Problems in Statistical Mechanics, Vol. 5, ed., E.G.D. Cohen (North-Holland, Amsterdam) p. 165.

Hénon, M. (1966a) Bull. Astron. (3) 1, fasc. 1, p. 57.

Hénon, M. (1966b) Bull. Astron. (3) 1, fasc. 2, p. 49.

Hénon, M. (1969) Quart. Appl. Math. 27, 291.

Hénon, M. (1974) Phys. Rev. B9, 1921.

Hénon, M. and C. Heiles (1964) Astron. J. 69, 73.

Hénon, M. and J. Wisdom (1983) Physica D, in press.

Iooss, G. and D.D. Joseph (1980) Elementary Stability and Bifurcation Theory (Springer, New York).

Jenkins, B.Z. and J.H. Bartlett (1972) Celest. Mech. 5, 407.

Knuth, D.E. (1973) The Art of Computer Programming, vol. 1 (2nd ed.) (Addison-Wesley, Reading, Massachusetts), end of Section 1.3.3 and Exercises.

Knuth, D.E. (1981) The Art of Computer Programming, vol. 2 (2nd ed.) (Addison-Wesley, Reading, Massachusetts), Section 3.2.1.

Kolmogorov, A.N. (1954) Dokl. Akad. Nauk SSSR 98, 527.

Manakov, S.V. (1974) Zh. Eksp. Teor. Fiz. 67, 543; english translation in Sov. Phys. JETP 40, 269.

Markus, L. and K.R. Meyer (1974) Mem. Am. Math. Soc. No. 144.

Martinet, L. and P. Magnenat (1981) Astron. Astrophys. 96, 68.

Moser, J. (1955) Nachr. Akad. Wiss. Göttingen, IIA, 87.

Moser, J. (1958) Astron. J. 63, 439.

Moser, J. (1962) Nachr. Akad. Wiss. Göttingen, Math.-Phys. Kl., 1.

Rannou, F. (1974) Astron. Astrophys. 31, 289.

Sinai, Ya.G. (1970) Usp. Mat. Nauk 25, 137; english translation in: Russ. Math. Surv. 25, 137.

Sinai, Ya.G. (1979) Funct. Anal. Appl. 13, 46.

Taylor, J.B. (1968) unpublished.

Tennyson, J.L., M.A. Lieberman and A.J. Lichtenberg (1979) in: Nonlinear Dynamics and the Beam–Beam Interaction, AIP Conf. Proc. No. 57, eds., M. Month, J.C. Herrera (American Institute of Physics, New York) p. 272.

Toda, M. (1967) J. Phys. Soc. Japan 22, 431.

COURSE 3

SEMICLASSICAL MECHANICS OF REGULAR AND IRREGULAR MOTION

Michael BERRY *

*Institute for Theoretical Physics, University of Utrecht
Utrecht, The Netherlands*

*Permanent address: H.H. Wills Physics Laboratory, Tyndall Avenue, Bristol BS8 1TL, UK.

G. Iooss, R.H.G. Helleman and R. Stora, eds.
Les Houches, Session XXXVI, 1981 − Comportement Chaotique des Systèmes Déterministes/
Chaotic Behaviour of Deterministic Systems
© *North-Holland Publishing Company, 1983*

Contents

1. Introduction

After half a century, the relation between quantum mechanics and classical mechanics is still not fully understood. Uncertainty extends even to the foundations of the subject: it is not known in general whether or how it is possible to specify uniquely the quantum mechanics of a system whose classical mechanics is given. This fundamental "problem of quantization" will not be discussed here. Instead, I shall concentrate on rather simple systems – noninteracting particles moving in low-dimensional spaces under external forces – for which it is known how to specify the quantum mechanics, in terms of Schrödinger's equation involving well-defined operators. For these systems, I shall study the *semiclassical limit*, i.e. the behaviour of wave functions, energy levels, etc. as Planck's constant $\hbar$ tends to zero (by comparison with classical quantities having the same dimensions). This is not the same as the classical limit, for which $\hbar$ is precisely equal to zero because, in general, quantal functions are nonanalytic in $\hbar$ as $\hbar \to 0$. So the semiclassical limit cannot be related to the classical limit by perturbation theory, but has a rich and interesting structure of its own. Nevertheless there must be some kind of correspondence principle, according to which the semiclassical limit reflects the nature of the underlying classical motion. We shall learn that what really affects the semi-classical mechanics is whether the classical motion is regular (predictable, integrable) or irregular (unpredictable, chaotic, non-integrable).

Of course these problems have a wide variety of actual and potential applications in quantum mechanics and indeed more generally throughout physics and applied mathematics wherever short waves are involved. Examples are the vibration spectra of non-symmetrical molecules, modes of acoustic oscillation in rooms with typical shapes, and the optics of waveguides. I shall not discuss any of these, but shall consider semiclassical mechanics as worth studying for its own sake, in order to understand the connection between two important branches of theoretical physics.

Many different sorts of scientists and mathematicians have studied semiclassical mechanics (or short-wave asymptotics) and the literature is correspondingly varied and scattered. I do not intend to attempt a complete survey, but shall concentrate on the qualitatively different sorts of semiclassical wave functions and spectra corresponding to qualitatively different sorts of classical motion. This is an aspect of quantum mechanics whose significance has been appreciated only recently, as advances in classical mechanics have unravelled the intricate varieties of predictability and chaos that orbits can display. To my knowledge this material has been reviewed only once before by Zaslavsky [118]; although Percival [1] and Duistermaat [2] have given reviews of aspects of it, from (respectively) physical and mathematical viewpoints. "Prehistoric" semiclassical mechanics was reviewed by Berry and Mount [3]. My treatment here will be nonrigorous, combining analytical and pictorial arguments, simple models and the results of computation, to bring out what I consider to be the essential points.

2. Regular and irregular classical motion

2.1. Two contrasting types of orbit

Here I discuss two contrasting types of Hamiltonian motion, that is motion governed by Newton's equations without dissipation. This is necessary in order to set the scene for the subsequent quantal treatment. More detailed expositions have been given by Ford [4], Arnol'd [5], Berry [6] and Helleman [7].

On the one hand, there is *regular motion*. This is exemplified by the one-dimensional oscillator (harmonic or anharmonic) which in its physical realization as a pendulum is the epitome of predictability ("as regular as clockwork"). Another example is the elliptical orbits of the planets when mutual perturbations are ignored. In systems with regular motion, trajectories with neighbouring initial conditions separate linearly.

On the other hand, there is *irregular motion*. This is exemplified by the (classical) motion of colliding molecules in a gas. If the molecules are confined to a plane, transformed into hard discs and all held fixed except one, then that one executes a motion with two freedoms idealising a pinball machine. Such motion is unpredictable in the sense that neighbouring trajectories separate exponentially, resulting in a sensitivity to initial conditions which has surprising consequences. For example, in order to accurately predict the motion for n collisions (after which the angle of emergence is in error by, say, $90°$), it is necessary to specify the initial position and momentum values to a number of digits D proportional to n, so that the capacity of any feasible calculator is exhausted for rather small numbers of collisions. (If the discs have radius r and mean separation l, a rough estimate for D when the variables are expressed in base-b notation is

$$D \approx n \log_b l/r, \tag{2.1}$$

as readers may verify. If $l/r = 10$, therefore, even specifying initial conditions with a relative error of one part in a million will ensure predictability for only six collisions.)

The distinction between regularity and irregularity is embodied in the geometry of typical trajectories in the system's phase space over infinitely long times. For N freedoms this is the $2N$-dimensional space q, p where $q = (q_1 \ldots q_n)$ are the coordinates and $p = (p_1 \ldots p_N)$ the momenta. There is a hierarchy of types of motion, increasingly chaotic, denoted by the terms integrable, quasi-integrable, ergodic, mixing, K-system, B-system, whose meaning is explained by Lebowitz and Penrose [8]. I shall not describe this hierarchy in detail, but give sufficient background for the later discussion of quantum mechanics.

For illustrative purposes it is not convenient to consider only smooth Hamiltonians, and I now introduce an important class of discontinuous systems with $N = 2$, namely the so-called *planar billiards*. These have Hamiltonian

$$\left. \begin{array}{l} H(x, y; p_x, p_y) = (p_x^2 + p_y^2)/2 \text{ inside a boundary } B \text{ in } q \text{ space} \\ \qquad\qquad = \infty \text{ outside } B \end{array} \right\}. \tag{2.2}$$

Motion consists of straight segments joined by specular reflections at B. As we shall see, the nature of the orbits is very sensitive to the form of B (for an elementary review of billiard dynamics, see Berry [9]).

2.2. Integrable systems

The simplest situation, corresponding to the regular case, is fully integrable motion. Here there exist N constants of motion in the form of functions $C_i(q, p)$ $(1 \le i \le N)$ in phase space, assumed to be "in involution" (i.e. all Poisson brackets between pairs of C_i vanish). One of these constants is the Hamiltonian itself if this is time-independent. If the C_i are independent of one another, their existence restricts motion to a surface Σ in phase space; Σ has dimensionality $N(=2N-N)$. Arnol'd [5] gives a clear formulation and proof of an important theorem, implicit in older literature, stating that if the C_i are "smooth enough", then

(i) Σ is an N-dimensional torus

(ii) the motion can be "integrated", i.e. the trajectories $q(t)$, $p(t)$ can be determined by elimination and integration. The proof (see also ref. [6]) involves constructing N nonsingular vector fields on Σ ("if a hairy Σ can be combed N ways without a singularity, then Σ is an N-torus"). (If the conditions of the theorem are not satisfied, and the vector fields do have singularities, then Σ need not have the topology of an N-torus, as will be shown in section 2.5 with a curious example.)

The simplest integrable systems are stationary harmonic or anharmonic oscillators with $N=1$ and mass μ, whose Hamiltonian is

$$H = \frac{p^2}{2\mu} + V(q), \tag{2.3}$$

where $V(q)$ (fig. 1a) is a potential well. The energy $E = H(q, p)$ is conserved, and since $N=1$ this one constant of motion suffices to make the system integrable. The "1-tori" on which motion occurs in phase space are simply the closed contours of H (fig. 1b).

All separable systems are integrable (they decouple into N one-dimensional systems). In particular, a particle moving in the plane ($N=2$) under a central potential $V(r)$ is integrable, the two constants of motion being E and the angular momentum L. Each choice of E and L labels a 2-torus Σ in the 4-dimensional phase space (fig. 2a). The trajectory winds

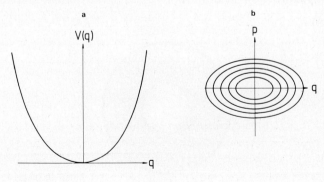

Fig. 1.

M. Berry

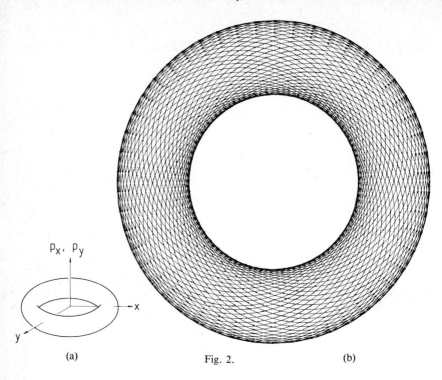

p_x, p_y

x

y

(a)   Fig. 2.   (b)

around Σ in a manner soon to be explained, and typically fills Σ as $t \to \infty$. When projected "down" p onto the coordinate plane x, y (fig. 2b), the trajectory is enveloped by a *caustic* in the form of two circles at the radii of closest approach and farthest recession from the centre of force. Caustics are the singularities of the projections of Σ. In the circular billiard (fig. 3) the outer circle corresponds to the boundary B and is not a caustic. In the

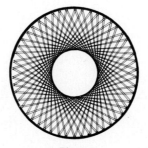

Fig. 3.

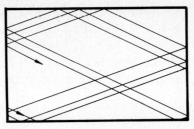

Fig. 4.

rectangular billiard (fig. 4), which is separable with p_x^2 and p_y^2 conversed, there are no caustics because the boundaries of the projection of Σ (a flat car tyre with four sheets above points q) coincide with B. Another separable billiard is the ellipse [9], whose orbits repeatedly touch an ellipse or hyperbola confocal with B.

The existence of tori throughout the phase space of an integrable system makes it natural to introduce an alternative set of coordinates and momenta known as *action-angle variables*, which will play an important part in the quantum mechanics of integrable systems. The *actions* $I = \{I_1...I_N\}$ are particular combinations of the C_i which label the tori Σ, defined in the most topologically natural way as

$$I_i \equiv \frac{1}{2\pi} \oint_{\gamma_i} \boldsymbol{p} \cdot \mathrm{d}\boldsymbol{q}, \tag{2.4}$$

where γ_i is the ith irreducible circuit of the torus. Now let $\boldsymbol{I}$ be the momenta of new phase-space variables. The conjugate coordinates are called the *angles* $\boldsymbol{\theta} = \{\theta_1...\theta_N\}$. Any point $\boldsymbol{q}, \boldsymbol{p}$ lies on a torus (fig. 5) labelled by

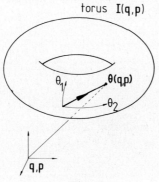

Fig. 5.

$I(q,p)$, and $\theta(q,p)$ locates the position of q, p on this torus.

The variables θ, I and q, p are related by a cononical transformation specified as follows. Let the torus Σ labelled by I be defined by a multivalued function with branches $p_j(q';I)$ in the phase (fig. 6). Then the generating function of the transformation is $S(q,I)$ with branches

$$S_j(q;I) = \int_{q_0}^{q} p_j(q';I) \cdot dq',$$
(2.5)

where q_0 is an arbitrary point "beneath" Σ. This gives the transformation

$$(q,p) \leftarrow S(q,I) \rightarrow (\theta,I)$$

$$p = V_q S, \quad \theta = V_I S.$$
(2.6)

It follows from these definitions that each angle variable θ_i changes by 2π during the corresponding circuit γ_i of Σ, thus justifying the term "angle". The fact that S is locally single-valued implies from eq. (2.6) that tori have the "Lagrangian" property

$$\partial p_i / \partial q_j = \partial p_j / \partial q_i.$$
(2.7)

Because there are only N independent constants of motion, each member of the original set C_i must be expressible in terms of the N I_i's. In particular, the Hamiltonian is conserved and can be written in the new variables as

$$H = H(I).$$
(2.8)

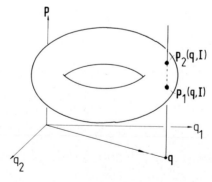

Fig. 6.

By one of Hamilton's equations, this means that the angle variables change at a constant rate, given by

$$\theta = \omega t + \theta_0, \tag{2.9}$$

where $\omega = \{\omega_1 ... \omega_N\}$ are the frequencies, given by

$$\omega(I) = V_r\, H(I). \tag{2.10}$$

The components ω_i give the rate at which a trajectory winds round Σ in directions labelled by the angles θ_i. In the general case, all ω_i will be mutually incommensurable and the orbit will eventually fill Σ densely. But if all ω_i are rationally dependent, i.e. if

$$\omega = M\omega_0, \tag{2.11}$$

where $M = \{M_1 ... M_N\}$ is a lattice vector (i.e. with integer components), then each orbit on Σ will be *closed*. In fact closure will occur after M_1 circuits of θ_1, M_2 circuits of $\theta_2 ..., M_N$ circuits of θ_N. These closed orbits are *nonisolated*, in the sense that each "rational" torus [i.e. satisfying eq. (2.11)] will contain infinitely many of them, related by parallel translation in θ (fig. 7 shows the case where $N = 2$, $M_1 = 1$, $M_2 = 0$). In the typical case, where $\omega(I)$ varies smoothly with I, tori made of closed orbits are of measure zero but nevertheless densely distributed. Finally, note that θ provides an invariant measure on Σ, in the sense that a cloud of phase-space points with uniform density θ preserves density as the cloud moves under time-evolution governed by $H(I)$.

2.3. Chaotic systems

The integrable systems so far considered are rare; amongst Hamiltonian systems, "almost all" are nonintegrable in the sense that there are no

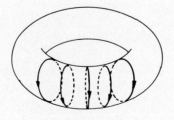

Fig. 7.

global constants of motion other than H. An important class of nonintegrable system is those with the property of *ergodicity:* almost all orbits eventually explore almost all points in a $2N-1$ dimensional energy surface, defined by $H(q,p) =$ constant, instead of being confined to an N-torus as with integrable systems. The case $N=1$ is degenerate, because $2N-1 = N$ and orbits are trivially ergodic as well as integrable (cf. fig. 1b) – provided of course that it is time-independent. When $N > 1$, ergodicity is usually associated with (but does not imply) motion which is chaotic or unpredictable as briefly described in section 2.1. The precise sense in which such motion is chaotic can be understood by dividing ("coarse-graining") phase space into cells ("Markov partitions") and studying the sequence ("Bernoulli shift") in which cells are explored; the sequences are often indistinguishable from random sequences (see for example Ornstein [105], and the elementary discussions in refs. [4] and [6]).

Are there any ergodic systems with $n=2$? Yes. It was shown by Sinai [10] that the billiard motion of a particle moving on the coordinate torus x, y (square with opposite sides identified) containing a circular reflecting obstacle with radius R (fig. 8) is ergodic if $R > 0$. In this case ergodicity is a consequence of exponential chaos (unpredictability) which in turn results from the defocusing of particle beams that hit the disc (fig. 9). Another effect of defocusing is that no orbits form caustics in coordinate space and indeed there are no phase space tori whose projections would have them as singularities. Sinai's billiard is the simplest example of a system of two or more hard spheres or discs moving in a compact space and colliding elastically. For the case where the number of particles is very large, Sinai's proof at last established the ergodicity of the hard-sphere gas.

Another ergodic billiard is the "stadium" of Bunimovich [11,12], consisting of two semicircles with radius R joined by parallel straight lines with length L (fig. 10) with $L > 0$. The semicircles cause a convergence of beams

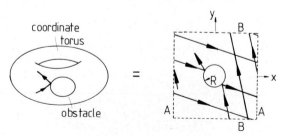

Fig. 8.

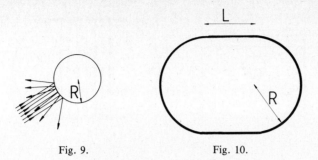

Fig. 9.   Fig. 10.

of trajectories but this is outweighed by the subsequent divergence before the next reflection. Again no caustics form, as the orbit in fig. 11 illustrates (cf. fig. 3).

In these ergodic billiards, a typical orbit passes through almost every point within B with almost every direction if followed long enough. The energy surface thus explored is 3-dimensional $(2N-1)$ in contrast to the 2-tori explored when there is an additional constant of motion. But not every orbit is typical: just as in integrable systems there is a dense-but-zero-measure set of closed orbits (each of which explores a 1-dimensional region in phase space). For integrable systems (with $N=2$) we have seen that these closed orbits are not isolated (fig. 7) but fill tori in a one-parameter family. And in some billiards (such as those of Sinai and Bunimovich) it is also possible to have one-parameter families of nonisolated closed orbits (fig. 12), although these do not fill tori. For ergodic billiards the nonisolated orbits drastically slow down the exploration of the energy surface [9], and are

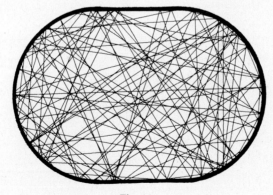

Fig. 11.

M. Berry

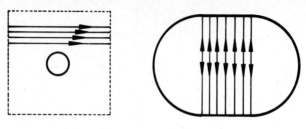

Fig. 12.

probably responsible for long tails in autocorrelation functions (Berry and Casati, unpublished). But vastly more numerous are the *isolated closed orbits*, some of which are shown in fig. 13. The distinction between isolated and nonisolated closed orbits has an important effect on the clustering of quantum-mechanical energy levels, to be explained in section 5.

These closed orbits, isolated and nonisolated alike, are unstable: the slightest error in position or momentum gives an orbit separating rapidly (exponentially, for isolated orbits) from the closed one and eventually exploring all the energy surface. By contrast, the closed orbits in an integrable system are only linearly unstable [cf. eq. (2.9)], and a perturbed orbit, far from filling the energy surface, merely fills a nearby torus. In this connection it is amusing to note that in the ideal roulette wheel and coin toss – two exemplars of unpredictable systems whose randomness arises largely from imperfect knowledge of initial conditions – the motion is in fact integrable (rigid disc rotating in or out of its plane), so that the instability is of a linear type and therefore much weaker than in, say, a pinball machine.

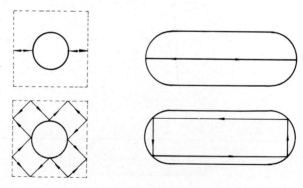

Fig. 13.

Another (non-billiard) ergodic system with $N=2$ is geodesic motion on a compact surface of everywhere negative curvature (which cannot be realized in Euclidean 3-space) [5]. No smooth Hamiltonian of the canonical type $H=$ kinetic + potential has been proved to be ergodic. However, Gutzwiller [13] gives strong evidence that the "anisotropic Kepler problem", with

$$H=ap_x^2+bp_y^2-\frac{C}{(x^2+y^2)^{1/2}} \qquad (a\neq b) \tag{2.12}$$

is ergodic, and Percival (private communication) has reason to believe that orbits in the potential

$$H=\frac{p_x^2+p_y^2}{2}+x^2y^2 \tag{2.13}$$

explore their (unbounded) energy surfaces chaotically.

2.4. Quasi-integrable systems

These are neither integrable nor ergodic. Excluding billiards for the moment, they have smooth Hamiltonians of the form

$$H(q,p)=H_0(q,p)+\varepsilon H_1(q,p), \tag{2.14}$$

where H_0 is integrable and ε is a small parameter that turns on a generic perturbation H_1. When $\varepsilon=0$, all trajectories lie on tori filling phase space. What happens to these tori when $\varepsilon\neq0$? The answer is given by the celebrated Kolmogorov [14], Arnol'd [15], Moser [16] (KAM) theorem (see also refs. [5,6]): under perturbation, most tori survive (albeit distorted). Therefore the motion is not ergodic. But for almost all H_1, some tori are destroyed, so the motion is not integrable either. The destroyed tori form a set of finite measure growing with ε. They are centred on those unperturbed tori whose frequencies ω_i are incommensurable, i.e. whose orbits are closed. The manner of their destruction is complicated but now fairly well understood [6]: of the continuous family of linearly unstable closed filling each rational torus, only a finite even number of closed orbits survives perturbation; half of these are unstable and half are stable. Motion near the unstable orbits is chaotic and fills regions with dimensionality $2N-1$. Near the stable closed orbits, most trajectories lie on "higher-order" tori, but these, like the original "parent" tori, have gaps near sites where "rational" tori would be, and the whole structure repeats in microcosm, recursively down to infinitely fine scales.

The solar system is quasi-integrable [15]. Planets unperturbed by their neighbours move (integrably) in Kepler ellipses. Including the perturbation results in a Hamiltonian of type (2.14) [6]. Some of the effects described in the last paragraph can be seen in the asteroids, which correspond, in effect, to an ensemble of zero-mass "test particles". Most of them move in approximately elliptical orbits, in spite of being perturbed by Jupiter, but there are gaps in the asteroid belt where orbital motion would be commensurate with (resonant with) Jupiter's; these gaps correspond to the destroyed tori. There are similar gaps in Saturn's rings, which may correspond to resonant perturbation of the orbits of ring particles by the satellite Mimas.

The tori whose existence the KAM theorem guarantees can be discovered "experimentally" by the caustics their orbits envelop when projected "down" onto coordinate space (cf. fig. 2). An example is shown in fig. 14; this was computed by Noid and Marcus [23] for a particle in the (nonintegrable) field of two anharmonically coupled harmonic oscillators. How can we deduce that the surface Σ in phase space, whose projection displays the caustics, is a torus? By the following procedure, devised by Ozorio de Almeida and Hannay [17] as part of a detailed study of the ways 2-tori can be embedded in 4-space and projected onto 2-space; traverse all branches of the caustic, keeping two-sheeted regions on one's right; one

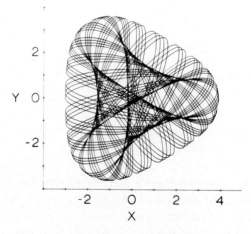

Fig. 14. Reprinted from ref. [77] with permission.

will rotate clockwise r times (not counting the necessary half-turns at cusps). The genus (number of handles) of the Σ which projects to give the caustic, is $1 - r$. For a torus, the genus must be unity, so r must be zero; it is easy to confirm that the caustic of fig. 14 passes the torus test. Generically, the forms of caustics are classified by catastrophe theory, but because I discussed this aspect in detail in last year's Les Houches lectures [18] I shall not elaborate it here.

Returning now to billiards, it is necessary in quasi-integrable cases to be very careful in specifying what class of deformation of the boundary B is being considered. For example we have already seen that if the integrable circle billiard is perturbed to the stadium (fig. 10), all tori are instantly destroyed (no matter how small L/R is) because the system becomes ergodic. However, Lazutkin [19] proved for billiards a theorem analogous to the KAM theorem but more general (in that it is nonperturbative): if B is convex and smooth enough, some orbits (of positive measure) will envelop caustics – i.e. there will be some tori in phase space. The sufficient condition for smoothness is that B's radius of curvature as a function of arc length must possess 553 continuous derivatives! Such a large number is almost certainly not necessary, but to have no continuous derivative is insufficient as the stadium example shows. Berry [9] illustrates Lazatkin's theorem with computations indicating quasi-integrable behaviour (i.e. some tori, some chaos) for a family of (analytic) oval billiards, and Bennet-tin and Strelcyn [20] study a generalization of the stadium made from four circular arcs, which has discontinuous curvature but nevertheless displays quasi-integrability.

2.5. Pseudo-integrable systems

Even if N constants of motion exist, independent and in involution, so that each orbit restricted to an N-dimensional surface Σ in phase space, it is not always the case that Σ is a torus. For billiards whose boundary B is a polygon with angles which are rational multiples of π (apart from the rectangle, equilateral triangle, 30°–60°–90° triangle and 45°–45°–90° triangle, which are integrable), Σ is not a torus but a *multiply-handled sphere* (i.e. genus $g > 1$) (for a reason to be explained, Arnol'd's theorem (section 2.2) does not apply). Richens and Berry [21] call these systems "pseudo-integrable"; they were previously studied by Zemlyakov and Katok [22].

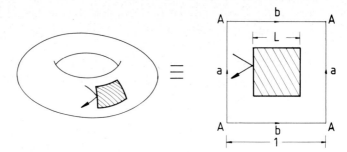

Fig. 15.

To illustrate this unexpected behaviour, consider the "square torus billiard" (fig. 15), which resembles Sinai's billiard except that the reflecting disc is replaced by a square. It is clear that the squares v_x^2 and v_y^2 of the velocity components are separately conserved, and that at most four directions are available to each trajectory ($v_x, v_y; -v_x, v_y; v_x, -v_y; -v_x, -v_y$). Therefore the phase-space surface Σ consists of four sheets, each being a copy of the accessible xy area, connected at the central square according to the identifications in fig. 16. To construct Σ, these four sheets must be sewn together. It is simplest first to join the inner squares to get a "skeleton" for Σ (fig. 17). Then the outer tori must be sewn onto each "square" (fig. 18). Finally, the two remaining ("l, q") holes must be joined

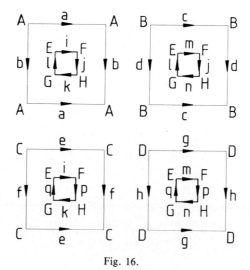

Fig. 16.

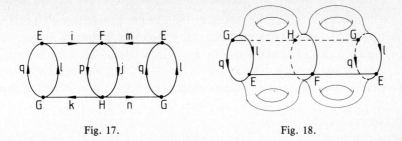

Fig. 17.   Fig. 18.

(fig. 19). It is clear from this construction that far from being a torus, Σ is a five-handled sphere.

Now, the constants v_x^2 and v_y^2 are smooth, independent and in involution. What then prevents Arnol'd's theorem (described in the first paragraph of section 2.2) from being applied? The answer, elaborated by Richens and Berry [21] is that certain vector fields on Σ, constructed from v_x^2 and v_y^2, possess singularities at the vertices of the reflecting square. Indeed it is possible to establish the topology of Σ from a study of these singularities. As an example, formulae given in ref. [21] yield the following equations for the genus g of Σ for billiards in a regular polygon with m sides:

$$g = \frac{(m-1)(m-2)}{2} \qquad (m \text{ odd})$$

$$g = \frac{(m-2)^2}{4} \qquad (m \text{ even}) \qquad (2.15)$$

As expected, this gives $g = 1$ for the triangle and square, which are integrable.

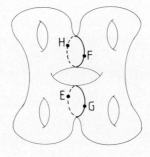

Fig. 19.

As discussed in ref. [22], rational billiards have a chaotic property resulting from the splitting of beams of trajectories hitting polygon vertices. According to Richens and Hannay (private communication) it is convenient to study this computationally in terms of the one-dimensional map produced by successive intersections of an orbit with a curve cutting Σ (for a torus this map is a simple rotation, but if $g > 1$ it is a "shuffling transformation").

For polygons with angles that are irrational multiples of π, each reflection of a typical trajectory produces a segment moving in a new direction, so these systems are not pseudo-integrable (unless we accept surfaces with $g = \infty$). Are they ergodic? Computations by Casati and Ford [24] suggest that the answer is yes, but that exploration of the three-dimensional energy surface is very slow and without exponential separation of orbits; to my knowledge nobody has proved a theorem demonstrating (or disproving) the ergodicity for irrational polygons.

2.6. Discrete area-preserving maps of the plane

The distinction between regular and irregular motion is very clearly exhibited by discrete maps M in the phase plane whose variables are a coordinate q and a momentum p. Under M (fig. 20), each point (q_{n+1}, p_{n+1}) is a deterministic function of its antecedent (q_n, p_n):

$$M: q_{n+1} = q_{n+1}(q_n, p_n)$$
$$p_{n+1} = p_{n+1}(q_n, p_n)$$

(2.16)

Area-preservation is ensured by requiring

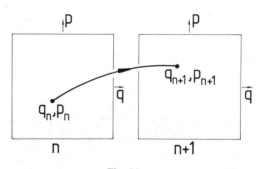

Fig. 20.

$$\det \begin{pmatrix} \dfrac{\partial q_{n+1}}{\partial q_n} & \dfrac{\partial q_{n+1}}{\partial p_n} \\[3mm] \dfrac{\partial p_{n+1}}{\partial q_n} & \dfrac{\partial p_{n+1}}{\partial p_n} \end{pmatrix} = \frac{\partial(q_{n+1}, p_{n+1})}{\partial(q_n, p_n)} = 1. \tag{2.17}$$

M can be regarded simply as an abstract dynamical system whose evolution takes place in discrete time steps according to the "Hamilton equations" (2.16) with (2.17) replacing Liouville's theorem on conservation of phase-space volume. But these maps also arise from continuous time dynamical systems, in two ways.

Firstly, one may have $N = 2$ and a time-independent Hamiltonian. In this case q and p are coordinates on a "surface of section" (see e.g. ref. [6]) through the three-dimensional energy surface. M is determined by the succession of points q_n, p_n in which a trajectory intersects the surface of section. For billiards, a convenient surface of section is given by successive bounces at B, and variables for which M is area-preserving (see e.g. ref. [9]) are $q = $ (arc length round B) and $p = $ (cosine of angle made by emerging trajectory with forward tangent to B).

Secondly, one may have $N = 1$ and a Hamiltonian periodic in time, i.e.

$$H(q, p, t + T) = H(q, p, t). \tag{2.18}$$

Then M is defined by a "stroboscopic phase portrait", i.e. by jumps of points q, p between snapshots of the motion at intervals T, i.e.

$$q_n \equiv q(nT), \qquad p_n \equiv p(nT), \tag{2.19}$$

and the motion between snapshots is ignored.

In terms of M, regularity or irregularity depends on the manner in which iterates (q_n, p_n) of some initial point (q_0, p_0) are distributed as $n \to \infty$. There are three possibilities. Firstly, iterates may lie on a zero-dimensional set in the plane (fig. 21), by forming a closed orbit, i.e. a *fixed point* of some finite power N of M:

$$q_{n+N} = q_n, \qquad P_{n+N} = p_n. \tag{2.20}$$

Secondly, iterates may fill a one-dimensional set in the plane (fig. 22), a so-called *invariant curve* which maps onto itself although its individual points do not. An example is the twist map which in polar coordinates is

$$M\colon r_{n+1} = r_n, \qquad \theta_{n+1} = \theta_{n+1} + 2\pi\alpha(r_n). \tag{2.21}$$

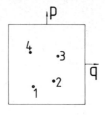

Fig. 21.

Here the invariant curves are circles. If the "rotation number" α is irrational, iterates eventually fill the circle, while if $\alpha = M/N$ (M, N mutually prime finite integers) the iterates form a closed orbit with period N. And thirdly, iterates may fill a two-dimensional set in the plane (fig. 23), a so-called *chaotic area*. I shall not discuss more exotic possibilities, such as orbits filling a "fractal" set with intermediate dimension, because there is no firm evidence that they occur in area-preserving maps (although they do occur – as "strange attractors" (see ref. [25] and other papers in the same volume) – in area-contracting maps corresponding to dissipative systems).

Integrable maps are those like fig. 22, where all points lie on invariant curves. These curves are the analogues of tori, and in the "surface of section" interpretation are precisely sections of 2-tori. Their existence throughout the phase plane implies a "constant of motion" in the form of a function $f(q, p)$ whose contours are the invariant curves and whose value is unchanged under M (for a twist map, $f = q^2 + p^2$).

Ergodic maps are those where almost every initial point explores almost all the phase plane, so that no finite measure is covered with invariant curves. An example is "Arnol'd's cat" [100], for which phase space is the unit torus and M the linear map corresponding to hyperbolic shear:

$$M: \begin{pmatrix} q_{n+1} \\ p_{n+1} \end{pmatrix} = \begin{pmatrix} 1 & 1 \\ 1 & 2 \end{pmatrix} \begin{pmatrix} q_n \\ p_n \end{pmatrix}. \tag{2.22}$$

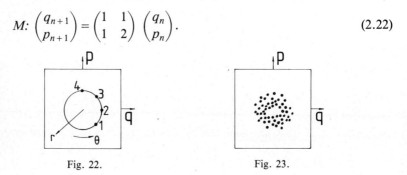

Fig. 22.   Fig. 23.

In the infinite plane this is trivially simple, but on the torus it winds area elements round and round in an irrational direction so that initially simple shapes quickly become unrecognizable (fig. 24). (Even after one hyperbolic mapping it can be hard to recognize shapes, and this idea is the basis of a Victorian amusement called the "anamorphic picture" (fig. 25), which can be comprehended only when viewed after an "inverse mapping" consisting of reflection from the exterior of a convex cylinder). There are closed orbits in Arnol'd's cat – they are executed by every point with rational

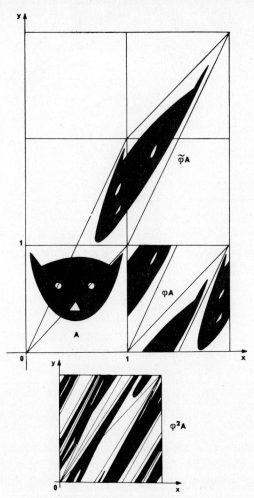

Fig. 24. Reprinted from ref. [100] with permission.

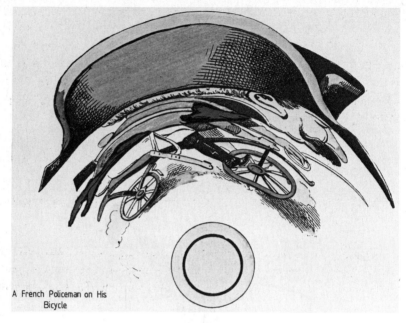

A French Policeman on His
Bicycle

Fig. 25.

q, p – but they are of measure zero and all isolated and unstable.

A quasi-integrable map, with a hierarchy of fixed points, invariant curves (tori) and chaotic areas, is generated by

$$H(q, p, t) = \frac{p^2}{2\mu} + TV(q) \sum_{n=-\infty}^{\infty} \delta(t - nT), \qquad (2.23)$$

corresponding to a particle with mass μ moving freely along a line between impulses at intervals T whose magnitude depends on the coordinate q. From Hamilton's equations, M is easily found to be

$$M: q_{n+1} = q_n + \frac{Tp_n}{\mu}$$

$$p_{n+1} = p_n - T \frac{\partial V}{\partial q}(q_{n+1}) \qquad (2.24)$$

The choice $V(q) = Aq^4/4$ (anharmonic potential well), followed by scaling, gives

$$M: q_{n+1} = q_n + p_n$$
$$p_{n+1} = p_n - q_{n+1}^3$$

(2.25)

Figure 26 shows the iterates of a number of initial points in the qp plane. Fixed points, invariant curves, and a continuous chaotic area are all clearly visible. Points outside the chain of eight stable "islands" escape rapidly [as $\exp(\exp 3n)$] to infinity under M.

3. Wave functions and Wigner functions

3.1. Maslov's association between phase-space surfaces and semiclassical wave functions

Now we turn to the main work of these lectures, which is to study how the different types of underlying classical motion manifest themselves in quantum mechanics. We begin by considering the morphologies of eigenfunc-

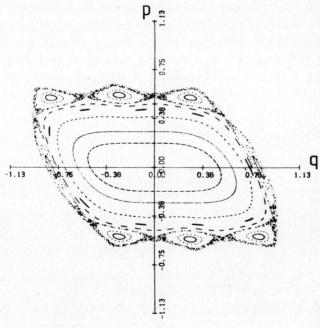

Fig. 26.

tions $\psi(q)$ of time-independent Hamiltonian operators whose analogous classical Hamiltonians generate regular or irregular bounded motion. The result will be strong evidence that the morphologies corresponding to the two sorts of motion are indeed very different. The suggestion that this might be the case goes back to Einstein [36], and was revided in a modern form by Percival [37].

I begin by describing an important semiclassical concept developed in full geometric detail in the 1960s by Maslov [26] but implicit in earlier work by Van Vleck [27] and Keller [28]. Maslov's ideas have been reviewed by Kravtsov [29] and (briefly) by me [18] in the context of propagating waves, by Duistermaat [2] and Guillemin and Sternberg [30] and Voros [44] from the mathematical viewpoint, and by Percival [1] and Eckmann and Sénéor [31] using simple examples.

The concept is an association between wave functons $\psi(q)$ and N-dimensional surfaces Σ in the $2N$-dimensional phase space q, p. ψ need not be an eigenstate of the Hamiltonian and Σ need not be one of the tori discussed in section 2. To start with, the association between ψ and Σ is purely geometric; dynamics is introduced later. Locally (fig. 27), Σ can be written as a function $p(q)$, and corresponds to an N-parameter ensemble of states (points) in classical phase space. We define a density on Σ in which these states are distributed uniformly in some coordinate $Q = \{Q_1 \dots Q_N\}$. A convenient definition of Q can be accomplished if we imagine an N-parameter family of surfaces filling phase space near Σ, labelled by $P = \{P_1 \dots P_N\}$, and then regard Q, P as alternative phase space variables in a canonical transformation from q, p. This is specified by a generating

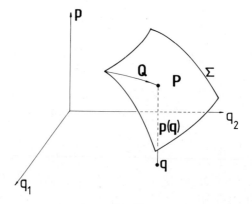

Fig. 27.

function $S(q;P)$ by

$$(q,p) \leftarrow S(q;P) \rightarrow (Q,P)$$
$$p = \nabla_q S, \; Q = \nabla_p S \tag{3.1}$$

(cf. the transformation (2.6) to action–angle variables, which is a special case). These equations imply that Σ has the "Lagrangian" property (2.7), and this will henceforth be assumed.

We want to associate with Σ a wave

$$\psi(q) = a(q)e^{ib(q)}. \tag{3.2}$$

Because there is as yet no dynamics, ψ is unrestricted by any wave equation, and we are free to choose any association with Σ. There is, however, a natural choice, based on two physical principles, one for $a(q)$ and one for $b(q)$.

For the amplitude $a(q)$ we require that the wave intensity $|\omega|^2$ is proportional to the density of points in classical q-space. This is obtained by projecting the density on Σ and using eq. (3.1), in terms of the Jacobian determinant

$$a^2(q) = K \left| \frac{dQ}{dq} \right| = K \det \left| \frac{\partial^2 S}{\partial q_i \partial P_j} \right|, \tag{3.3}$$

where K is a constant.

For the phase $b(q)$ we use de Broglie's rule relating the classical momentum $p(q)$ to the wave vector $k(q)$ of a locally plane wave, so that, in terms of the phase difference between two points separated by δq,

$$b(q + \delta q) - b(q) = \nabla b \cdot \delta q = k \cdot \delta q = \frac{p \cdot \delta q}{\hbar}, \tag{3.4}$$

thereby introducing Planck's constant $\hbar$. Thus

$$\nabla b(q) = p(q)/\hbar \tag{3.5}$$

and

$$b(q) = \frac{1}{\hbar} \int_{q_0}^{q} p(q') \cdot dq' = \frac{S(q;P)}{\hbar}, \tag{3.6}$$

again using eq. (3.1), where q_0 is a fixed position where S is defined to vanish. This phase is uniquely defined locally, independent of the integration path between q_0 and q, because the Lagrangian property (2.7) makes

S a (locally) single-valued function of q.

In this way we construct the local plane wave

$$\psi(q) = K \left| \det \frac{\partial^2 S(q;P)}{\partial q_i \partial q_j} \right|^{1/2} \exp\left(\frac{i}{\hbar} S(q;P)\right) \tag{3.7}$$

associated with the surface Σ labelled by P. Why should this association be a useful one? The answer lies in regarding ψ as an initial quantum state and letting it evolve according to the Schrödinger equation under dynamics governed by a Hamiltonian H. After a time t, ψ will have evolved into a new wave ψ'. In addition, Σ will have evolved into a new surface Σ' by virtue of the classical Hamiltonian motion of each of its points. Solving the time-dependent Schrödinger equation asymptotically, i.e. to lowest order in $\hbar$, it can be shown (see e.g. ref. [27] or Dirac [32]) that the ψ' can be constructed from Σ' by precisely the recipe (3.7). Therefore the construction persists in time, at least at the level of semiclassical approximations, and so represents a natural association between evolving quantal waves and N-parameter families of classical orbits (i.e. evolving classical surfaces). As $t \to \infty$, the association breaks down in a most interesting way, to be discussed in section 6. Now we pursue the purely geometric aspects of the association.

3.2. Globalization

The procedure based on fig. 27 and eq. (3.7) is well defined only if $p(q)$ is singlevalued. But what if Σ is curved (fig. 28) in such a way that a fibre drawn "upwards" from q intersects it at several momenta $p_i(q)$? It is natural to invoke the principle of superposition and extend the association

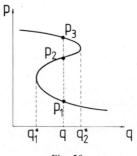

Fig. 28.

by writing $\psi(q)$ as a sum of terms like eq. (3.7), one for each branch $p_i(q)$. This raises the question of how these contributions are to be joined at caustics, where the projection of Σ "down" onto q is singular (e.g. at the points q_1^* and q_2^* on fig. 28). There are two difficulties. Firstly, it is not clear how to relate the phases of the different contributions. Secondly, eq. (3.7) has the undesirable property that it becomes infinite on caustics, because of the divergence of the "projection" Jacobian dQ/dq [eq. (3.3)] determining the amplitude $a(q)$.

Both difficulties can be overcome by Maslov's ingenious procedure [26] of requiring the association between waves and phase-space surfaces to hold for momentum as well as position coordinates. Therefore the semiclassical momentum wave function $\bar{\psi}(p)$ will be given by a formula like eq. (3.7) with Σ specified by the function $q(p;P)$ and the generating function $\bar{S}(p;P)$ constructed from the line integral $\int q \cdot dp$ rather than $\int p \cdot dq$. The beauty of this method is that Σ is a smooth surface and so cannot have points where the projection is singular in both q and p. So when eq. (3.7) gives problems near q-caustics its momentum analogue is well-behaved. But $\psi(q)$ and $\bar{\psi}(p)$ are related by Fourier transformation, so that Maslov was able to employ the momentum analogue of eq. (3.7) to obtain representation of $\psi(q)$ in terms of an *oscillatory integral* which is well-behaved at caustics. I shall not discuss here the form of this wave close to caustics (this aspect was reviewed at Les Houches in 1980 [18] and by Berry and Upstill [33] under the heading "catastrophe optics") except to state that near a typical caustic (a smooth surface with dimensionality $N-1$ in q-space), eq. (3.7) is replaced by an Airy function. When q is not close to a caustic, the Fourier integral for $\psi(q)$ can be evaluated by the method of stationary phase, because when $\hbar \to 0$ the integrand is a rapidly-oscillating function of its variables p. The result is that the representation of ψ as a sum of terms like eq. (3.7) is regained, but with a definite phase relationship, which I now describe, between the terms corresponding to the branches that join on the caustic.

In constructing the correct superposition of terms (3.7), the phases S_i must be chosen to be the branches of a single action function $S(q)$ on Σ, so that the value of S on a caustic must be the same irrespective of the branch on which the caustic is approached. Corresponding to fig. 28, for example, S must be chosen as follows:

$$\text{branch 1: } S = \int_{q_1^*}^{q} p_1(q')dq', \tag{3.8}$$

branch 2: $S = \displaystyle\int_{q_1^*}^{q} p_2(q')dq'$,

branch 3: $S = -\displaystyle\int_{q}^{q_2^*} p_3(q')dq' + \int_{q_1^*}^{q_2^*} p_2(q')dq'$.   (3.8 cont'd)

In addition, the modulus signs must be removed from the determinantal amplitude factor, so that during passage through a typical caustic (smooth $N-1$ dimensional surface), where $|dq|$ has a simple zero, the determinant changes phase by π and its square root by $\pi/2$. The only problem lies in deciding the sign of this change. The solution is that the phase change of ψ is $-\pi/2$ if the caustic is transversed with the (locally) two-sheeted region of Σ on the right. This is the well-known "phase advance of $\pi/2$ on crossing a caustic" [30]. Thus in fig. 28 the phase changes by $-\pi/2$ between p_3 and p_2 and by a further $+\pi/2$ (back to its original value) between p_2 and p_1.

The final step in generalizing the construction (3.7) is to consider the case where Σ is a closed surface in the form of an N-torus. This is not the only topology compatible with the Lagrangian conditions (2.7): for $N=2$, Hayden and Zeeman [34] have constructed a Lagrangian Klein bottle, but its quantal (or even classical) meaning is still obscure.

For an N-torus it is possible to return to any original point q in N essentially different ways, corresponding to the irreducible circuits γ_i of Σ. There is no global action function S which is single valued on Σ, because around γ_i there must be a change $\Delta_i S$ given by

$$\Delta_i S = \oint_{\gamma_i} p \cdot dq,$$ (3.9)

and equal to the sum of the areas of the projections of γ_i onto the N qp planes. During such a circuit, the phase of the wave constructed by eq. (3.7) will change by $\Delta_i S/\hbar$ plus a multiple α_i of $-\pi/2$ equal to the number of caustics encountered during γ_i. But the wave function $\psi(q)$ must be single valued under continuation, implying

$$\frac{1}{\hbar} \oint_{\gamma_i} p_i \cdot dq_i - \tfrac{1}{2}\alpha_i\pi = 2m_i\pi \qquad (1\le i\le N),$$ (3.10)

i.e.

$$\frac{1}{2\pi} \oint_{\gamma_i} p_i \cdot dq_i = (m_i + \tfrac{1}{4}\alpha_i)\hbar,$$ (3.11)

where the m_i are integers. This is a set of N *quantum conditions* on the geometry of Σ. A typical torus will not satisfy these conditions, but recall that Σ is embedded in an N-parameter family labelled by P, and if (as is true semiclassically) $\hbar$ is small compared with $\Delta_i S$ it will always be possible close to any given Σ to find a surface for which eq. (3.11) is satisfied and ψ is a single-valued semiclassical wave function. The numbers α_i, often called the "Maslov indices", depend on how Σ is embedded in phase space; a general topological discussion of α_i was given by Arnol'd [35] (see also refs. [30,1,2]). The Maslov indices embody the solution in N dimensions of what in one dimension is the "connection problem" [3,62] for WKB solutions of Schrödinger's equation.

The wave ψ thus obtained, based on a quantized torus Σ, will not in general be an eigenfunction of the Hamiltonian H, because it will change with time as Σ deforms under H. Only when Σ is an *invariant torus of motion* under H as described in section 2 will the surface remain fixed as its individual points q, p wind around it, and then ψ will be an energy eigenfunction, labelled by N quantum numbers m_i. The energy spectrum implied in this case by the conditions (3.11) will be studied in section 4.

It is obvious that this method of constructing semiclassical energy eigenfunctions can succeed only if tori exist, i.e. if the motion is regular as in integrable systems or throughout most of the phase space in quasi-integrable systems. For irregular and in particular ergodic motion, no tori exist and Maslov's method fails completely. At present there exists no asymptotic theory for the eigenfunctions corresponding to irregular motion. It is possible, however, to make conjectures in this case, on the basis of a general quantal phase-space formalism now to be described.

3.3. Wigner's function and the semiclassical eigenfunction hypothesis

In 1932, Wigner [38], introduced a phase-space distribution function $W(q,p)$ corresponding to a quantum state $\psi(q)$. This is defined by the N-fold integral

$$W(q,p) = \frac{1}{(2\pi\hbar)^N} \int \cdots \int d^N X \, \exp(-i p \cdot X/\hbar) \psi^*(q - X/2) \psi(q + X/2),$$

(3.12)

namely the Fourier transform of the product of ψ and ψ^* at positions separated by X. W is a quantal generalization of the classical density of points in phase space. It is also possible to generalize other classical functions to get phase-space representations of quantal operators, and to generalize Liouville's equation to get a phase-space representation of the Schrödinger equation. Groenewold [39], Moyal [40], Takabayasi [41] and Baker [42] have shown that the resulting phase-space picture gives a complete representation of quantum mechanics, alternative to more familiar approaches in terms of wave functions, Hilbert space operators, or functional integrals. Wigner's picture is peculiarly well suited to the present problem, because it is in phase-space that the distinction between classical regular and irregular motion manifests itself most clearly, and so one can hope that the analogous quantal distinctions will reveal themselves with corresponding clarity in the form of $W(q,p)$. This idea was first expressed by Nordholm and Rice [43] and developed by Voros [44], Berry [45] and Berry and Balazs [46]. Other semiclassical aspects of Wigner's function were studied by Balazs and Zipfel [47], Heller [48] and Korsch [49].

Despite appearances, $W(q,p)$ as defined by eq. (3.12) has complete formal symmetry in q and p. It contains all the information about the quantum state. In particular, the coordinate probability density is obtained by projection "down" p onto q:

$$|\psi(q)|^2 = \int \ldots \int d^N p \, W(q,p),\tag{3.13}$$

and the momentum probability density is obtained by projection "across" q onto p:

$$|\bar{\psi}(p)|^2 = \int \ldots \int d^N q \, W(q,p).\tag{3.14}$$

W can also be employed more generally, to describe statistical mixtures of pure states ψ_i with weights a_i, by adding terms of the pure-state form (3.12), one for each state ψ_i in the mixture, with coefficients a_i, to get a phase-space representation of the quantal density matrix.

It is natural to ask what $W(q,p)$ looks like for a semiclassical state of the form (3.7), associated with a surface Σ. To find out, simply substitute eq. (3.7) into eq. (3.12), to get

$$W(q,p) = \frac{K^2}{(2\pi\hbar)^N} \int \ldots \int d^N X$$
$$\left| \det\left(\frac{\partial^2 S(q+X/2;P)}{\partial q_i \partial P_j}\right) \det\left(\frac{\partial^2 S(q-X/2;P)}{\partial q_i \partial P_j}\right) \right|^{1/2}$$

$$\times \exp\left(\frac{i}{\hbar}\left[\int_{q-X/2}^{q+X/2} p(q', P) \cdot dq' - p \cdot X\right]\right). \tag{3.15}$$

As $\hbar \to 0$, the integrand oscillates rapidly and is dominated by the region near $X = 0$. Expanding the phase for small X gives

$$\int_{q-X/2}^{q+X/2} p(q', P) \cdot dq' \approx X \cdot p(q). \tag{3.16}$$

On setting $X = 0$ in the determinants, the integral can be evaluated to give the purely classical result

$$W(q,p) \approx K^2 \left| \det \frac{\partial^2 S(q;P)}{\partial q_i \partial q_j} \right| \delta[p - p(q,P)]. \tag{3.17}$$

In this approximation, therefore, W is nonzero only on the surface Σ employed in the construction of ψ. This satisfying result can be expressed in a more illuminating way if we define Σ as the memeber P^* of the N-parameter family of surfaces labelled by P, and use the fact that

$$\left| \det \frac{\partial^2 S(q;P)}{\partial q_i \partial P_j} \right| = \left| \frac{\partial P}{\partial P} \right| \tag{3.18}$$

to change variables in the delta function in eq. (3.17). This gives

$$W(q,p) \approx K^2 \delta[P(q,p) - P^*], \tag{3.19}$$

where $P(q,p)$ is the label of the particular surface that passes through q, p. It is clear from this representation that W is of uniform strength in the variable Q conjugate to P.

It is possible to go much deeper into the asymptotics of W and give a full stationary-phase evaluation of eq. (3.15). This reveals ([45,17,50,46,119]) that in the semiclassical limit the purely classical W given by eq. (3.19) softens its delta function and develops an intricate fringe pattern (whose details depend on the geometry of Σ) in phase space near Σ. Moreover this more refined semiclassical W projects down onto q by eq. (3.13) to give a probability density with the correct non-diverging behaviour on caustics, and this approach leads to an infinite hierarchy of nonlinear identities between the caustic wave functions classified by catastrophe theory (Berry and Wright [51], summarized in ref. [18]). But for our present purposes the crude approximation eq. (3.19) is sufficient.

We wish to apply eq. (3.19) to the energy eigenstates of a system (integrable or quasi-integrable) for which some orbits trace out phase-space tori. Such tori must satisfy the quantization condition (3.11) which according to eq. (2.4) restricts the action variables to

$$I_m = (m + \alpha/4)\hbar, \tag{3.20}$$

where $m \equiv \{m_1 ... m_N\}$ is the set of quantum numbers and $\alpha \equiv \{\alpha_1 ... \alpha_N\}$ are the Maslov indices. Then the Wigner function W_m representing the state labelled by m is given by eq. (3.19) as the following (correctly normalized) expression:

$$W_m(q,p) \approx \frac{1}{(2\pi)^N} \delta[I(q,p) - I_m], \tag{3.21}$$

where $I(q,p)$ is the action of the torus passing through q, p. The N-dimensional delta function implies that the Wigner function for an eigenstate is concentrated on the region that an orbit explores over infinite time – i.e. on the torus.

It is natural to extend this idea to cases where the motion is irregular and so not confined to tori. The resulting "semiclassical eigenfunction hypothesis" can be expressed as follows:

> Each semiclassical eigenstate has a Wigner function concentrated on the region explored by a typical orbit over infinite times.

The stipulation "typical" excludes the measure-zero closed orbits exploring one-dimension regions, which (except for certain degenerate cases) are too small to support quantum states (see also section 5.3). It was realized by Berry [52] and Voros [53] that this plausible hypothesis has powerful predictive force, as will now be explained.

First of all, let us apply the hypothesis to the extreme case of an ergodic system, whose orbits fill whole energy surfaces in phase space. Each quantum state corresponds to one energy surface, selected by a quantum condition. What these eigenenergies are is unknown, because nobody has so far discovered how to associate a wave with an energy surface in such a way that quantization follows from single-valuedness (the structure of the spectrum, and implicit quantum conditions for a particular ergodic system, will be discussed in sections 4 and 5). For an ergodic system, then, the hypothesis gives for the correctly normalized Wigner function representing an eigenstate with energy E,

$$W(q,p) \approx \frac{\delta[E - H(q,P)]}{\int ... \int d^N q \, d^N p \, \delta[E - H(q,p)]}. \tag{3.22}$$

In contrast to eq. (3.21), this is a one-dimensional delta function, reflecting the fact that W is spread over a much larger region of phase space.

The prediction of different morphologies for W is supported by computations of Hutchinson and Wyatt [54] for a Hamiltonian with $N = 2$ corresponding to motion of a particle in a potential (of "Hénon–Heiles" type) giving predominantly regular motion at low energies and irregular motion at high energies.

3.4. Regular and irregular quantum states

By employing the "integrable" or "ergodic" Wigner functions (3.21) and (3.22), together with the definition (3.12), it is possible to obtain morphological information about two aspects of $\psi(q)$: its local average strength and its pattern of local oscillations.

Consider first the probability density $|\psi(q)|^2$, obtained according to eq. (3.13) by projecting $W(q,p)$ "down" p. For a system with tori, eq. (3.21) simply gives the particular case of eq. (3.3) appropriate for this form of Σ, namely (fig. 29) the sum over branches

$$|\psi(q)|^2 \approx \frac{1}{(2\pi)^N} \sum_i \left| \frac{\mathrm{d}\theta}{\mathrm{d}q} (q, p_i(q)) \right| . \tag{3.23}$$

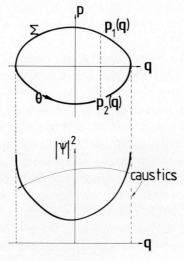

Fig. 29.

As already discussed, there are caustics at the singularities of the projection of the torus onto q, i.e. on local boundaries of the region explored by the orbit in q-space. Since the torus is usually locally parabolic near a caustic, we can extract the analytic form of the divergence of the approximate intensity $|\psi(q)|^2$ by taking a coordinate x perpendicular to the caustic and evaluating

$$|\psi(x)|^2 \sim \int dp\, \delta(x-p^2) \sim \frac{\Theta(x)}{x^{1/2}}, \tag{3.24}$$

where Θ denotes the unit step function. In general the caustic has cusps, swallowtails, umbilics and other morphologies classified by catastrophe theory as singularities of projections [18,17] and becomes very complicated as N increases.

On the other hand, for an ergodic system (3.22) gives

$$|\psi(q)|^2 \approx \int \ldots \int d^N p\, \delta[E - H(q,p)], \tag{3.25}$$

i.e. the projection of the energy surface. In ref. [52] I evaluated this for the case of a Hamiltonian

$$H(q,p) = \frac{|p|^2}{2\mu} + V(q), \tag{3.26}$$

with $V(q)$ corresponding to ergodic motion (e.g. the $N=2$ stadium billiard potential discussed in section 2.3), to get the result

$$|\psi(q)|^2 \sim [E - V(q)]^{N/2 - 1} \Theta[E - V(q)], \tag{3.27}$$

where the step function confines $|\psi|^2$ to the classically allowed part of q-space. In the trivial case $N=1$, when the ergodic system is also integrable, this reproduces the caustic divergence (3.24) on the classical boundary. But when $N>1$ the result shows that $|\psi(q)|^2$ does not diverge on the boundaries of the classical region: instead, it displays what I call *anticaustics*. Geometrically, this unexpected result is made plausible by a one-dimensional analogy in which instead of projecting a closed curve as in fig. 29 we project the patch of phase space enclosed by the curve (fig. 30). In section 6 we shall see evidence for the existence of anticaustics.

If the dynamical system consists of geodesic motion on a compact Riemannian manifold M, and is such that the geodesics are ergodic (e.g. if M has everywhere negative curvature), then there are no q-space boundaries. The semiclassical hypothesis predicts that for the corresponding

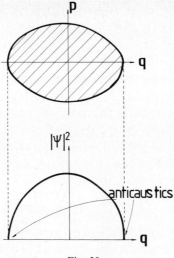

Fig. 30.

"quantal" problem, i.e. determination of eigenvalues and eigenfunctions $\psi_j(q)$ of the Laplace operator, $|\psi_j|^2$ should spread over the whole q space. That this is the case has been proved by Shnirelman [55], who shows that, for any smooth function $a(q)$,

$$\int_{\dot{M}} \cdots \int \mathrm{d}q\, a(q)|\psi_j(q)|^2 = \int_{\dot{M}} \cdots \int \mathrm{d}q\, a(q)/\int_{\dot{M}} \cdots \int \mathrm{d}q \qquad \text{as } j \to \infty \qquad (3.28)$$

for almost all j, where j labels increasing eigenenergies.

At this point we should pause to ask exactly what sort of semiclassical |wave function|^2 is being calculated by this procedure involving the purely classical Wigner functions (3.21) and (3.22). After all, we know that the exact $|\psi|^2$ will have oscillations and moreover will behave smoothly across classical boundaries. The probable answer [which I cannot prove except for simple cases but which is supported in Shnirelman's [55] special case by the result (3.28)] is that what is being calculated is the limit as $\hbar \to 0$ of a probability density $|\psi|^2_{\mathrm{sm}}$ which has been *smoothed* over a distance $\varDelta$ in each direction in q space, where $\varDelta$ vanishes as $\hbar \to 0$ more slowly than $\hbar$, so that oscillatory detail on the scale of the de Broglie wavelength is smoothed away. Mathematically, for any function $f(q)$ this procedure can be defined by

$$f(q) \to f_{\mathrm{sm}}(q) \equiv \frac{1}{\varDelta^N} \prod_{i=1}^{N} \int_{q_i - \varDelta/2}^{q_i + \varDelta/2} \mathrm{d}Q_i f(Q), \qquad (3.29)$$

where

$$\lim_{\hbar \to 0} \Delta = 0 \qquad \text{but } \lim_{\hbar \to 0} \frac{\hbar}{\Delta} = 0. \tag{3.30}$$

The vanishing of Δ ensures that the smoothed functions are "semiclassically sharp". On this interpretation, the classical Wigner functions (3.21) and (3.22) should be considered as the result W_{sm} of smoothing over q.

In view of this, it might be thought that eqs. (3.21) and (3.22) destroy all information about the oscillations of $\psi(q)$. But this is, surprisingly, not the case. Such information can in fact be obtained [52,53] from the *local autocorrelation function* of ψ, defined as

$$C(X;q) = [\psi(q+X/2) \, \psi^*(q-X/2)]_{\text{sm}} / |\psi(q)|^2_{\text{sm}}, \tag{3.31}$$

where the bars denote local semiclassical averaging as in (3.29). C contains information about the oscillations on the scale X near q, with irrelevant overall phases averaged away. For example, if (in one dimension),

$$\psi(q) = \exp\left[i\left(\frac{p_1 q}{\hbar} + \alpha \right) \right] \cos\left(\frac{p_2 q}{\hbar} + \beta \right) \tag{3.32}$$

then

$$|\psi(q)|^2_{\text{sm}} = \tfrac{1}{2} \tag{3.33}$$

and

$$C(X) = \exp(i p_1 X/\hbar) \cos(p_2 X/\hbar). \tag{3.34}$$

From the definition (3.12), C is given in terms of W_{sm} by

$$C(X;q) = \int \ldots \int \mathrm{d}^N p \, W_{\text{sm}}(q,p) \exp(i p \cdot X/\hbar) / |\psi(q)|^2_{\text{sm}}. \tag{3.35}$$

For states associated with tori, eq. (3.21) gives

$$C(X;q) \sim \sum_i \left| \frac{\mathrm{d}\theta}{\mathrm{d}q}(q, p_i(q)) \right| \exp\left(\frac{i}{\hbar} p_i(q) \cdot X \right), \tag{3.36}$$

whose simple interpretation is of local de Broglie waves interfering at q. Each wave [of the type (3.7)] has a wave vector $p_i(q)/\hbar$ corresponding to one of the branches of Σ lying "over" q. For a torus there can be only a finite number of branches and so ψ is the local superposition of finitely many plane waves, which gives an anisotropic pattern of interference fringes.

Now contrast this with what happens for an ergodic system with Hamiltonian (3.26). The Wigner function (3.25) and formula (3.35) for the autocorrelation function give the following result, obtained in refs. [52] and [53]:

$$C(X;q) \approx \frac{\Gamma(\tfrac{1}{2}N) J_{N/2-1}(|X|\{2\mu[E-V(q)]\}^{1/2}/\hbar)}{\{|X|\{2\mu[E-V(q)]\}^{1/2}/\hbar\}^{N/2-1}}, \tag{3.37}$$

where J denotes standard Bessel functions. Here, in contrast to eq. (3.36), C depends only on the length $|X|$ of the vector X, so that the local oscillations of ψ are, on the average, isotropic. This is because they result from the interference of de Broglie waves with the same wavelength but all possible directions, corresponding for the exploration by a trajectory of all momenta with the same energy $(p^2/2\mu)(q) + V(q)$, rather than the finite set $p_i(q)$ in the case of tori.

How can a function have locally isotropic autocorrelations about every point q? Only be being in some sense a *random* function of q. The randomness of $\psi(q)$ is governed by $W(q,p)$, which corresponds to a local spectral function specifying the distribution of local wave vectors $k = p/\hbar$ combining to interfere at q. For an ergodic Hamiltonian of type (3.26) the set of wave vectors consists of those with all possible directions but the same length – for $N=2$ this would be called a "ring spectrum".

To define the statistics of a random function ψ, it is necessary to specify more than the bilinear averages $|\psi|^2$ and C, in order to describe the fluctuations about these averages. In ref. [52] I suggested that the large number of interfering de Broglie waves would have random phases. This would make ψ a complex Gaussian random function of q, with local intensity fluctuations governed by the probability distribution

$$P(q,|\psi|^2) = (|\psi|_{sm}^2)^{-1} \exp(-|\psi|^2/|\psi|_{sm}^2), \tag{3.38}$$

where $|\psi|_{sm}^2$ is the "torus-projection" formula (3.27). A convenient measure of the fluctuations is the set of moments I_n, given by

$$I_n \equiv |\psi|_{sm}^{2n} = n!. \tag{3.39}$$

These grow with n, but not as rapidly as moments associated with waves dominated by caustics, which (as I have shown for the optical case in a study [56,33] of the statistics of twinkling starlight) diverge as $\hbar \to 0$ according to power laws determined by catastrophe theory. Ozorio [50] has made a preliminary study suggesting that the Gaussian random nature of

M. Berry

a wave function is preserved under "metaplectic transformations", i.e. linear phase-space-transformations. For a detailed semiclassical study of the moments (3.39), see ref. [120].

Evidently the simple idea underlying the semiclassical eigenfunction hypothesis has led to dramatic predictions about the morphology of wave functions. It implies that as $\hbar \to 0$ wave functions separate into two universality classes, associated with regular and irregular classical motion. In the regular case, ψ is associated with tori, and has vivid anisotropic interference oscillations rising to high intensities $|\psi|^2$ on caustics. In the irregular case, ψ is associated with chaotic regions in phase space (e.g. the whole energy surface in the case of ergodic systems), and has a random pattern of oscillations (isotropic for ergodic systems) with anticaustics at classical boundaries. I emphasize that these universality classes are *emergent* properties as $\hbar \to 0$; away from the semiclassical limit, it may often be impossible to unambiguously categorize a state as being regular or irregular.

For the regular states, this description merely reformulates what is already known. For irregular states it predicts an unfamiliar structure. In section 6 I shall display some irregular wave functions in one dimension, generated by a time-dependent Hamiltonian. Here I show (fig. 31) the nodal lines of a high-lying eigenstate (the 157th) of the Laplace operator with the condition that ψ vanish on the boundary of a stadium (cf. fig. 10) and on its diameters as computed by McDonald and Kaufman [57]. Recall from section 2.3 that the stadium is classically ergodic, so that the quantal eigenstates should be irregular. And it is clear that the nodal lines do wander irregularly with no systematic well-defined direction. Compare this with fig. 32, showing nodal lines for an eigenfunction of a circle (with radial quantum number 9 and angular quantum number 28), which form a regular pattern as expected because the circle is an integrable billiard (cf.

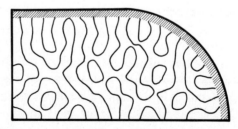

Fig. 31. Reprinted from ref. [57] with permission.

Fig. 32.

section 2.2). A similar contrast between the nodal patterns of states corresponding to regular and irregular classical motion was found by Stratt et al. [58]; their computations were for a smooth potential where some orbits filled neither tori nor whole energy surface, and indeed they found, roughly corresponding to these, states whose nodal pattern could not be classified as clearly regular or clearly irregular.

3.5. Crossing of nodal lines when $N=2$

The most striking difference between the nodal patterns for the irregular and regular wave function in figs. 31 and 32 is that in the irregular case the nodal lines do not intersect (except at the stadium boundary), while in the regular case they do. Is this a general property, which could be used to distinguish regular from irregular states in systems with $N=2$? The answer is no.

To see why, consider first the theorem published by Uhlenbeck [59] in 1976, which states that it is a generic property of eigenfunctions that their nodal lines do not intersect. "Generic" means here that the property fails to hold only for Hamiltonians forming a set of zero measure among all Hamiltonians. The idea is very simple. Suppose two nodal lines do cross. Near such a crossing the graph of a (real) eigenfunction $\psi(x, y)$ must be saddle-shaped, with contours in the form of hyperbolae near the nodal lines (it is easy to prove from the Schrödinger equation – with or without a potential but with isotropic dependence of H on the momentum components – that the nodal lines cross at right angles). At the crossing, not only ψ but its gradients $\partial\psi/\partial x$ and $\partial\psi/\partial y$ vanish. This gives three conditions on two coordinates xy and so generically no point can be found at which they are all satisfied. It could be objected that $\psi(x, y)$ is not a generic function because it is restricted by being a solution of the wave equation; but this relates ψ to its second derivatives and so does not spoil any genericity argument involving the gradient.

The form of this argument suggests (cf. ref. [57]) that node-crossing *will* occur generically in a family of Hamiltonians depending on a single parameter A, i.e. it is possible that for isolated values of A the nodes of a given eigenfunction may cross. If this is correct, then nodal lines behave differently from energy levels, whose crossing (degeneracy) requires two parameters, as we shall see in section 5.

There is no implication that "generic" here means "being associated with irregular classical motion", and indeed it cannot mean this, because the KAM theorem (section 2.4) establishes the persistence of tori in large classes of nonintegrable systems. Therefore even among regular states, associated with tori, the overwhelming majority will not show nodal line crossings.

So what is the special property allowing nodal lines to cross? It is not the existence of some tori throughout some but not all phase space, as in quasi-integrable systems. Nor is it the existence of tori throughout the whole phase space, as in integrable systems. In fact (as realized by Pechukas [60] in 1972) it is *separability* of the Schrödinger equation in an orthogonal coordinate system. This is not implied by integrability: even at the classical level the canonical transformation to action-angle variables does not imply that separation can be effected by coordinate transformation alone – see ref. [17] for a discussion of nonseparable tori.

If the system is separable, with orthogonal coordinates ζ, η, then eigenfunctions can be found with the special form

$$\psi(\zeta,\eta) = F(\zeta)\,G(\eta) \tag{3.40}$$

for which the nodal lines do cross. But, though separability allows nodal lines to cross, even this is not sufficient to *guarantee* that they do so. Because if the spectrum has *degeneracies* it is possible to add solutions of the form (3.40) to get new eigenfunctions whose nodal lines do not cross. Courant and Hilbert [61] give a dramatic example of this using the following eigenfunctions of a square with side π:

$$\psi_1(x,y) = \sin 2rx \sin y; \qquad \psi_2(x,y) = \sin x \sin 2ry. \tag{3.41}$$

The combination

$$\psi(x,y) = \psi_1(x,y) + (1+\varepsilon)\,\psi_2(x,y) \tag{3.42}$$

is also an eigenfunction. Figure 33a shows the nodal lines for $r = 12$ and $\varepsilon = 0$; there are 10 crossings, dividing the square into 12 regions. Figure 33b

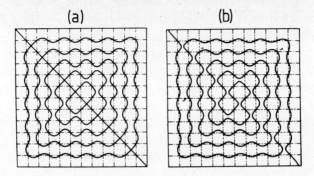

Fig. 33. Reprinted from ref. [61] with permission.

shows the nodal structure for a small value of ε; all the crossings have been destroyed, and the single nodal line divides the square into just two regions.

It is clear that the crossing of nodal lines is a very fragile property indeed, and certainly not as robust as the tori on which the existence of regular wave functions depend. Therefore although irregular wave functions must surely have no node crossings (except at isolated values of parameters in H), no implication about the nature of the associated classical motion can be drawn from the observation that nodal lines do not cross.

4. Eigenvalues: spectra on the finest scales

4.1. Quantum conditions

Having discussed the morphologies of semiclassical quantum states associated with regular and irregular motion, it is now necessary to consider how these classical distinctions affect quantal energy spectra. We shall find that these connections are quite subtle and depend on the scale on which the spectrum is studied. In my opinion the importance of understanding the rich variety of spectra on different scales (ranging from the locating of individual eigenvalues to the average density of states) has not hitherto been sufficiently appreciated.

On the finest scale, the most complete description of the spectrum would be explicit exact analytical formulae generating the energies E of all eigenstates. Obviously, this ideal cannot be achieved except in very special cases, and so we seek semiclassical formulae, giving the energies with an accuracy that increases as $\hbar \to 0$. This more limited programme has been carried out only for quantum states associated with regular motion, i.e. with phase-space tori. The resulting semiclassical quantum condition is known by various names in different sects of physics or mathematics: Bohr, Sommerfeld, Einstein [36], Brillouin, Keller [28], Maslov [26]... I shall refer to the procedure simply as "torus quantization".

According to arguments given in sections 3.2 and 3.3, a quantal energy state labelled by quantum numbers $m \equiv \{m_1 ... m_N\}$ is associated with a torus whose actions I_m are given by eq. (3.20) [cf. also eq. (3.11)]. The energy E_m of this state can be found in terms of the Hamiltonian (2.8) written as a function of the actions, and is simply

$$E_m = H(I_m). \tag{4.1}$$

This explicit formulae is exact as $\hbar \to 0$. For fixed E this corresponds to highly excited states, but the formula often gives results accurate to a few per cent even for the ground state (e.g. when $N = 1$, in which case (4.1) is the familiar WKB level formula [3]); in a few cases (harmonic oscillator, Coulomb potential, rectangular boxes) it gives all levels exactly. For integrable systems, the whole phase space is filled with tori and eq. (4.1) approximates all the levels (for applications to quantum billiards, see Keller and Rubinow [113]). For nonintegrable systems which are quasi-integrable (section 2.4), the KAM theorem guarantees the existence of tori filling some regions of phase space, and so eq. (4.1) can be employed to find a finite proportion of the levels, provided the actions and energies of the tori can be determined. This is a difficult problem of classical mechanics, which has been tackeld analytically by perturbation and iteration methods, and numerically by studying caustics and the Poincaré surface of section. By such techniques, Marcus and his colleagues [23,63–65], Percival and Pomphrey [66], Jaffé and Reinhardt [67], and Chapman, Garrett and Miller [68], obtained, for some chemically interesting quasi-integrable systems with $N = 2$ and 3, eigenvalues in very good agreement with "exact" computations.

What determines energy levels if the classical motion is irregular? In quasi-integrable systems with irregular regions whose phase-space volume is not large in comparison with h^N (see refs. [45,6] and section 5.1), the levels can be found from eq. (4.1) using low-order perturbation theory to approximate the motion in terms of tori even though the actual motion is not entirely confined to tori. But when large regions of phase space are filled by truly chaotic motion, torus quantization must fail. In the extreme case of an ergodic system, there are no tori, no actions and so no quantum numbers, and eq. (4.1) cannot give any of the levels. Nobody has yet succeeded in finding an explicit quantum condition, analogous to eq. (4.1), giving the levels of the "irregular spectrum" (to use Percival's terminology [37]) associated with chaotic trajectories.

It is, however, always possible to write *implicit* quantum conditions, by the conventional procedure of expanding the unknown eigenstate $|\psi>$ of the Hamiltonian $\hat{H}$ in terms of a complete set $|\phi_m>$ of basis states labelled by the N-fold index m. Then Schrödinger's equation,

$$\hat{H}|\psi> = E|\psi>, \tag{4.2}$$

leads to the following secular determinant:

$$\det_{m,n}\{<\phi_m|\hat{H}|\phi_n> - E\delta_{m,n}\} = 0. \tag{4.3}$$

Although formally exact, this procedure, involving matrix elements labelled by $2N$ integers, is not suitable for understanding the semiclassical limit.

For general Hamiltonians I do not know how to improve on eq. (4.3), but for billiards with $N=2$ it is possible to derive a determinantal eigenvalue condition [69] which is much more compact in that each element is labelled by two integers instead of four. Moreover this new determinant is rapidly convergent and well suited to semiclassical studies of the spectrum. I shall illustrate the procedure by outlining its application to Sinai's billiard (section 2.3 and fig. 8); full details can be found in ref. [70]. The motivation for studying Sinai's billiard is that the classical motion is ergodic so that this is a system with no tori at all, so one knows from the outset that torus quantization is meaningless.

Because of periodicity with respect to the coordinate torus in Sinai's billiard, this system can be represented in terms of wave propagation amongst hard discs centred on the points $\varrho = \{\varrho_1, \varrho_2\}$ of the unit square lattice (fig. 34). Eigenvalues E are determined by Schrödinger's equation,

$$\frac{\partial^2 \psi}{\partial x^2} + \frac{\partial^2 \psi}{\partial y^2} + k^2 \psi = 0, \qquad k^2 \equiv \frac{2\mu E}{\hbar^2}, \tag{4.4}$$

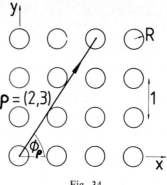

Fig. 34.

with boundary conditions

$$\psi(r+\varrho)=\psi(r) \qquad \text{(torus periodicity)},$$

$$\psi(|r|=R)=0 \qquad \text{(hard discs)},$$

(4.5)

where $r \equiv \{x, y\}$.

Thus the problem is transformed into one concerning waves in a periodic structure, and so the methods of solid-state electron band-structure theory can be brought to bear. Of several possible techniques (developed primarily for soft potentials and low energies) the one most suited to this problem (which has hard potentials and high – semiclassical – energies) is that devised by Korringa, Kohn and Rostoker (KKR) [71], where Green's identity plus periodicity is used to rewrite the Schrödinger equation as an integral equation round the boundary of one disc. In this equation the unknown function (normal derivative of ψ) is expanded as an angular Fourier series, i.e. in an angular momentum representation. After some (nontrivial) manipulation [70] this leads to the following determinantal condition for the energies E:

$$\det_{l, l'}[\delta_{ll'} + \sin \eta_l(E)\,e^{i\eta_l(E)}S_{l-l'}(E)] = 0 \qquad (-\infty < l, l' < +\infty). \quad (4.6)$$

In this equation, η_l is the lth scattering phase shift from a single disc, given in terms of real Bessel functions J and Y [72] by

$$\tan \eta_l(E) = J_l(kR)/Y_l(kR) \tag{4.7}$$

and $S_{l-l'}$ are the "structure constants", given in terms of Hankel functions $H^{(1)}$ [72] by the lattice sum

$$S_l(E) = -\mathrm{i} \sum_{\varrho}{}' H_l^{(1)}(k\varrho)\mathrm{e}^{\mathrm{i}l\phi_\varrho}, \tag{4.8}$$

where the prime denotes exclusion of $\varrho = 0$ and where ϱ, ϕ_ϱ are the plane polar coordinates of ϱ (fig. 34).

In physical terms, eq. (4.6) is the condition that the wave scattered from a single disc interferes constructively with the waves scattered from and amongst all the other discs. It differs in two essential ways from the secular determinant (4.3): E appears non-linearly (as k in the Bessel functions) rather than linearly, and each matrix element is labelled by two integers rather than four.

The formally exact equation (4.6) involves an infinite determinant, but this can be truncated according to a simple semiclassical rule, as follows. When $\hbar$ is small, the argument kR of the Bessel functions in the phase shifts eq. (4.7) is large. Then as $|l|$ increases from zero the Bessel functions oscillate until $|l| \approx kR$. For $|l| > kR$, J_l gets exponentially small and Y_l exponentially large, so that the phase shifts can be set equal to zero and the determinant truncated at $l = kR$ [the full justification [70] of this procedure involves intricate asymptotics on $S_l(E)$]. Semiclassically, then, the size of the determinant is given by $l_{\max}$, defined as

$$l_{\max} = kR = \frac{2\mu E}{\hbar} = \frac{2\pi R}{\lambda}$$

$$= \frac{\text{perimeter of billiard boundary}}{\text{de Broglie wavelength of state being studied}}. \tag{4.9}$$

My opinion is that the determinant (4.6), truncated in this way and approximated by replacing the Bessel functions by their "Debye" asymptotic forms [72], is the semiclassical quantum condition for Sinai's billiard, analogous to the torus quantization rule (4.1). Of course eq. (4.6) is an implicit equation, and more complicated than eq. (4.1). Moreover the work of finding the levels increases as $\hbar \to 0$ (because $l_{\max} \to \infty$), but less rapidly than in any other method I know (a further discussion of this point will be given in section 5.4). A determinant analogous to eq. (4.6) can be written down [69] for any billiard (integrable, quasi-integrable or pseudointegrable as well as ergodic). Its size is always given by eq. (4.9).

4.2. Degeneracies

We shall learn that there is a great deal more structure in the spectrum than can be immediately appreciated from a listing of the energies of all the eigenstates. This structure is embodied in correlations and clusterings of the levels. On the finest scale, such correlations concern neighbouring levels, and the first question to ask is: under what circumstances do these coincide? In other words: when do degeneracies occur?

If the Hamiltonian operator $\hat{H}$ has any symmetry, this may produce degeneracies, whose nature can be studied using group theory. It is not my intention to consider degeneracies of this type, and so when $\hat{H}$ does have symmetry I shall consider only states which all have the same symmetry class. This procedure is equivalent to considering all the states in a suitably "desymmetrized" Hamiltonian. For example, in the case of Sinai's billiard there will be degeneracies between states related by reflection about the axes $x=0$ or $y=0$ (fig. 8) or the diagonal $x=y$. We can eliminate these by considering, for example, only states which are odd under these reflections, and so effectively studying the modes of vibration of a membrane whose boundary is the billiard in fig. 35. Henceforth I shall consider all Hamiltonians as having been desymmetrized in this way.

For a typical (generic) such Hamiltonian it seems clear that degeneracy is infinitely improbable. But in a one-parameter family of Hamiltonians:

$$\hat{H} = H(\hat{q}, \hat{p}; A), \qquad\qquad (4.10)$$

it might be expected that levels E can degenerate for isolated values of the parameter A. The reason would be that the jth level $E_j(A)$ is a curve in E, A space, and the crossing of curves in the plane (fig. 36) is a geometric occurrence which is stable under perturbation. But the surprising fact is that this picture is not correct: for typical systems with real eigenfunctions (the only ones considered here), it is necessary to vary two parameters, not one, in order to make two levels degenerate. This is the content of a theorem due originally to Von Neumann and Wigner [73] and Teller [74] and later

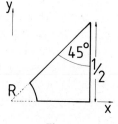

Fig. 35.

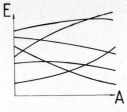

Fig. 36.

generalized by Arnol'd (ref. [5], appendix 10). The proof is based on a simple argument which I now give.

Consider Hamiltonians $\hat{H}$ close to a Hamiltonian $\hat{H}^*$ at which two orthogonal states $|u>$ and $|v>$ have the same energy E^*, i.e.

$$\hat{H}^*|u> = E^*|u>, \quad \hat{H}^*|v> = E^*|v>, \quad <u|v> = 0. \tag{4.11}$$

We wish to study the eigenvalues $|\psi>$ of $\hat{H}$, with energy E. Define

$$\hat{H} \equiv \hat{H}^* + \Delta\hat{H}, \quad E \equiv E^* + \Delta E, \tag{4.12}$$

and consider $\Delta\hat{H}$ and ΔE as small. Then the Schrödinger equation,

$$\hat{H}|\psi> = E|\psi>, \tag{4.13}$$

can be studied by degenerate perturbation theory. $|\psi>$ can be approximated as a normalized linear combination of $|u>$ and $|v>$, namely

$$|\psi> \approx \cos\chi|u> + \sin\chi|v>. \tag{4.14}$$

When substituted into eq. (4.13) using eqs. (4.11) and (4.12), this gives two homogeneous linear equations for $\cos\chi$ and $\sin\chi$, whose consistency gives a condition on the eigenvalue deviations $\Delta E_\pm$ which can be solved to give

$$\Delta E_\pm = \{<u|\Delta\hat{H}|u> + <v|\Delta\hat{H}|v>$$
$$\pm [(<u|\Delta\hat{H}|u> - <v|\Delta\hat{H}|v>)^2 + 4(<u|\Delta\hat{H}|v>)^2]^{1/2}\}. \tag{4.15}$$

The discriminant is a sum of squares, so that coincidence of the eigenvalues ($\Delta E_+ = \Delta E_-$) requires the (real) matrix elements to satisfy two conditions and not one. Generically, this can be accomplished by varying two parameters in $\hat{H}$, as asserted.

Let the two parameters be A and B. Then eq. (4.15) implies that the connection of eigenvalues surface $E = E_\pm(A, B)$ in the space E, A, B takes the form of a double cone (diabolo) (fig. 37) whose sheets are joined at the "diabolical point" E^*, A^*, B^*, where A^*, B^* are the parameters for which

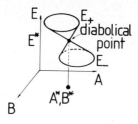

Fig. 37.

the degeneracy occurs. If only one parameter A is varied, the diabolical point will almost surely be missed and the curves $E_\pm(A)$ will avoid each other like branches of a hyperbola obtained by slicing the cone rather than crossing. So, instead of fig. 36 we expect curves like fig. 38, with near-degeneracies rather than actual crossings.

In testing the resulting picture of the spectrum, it is plausible to assume that a one-parameter family of non-symmetric classically ergodic systems will be generic in the quantal sense, and so will produce no level crossings. Such a family is the desymmetrized Sinai billiard (fig. 35) with the hard-disc radius R acting as the parameter. The levels $E_j(R)$ were computed "exactly" [70] using a simple modification of the determinant (4.6) which selects only states with the required symmetry. Figure 39 shows the spectrum. There are many near-degeneracies but careful examination shows that no two levels actually cross, so this test is successful.

The conclusion is that in a generic two-parameter family of Hamiltonian systems (e.g. Sinai's billiard with the discs replaced by ellipses of variable eccentricity and radius), we expect degeneracies at diabolical points A^*, B^* as in fig. 37. But how can be sure in any practical case that there really is a degeneracy at A^*, B^*, and not merely a near-touching of blunted cones, analogous to the near-crossing of curves in fig. 38? The answer lies in a little-known theorem, derived in ref. [70] by extending the algebra leading to eq. (4.15), concerning the two (real) eigenfunctions $\psi_\pm(q; A, B)$ whose energies degenerate at A^*, B^*: when taken round a circuit C in parameter

Fig. 38.

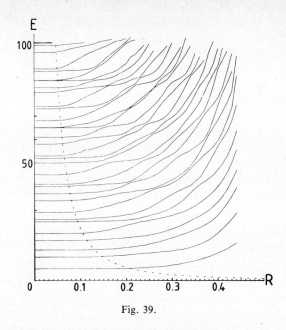

Fig. 39.

space, then, if (and only if) C encloses a degeneracy, ψ_+ and ψ_- will change sign. The theorem is mentioned by Arnol'd (appendix 10 of ref. [5]) and its converse was proved by Longuet–Higgins [75]. It is of course well known that wave functions must be single-valued under continuation with respect to their variables (e.g. q), but this theorem shows, surprisingly, that they need not be single-valued under continuation with respect to parameters in the Hamiltonian.

For systems whose levels are given by the torus quantization formula (4.1), degeneracies do not obey the generic two-parameter rule, but typically occur on varying only one parameter A. Therefore the picture of level crossings given by fig. 36, which we discarded in favour of fig. 38, is reinstated for the special class of systems with torus quantization. To see why this is so, regard the state with quantum numbers m^* as arising from the intersection of the $N-1$ dimensional hypersurface $E = H(I_{m^*}; A)$, in the N-dimensional space of quantum numbers m (fig. 40), with the lattice point m^*. Typically (i.e. for a fixed value of A) this hypersurface will not intersect any other lattice point, so the state m^* will be non-degenerate. But on varying A the hypersurface through m^* will smoothly change its orientation and there will typically be values A^* where it cuts another lattice

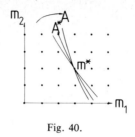

Fig. 40.

point, so that for such a value $A*$ there is a pair of degenerate states.

To summarize so far, we can say that, in the space of all Hamiltonians, those Hamiltonians with degenerate levels form a set of codimension two (connected cones), while in the space of Hamiltonians with all or part of the spectrum given by torus quantization, those Hamiltonians with degenerate levels form a set of codimension one (crossing curves). For the class of separable systems (all of which are integrable and so have torus-filled phase spaces), it is simple to show exactly that degenerate levels form a set of codimension one as torus quantization predicts (even through torus quantization is usually approximate rather than exact in these cases). But for nonseparable systems, even those which are integrable, the one-parameter degeneracies that torus quantization predicts will probably be split by the multi-dimensional analogue of barrier penetration, and I expect the exact spectra for this class of systems to have two-parameter degeneracies. On varying only one parameter, they should exhibit avoided crossings with exceedingly narrow separations, of order $\exp(-\text{constant}/\hbar)$, to be compared with the mean level separation which will be shown in section 5.1 to be of order $\hbar^N$. Semiclassically, if the spectrum is studied only to an accuracy described by power-law asymptotics in $\hbar$, such systems (i.e. all those with tori throughout regions of phase space), will seem to posses one-parameter degeneracies, but these are fragile and turn into two-parameter degeneracies when studied with exponential precision.

There are some special integrable systems whose levels are all proportional to integers, and these are not naturally embedded in continuous families. Their degeneracy structure can be very strange as I now indicate with two examples.

Consider first the equal-frequency harmonic oscillator with $N=2$, whose Hamiltonian is

$$H = \frac{p_1^2 + p_2^2}{2\mu} + \tfrac{1}{2}\mu\omega^2(q_1^2 + q_2^2). \tag{4.16}$$

The exact levels are

$$E_{m_1, m_2} = \hbar\omega(m_1 + m_2 + 1) \qquad (0 \le m_1, m_2 < \infty). \tag{4.17}$$

By considering contours of constant E in the m_1, m_2 plane (fig. 41), we see that levels occur at energies

$$E = \hbar\omega p \qquad (p = 1, 2 \ldots) \tag{4.18}$$

with degeneracies p. The mean density of states (cf. section 5.1) is

$$\bar{d}(E) = \overline{\sum_1^\infty \delta(E - \hbar\omega p)p} = \int_0^\infty dp\, p\delta(E - \hbar\omega p) = \frac{E}{\hbar^2\omega^2}. \tag{4.19}$$

Therefore the mean spacing irrespective of degeneracies is of order $\hbar^2/E$, and in terms of this mean spacing the levels (4.18) arrive at ever-increasing intervals in ever-more-degenerate groups.

The second example is the 45° right triangle billiard with hypotenuse $2^{-1/2}$, i.e. the desymmetrized Sinai billiard (fig. 35) with $R = 0$. This is a separable system whose exact levels are

$$E_{m_1, m_2} \equiv \frac{2\pi^2\hbar^2}{\mu}\, \mathscr{E}_{m_1, m_2} = \frac{2\pi^2\hbar^2}{\mu}(m_1^2 + m_2^2) \qquad (1 \le m_1 < m_2 < \infty). \tag{4.20}$$

Thus the levels $\mathscr{E}$ are all those integers which can be written as the sum of two squares. For small m_1, m_2, this square decomposition can be carried out in only one way. But when $\mathscr{E} = 65$ $(= 7^2 + 4^2 = 8^2 + 1^2)$ the first degeneracy occurs (it can be seen in fig. 39 for $R = 0$). When $\mathscr{E} = 325$ $(= 15^2 + 10^2 = 18^2 + 1^2 = 17^2 + 6^2)$ the first triple degeneracy occurs. It seems that degeneracies are rare. But as $\mathscr{E}$ increases, they come to dominate

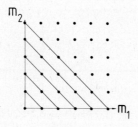

Fig. 41.

the spectrum. This curious conclusion, obtained in ref. [70], follows from two facts. Firstly, the average density of states (in $\mathscr{E}$) can be found from eq. (4.20) by counting lattice points, or from formulae to be derived in section 5.1, to be

$$\bar{d}(\mathscr{E}) = \pi/8, \tag{4.21}$$

so that the asymptotic mean level spacing irrespective of degeneracy is $8/\pi$. Secondly, it is a number-theoretic result that for large integers $\mathscr{E}$ the probability that $\mathscr{E}$ can be written as the sum of two squares decreases as $(\ln \mathscr{E})^{-1/2}$, so that the levels are separated by widening gaps of size $\sim (\ln \mathscr{E})^{1/2}$. These two facts are consistent only if the degeneracies increase as $(\ln \mathscr{E})^{1/2}$ to keep the net level density constant. In the original energy units, the resulting semiclassical spectrum has levels with degeneracies of order $[\ln(\text{constant} \times E/\hbar^2)]^{1/2}$ separated by gaps of order $\hbar^2[\ln(\text{constant} \times E/\hbar^2)]^{1/2}$. Pinsky [76] has found similar strange behaviour for the levels of the equilateral triangle billiard.

It is instructive to see how pseudointegrable systems fit into this general picture of degeneracy structure. Recall from section 2.5 that these are delicate "marginal" systems with $N = 2$ and two constants of motion, whose nonintegrability consists in the fact that the two-dimensional phase-space surfaces that their orbits fill are not tori but multiply-handled spheres.

To study these we consider the square torus billiard of fig. 15, parameterized by the side length L of the reflecting square. Quantum-mechanically, this was desymmetrized so as to become effectively the billiard shown in fig. 42. Richens and Berry [21] computed some of the levels $E_j(L)$ "exactly", with results shown in fig. 43. There are no degeneracies except at $L = 0$, suggesting that all the levels avoid each other which would mean that this class of systems behave like typical Hamiltonian systems and require two parameters to produce degeneracy. It can be

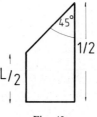

Fig. 42.

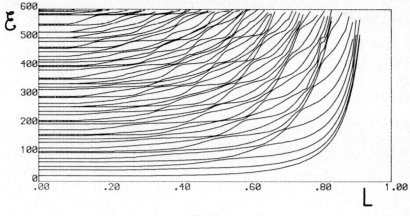

Fig. 43.

shown [21] that this is not quite correct: for every rational value of L there is a subset of all the levels (which gets smaller as the order of rationality increases) for which an exact torus-like quantization formula can be found, and which exhibit the increasing number-degeneracies discussed in connection with the 45° right triangle. None of these degeneracies (except for three at $L = 0$) appears in fig. 43 because the energy range is not large enough.

Marcus [77] suggests that the presence of many near-degeneracies (overlapping avoided crossings), in curves of energy levels as functions of a single parameter, is an indication that the corresponding quantum states are irregular in the sense described in section 3.4, i.e. associated with chaotic classical orbits filling large fractions of the $2N - 1$ dimensional energy surface. Figure 43 shows this implication to be unjustified: there are many near-degeneracies, but for the square torus billiard all orbits lie on $N - $ (i.e. $2 - $) dimensional surfaces and explore them without exponential chaos. Marcus's further suggestion that the implication holds in the reverse direction, i.e. that the energy levels of quantum states associated with classically chaotic motion will show many avoided crossings, is almost certainly correct. I conclude by drawing attention again to the extreme fragility of one-parameter crossings; these are analogous to the nodal-line intersections of wave functions (section 3.5), in that any departure from separability (which need not amount to classical chaos or even nonintegrability) may destroy them. For a detailed study of diabolical points in a quantum billiard system, see ref. [69].

4.3. Level spacings

The next step in understanding spectra is to study their finescale texture as embodied by the distribution of spacings between neighbouring levels, that is, to study spectra on the scale of the mean level spacing, which as we shall see in section 5.1 is of order $\hbar^N$. The spacings distribution $P(S)$ is defined by

$$P(S)\mathrm{d}S = \text{probability that the spacings of a randomly}$$
$$\text{chosen pair of neighbouring levels will}$$
$$\text{between } S \text{ and } S+\mathrm{d}S, \qquad (4.22)$$

where S is measured as a fraction of the mean level spacing $[\bar{d}(E)]^{-1}$ at the energy considered, so that the jth spacing S_j is defined as

$$S_j \equiv \bar{d}(E_j)(E_{j+1} - E_j). \qquad (4.23)$$

The function $P(S)$ was devised by Wigner and Landau and Smorodinsky (see e.g. ref. [78] in nuclear physics, where the many-body levels of nuclei are modelled by the eigenvalues of matrices in an ensemble (e.g. all matrices whose elements are Gaussian-distributed). How can an ensemble be defined in the case of a single Hamiltonian, for which the levels are deterministic rather than random? The answer is: by taking the semiclassical limit $\hbar \to 0$, so that infinitely many levels lie near any given E and eq. (4.22) is the average over their spacings. For billiards this procedure is equivalent to averaging over all levels in the spectrum with $\hbar$ fixed, i.e.

$$P(S) = \lim_{J \to \infty} \frac{1}{J} \sum_{j}^{j=1} \delta(S - S_j). \qquad (4.24)$$

We shall study the behaviour of $P(S)$ as $S \to 0$, because this tells us about the finest scales of level *clustering*. If $P(S) \to 0$ as $S \to 0$, neighbouring levels can be considered to "repel" each other, leading to a degree of regularity in the arrangement of levels, which can be quantified by the manner in which $P(S)$ vanishes. If on the other hand $P(S) \to$ constant as $S \to 0$, neighbouring levels cluster rather than repel.

The form of $P(S)$ as $S \to 0$ for a given Hamiltonian H depends on the degeneracy structure of families of similar Hamiltonians in which H can be embedded. In the generic case, we saw in the last section how H can be embedded in a family with two parameters A, B which has degeneracies at diabolical points in E, A, B space as in fig. 37. Let the actual Hamiltonian under study have parameters A_0, B_0. Then the line with $A = A_0$, $B = B_0$ (fig. 44) will thread its way among cones in E, A, B space, and the cones will presumably be distributed thickly if $\hbar$ is small. In eq. (4.24) the sum over j corresponds to an "energy average"

$$P(S) = \overline{\delta[S - S_j(A_0, B_0)]} \tag{4.25}$$

over the spacings S_j "above" A_0, B_0. For small S the only contributions to the average come from spacings for which the line in fig. 44 passes near diabolical points (labelled by k) whose parameters A_k^*, B_k^* lie close to A_0, B_0. The corresponding spacing S_k takes the positive-definite "conical" form [cf. eq. (4.15) with the matrix elements expanded in $A - A^*$, $B - B^*$] as follows:

$$S_k(A_0, B_0) =$$
$$[a_k(A_0 - A_k^*)^2 + 2b_k(A_0 - A_k^*)(B_0 - B_k^*) + c_k(B_0 - B_k^*)^2]^{1/2}, \tag{4.26}$$

where a_k, b_k and c_k describe the geometry of the kth cone.

Now, by hypothesis, there is nothing special about the system with parameters A_0, B_0, and so the energy average in eq. (4.25) can be augmented by an ensemble average over a region of A, B near A_0, B_0. Let $\varrho(A, B, E)$ be the density of diabolical points, and let $\pi(a, b, c)$ be the probability distribution of cone geometry parameters; the forms of the functions ϱ and π are unknown. Then the average (4.25) becomes, on using eq. (4.26) because S is small,

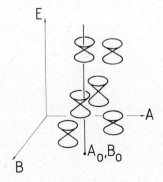

Fig. 44.

$$P(S) = \frac{\varrho}{\bar{d}(E)} (A_0, B_0, E) \int da \int db \int dc \ \pi(a, b, c)$$

$$\int dA \int dB \ \delta[S - (aA^2 + 2bAB + cB^2)^{1/2}]. \tag{4.27}$$

The integral over A, B can be extended to infinity because the positive definiteness of the quadratic form ensures convergence. Now change variables to

$$\alpha \equiv A/S, \qquad \beta \equiv B/S, \tag{4.28}$$

and eq. (4.27) becomes

$$P(S) = \left(\frac{\varrho(A_0, B_0, E)}{\bar{d}(E)} \int da \int db \int dc \ \pi(a, b, c) \right.$$

$$\left. \int d\alpha \, d\beta \ \delta[1 - (a\alpha^2 + 2b\alpha\beta + c\beta^2)^{1/2}] \right) \times S. \tag{4.29}$$

The factor in the two-line round brackets is a purely geometric average not involving S, so that

$$P(S) \sim S \qquad \text{as } S \to 0. \tag{4.30}$$

This argument therefore leads us to expect level repulsion for generic systems. The essential feature of the degeneracy structure which gives rise to this conclusion is the fact that the total length of intersection of a diabolo by two parallel sheets separated by S, where the diabolical point lies between the sheets, is proportional to S as $S \to 0$.

Figure 45 shows a test of this prediction of "linear" level repulsion, obtained by calculating eigenvalues of the desymmetrized Sinai billiard (fig. 42) and making a histogram of all spacings (several hundred in all) with $0.20 \le R \le 0.44$. It is clear that the levels to repel, and that the linear law gives a good fit. McDonald and Kaufman [57] and Casati, Valz-Griz and Guarneri [79], in computations of $P(S)$ for the desymmetrized stadium billiard, also obtain level repulsion, but their histograms do not show sufficient resolution to say whether the linear law is obeyed. Figure 46 shows another test, this time for the more delicate case of the (pseudo-integrable) desymmetrized square torus billiard (fig. 42). Once again the linear law of level repulsion gives a good fit, and it appears that pseudointegrable systems behave generically as far as $P(S)$ is concerned. (Because fig. 46 is

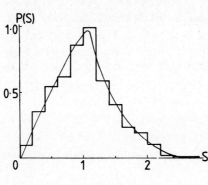

Fig. 45.

a compilation for different values of L, there might be a delta-peak in $P(S)$ at $S = 0$ because of the increasing number-theoretic degeneracies for each rational L, discussed in section 4.2, but it is probable that this delta-peak has zero height because of the zero measure of rationals.)

Zaslavsky [80] predicts that for classically chaotic systems $P(S)$ will display not the limiting form (4.30) but a nonlinear repulsion S^γ where γ depends on the rate of exponential separation of trajectories. I have criticised Zaslavsky's argument elsewhere [70], but point out that my own argument leading to linear repulsion is not watertight: it could fail if the cone-

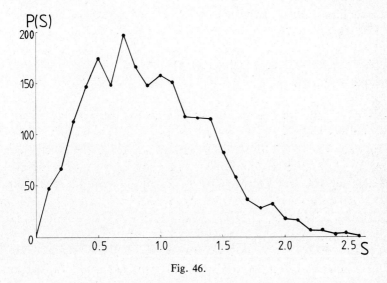

Fig. 46.

shape distribution $\pi(a, b, c)$ becomes singular as $\hbar \to 0$ in such a way that the previously mentioned two-line geometric factor in eq. (4.29) diverges.

Now let us consider $P(S)$ for some classes of non-generic system. Suppose we have a system so special that when embedded in an ensemble of similar systems it exhibits m-parameter degeneracies, i.e. degeneracies with codimension m, where m differs from the generic value of two. Then an argument precisely analogous to that based on cones when $m = 2$, employing the fact that S as a function of parameters is given not by eq. (4.26) but by the square root of a quadratic form in m variables, leads instead of eq. (4.30), to

$$P(S) \sim S^{m-1} \qquad \text{as } S \to 0. \tag{4.31}$$

For systems with torus quantization we have seen in eq. (4.2) that degeneracies are produced by varying only one parameter, so $m = 1$ and eq. (4.31) predicts that $P(S) \to \text{constant}$ as $S \to 0$. Therefore systems with tori should show level clustering rather than level repulsion. This is consistent with a more elaborate argument by Berry and Tabor [81], indicating that for integrable systems (with $N > 1$), $P(S)$ generically has the universal form

$$P(S) = e^{-S}, \tag{4.32}$$

corresponding to levels arriving irregularly. Figure 47 shows tests of this prediction for (a) the spectrum of a rectangle with side ratio 2, and (b) a

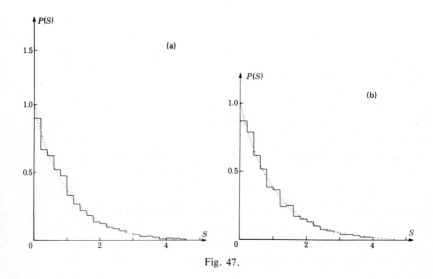

Fig. 47.

two-dimensional system whose potential is a square well in one direction and a harmonic oscillator in the other. Evidently the negative exponential is a very good fit to the computed histograms.

To summarize, we have found that for a generic system, e.g. one whose classical motion is irregular, the quantum levels exhibit linear repulsion and so are fairly regularly distributed. For a typical system with tori, whose classical motion is regular, the energy levels are irregularly distributed [and indeed from eq. (4.32) arrive at random as in a Poisson process such as radioactive decay].

I conclude with two curiosities. Consider first a class of systems in which degeneracies are strictly forbidden, so that even infinitely many parameters will not produce them. Then we must take $m \to \infty$, and eq. (4.31) suggests that $P(S)$ will vanish faster than any power of S as $S \to 0$. One such class of system is the eigenstates of the one-dimensional Schrödinger equation in a finite range with vanishing wave function at the ends. And indeed Pokrovski [82] showed that for electron states in a disordered linear chain of N delta-function potentials,

$$P(S) \sim \exp(-C/S^2)/S^2 \qquad \text{as } S \to 0, \tag{4.33}$$

where C depends on N but not S) in conformity with this prediction. (Recently, Molcanov [83], in a rigorous analysis, obtained the exponential distribution (4.32) for a one-dimensional potential – in the form of a non-differentiable ("fractal") curve derived from Brownian motion – in apparent contradiction with the prediction. But an essential feature of Molcanov's argument is the taking of a limit in which the potential exists over an infinite range, so that localized wave functions separated by great distances can have essentially independent eigenvalues and the basis of the prediction of level repulsion breaks down).

The second curiosity concerns two-dimensional harmonic oscillators with frequencies ω_1 and ω_2. Recall fig. 41 and eqs. (4.18) and (4.19), which show that if $\omega_1 = \omega_2$ degeneracies increase with energy, so that on the scale $\bar{d}^{-1}$ of the mean level spacing the gaps between neighbouring levels increase. Therefore the function $P(S)$ does not exist: the levels never settle down to a limiting distribution. A similar result holds for all rational values of ω_1/ω_2. What about irrational values? These are classically integrable systems and so might be expected to obey the negative-exponential rule (4.32). But for irrational harmonic oscillators the basis of the rule breaks down, because it can be shown quite easily that the levels, given by

$$E_{m_1 m_2} = \hbar[\omega_1(m_1 + 1/2) + \omega_2(m_2 + 1/2)], \tag{4.34}$$

are never degenerate. A lengthy number-theoretic analysis [81] of ways levels can come close together strongly suggests that $P(S)$ should show level repulsion, and indeed computation confirms this, as fig. 48 (for $\omega_1/\omega_2 = 2^{1/2}$) shows. The decay as $S \rightarrow 0$ is faster than for generic systems (cf. figs. 45 and 46), and suggest that $P(S)$ is more like eq. (4.33) – i.e. vanishing faster than any power of S – than the merely linear repulsion (4.30).

5. Eigenvalues: spectra on larger scales

5.1. Mean level density

Now we study spectra on scales large compared with the mean level spacing. It will be convenient to work with two functions: the level density $d(E)$ and the *mode number* $\mathcal{N}(E)$, defined in terms of the Hamiltonian operator $\hat{H}$ or the levels $E_1...E_j...$ (in increasing E with degenerate states counted separately) by

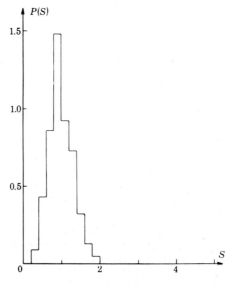

Fig. 48.

$$d(E) \equiv \sum_{j=1}^{\infty} \delta(E - E_j) = \mathrm{Tr}\,\delta(E - \hat{H}) \tag{5.1}$$

and

$$\mathcal{N}(E) \equiv \sum_{j=1}^{\infty} \Theta(E - E_j) = \mathrm{Tr}\,\Theta(E - \hat{H}). \tag{5.2}$$

These functions have of course been chosen so that their singularities (spikes for d and steps for $\mathcal{N}$) give the positions of the levels. $\mathcal{N}(E)$ is simply the number of states with energies below E. Obviously,

$$d(E) = \mathrm{d}\mathcal{N}(E)/\mathrm{d}E. \tag{5.3}$$

Sometimes $\mathcal{N}(E)$ is the more convenient function to work with, and sometimes $d(E)$ is more convenient.

There is a beautiful semiclassical theory for d and $\mathcal{N}$, whose principal architects were Gutzwiller [84–88] and Balian and Bloch [89–92]. This is based on representing $d(E)$ in the form

$$d(E) = \bar{d}(E) + d_{\mathrm{osc}}(E), \tag{5.4}$$

where $\bar{d}$ is the mean level density and d_{osc} is a series of oscillatory corrections. There is a similar representation for $\mathcal{N}(E)$. The terms on the right side of eq. (5.4) correspond to successive *smoothings* of the singular function $d(E)$. On the coarsest scale, that is after smoothing over energy ranges ΔE large enough to obliterate all traces of individual levels and all scales of level clustering, only $\bar{d}(E)$ survives. As ΔE is made smaller, more and more terms $d_{\mathrm{osc}}(E)$ contribute, with faster oscillations, until eventually they sum up to give a series of delta functions at the exact level positions. In simple terms, the semiclassical representation (5.4) is a generalization of the following expansion for a series of equally spaced unit delta functions:

$$\sum_{n=-\infty}^{\infty} \delta(E - n) = 1 + 2\sum_{m=1}^{\infty} \cos 2\pi m E. \tag{5.5}$$

We begin in this section by studying the mean level density $\bar{d}(E)$ and the mean mode number $\bar{\mathcal{N}}(E)$. These are given by the simple semiclassical rule (Landau and Lifshitz [93]) that each quantum state is associated with a phase-space volume h^n. Therefore $\bar{\mathcal{N}}(E)$ is h^{-N} times the volume of phase space for which $H(q, p)$ is less than E, i.e.

$$\bar{\mathcal{N}}(E) \approx \frac{1}{h^n} \int \mathrm{d}q \int \mathrm{d}p\, \Theta[E - H(q, p)], \tag{5.6}$$

and

$$\bar{d}(E) \approx \frac{1}{h^N} \int dq \int dp \, \delta[E - H(q,p)], \qquad (5.7)$$

so that the density is proportional to the size of the energy surface.

It is instructive to obtain these formulae in several different ways. The simplest is via the Weyl association [39–42] of a phase-space function $a_w(q,p)$ with any quantal operator $\hat{a}$ which is a function of the $\hat{q}$ and $\hat{p}$ operators:

$$a_w(q,p) = \frac{1}{h^N} \text{Tr}\left(\hat{a} \int dQ \int d\Pi \exp\left\{\frac{i}{\hbar}\left[(\hat{p}-p)\cdot Q + (\hat{q}-q)\cdot \Pi]\right\}\right), \qquad (5.8)$$

and implies

$$\text{Tr}\,\hat{a} = \frac{1}{h^N} \int dq \int dp \, a_w(q,p). \qquad (5.9)$$

It is not hard to verify that if $\hat{a}$ is the sum of an operator depending only on $\hat{q}$ and an operator depending only on $\hat{p}$, $a_w(q,p)$ is the function obtained by replacing $\hat{q}$ q and $\hat{p}$ by p , but that in general a_w $(q, p$ does not equal the classical function obtained in this way. But, by the correspondence principle this classical replacement must hold *in the semiclassical limit*. We therefore choose for $\hat{a}$ the operator in eq. (5.1), namely

$$\hat{a} = \delta(E - \hat{H}), \qquad (5.10)$$

so that semiclassically a_w is simply obtained by replacing $\hat{H}$ by the classical Hamiltonian function, and eq. (5.7) follows at once.

The second method, which we shall also use to study d_{osc} in section 5.2, is based on writing eq. (5.1) as

$$d(E) = -\frac{1}{\pi} \lim_{\varepsilon \to 0} \text{Im}\,\text{Tr}\left(\frac{1}{E + i\varepsilon - \hat{H}}\right)$$

$$= -\frac{1}{\pi} \text{Im} \int dq \, G^+(q, q'; E)_{q=q'}, \qquad (5.11)$$

where G^+ is the outgoing energy-independent *Green function*, defined by

$$G^+(q, q'; E) \equiv \left\langle q' \left| \frac{1}{E + i\varepsilon - \hat{H}} \right| q \right\rangle \qquad (5.12)$$

and satisfying

$$[E + i\varepsilon - H(q, -i\hbar\nabla_q)]G^+(q, q') = \delta(q - q'). \tag{5.13}$$

For simplicity let us restrict ourselves to Hamiltonians of the "kinetic + potential" type (3.26). Then the Green function satisfies a Helmholtz equation

$$\left(\nabla_q^2 + \frac{p^2(q)}{\hbar^2}\right)G^+(q, q') = \frac{2\mu}{\hbar^2}\delta(q - q') \tag{5.14}$$

with varying local momentum

$$p^2(q) = 2[E - V(q)]. \tag{5.15}$$

Let

$$\varepsilon = |q - q'|. \tag{5.16}$$

Then from eq. (5.11), $d(E)$ depends on G as $\varepsilon \to 0$. The mean density $\bar{d}$ can be obtained by a local approximation in which eq. (5.14) is solved as though $p(q)$ had everywhere the value it has at q. The resulting outgoing "free-particle Green function with momentum $p(q)$" is

$$G^+(q, q'; E)_{q \approx q'} \approx \frac{-2i\mu}{(2\pi)^{N/2+1}}\left(\frac{p(q)}{\varepsilon}\right)^{N/2-1}$$

$$H_{N/2-1}^{(1)}\left(\frac{p(q)\varepsilon}{\hbar}\right), \tag{5.17}$$

where $H^{(1)}$ is a Hankel function [72]. This diverges as $\varepsilon \to 0$ but its imaginary part remains finite. The small-argument limiting form for the Hankel functions [72], together with eq. (5.11), now gives $\bar{d}$ as an integral over the classically accessible space, namely

$$\bar{d}(E) \approx \left(\frac{\mu}{2\pi\hbar^2}\right)^{N/2}\frac{1}{\Gamma(N/2)}\int dq [E - V(q)]^{N/2-1}\Theta(q). \tag{5.18}$$

This is precisely what eq. (5.7) gives for Hamiltonian of type (3.26) (after integrating away the quadratic p-dependence).

For billiard Hamiltonians with $N = 2$, whose boundary encloses area $\mathscr{A}$ eqs. (5.6) and (5.7) give

$$\bar{\mathcal{N}}(E) \approx \frac{\mathscr{A}\mu E}{2\pi\hbar^2} = \frac{\mathscr{A}k^2}{4\pi} \tag{5.19}$$

and

$$\bar{d}(E) \approx \frac{\mathcal{A}\mu}{2\pi\hbar^2}. \tag{5.20}$$

These results are known as "Weyl formulae" (for a review, see Baltes and Hilf [94]).

Of course it is also possible to obtain $\bar{d}(E)$ from the quantum conditions discussed in section 4.1. For integrable systems, the torus quantization rule (4.1), together with eq. (5.1), gives

$$d(E) = \sum_m \delta[E - H(I_m)]. \tag{5.21}$$

To find $\bar{d}$ we simply replace the sum by an integral and change variables from m to I using eq. (3.20), giving

$$\bar{d}(E) \approx \frac{1}{\hbar^N} \int dI \delta[E - H(I)]. \tag{5.22}$$

Introducing the torus angles θ as dummy integration variables, we obtain an integral over phase space I, θ:

$$\bar{d}(E) \approx \frac{1}{\hbar^N} \int dI \int \frac{d\theta}{(2\pi)^N} \delta[E - H(I)]. \tag{5.23}$$

After a canonical transformation from I, θ back to the original q, p variables (a transformation whose Jacobian is unity), this gives precisely eq. (5.7).

It is more tricky to extract $\bar{d}(E)$ from the implicit quantization condition (4.6) for Sinai's billiard, and I simply outline the procedure (details are given in ref. [70]). A slight manipulation of eq. (4.6) gives

$$F(E) \equiv \det\nolimits_{l, l'} \left(\frac{e^{-i\eta_l(E)}}{\sin \eta_l(E)} \delta_{l, l'} + S_{l-l'}(E) \right) = 0, \tag{5.24}$$

where $F(E)$ is a function which can be shown to be real for real E. At each zero E_j (energy level of Sinai's billiard), the phase of F jumps by π. F also has poles E_p (where the structure constants S_l diverge and where $\sin \eta_l$ vanishes), and at E_p the phase of F also jumps by π. This leads to the following representation for $\mathcal{N}(E)$:

$$\mathcal{N}(E) = \lim_{\varepsilon \to 0} -\frac{1}{\pi} \text{Im} \ln[F(E + i\varepsilon)] + \sum_p \Theta(E - E_p). \tag{5.25}$$

After some algebra this gives the exact formula

$$\mathcal{N}(E) = \sum_m \Theta(k^2 - 4\pi^2 m^2) + \frac{1}{\pi} \sum_{l=-\infty}^{\infty} \eta_l(E)$$

$$- \frac{1}{\pi} \operatorname{Im} \operatorname{Tr} \ln[\delta_{ll'} + \sin\eta_l(E) e^{i\eta_l(E)} S_{l-l'}(E)]. \tag{5.26}$$

The first term is the "unperturbed" mode number corresponding to the integrable billiard with $R = 0$, with steps at each level of the "empty" torus. The other terms embody scattering by the discs. Only the first two terms contribute to the mean mode number in lowest approximation, and a little analysis [70] leads to

$$\bar{\mathcal{N}}(E) = \frac{(1 - \pi R^2) k^2}{4\pi}, \tag{5.27}$$

which is precisely the "Weyl" result (5.20) for this case.

In order to test numerically these theories for $\mathcal{N}(E)$, it is necessary to compute the levels E_j, construct the exact stepped curve $\mathcal{N}(E)$ using eq. (5.2) and compare its trend with eq. (5.6). Such a test has been carried out for the desymmetrized Sinai billiard (fig. 35) and is very instructive. Figure 49 shows the comparison of $\mathcal{N}(E)$ (stepped curve) with the Weyl formula (5.27) (divided by 8 because of the reduced area after desymmetrization) (full curve), for values of R from zero (integrable) to $R = 0.4$. Evidently the agreement is poor. The reason is that eq. (5.27) is an *asymptotic* formula which is here being tested on *lowlying* states. It is necessary to include *corrections* to eq. (5.27). These are not contributions to $\mathcal{N}_{\text{osc}}$ (which will be discussed in the next section), but are smooth terms of lower order in k, depending on aspects of the billiard geometry other than its area $\mathcal{A}$. For the desymmetrized Sinai billiard, general formulae [94] give

$$\bar{\mathcal{N}}(E) \approx \frac{1}{32\pi} (1 - \pi R^2) k^2 - \frac{1}{4\pi} \left[1 + 2^{-1/2} - R\left(2 - \frac{\pi}{4}\right) \right] k + \frac{31}{96}. \tag{5.28}$$

$$\text{area} \qquad\qquad \text{perimeter} \qquad\qquad\quad \text{curvature}$$
$$\text{+ corners}$$

This corrected formula is shown in fig. 49 as dashed curves; evidently the agreement is dramatically improved. Computations of the spectrum of the (pseudo-integrable) desymmetrized square torus billiard [21] give essentially the same result.

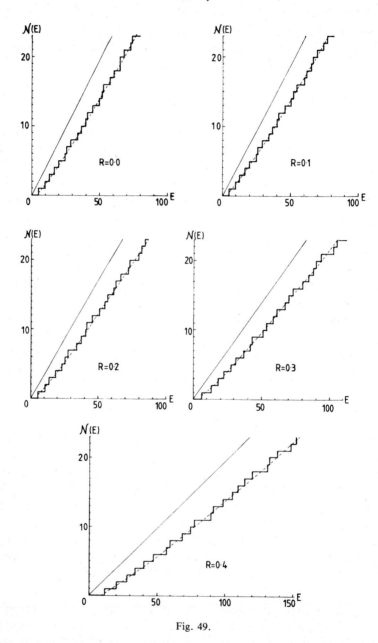

Fig. 49.

This study of the spectrum on the coarsest scales, embodied in $\bar{d}$ and $\bar{\mathcal{N}}$, has revealed that with this degree of smoothing all evidence of regularity of the underlying classical motion is lost.

5.2. Oscillatory corrections and closed orbits

The results of the last section confirm what was asserted in section 4, that the mean level spacing $\bar{d}^{-1}$ is of order $\hbar^N$. This might lead us to expect that the mean level density $\bar{d}(E)$ would be obtained simply by smoothing the exact $d(E)$ over an energy range of order $\hbar^N$. But such an expectation is mistaken, because as we shall see the corrections d_{osc} oscillate with energy "wavelength" of order $\hbar$, which when $N > 1$ (i.e. in nontrivial cases) is infinitely larger than the mean spacing as $\hbar \to 0$. In this important respect the spectrum in the general case contrasts with those special cases (which actually correspond to $N = 1$ as analyzed for example in ref. [3]) represented by eq. (5.5), in which the slowest-oscillating corrections have energy wavelength equal to the mean spacing.

The clearest route to understanding the nature of the contributions to d_{osc} is the Green function method based on eqs. (5.12)–(5.15). Because full details can be found in the original papers by Gutzwiller [84–88] and Balian and Bloch [89–92], which have been partially reviewed before [3]. I confine myself here to using the very simplest arguments leading to the essential results.

As defined by eq. (5.12) the Green function $G^+(q, q'; E)$ gives the amplitude at q of waves continuously emitted from q' with energy E. The corresponding classical paths form an N-parameter family (N components of momentum p' of the emitted particles, minus 1 because p' must lie on the energy surface $H(q', p') = E$ with q' and E fixed, plus 1 because particles are emitted at all times). Therefore the method described in section 3.1 can be employed to construct the semiclassical Green function as a series of contributions from *all classical paths leading from q' to q with energy E*.

For the jth such path, let the momentum at q be $p_j(q; q', E)$. Then the phase of this path's contribution to G^+ is $S_j(q, q'; E)/\hbar$, where

$$S_j(q, q'; E) = \int_{q'}^{q} p_j(q_1, q'; E) \cdot dq'. \tag{5.29}$$

Derived from S_j is an amplitude factor $a_j(q, q'; E)$ whose form will not be specified here (see ref. [84]), and this contributes to G^+ after multiplication by $\hbar^{-(N+1)/2}$ (as can be seen for example from the "free-particle" formula (5.17) on replacing the Hankel function by its asymptotic approximation). Therefore

$$G^+(q,q';E) \approx \frac{1}{\hbar^{(N+1)/2}} \sum_j a_j(q,q';E) \exp\left(\frac{i}{\hbar} S_j(q,q';E)\right). \qquad (5.30)$$

To find $d(E)$ it is necessary to let $q' \to q$ and take the trace (5.11) by integrating over q. When $q' \to q$, eq. (5.32) includes all classical paths beginning and ending at q. These are of two kinds: the "paths of zero length" – i.e. the limit as $\varepsilon \to 0$ of the direct path from q to $q + \varepsilon$—and the paths looping back to q after a finite excursion. The paths of zero length were discussed in the last section, and shown to give rise to the smooth term $\bar{d}$ [cf. eqs. (5.17) and (5.18)]. The looping paths contribute to d_{osc} as I now show.

A looping path need not be a closed orbit, because it may (and usually does) return to q with a momentum p different from its initial momentum p'. But in the integration (5.11) over q such non-closed looping paths give negligible contributions. The reason is that $S_j(q,q;E)$ must be stationary under local variations of q if the contributions from nearby positions are not to cancel by destructive interference, and this implies

$$\nabla_q S(q,q;E) = \lim_{q' \to q} [\nabla_q S(q,q';E) + \nabla_{q'} S(q,q';E)]$$

$$= \lim_{q' \to q} [p(q) - p'(q')] = 0, \qquad (5.31)$$

so that the initial and final momenta must in fact be equal.

We have now reached the following central conclusion: *only closed orbits with energy E contribute to d_{osc} (E)*; these include repetitions of primitive orbits, i.e. closed orbits traversed once. To study the form of these contributions, let j now label the primitive closed paths, and let the action round the jth path be

$$S_j(E) = \oint p_j(q;E) \cdot dq. \qquad (5.32)$$

Then it is clear that the phase of the contribution to d_{osc} from the jth primitive closed orbit, when traversed p times, is $pS_j(E)/\hbar$.

The amplitude is found by performing the integral over q in eq. (5.11), and the manner of doing this depends on two aspects of the primitive closed orbits: whether they are *isolated or nonisolated*, and whether they are *stable or unstable*.

Let the jth primitive closed orbit be embedded in an l_j-parameter family of closed orbits. l_j can vary from zero (for an isolated orbit) to $N-1$ (for the torus-filling orbits of an integrable system – section 2.2 – taking account of the fact that each closed orbit occupies one dimension). Then in eq. (5.11) l_j+1 dimensions of q integration (one along the orbit and l_j "across" the family) can be performed easily because the phase $S_j/\hbar$ is constant, and gives a factor corresponding to the measure of the family. The remaining $N-l_j-1$ dimensions, over the looped non-closed paths in the neighbourhood of the family, must be performed by stationary phase; each dimension gives a factor $\hbar^{1/2}$. Using eq. (5.30) and eq. (5.11), the resulting formula for the oscillatory level density corrections is

$$d_{\mathrm{osc}}(E) = \sum_j \sum_{p=1}^{\infty} \frac{A_{j,p}(E)}{\hbar^{1+l_j/2}} \sin\left(\frac{pS_j(E)}{\hbar} + p\alpha_j\right). \tag{5.33}$$

This is a sum over all primitive orbits j and repetitions p. The phases α_j are analogous to the Maslov indices discussed in section 3.2, and depend on the focusing of trajectories close to the closed orbit. The behaviour of the amplitudes $A_{j,p}$ as function of repetition number p depends on the stability of the primitive orbit j. For isolated orbits, $A_{j,p}$ oscillates with p if the orbit is stable, and decays exponentially if the orbit is unstable – types of behaviour to be expected in view of the repeated focusing or continued defocusing of beams of trajectories. For integrable systems, where all orbits are embedded in $N-1$ parameter families, Berry and Tabor [95,96] showed that $A_{j,p}$ decreases as $p^{-(N-1)/2}$. The Green function theory leading to eq. (5.33) is reviewed in detail by Rajaraman [97] and de-Witt-Morette et al. [98].

Just as for $\bar{d}(E)$ it is instructive to see how the general formula (5.33) for $d_{\mathrm{osc}}(E)$ is implicit in the quantization conditions obtained in section 4.1.

Consider first the integrable case, where torus quantization gave eq. (5.21) for $d(E)$. This is an N-dimensional sum over quantum numbers m, which can be transformed exactly into a sum of integrals in action space I by using the Poisson summation formula and the relation (3.20) between I and m. Recall first that the Poisson formula transforms sums over a unit lattice m into sums over another unit lattice M: for any function $f(m)$ defined on the lattice,

$$\sum_m f(m) = \sum_M \int \mathrm{d}m \exp(2\pi \mathrm{i}m \cdot M)f(m), \tag{5.34}$$

where the integrals involve any interpolation of f for continuous m. Applied to eq. (5.21) this procedure gives

$$d(E) = \frac{1}{\hbar^N} \sum_M \exp\left(-i\frac{\pi}{2}\alpha \cdot M\right) \int dI \delta[E - H(I)] \exp\left(\frac{2\pi i}{\hbar} M \cdot I\right). \quad (5.35)$$

The term $M = 0$ corresponds to replacing the original m-sum by an integral, which as we saw in eq. 5.1 gives the mean level density $\bar{d}$. The terms $M \neq 0$ give the oscillatory corrections we want to study now.

Action space I, that is the space whose points are tori, is stratified by the surfaces of constant E, which have $N - 1$ dimensions. The delta function in eq. (5.35) restricts integration to the surface with energy E. Introduce coordinates $\xi = \{\xi_1 \ldots \xi_{N-1}\}$ on this surface (fig. 50). Then for small $\hbar$ eq. (5.35) involves an integration with respect to ξ over a rapidly-oscillating exponential whose phase is $2\pi M \cdot I(\xi)/\hbar$. The principal contributions will come from those values ξ^M for which this phase is stationary, i.e. for which

$$M \cdot \frac{\partial I}{\partial \xi_i} = 0 \qquad \text{when } \xi = \xi^M \qquad (i = 1, 2 \ldots N-1). \quad (5.36)$$

Because $\partial I / \partial \xi_i$ are tangent vectors in the energy surface, this condition for ξ^M has the geometric meaning that the tori I^M contributing to d_{osc} are those lying on the energy surface at places where it is perpendicular to M. But the frequency vector ω[eq. (2.10)] is also perpendicular to the energy surface, so that ω and M are parallel in action space. Now, the components M_i are integers, so that the ω_i are mutually commensurate. Therefore the tori I^M are just those on which orbits are *closed* [cf. the discussion follow-

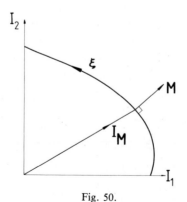

Fig. 50.

ing eq. (2.11)]. Moreover the lattice vector M specifies the topology of the orbit in terms of its winding numbers round the cycles of the torus. Therefore we have shown that the terms $M \neq 0$ in eq. (5.35) correspond to all the topologically distinct closed orbits with energy E. The phase of each contribution is $2\pi M \cdot I^M/\hbar$, which according to the definition (2.4) is precisely the action S round the orbit. This argument shows that for integrable systems torus quantization gives terms for d_{osc} precisely in accord with the general formula (5.33). For further details see refs. [95,96].

It is at first surprising to find that d_{osc} depends on the closed orbits in this way, because the representation (5.23) from which we started involved the actions I_m corresponding to the quantized tori, and not the actions I^M of the "rational" tori supporting closed orbits. These actions are different, and in particular the actions I_m corresponding to individual quantum states will not generally have commensurable frequencies and so will not support closed orbits. I shall discuss this point further in the next section, but remark that the apparently paradoxical result that all details of $d(E)$, including positions of individual levels, can be obtained by summing closed orbits is in fact analogous to, and no more surprising than, representing an irrational number as a sum over infinitely many rationals.

In the case of Sinai's billiard, extraction of the oscillatory contributions from the implicit quantization condition (4.6) is a much more difficult business and I only indicate the results (a full derivation is given in ref. [70]). The oscillatory contributions come from the first and third terms in the exact representation (5.26). The first term (which corresponds to the spectrum of the integrable "empty torus" or "empty lattice") can be transformed by the Poisson formula as in the general integrable case just discussed. This gives terms $\mathcal{N}_{osc}$ corresponding to each closed orbit belonging to the *nonisolated* set, which do not strike a disc (cf. fig. 12). Two examples are shown in fig. 51. But also included in this set of contributions are terms corresponding to nonisolated orbits that would pass through "ghost" discs, such as those in fig. 52. These are obviously un-

Fig. 51.

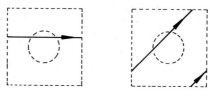

Fig. 52.

physical and in fact cancel as will soon be explained.

The contributions $\mathcal{N}_{osc}$ from the third term in eq. (5.26) are extracted as follows. First the Tr *ln* is expanded using the formula

$$\text{Tr}\,ln(\delta_{ll'} + M_{ll'}) = \sum_{n=1}^{\infty} \frac{(-1)^n}{n} \left(\sum_{l_1} \cdots \sum_{l_n} M_{l_1 l_2} M_{l_2 l_3} \cdots M_{l_n l_1} \right). \tag{5.37}$$

Next, the phase shift factors are written as

$$\sin \eta_l e^{i\eta_l} = \frac{1}{2i}(e^{2i\eta} - 1) \tag{5.38}$$

and the "-1" neglected for the moment. Then the sums over angular momenta l are replaced by integrals. Next, all Bessel functions are replaced by their asymptotic approximations. Finally the multiple l-integrals are evaluated by the method of stationary phase. The result of this lengthy procedure is that the nth term in eq. (5.37) gives contributions to $\mathcal{N}_{osc}$ from *isolated* closed orbits with n bounces from discs (cf. fig. 13). Two examples with $n = 4$ are shown in fig. 53. But also included in this set of contributions are oscillatory terms corresponding to closed orbits that would pass through "ghost" discs, as shown in fig. 54. Just like their nonisolated counterparts, these are obviously unphysical.

Both sorts of impossible path are cancelled by contributions from the "-1" terms in eq. (5.38). This occurs through a subtle diagrammatic identity [70] whose action can be glimpsed from the equation represented sym-

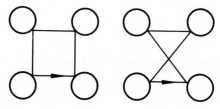

Fig. 53.

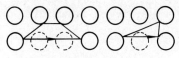

Fig. 54.

bolically in fig. 55, which refers to the impossible leg of the first diagram in fig. 54.

After all this, the final result is that d_{osc} contains contributions from all topologically distinct physically possible closed orbits. These are of two sorts. Firstly, terms from nonisolated orbits, which give contributions of order $\hbar^{-3/2}$ in precise accord with eq. (5.33) for this case where the orbits form one-parameter families. Secondly, terms from isolated orbits, which give contributions of order $\hbar^{-1}$, again in accord with eq. (5.33). The isolated orbits are all unstable and their contributions decay with increasing repetition as Δ_j^{-p}, where Δ_j (whose value exceeds unity) is a "determinant of bounces" corresponding to the jth primitive closed orbit; the order of Δ_j is the number n_j of bounces in the primitive orbit.

5.3. Comments on the closed-orbit sum

The most important feature of eq. (5.33) is its implication that each closed orbit (labelled by j and p) contributes an oscillation to the level density $d(E)$. The oscillation has energy "wavelength" ΔE given by

$$\frac{p}{\hbar}\frac{\mathrm{d}S_j}{\mathrm{d}E}\Delta E = 2\pi. \qquad (5.39)$$

Now from eq. (5.34),

$$\frac{\mathrm{d}S_j}{\mathrm{d}E} = \oint \frac{\mathrm{d}p_j}{\mathrm{d}E}\cdot\mathrm{d}q = \int \mathrm{d}t\,\dot{\boldsymbol{q}}\cdot\frac{\mathrm{d}p_j}{\mathrm{d}E} = \int \mathrm{d}t\,\nabla_p H\cdot\frac{\mathrm{d}p_j}{\mathrm{d}E} = T_j(E), \qquad (5.40)$$

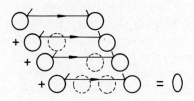

Fig. 55.

where $T_j(E)$ is the period of the jth closed orbit with energy E, so that

$$\Delta E = h/pT_j(E), \tag{5.41}$$

so that the closed orbit describes level clustering on scales of order $\hbar$. Now recall from eq. (5.7) that the mean level spacing of order $\hbar^N$. Therefore if $N > 1$ the closed orbits do indeed describe level clustering on scales much larger than the mean level spacing, as asserted at the beginning of section 5.2.

For billiards,

$$\frac{S_j}{\hbar} = \oint \frac{p_j \cdot dq}{\hbar} = kL_j, \tag{5.42}$$

where L_j is the spatial length of the jth orbit, implying that the corresponding k-wavelength Δk of the oscillations is

$$\Delta k = 2\pi/L_j. \tag{5.43}$$

Now I want to dispose of a fallacy based on a misinterpretation of eq. (5.33). Consider the terms with given j, i.e. those corresponding to all repetitions of a single primitive closed orbit. The terms will interfere constructively if

$$S_j(E_m) = (2\pi m - \alpha_j)\hbar, \tag{5.44}$$

defining a series of energies E_m corresponding to integers m. At these energies, the sum of the contributions to d_{osc} depends on the amplitude factors $A_{j,p}(E)$: for stable isolated orbits, the sum gives delta-functions at E_m [87], for unstable isolated orbits, the sum gives Lorentzian peaks at E_m [87], while for nonisolated orbits the sum gives other singularities at E_m [91,95], e.g. of logarithmic or inverse square root type. The fallacy is to suppose that E_m are eigenvalues of the Hamiltonian, i.e. that eq. (5.44) is a quantization condition associating individual quantum states with repetitions of individual closed orbits.

Why is this a fallacy? For a start, eq. (5.44) gives energies separated by distance of order $\hbar$, whereas the energy levels have separation $\hbar^N$. But why not superpose the level sequences obtained from eq. (5.44) with all topologically diffeent orbits j? Because this would give too many levels! An instructive demonstration of this is provided by free motion of a particle with mass μ on a coordinate 2-torus, represented by a rectangle with sides a, b and periodic boundary conditions. This is an integrable system, whose exact levels E_m are labelled by $m = (m_x, m_y)$ and given by

$$E_m = \frac{2\pi^2 \hbar^2}{\mu} \left(\frac{m_x^2}{a^2} + \frac{m_y^2}{b^2} \right). \tag{5.45}$$

Compare this with what eq. (5.44) would give: there are closed orbits labelled by $M = (M_x, M_y)$, corresponding to lattice translations $(M_x a, M_y b)$ (windings round the torus), with length

$$L_M = (M_x^2 a^2 + M_y^2 b^2)^{1/2}. \tag{5.46}$$

Then eq. (5.44) would give, on using eq. (5.42),

$$E_{m,M} = \frac{\hbar^2}{2\mu} \frac{(2\pi)^2 m^2}{L_M^2} = \frac{2\pi^2 \hbar^2 m^2}{\mu (M_x^2 a^2 + M_y^2 b^2)}. \tag{5.47}$$

This is clearly nonsense: it describes "levels" labelled by three quantum numbers rather than two, with an "infrared catastrophe" of levels with arbitrary small E corresponding to slowly-traversed closed orbits which are nevertheless long enough for their action to exceed $\hbar$.

Nevertheless, there are two circumstances where eq. (5.44) does give semiclassical quantal levels correctly. The first (and trivial) case is potential wells with $N = 1$, when there is only one topology of closed orbit and the levels do have separation $\hbar$ (see for example ref. [3] and references therein). The second case occurs when the closed orbits are isolated and stable. Then Miller [99] showed that by considering lowest-order fluctuations about the closed orbit it was possible to generalize eq. (5.44) into a condition with a full set of N quantum numbers. But Voros [44] pointed out that isolated stable orbits are always surrounded by tori, and explained how this quantum condition is really an approximate version of torus quantization, appropriate for the thin tori surrounding isolated stable orbits (when applied to a circularly symmetric potential with $N = 2$, for example, the approximation is accurate for levels near the bottoms of the wells in the effective potentials for each angular momentum).

In general, though, a single closed orbit gives not individual levels but a collective property of the spectrum, namely an oscillatory clustering with scale ΔE given by eq. (5.41). Conversely the determination of individual levels from of eq. (5.33) involves the closed orbits collectively, and would require the summation over sufficiently many closed orbits for individual delta functions to emerge as a result of constructive interference at certain energies and destructive interference at all other energies. Is this a feasible procedure for calculating the individual levels? I shall now argue that it is not.

To begin to see delta functions emerging from the path sum (5.33), it is necessary to include at least all orbits giving oscillations whose energy wavelength ΔE, given by eq. (5.41), exceeds the mean spacing $\delta E = \bar{d}^{-1}$ between neighbouring levels. Since the lower orbits give faster oscillations, it is necessary to sum over all closed orbits whose periods $pT_j(E)$ are less than T_{max}, given by

$$T_{max} = \frac{\int dq \int dp \, \delta[E - H(q,p)]}{h^{N-1}}.$$ (5.48)

For billiards with $N = 2$, it is necessary to sum over all closed orbits with length less than L_{max}, where [using eq. (5.19)]

$$L_{max} = k \mathscr{A},$$ (5.49)

where $\mathscr{A}$ is the area of the billiard.

Semiclassically, i.e. as $\hbar \to 0$ or $k \to \infty$, T_{max} and L_{max} increase, and it is necessary to include ever more closed orbits in the sum. How many? This depends on the behaviour as $T \to \infty$ of the function

$$v(T) = \text{number of closed orbits with periods less than } T, \quad (5.50)$$

or its billiard analogue $v(L)$. For integrable systems, closed orbits are classified by the N-dimensional winding number M, eq. (2.11), and T is proportional to the components of M, so

$$v(T) \sim T^N \qquad \text{as } T \to \infty \quad (5.51)$$

(this is just the number of unit lattice points within an N-dimensional sphere whose radius is proportional to T). Therefore the number v_s of steps required to determine the levels of an integrable system by direct summation over closed orbits is

$$v_s \sim v(T_{max}) \sim \hbar^{-N(N-1)}.$$ (5.52)

Of course this would be a foolish way to calculate the levels of an integrable system, because the torus quantization formula (4.1) gives the levels explicitly. Nevertheless it is instructive to see the delta functions emerging as more topologies of closed orbit are included, and I now illustrate this with a two-dimensional centralforce Hamiltonian (cf. fig. 2). Orbits classified by (M_1, M_2) close after M_1 librations and M_2 rotations;

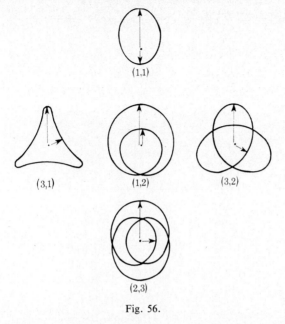

Fig. 56.

fig. 56 shows some of the simplest topologies. The Hamiltonian is a particle moving in a Morse potential, namely

$$H = \frac{p_x^2 + p_y^2}{2\mu} + V_0(e^{-2\delta(r-r_0)} - 2e^{-\delta(r-r_0)}) \qquad (5.53)$$

($\mu = 1$ proton mass, $V_0 = 0.2$ eV, $r_0 = 0.25$ nm, $\delta = 10$ nm^{-1}) which has a total of 166 bound levels. Figure 57 shows the effect of including increasing numbers of topologies M. In the last frame, which includes several hundred closed orbits, the delta functions are beginning to emerge very clearly (arrows mark exact levels, and $\bar{d}$ is given by the chain curves). For more details, see ref. [95].

What about ergodic systems? These will be discussed with reference to Sinai's billiard. Simple arguments [70] indicate that there is only a finite number of nonisolated closed orbits, but that the isolated orbits proliferate exponentially, i.e.

$$\dot{v}(L) \sim \exp(\text{constant} \times L) \qquad \text{as } L \to \infty, \qquad (5.54)$$

so that the isolated orbits dominate the sum (5.31) in spite of their individually weaker contributions. As mentioned at the end of section 5.2,

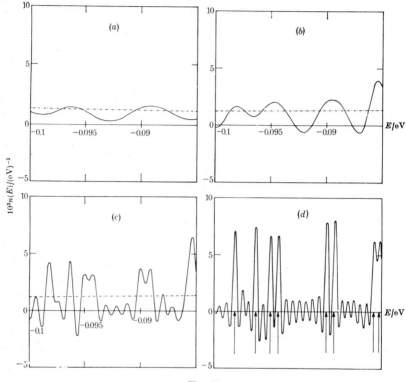

Fig. 57.

the contribution of each such orbit involves the $n_j \times n_j$ determinant of bounces. The computation of such a determinant (by Gaussian elimination) takes a number of steps of order n_j^3. As L increases, the majority of new orbits are traversed only once, and so the bounce number n_j is proportional to L. Therefore the number ν_s of steps required to evaluate the path sum for this ergodic system is

$$\nu_s \sim \nu(L_{max}) \times L_{max}^3 \sim (k \mathscr{A})^3 \exp(\text{constat} \times k \mathscr{A}),$$

$$\sim \hbar^{-3} \exp(\text{constant}/\hbar) \tag{5.55}$$

and greatly exceeds the corresponding number (5.51) for an integrable system.

However, there is in this case no torus quantization procedure to give the levels, and the closed orbit sum must compete instead with the determinant (4.6), whose zeros give the levels. How many steps ν_s are required to evaluate this? The answer lies in eq. (4.9), which gives the effective size of the determinant and leads [70] to

$$v_s \sim \hbar^{-3}. \tag{5.56}$$

It is obvious, therefore, that in the semiclassical limit the "compact" representation in the form of a determinant is vastly (i.e. exponentially) more efficient than the "expanded" representation as a closed path sum, as a means for calculating individual levels.

In a tour de force, Gutzwiller [101] has recently evaluated the path sum for an ergodic system, namely the anisotropic Kepler problem with Hamiltonian (2.12). The exponentially large number of closed orbits below any given length are classified in terms of binary sequences, which enables the path sum (5.33) to be interpreted as the partition function for a system of interacting spins, and this can be approximated by a determinant in terms of whose eigenvalues the energy levels can be determined. In this way the first 18 levels are computed with an accuracy of a few per cent. But a 50×50 determinant must be diagonalized in order to achieve this, so that [as expected on the basis of eq. (5.55)] the closed path sum is hardly a practical method for calculating the levels.

The enormous labour involved in the evaluation of closed orbit sums for nonintegrable systems does not mean that they are useless in understanding the spectrum. Suppose for example that what is required is not a complete knowledge of all the levels, but the level density *smoothed* over energy ranges $\Delta E \gg \delta E$. Then the path sum need only be taken up to paths with period pT_j given by eq. (5.41), which will involve only a small number of paths if ΔE is large. In this way Richens and Berry [21] obtained with 225 closed orbits a good fit to the smoothed $d(E)$ in the case of the (pseudointegrable) square torus billiard. A physical problem for which the closed path sum is a promising approach is the acoustics of auditoriums, where absorption at the walls causes individual eigenfrequencies to be broadened into resonances much wider than their separation, thus effecting a natural smoothing.

Finally, although I have presented these closed orbit expansions as semiclassical approximations, and this is how they arise in the general theories of Gutzwiller [84–88] and Balian and Bloch [89–91], it is encouraging that there are two *exact* connections known between the spectrum and the closed orbits for nonintegrable systems. Firstly, for the wave equation on a smooth closed Riemannian manifold (which need not have constant curvature), if the eigenwavenumbers are k_j $[=(2\mu E_j)^{1/2}/\hbar]$, then the function

$$Q(\lambda) \equiv \sum_{j=1}^{\infty} e^{ik_j\lambda} \qquad (5.57)$$

was proved by Chazarain [102] to have singularities for each λ equal to the length of a closed geodesic $[Q(\lambda)$ is the Fourier transform of the k density of states]. Balian and Bloch [92] obtain a similar exact result for the Schrödinger equation: the transform with respect to $\hbar^{-1}$ is singular at the action values of classical closed orbits. And secondly. the "Selberg trace formula" (reviewed by McKean [103] and Hejhal [104]) shows that for the ergodic system consisting of geodesic motion on a compact surface of constant negative curvature a sum over eigenvalues of the Laplacian is exactly given by a sum over (isolated, unstable) closed orbits. Unfortunately, neither side in the (exact or asymptotic) equality

$$\sum \text{ (eigenvalues)} \approx \sum \text{ (closed orbits)} \qquad (5.58)$$

is exactly known for any nonintegrable system.

6. Quantum maps

6.1. Evolving states

Recall the association described in sections 3.1 and 3.2, between quantal wave functions $\psi(q)$ and N-dimensional surfaces Σ in phase space. The association, embodied in eq. (3.7), was natural because under semiclassical conditions it persists as the state ψ and surface Σ evolve, one classically and the other quantally, under the action of the Hamiltonian H. Now I want to consider the largely unsolved problem of what happens to this association over long times t. The following argument indicates why we can expect this problem to be difficult. In almost all cases (i.e. when Σ is not an invariant surface and H is not linear or quadratic) Σ develops, as will be explained, an infinitely complicated morphology as $t \to \infty$, with foldings and convolutions down to arbitrarily fine scales. But it is hard to see how phase-space fine structure in volumes smaller than h^N can have quantal significance. This suggests that the asymptotic association between ψ and Σ holds only up to a time $t_{\max}(\hbar)$ (which increases with $\hbar$), after which ψ

and Σ evolve differently. Alternatively stated, we expect that the limits $\hbar \to 0$ and $t \to 0$ cannot be interchanged.

A wide-ranging and perceptive discussion of the problem is given by Chirikov et al. [106], especially in the context of classically chaotic systems (although even the semiclassical approximation as $t \to \infty$ of nonstationary states in classically *integrable* systems presents difficulties). Possible meanings that can be assigned to the term "quantum stochasticity" for evolving states have given rise to an extensive chemical literature which I shall not attempt to summarize; it is well discussed by Hutchinson and Wyatt [107].

A direct analytical attack on the problem appears to be extremely difficult, and so it is natural to seek simple models where it is possible to compute the evolution of both ψ and Σ and see if and when the association between them breaks down. This is where area-preserving maps M of the phase plane q, p (discussed in section 2.6) become useful (and indeed almost indispensible, given the limitations of present computers). There are four main reasons for using such maps as models. Firstly, they can easily be quantized as I show in a moment. Secondly they show all varieties of regular and chaotic behaviour. Thirdly, phase space is a plane, so that "surfaces" Σ are curves whose evolution is easy to depict. And fourthly the coordinate space q is one-dimensional, so that probability densities $|\psi(q)|^2$ can be depicted as graphs. The study of such "quantum maps" was pioneered by Casati et al. [108] in a particular case. Here I present some theory for a rather more general class of maps, as independently developed later by Berry et al. [109]. I shall emphasize, however, that the most interesting questions are still open.

The map to be quantized is eq. (2.24), derived from the T-periodic Hamiltonian (2.23), and some results will be presented for the special case, eq. (2.25). For quantization, we seek the unitary evolution operators $\hat{U}$ corresponding to the action of $\hat{H}$ over time T. This operator transforms states $|\psi_n\rangle$ into states $|\psi_{n+1}\rangle$ according to the quantum map

$$|\psi_{n+1}\rangle = \hat{U}|\psi_n\rangle. \tag{6.1}$$

To find $\hat{U}$ explicitly, we notice that in the time $0 \le t < T$, $\hat{H}$ can be written as the limit $\tau \to 0$ of

$$\begin{aligned} H(\hat{q}, \hat{p}, t) &= \hat{p}^2/2\mu & 0 \le t < T - \tau \\ &= \frac{TV(\hat{q})}{\mu} & T - \tau \le t < T \end{aligned} \tag{6.2}$$

In the first time interval, $0 \leq t < T - \tau$, states evolve under

$$\hat{U}_1 = \exp\left(\frac{-i}{2\mu\hbar}\hat{p}^2(T-\tau)\right),$$ (6.3)

and in the second interval, $T - \tau \leq t < T$, the evolution operator is

$$\hat{U}_2 = \exp\left(\frac{-i}{\hbar}TV(\hat{q})\right).$$ (6.4)

The full evolution operator $\hat{U}$ is therefore given by

$$\hat{U} = \hat{U}_2\hat{U}_1 = \exp\left(\frac{-i}{\hbar}TV(\hat{q})\right)\exp\left(\frac{-i}{2\hbar\mu}\hat{p}^2 T\right),$$ (6.5)

after letting $\tau \to 0$.

In the position representation, this operator inserted into eq. (6.1) gives the following discrete-time Schrödinger equation giving the quantum map transforming wave functions $\psi_n(q)$ into wave functions $\psi_{n+1}(q)$:

$$\psi_{n+1}(q) = \left(\frac{\mu}{2\pi\hbar T}\right)^{1/2} \exp\left[\frac{\pi}{4} + \frac{TV(q)}{\hbar}\right)\right]$$

$$\times \int_{-\infty}^{\infty} dq' \exp\left(\frac{i\mu}{2\hbar T}(q-q')^2\right)\psi_n(q').$$ (6.6)

This innocent-looking equation can, as we shall see, generate a wealth of interesting structure.

To explain how this was calculated, first note that in the limit $T \to 0$ the Hamiltonian (2.23) can be replaced by its time-average, which is the (integrable) stationary Hamiltonian $\bar{H}$ given by

$$\bar{H}(q, p) = \frac{p^2}{2\mu} + V(q),$$ (6.7)

corresponding to an oscillator. Therefore T is a parameter which turns on nonintegrability. For power-law anharmonic oscillators T can be scaled away as in the quartic case (2.25), and this simply means that a point in the original q, p plane [mapping under eq. (2.24)] is associated with a point in the scaled q, p plane which is closer to the origin the smaller T is.

For the initial curve Σ we [109] chose a (closed) contour of the average Hamiltonian (6.7). Of course Σ is not an invariant curve of M when $T > 0$. Nearby points on Σ will map apart under repeated application of M and so Σ will become infinitely long as $n \to \infty$. On the other hand, because of eq. (2.17) the area inside Σ must remain the same. Therefore Σ will develop infinite complexity. This complexity is due to the development of two principal morphologies in Σ, which are called "whorls" and "tendrils" [109,46] and which are associated with phase plane regions where motion is, respectively, regular and irregular. As can be seen from fig. 26, both sorts of motion are generated by M given by eq. (2.25).

Whorls are associated with invariant curves surrounding a stable fixed point of M. They can arise, for example in the twist map (2.21) (see fig. 22), provided the rotation number α is not independent of radius. What happens is that points at different radii rotate around the central fixed point at different rates. Therefore radii map to spirials, and parts of Σ passing close to stable fixed points will wrap around them as indicated in fig. 58. For the particular M being studied now, we expect from fig. 26 that whorls will occur whenever Σ passes through the central area of the qp plane (covered with invariant curves) or through the surrounding island chain.

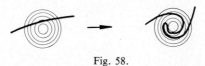

Fig. 58.

Tendrils are associated with the chaotic areas surrounding unstable fixed points. Iterates of curves Σ passing through such areas flail wildly back and forth as the individual points of Σ map exponentially apart, as indicated in fig. 59. From fig. 26 we expect tendrils whenever Σ passes through chaotic outer areas of the qp plane.

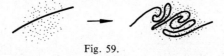

Fig. 59.

Figure 60a and 61 show how these morphologies develop for two initial curves Σ (contours of $\bar{H}$). The first curve ($n = 0$ on fig. 60a) lies mainly in the central area, and so begins to wrap into a whorl around the origin. By $n = 20$ the "spiral galaxy" structure is clearly visible and two subwhorls have started to appear, associated with other islands. The second curve ($n = 0$ on fig. 61) lies largely in the outer, escaping region of the plane and rapidly shoots tendrils out towards infinity.

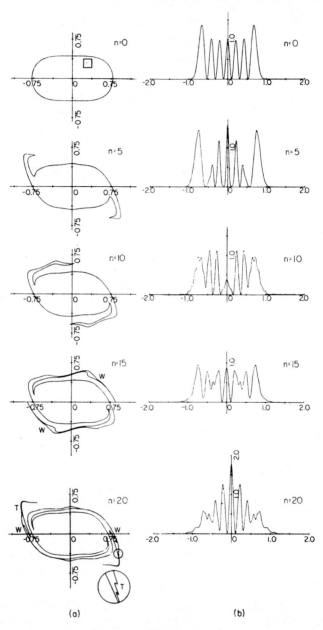

Fig. 60.

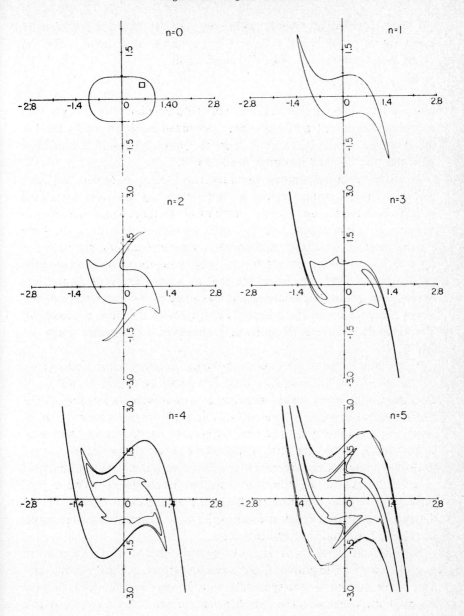

Fig. 61.

The development of a whorl can be seen very clearly in computations by Lewis (reported by Berry [110] of the evolution of a curve under the "nonlinear Harmonic oscillator" Hamiltonian

$$H = f(q^2 + p^2),$$ (6.8)

whose points map continuously in circles with angular velocity depending on radius. Beautiful tendrils have been calculated by Shepelyansky [111] in the evolution of the curve $p = 0$ under the map (2.24) with a sinusoidal potential (the "kicked quantum pendulum").

To study the corresponding quantum map (i.e. eq. (6.6) with a quartic potential $V(q)$, we must choose as initial state $\psi_0(q)$ a wave associated with Σ. Because Σ is an invariant curve of $\bar{H}$, ψ_0 must be an eigenstate of $\bar{H}$. For the curve mapped on fig. 60a, ψ_0 was chosen to be the sixth eigenstate of $\bar{H}$. This state, and its iterates under eq. (6.6), are shown in fig. 60b. The graph of $|\psi|^2$ for $n = 0$ clearly shows the association with the initial curve Σ, with maxima at the caustics of projection of Σ, and, between these, quasisinusosidal oscillations caused by the interference of two waves associated with the intersections $p_i(q)$ with the fibre $q = $ constant. The value of $\hbar$ corresponding to ψ_0 is the area of the square drawn inside Σ.

By the 20th iteration the classical curve has developed complexity (arms of the whorl) on scales smaller than $\hbar$, and so we expect ψ_n not to be associated with every detail. Instead, it is reasonable according to the "semiclassical hypothesis" stated in section 3.3 (in the context of eigenfunctions) and developed in section 3.4, to associate ψ with the area filled by the whorl, and on the basis of fig. 30 we expect the projection $|\psi(q)|^2$ to display anticaustics. Moreover, associated with the quasicontinuum of p-values intersected by fibres $q = $ constant (each giving a wave with wavelength h/p contributing to the interference pattern making up the total ψ) is an obvious loss of spectral purity of ψ. Therefore the predictions of section 3 are confirmed in this case.

For the initial curve of fig. 61, ψ_0 was chosen as the eighteenth eigenstate of $\bar{H}$. The graph of $|\psi_0|^2$, and its iterates, is shown in fig. 62a. The waves ω_0 and ψ_1 are spectrally pure (2-wave interference), because for $n = 1$ the classical curve (fig. 61) has not yet "curled over" to give multiple intersections. But it does curl over when $n = 2$, with immediate and dramatic effect on $|\psi_2(q)|^2$.

For a more careful classical-quantum comparison, the classical curves of fig. 61 are projected onto q, and the resulting "classical $|\psi|^2$" graphs are shown in fig. 62c. The most striking features are the caustic spikes, which proliferate as n increases and the classical curve throws off tendrils. The caustics are evidently related to features of the wave functions (fig. 62a) until $n=2$ but thereafter there is no obvious association. This corresponds precisely with the throwing-off for $n=3$ of a fine tendril, smaller than $\hbar$. Related to this is the fact that for large n the neighbouring caustics in fig. 62c are closer than the de Broglie wavelength and so cannot be directly associated with features of ψ. This suggests, in the spirit of the smoothing procedure described in section 3.4 and embodied in eq. (3.29), that a better comparison between classical and quantal calculations will be obtained if we smooth both the graphs of $|\psi|^2$ (fig. 62a) and the curve projections (fig. 62c), on the scale of the average de Broglie wavelength. The smoothed curves are shown in fig. 62b (quantal) and fig. 62d (classical). Clearly the agreement is much better, suggesting a continuing association between ψ and Σ in some average sense if not in detail.

The breakdown of the detailed association between classical and quantum mechanics for this map can be followed very nicely in terms of the Wigner function defined by eq. (3.12). By employing this definition in conjunction with the quantum map equation (6.6), it is not hard [109] to obtain the following quantum map transforming Wigner functions $W_n(q,p)$ into their iterates $W_{n+1}(q,p)$:

$$W_{n+1}(q,p) = \int_{-\infty}^{\infty} dq' \int_{-\infty}^{\infty} dp' K(q,p;q',p') W_n(q,p), \qquad (6.9)$$

where for the potential $V = Aq^4/4$ the Wigner propagator K is

$$K(q,p;q',p') = \frac{2\delta(q'-q+p'T/\mu)}{\hbar^{2/3}(6AT|q|)^{1/3}} \mathrm{Ai}\left(\frac{-2\mathrm{sign}(q)(p'-p-TAq^3)}{\hbar^{2/3}(6AT|q|^{1/3}} \right),$$

$$(6.10)$$

Ai being the Airy function [72]. As $\hbar \to 0$, the Airy function turns into a delta function, and eq. (6.10) becomes the Liouville propagator transforming classical densities pointwise in accordance with the classical map M (2.25).

For the initial Wigner function, Korsch and Berry [112] chose a Gaussian centred on $q=p=0$, with contours in the form of concentric circles. The half-width of the Gaussian corresponds to a substantial fraction of W lying in the chaotic escaping region of fig. 26. Contours of the first three iterates

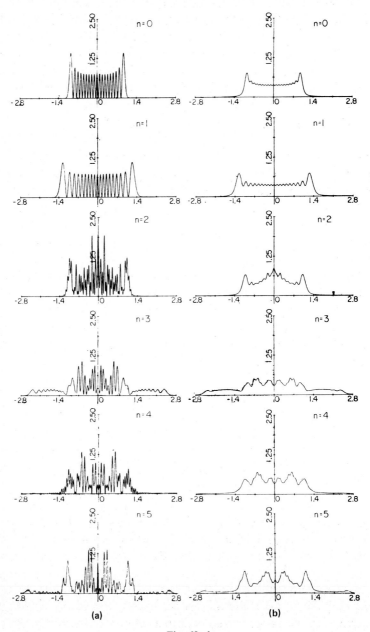

Fig. 62a,b.

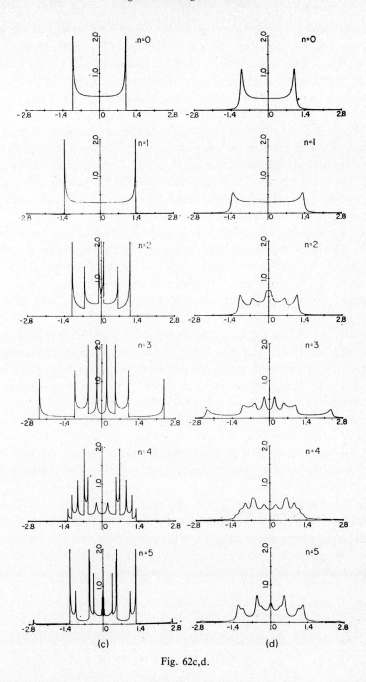

Fig. 62c,d.

W_1, W_2 and W_3 are shown in fig. 63 for values of $\hbar$ shown as the areas of squares on the right. The classical Wigner maps ($\hbar = 0$) are shown in fig. 63e. As $\hbar$ increases (figs. 63d,c,b,a), more and more of the classical complexity is smoothed away, and the extreme quantal functions (fig. 63a) show little resemblance to their classical counterparts (fig. 63e).

These classical curve maps and Wigner contour maps show strong resemblances to the morphologies developed when cream spreads in coffee, and indeed both have a common origin in generic measure-preserving maps. In quantum mechanics the development of complexity on infinitely fine scales is inhibited by the finite value of $\hbar$; in fluid mechanics the fine scale of turbulent mixing is determined by viscosity (see for example the vorticity contour maps computed by Zabusky [114]).

It is worth pointing out that these computations of the evolution of states under quantum maps show Planck's constant playing an unfamiliar role. Instead of adding quantum detail to a smooth underlying classical structure (e.g. the initial curve Σ), $\hbar$ acts in the opposite way when $t \rightarrow \infty$, to impose a quantum smoothing onto classical structures with infinitely fine detail (e.g. fully-developed whorls and tendrils).

To my knowledge there are no analytical results giving information about $\psi(q)$ for small $\hbar$ as $t \rightarrow \infty$, i.e. in the regime where the $\Sigma - \psi$ association (3.7) no longer holds (although it is plausible that ψ will resemble the irregular wave functions discussed in section 3.4). But by estimating the corrections to eq. (3.7) in the case of a classically chaotic system, Shepelyansky [111] (see also ref. [106]) concludes that the new regime begins at a transition time t_{max} of order $\hbar^{-1}$; for $t < t_{max}$, therefore, ψ should be associated with Σ, even though this curve can be very complicated and give rise to many momenta $p_i(q)$ giving contributions to ψ. Associated with these many contributions are many caustics (cf. fig. 62d for $n = 5$), near which eq. (3.7) fails and must be modified as explained in section 3,2 and references therein. Shepelyansky [111] also estimates that for chaotic systems these caustics, although numerous, are greatly outnumbered by non-degenerate contributions P_i for which eq. (3.7) is valid, so that caustics give an insignificant net contribution to ψ.

6.2. Stationary states

So far we have used quantum maps to study evolving states. Can they be used to study stationary states? Unitary operators have eigenvalues on the

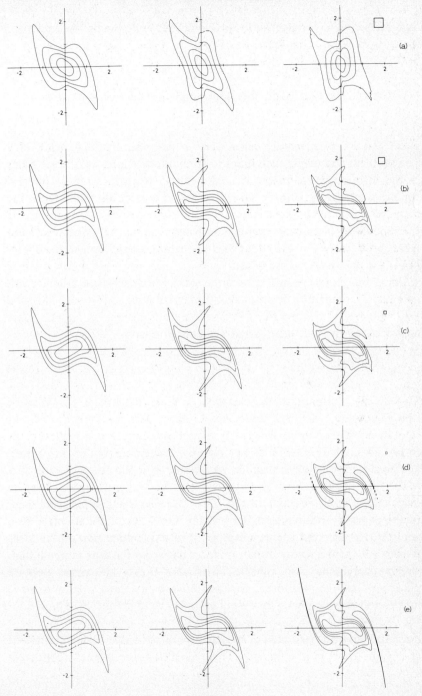

Fig. 63.

unit circle, and so the eigenstates $|\phi>$ must iterate to themselves under eq. (6.1), apart from a phase factor. Thus

$$\hat{U}|\phi> = e^{i\alpha}|\phi>. \tag{6.11}$$

For maps of the type (6.5), the "eigenangle" α may be written as

$$\alpha = \tilde{E}T/\hbar, \tag{6.12}$$

where $\tilde{E}$ is a "quasi-energy" eigenvalue (as described by Zeldovich [115]). The problem with using such maps to study the asymptotics of eigenstates is that little seems to be known about the spectra generated by integral equations of the type (6.6). Are they continuous or discrete? Does the nodal structure of the eigenfunctions (or their real and imaginary parts if they should be essentially complex) bear any relation to the spectrum? For what sort of potential $V(q)$ does eq. (6.6) exhibit bound states? We do not know the answers to these questions.

But there is another sort of quantum map whose spectra can be studied analytically, namely those obtained by quantizing maps M of the "Arnol'd cat" type [eq. (2.22) and fig. 24]. These are linear maps on a phase space consisting of the unit torus; their general form is

$$M: \begin{pmatrix} q_{n+1} \\ p_{n+1} \end{pmatrix} = \begin{pmatrix} T_{11} & T_{12} \\ T_{21} & T_{22} \end{pmatrix} \begin{pmatrix} q_n \\ p_n \end{pmatrix}_{\text{mod. 1}}. \tag{6.13}$$

For M to be continuous, the elements T_{ij} must be integers; for M to be area-preserving, $|\det T_{ij}|$ must equal unity; and if $|T_{11} + T_{22}| > 2$ the transformation of area elements is hyperbolic and M is ergodic on the torus. This is a rather artificial class of system (although an optical analogue is in principle realizable) whose interest in the present context lies in the fact that it is the only ergodic system for which the exact quantum mechanics is at all well-understood. The following treatment is a summary of the paper by Hannay and Berry [116], but I also draw attention to a study by Izraelev and Shepelyansky [117] of a quantum map on the torus arising not from a linear transformation but from the kicked pendulum.

Quantization has two aspects. The first is purely kinematic. Because phase space is the unit torus, coordinate wave functions $\psi(q)$ must have unit period in q, implying that momentum wave functions $\bar{\psi}(p)$ must consist of a series of delta functions at $p = nh$, where n is an integer. But $\bar{\psi}(p)$ must also have unit period in p, so that $\psi(q)$ must consist of delta functions at $q = mh$, where m is an integer. These conditions are mutually consistent only if the torus area is an integer multiple of Planck's constant, which in this case (unit torus) gives the strange condition

$$h = 1/N,$$ (6.14)

where N is an integer. Measurements of q and p can yield only values in the "quantum lattice"

$$q = Q/N, \ p = P/N \qquad (1 \le Q, P \le N),$$ (6.15)

with Q and P integers. The semiclassical limit is $N \to \infty$; as it is reached, h vanishes by discontinuous steps and q and p become continuous variables. For a technical reason [116], N will henceforth be restricted to be odd.

The second aspect of quantization is the introduction of dynamics, in the form of a unitary operator $\hat{U}$ corresponding to M. In the discrete representation (6.15), $\hat{U}$ propagates states $\psi(Q)$ according to the analogue of eq. (6.6), namely

$$\psi_{n+1}(Q) = \sum_{Q'=1}^{N} U_{QQ'} \psi_n(Q').$$ (6.16)

$\hat{U}$ is constructed by "periodizing" the (simple) unitary operator corresponding to M acting as a map in the original "untorized" infinite q, p plane. It turns out that $\hat{U}$ preserves periodicity in both $\psi(Q)$ and $\bar{\psi}(P)$ only if the elements T_{ij} in eq. (6.13) have the form

$$M: \begin{pmatrix} \text{even} & \text{odd} \\ \text{odd} & \text{even} \end{pmatrix} \text{ or } \begin{pmatrix} \text{odd} & \text{even} \\ \text{even} & \text{odd} \end{pmatrix}.$$ (6.17)

This excludes the familiar cat map (2.22) but includes for example, the simple ergodic map with matrix

$$M: \begin{pmatrix} 2 & 1 \\ 3 & 2 \end{pmatrix},$$ (6.18)

The explicit construction of $\hat{U}$ for an arbitrary matrix in the "permitted" class (6.17) is an intricate process [116] involving the Gauss sums of number theory, but for the particular case (6.18) the evolution operator takes the simple form

$$U_{QQ'} = \frac{1}{N^{1/2}} \exp\left[\frac{-i\pi}{4} + \frac{2\pi i}{N} \left(Q^2 + Q'^2 - QQ' \right) \right].$$ (6.19)

According to eq. (6.11), eigenstates of $\hat{U}$ map into themselves under eq. (6.16) apart from phase factors $\exp(i\alpha)$, given by

$$\det_{QQ'}(U_{QQ'} - e^{i\alpha}\delta_{QQ'}) = 0. \tag{6.20}$$

Because $U_{QQ'}$ is an $N \times N$ matrix, there are N eigenvalues $\alpha_j (1 \le j \le N)$, and it is interesting to ask how they are distributed round the unit circle, especially in the semiclassical limit $N \to \infty$.

To answer this question, consider the effect of the *classical* map M on points in the *quantum* lattice (6.15). Being rational, these points map around closed orbits ("cycles") in contrast to generic points on the torus, which have irrational coordinates and so never return to their starting points. For each N, there will be a number $n(N)$ of iterations after which every rational point with denominator N will have completed at least one cycle. $n(N)$ is the *period* of the map, defined as the smallest number satisfying

$$\begin{pmatrix} T_{11} & T_{12} \\ T_{21} & T_{22} \end{pmatrix}^{n(N)} = \begin{pmatrix} 1 & 0 \\ 0 & 1 \end{pmatrix}_{\mathrm{mod},\, N}, \tag{6.21}$$

and given by the lowest common multiple of the lengths of cycles of points in the quantum lattice.

Now, for these linear maps the corresponding Hamiltonians are quadratic. For example, eq. (6.18) corresponds to

$$H = \frac{\operatorname{ar sinh} 3^{1/2}}{2} \left(\frac{p^2}{3^{1/2}} - q^2 \cdot 3^{1/2} \right), \tag{6.22}$$

acting for unit time. It follows from this that certain quantum-mechanical quantities (e.g. Wigner's function) evolve classically. Therefore after $n(N)$ iterations of the quantum map, wave functions will have returned to their original values, apart from a possible phase factor, i.e.

$$\hat{U}^{n(N)} = \hat{I}\, \exp[i\sigma(N)], \tag{6.23}$$

where σ is in general unknown. Thus the eigenvalues α must be multiples of $2\pi/n(N)$, apart from a shift, i.e.

$$\alpha_j = \frac{2\pi m_j}{n(N)} + \frac{\sigma(N)}{n(N)} \quad \begin{pmatrix} 1 \le j \le N \\ 1 \le m_j \le n(N) \end{pmatrix}. \tag{6.24}$$

The spectrum of $\hat{U}$ therefore consists of N eigenangles distributed among $n(N)$ possible eigenlevels. $n(N)$ is an extremely erratic function, defined number-theoretically by eq. (6.21). Sometimes $n < N$, in which case some levels must be multiply occupied, and sometimes $n > N$, in which case some levels must be empty. In the map (6.18), for instance,

$$n(1) = 1, \quad n(3) = 6, \quad n(5) = 3, \quad n(7) = 8, \quad n(9) = 18, \quad n(11) = 10. \qquad (6.25)$$

Number-theoretic arguments, together with numerical experiments, suggest that in a suitably-defined asymptotic sense

$$n(N) \sim CN \qquad \text{as } N \to \infty. \qquad (6.26)$$

where C is constant, but the limit is approached very slowly and erratically. Whatever the value of $n(N)$, the angles seem to be fairly uniformly distributed over their possible sites, behaviour reminiscent of the level repulsion discussed in section 4.3 for generic systems of more conventional type. Computations also show that the Wigner functions for individual eigenstates spread all over the quantum lattice in accordance with the semiclassical eigenfunction hypothesis (section 3) applied to these ergodic systems, and do not concentrate about individual cycles (closed orbits).

The erratic behaviour of $n(N)$ as $N \to \infty$ shows that the semiclassical mechanics of this chaotic system is very different from that of an integrable system whose eigenvalues are given by torus quantization (4.1) as smooth functions of $\hbar$. In fact decreasing $\hbar$, and hence increasing N, causes the spectrum to depend on the iteration of points in an ever-finer quantum lattice, i.e. on an increasingly intricate cycle structure. Here we again see $\hbar$ playing the same role as in the evolution of nonstationary states over long times, namely obscuring an underlying classical structure which has infinite complexity.

Acknowledgments

I thank the Institute for Theoretical Physics in the University of Utrecht for hospitality whilst these lectures were written. The work was not supported by any military agency.

References

[1] I.C. Percival, Adv. Chem. Phys. 36 (1977) 1–61.
[2] J.J. Duistermaat, Commun. Pure. Appl. Math. 27 (1974) 207–281.

[3] M.V. Berry and K.E. Mount, Rep. Prog. Phys. 35 (1972) 315–397.

[4] J. Ford, in: Fundamental Problems in Statistical Mechanics, ed., E.G.D. Cohen, vol. III (North-Holland, Amsterdam, 1975) pp. 215–255.

[5] V.I. Arnol'd, Mathematical Methods of Classical Dynamics (Springer, New York, 1978).

[6] M.V. Berry, in: Topics in Nonlinear Dynamics, ed., S. Jorna, Am. Inst. Conf. Proc. 46 (1978) 16–120.

[7] R. Helleman, in: Fundamental Problems in Statistical Mechanics, ed., E.G.D. Cohen, vol. V (North-Holland, Amsterdam, 1980) pp. 165–233.

[8] J.L. Lebowitz and O. Penrose, Physics Today 26 (February 1973) 23–29.

[9] M.V. Berry, Eur. J. Phys. 2 (1981) 91–102.

[10] Ya. G. Sinai, Russ. Math. Surv. 25 (1970) 137–189.

[11] L.A. Bunimovich, Funct. Anal. Appl. 8 (1974) 254–255.

[12] L.A. Bunimovich, Commun. Math. Phys. 65 (1979) 295–312.

[13] M.C. Gutzwiller, J. Math. Phys. 14 (1973) 139–152; 18 (1977) 806–823.

[14] A.N. Kolmogorov, Dokl. Akad. Nauk 98 (1954) 527–530.

[15] V.I. Arnol'd, Russ. Math. Surv. 18, No. 5 (1963) 13–39; No. 6, 61–196.

[16] J. Moser, Nachr. Akad. Wiss. Göttingen 1 (1962) 1–20.

[17] A.M. Ozorio de Almeida and J.H. Hannay, Ann. Phys. (N.Y.) 138 (1982) 115–154.

[18] M.V. Berry, Singularities in Waves and Rays, Lectures at Les Houches 1980, Session XXXV, Physics of Defects, eds., R. Balian, M. Kléman, J.P. Poirier (North-Holland, Amsterdam, 1981) p. 453.

[19] V.F. Lazutkin, Math. Izv. USSR. 37 (1973) 186–216.

[20] G. Benettin and J.M. Strelcyn, Phys. Rev. A17 (1978) 773–785.

[21] P.J. Richens and M.V. Berry, Physica 2D (1981) 495–512.

[22] A.N. Zemlyakov and A.B. Katok, Math. Notes (USSR) 18 (1975) Nos. 1–2, 760–764.

[23] D.W. Noid and R.A. Marcus, J. Chem. Phys. 67 (1977) 559–567.

[24] G. Casati and J. Ford, J. Comp. Phys. 20 (1976) 97–109.

[25] B.B. Mandelbrot, in: Nonlinear Dynamics, ed., R.H.G. Helleman, Ann. N.Y. Acad. Sci. 357 (1980) 249–259.

[26] V.P. Maslov and M.V. Fedoriuk, Semiclassical Approximation in Quantum Mechanics (Reidel, Dordrecht, 1981). (original Russian edition 1965).

[27] J.H. van Vleck, Proc. Natl. Acad. Sci. USA 14 (1928) 178–188.

[28] J.B. Keller, Ann. Phys. N.Y. 4 (1958) 180–188.

[29] Yu. A. Kravtsov, Sov. Phys. Acoust. 14 (1968) 1–17.

[30] V. Guillemin and S. Sternberg, Geometric Asymptotics, Am. Math. Soc. Surv. No. 14 (Providence, USA, 1977).

[31] J.P. Eckmann and R. Sénéor, Arch. Rat. Mech. Anal. 61 (1976) 153–173.

[32] P.A.M. Dirac, The Principles of Quantum Mechanics (Clarendon Press, Oxford, 1947).

[33] M.V. Berry and C. Upstill, Prog. Opt. 18 (1980) 257–346.

[34] J. Hayden and E.C. Zeeman, Math. Proc. Camb. Phil. Soc. 89 (1981) 193–200.

[35] V.I. Arnol'd, Funct. Anal. Appl. 1 (part 1 1967) 1–13.

[36] A. Einstein, Ver. Deut. Phys. Ges. 19 (1917) 82–92.

[37] I.C. Percival, J. Phys. B6 (1973) L229–232.

[38] E.P. Wigner, Phys. Rev. 40 (1932) 749–759.

[39] H.J. Groenewold, Physica 12 (1946) 405–460.

[40] J.E. Moyal, Proc. Camb. Phil. Soc. 45 (1949) 99–124.

[41] T. Takabayasi, Proc. Theor. Phys. Jap. 11 (1954) 341–373.

[42] G.A. Baker, Jr. Phys. Rev. 109 (1958) 2198–2206.

[43] K.S.J. Nordholm and S.A. Rice, J. Chem. Phys. 61 (1974) 203–223, 768–779.

[44] A. Voros, Ann. Inst. H. Poincaré XXIV (1976) 31–90.

[45] M.V. Berry, Phil. Trans. Roy. Soc. Lond. 287 (1977) 237–271.

[46] M.V. Berry and N.L. Balázs, J. Phys. A:12 (1979) 625–642.

[47] E.J. Heller, J. Chem. Phys. 65 (1976) 1289–1298; 67 (1977) 3339–3351.

[48] N.L. Balázs and G.G. Zipfel, Jr. Ann. Phys. N.Y. 77 (1973) 139–156.

[49] H.J. Korsch, J. Phys. A:12 (1979) 811–823.

[50] A.M. Ozorio de Almeida, Physica 110 A (1982) 501–517.

[51] M.V. Berry and F.J. Wright, J. Phys. A:13 (1980) 149–160.

[52] M.V. Berry, J. Phys. A:10 (1977) 2083–2091.

[53] A. Voros, in: Stochastic Behaviour in Classical and Quantum Hamiltonian Systems, eds., G. Casati, J. Ford, Lecture notes in Physics 93 (Springer, Berlin, 1979) pp. 326–333.

[54] J.S. Hutchinson and R.E. Wyatt, Chem. Phys. Lett. 72 (1980) 378–384.

[55] A.I. Shnirelman, Usp. Mat. Nauk 29, No. 6 (1974) 181–182.

[56] M.V. Berry, J. Phys. A:10 (1977) 2061–2081.

[57] S.W. McDonald and A.N. Kaufman, Phys. Rev. Lett. 42 (1979) 1189–1191.

[58] R.M. Stratt, N.C. and W.H. Miller, J. Chem. Phys. 71 (1979 3311–3322.

[59] K. Uhlenbeck, Am. J. Math. 98 (1976) 1059–1078.

[60] P. Pechukas, J. Chem. Phys. 57 (1972) 5577–5594.

[61] R. Courant and D. Hilbert, Methods of Mathematical Physics, vol. I (Interscience, New York, 1953) pp. 451–456.

[62] N. Fröman and P.O. Fröman, JWKB approximation (North-Holland, Amsterdam, 1965).

[63] W. Eastes and R.A. Marcus, J. Chem. Phys. 61 (1974) 4301–4306.

[64] D.W. Noid and R.A. Marcus, J. Chem. Phys. 62 (1975) 2119–2124.

[65] D.W. Noid, M.L. Koszykowski and R.A. Marcus 71 (1979) 2864–2873.

[66] I.C. Percival and N. Pomphrey, Mol. Phys. 31 (1976) 97–114.

[67] C. Jaffé and W.P. Reinhardt, J. Chem. Phys. 71 (1979) 1862–1869.

[68] S. Chapman, B. Garrett and W.H. Miller, J. Chem. Phys. 64 (1976) 502–509.

[69] M.V. Berry and M. Wilkinson, submitted to Proc. Phys. Soc. Lond. (1983).

[70] M.V. Berry, Ann. Phys. N.Y. 131 (1981) 163–216.

[71] J.M. Ziman, Solid State Phys. 26 (1971) 1–101.

[72] M. Abramowitz and I.A. Stegun, Handbook of Mathematical Functions (Washington, U.S. Nat. Bur. Standards, 1964).

[73] J. von Neumann and E.P. Wigner, Phys. Z. 30 (1929) 467–470.

[74] E. Teller, J. Phys. Chem. 41 (1937) 109–116.

[75] H.C. Longuet-Higgins, Proc. Roy. Soc. A344 (1975) 147–156.

[76] M. Pinsky, SIAM J. Math. Anal. 11 (1980) 819–827.

[77] R.A. Marcus, in: Nonlinear Dynamics, ed., R.H.G. Helleman, Ann. N.Y. Acad. Sci. 357 (1980) pp. 169–182.

[78] C.F. Porter, ed., Statistical Theory of Spectra: Fluctuations (Academic Press, New York, 1965).

[79] G. Casati, F. Valz-Gris and I. Guarneri, Nuovo Cimento Lett. 28 (1980) 279–182.

[80] G.M. Zaslavskii, Sov. Phys. JETP 46 (1977) 1094–1098.

[81] M.V. Berry and M. Tabor, Proc. Roy. Soc. A356 (1977) 375–394.

[82] V.L. Pokrovskii, JETP Lett. 4 (1966) 96–99.

[83] S.A. Molcanov, Commn. Math. Phys. 78 (1981) 429–446.

[84] M.C. Gutzwiller, J. Math. Phys. 8 (1967) 1979–2000.

[85] M.C. Gutzwiller, J. Math. Phys. 10 (1969) 1004–1020.

[86] M.C. Gutzwiller, J. Math. Phys. 11 (1970) 1791–1806.

[87] M.C. Gutzwiller, J. Math. Phys. 12 (1971) 343–358.

[88] M.C. Gutzwiller, in: Path Integrals and Their Application in Quantum Statistical and Solid State Physics, eds., G.J. Papadopoulos, J.T. Devreese (Plenum, New York, 1978) pp. 163–200.

[89] R. Balian and C. Bloch, Ann. Phys. N.Y. 60 (1970) 401–447.

[90] R. Balian and C. Bloch, Ann. Phys. N.Y. 64 (1971) 271–307.

[91] R. Balian and C. Bloch, Ann. Phys. N.Y. 69 (1972) 76–160.

[92] R. Balian and C. Bloch, Ann. Phys. N.Y. 85 (1974) 514–545.

[93] L.D. Landau and E.M. Lifshitz, Quantum Mechanics (Nonrelativistic Theory) Pergamon, Oxford, 1965).

[94] H.P. Baltes and E.R. Hilf, Spectra of Finite Systems (B – I Wissenschaftsverlag, Mannheim, 1978).

[95] M.V. Berry and M. Tabor, Proc. Roy. Soc. A349 (1976) 101–123.

[96] M.V. Berry and M. Tabor, J. Phys. A:10 (1977) 371–379.

[97] R. Rajaraman, Phys. Rep. 21 (1975) 227–313.

[98] C. DeWitt-Morette, A. Maheshwari and B. Nelson, Phys. Rep. 50 (1979) 255–372.

[99] W.H. Miller, J. Chem. Phys. 63 (1975) 996–999.

[100] V.I. Arnol'd and A. Avez, Ergodic Problems of Classical Mechanics (Benjamin, Reading, MA, 1968).

[101] M.C. Gutzwiller, Phys. Rev. Lett. 45 (1980) 150–153.

[102] J. Chazarin, Inv. Math. 24 (1974) 65–82.

[103] H.P. McKean, Comm. Pure Appl. Math, 25 (1972) 225–246.

[104] D. Hejhal, The Selberg Trace Formula for PSL(2,R), Lecture Notes in Mathematics, Vol. 548 (Springer, New York, 1976).

[105] D.S. Ornstein, in Proc. 6th IUPAP conf on Stat. Mech., eds., S.A. Rice, K.F. Freed, J.C. Light (University of Chicago Press, 1972).

[106] B.V. Chirikov, F.M. Izraelev and D.L. Shepelyansky, Preprint from Institute of Nuclear Physics, Novosibirsk.

[107] J.S. Hutchinson and R.E. Wyatt, Phys. Rev. A23 (1981) 1567–1584.

[108] G. Casati, B.V. Chirikov, J. Ford and F.M. Izraelev, in: Stochastic Behaviour in Classical and Quantum Hamiltonian Systems, eds., G. Casati, J. Ford, Lecture Notes in Physics 93 (Springer, Berlin, 1979) pp. 334–352.

[109] M.V. Berry, N.L. Balázs, M. Tabor and A. Voros, Ann. Phys. N.Y. 122 (1979) 26–63.

[110] M.V. Berry, in: Nonlinear Dynamics, ed., R.H.G. Helleman, Ann. N.Y. Acad. Sci. 357 (1980) 183–203.

[111] D.L. Shepelyansky, Sov. Phys. Dokl. 26 (1981) 80–82.

[112] H.J. Korsch and M.V. Berry, Physica 3D (1981) 627–636.

[113] J.B. Keller and S.I. Rubinow, Ann. Phys. N.Y. 9 (1960) 24–75.

[114] N.J. Zabusky, in: Proc. Orbis Scientiae (Miami) on the Significance of Nonlinearity in the Natural Sciences, eds., A. Perlmutter, L.F. Scott (Plenum, New York, 1977) pp. 145–205.

[115] Ya.B. Zel'dovich, Sov. Phys. JETP 24 (1967) 1006–1008.

[116] J.H. Hannay and M.V. Berry, Physica 1D (1980) 267–290.

[117] F.M. Izraelev and D.L. Shepelyansky, Sov. Phys. Dokl. 24 (1979) 996–8.

[118] G.M. Zaslavsky, Phys. Rep. 80 (1981) 157–250.

[119] A.M. Ozorio de Almeida, Ann. Phys. (N.Y.) (1983) in press.

[120] M.V. Berry, J.H. Hannay and A.M. Ozorio de Almeida, Physica D (1983) in press.

COURSE 4

BIFURCATIONS DE DIFFEOMORPHISMES DE $\mathbb{R}^2$ AU VOISINAGE D'UN POINT FIXE ELLIPTIQUE

A. CHENCINER

*Departement de Mathématiques, Université Paris VII,
2, place Jussieu, 75251 Paris Cedex 05, France*

G. Iooss, R.H.G. Helleman and R. Stora, eds.
*Les Houches, Session XXXVI, 1981 — Comportement Chaotique des Systèmes Déterministes/
Chaotic Behaviour of Deterministic Systems*
© *North-Holland Publishing Company, 1983*

Table des matières

Note liminaire. Le texte qui suit ne contient pas de démonstration précise; il doit être considéré comme l'exposition d'un courant d'idées et un guide pour la littérature (un peu trop) abondante sur le sujet.

Par rapport aux approches existantes la nouveauté consiste en l'exhibition dès la codimension deux de phénomènes du type conservatif (déployés dans l'espace des paramètres) par opposition aux phénomènes dissipatifs étudiés dans la théorie classique des bifurcations.

Melampyrum silvaticum

0. Introduction: Equations différentielles et diffeomorphismes; exemples autour de Van der Pol et de Liénard

> *Once in possession of the general theory two roads lay open before us. We could follow Poincaré, Levi-Civita, Birkhoff, and study "two degrees of freedom" and the extensive doctrine centering around the three-body problem. The second and more modest road which we have selected led to non-linear differential equations of the second order: the group of problems stirred up a generation ago by Van der Pol.*
>
> Solomon Lefschetz 1957

0.1. Equations de Van der Pol et de Liénard

Le phénomène des oscillations non-linéaires a été particulièrement étudié sur les équations du type

$$\frac{d^2x}{dt^2} + f(x)\frac{dx}{dt} + x = 0. \tag{1}$$

En posant

$$y = \frac{dx}{dt} + F(x), \qquad F(x) = \int_0^x f(\zeta)d\zeta,$$

on obtient le système (champ de vecteurs,...) equivalent

$$dx/dt = y - F(x),$$
$$dy/dt = -x. \tag{2}$$

Soit

$$u(x, y) = x^2/2 + y^2/2;$$

le long d'une solution (orbite, courbe intégrale, trajectoire,...) du système (2) on a a $du/dt = -xF(x)$; u est donc une fonction de Liapunov (décroissante sur les orbites) dès que $xF(x) > 0$ pour $x \neq 0$, ce qui force alors l'ori-

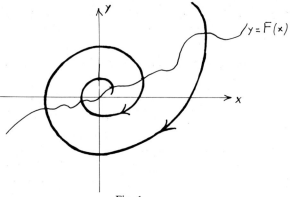

Fig. 1.

gine à être un attracteur global (fig. 1). Lorsque $xF(x)$ n'est pas de signe constant pour $x \neq 0$ des orbites périodiques peuvent apparaître [puisque (0,0) est la seule singularité (= position d'équilibre = point singulier) du système (2) il s'agit forcément de courbes fermées entourant l'origine]; l'archétype de cette situation est l'équation de Van der Pol $[f(x) = \mu(x^2 - 1), \mu > 0]$ dont l'étude et l'origine, qui remontent aux années 20, sont résumées respectivement sur les figs. 2 et 3 (on remarquera que $xF(x) < 0$ pour $x \neq 0$ assez petit: l'origine est donc un répulseur). Un bon article de revue, contenant une abondante bibliographie, est Van der Pol (1934). Pour un bon exposé mathématique, voir Lefschetz (1977) chapitre 11.

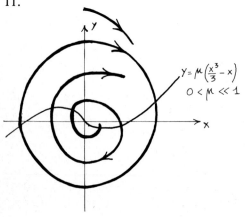

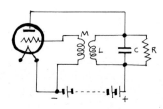

Fig. 2.   Fig. 3.

Remarque. Les figs. 2 et 3 correspondent au cas où $\mu \ll 1$; le cas $\mu \gg 1$ est assez différent phénoménologiquement, mais pas topologiquement.

Plusieurs techniques sont disponibles pour prouver l'existence d'orbites périodiques d'un champ de vecteurs dans le plan; dans le cas de l'équation de Van der Pol il suffit, grâce à la symétrie par rapport à l'origine du champ de vecteurs, d'étudier l'application de "demi-premier retour" qui à un point A situé sur le demi-axe Oy positif fait correspondre le premier point C où l'orbite issue de A coupe le demi-axe Oy négatif (fig. 2): une orbite périodique correspond à $OA = -OC$; un argument de monotonie montre alors l'existence et l'unicité de l'orbite périodique, ainsi que son caractère attractant.

Plus généralement, on peut chercher des anneaux dans lesquels rentre (ou sort) le champ de vecteurs; s'il n'y a pas de singularité dans un tel anneau, le théorème de Poincaré–Bendixson [voir Lefschetz (1977) ou Hirsch and Smale (1974)] fournit une orbite périodique, mais l'unicité reste un problème difficile (voir fig. 4). On trouvera dans Lins et al. (1977) une étude du nombre d'orbites périodiques du système (2) en liaison avec le degré du polynôme $F(x)$.

Enfin, et c'est notre thème principal, on peut chercher à voir apparaître de petites orbites périodiques dans une famille à plusieurs paramètres d'équations différentielles: de global le problème devient local et donc justiciable de calculs de perturbation, ce qui ne peut que plaire à des physiciens.

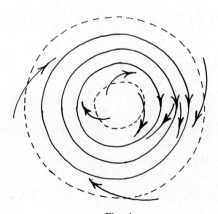

Fig. 4.

Dans le paragraphe suivant nous indiquons l'esprit de cette approche dans le cas de l'équation de Van der Pol avec μ petit.

0.2. Bifurcations de familles à 1 paramètre

Pour $\mu > 0$ le changement de variables

$$X = x\mu^{1/2},$$
$$Y = y\mu^{1/2},$$

$$(3)$$

transforme le système (2) [où $F(x) = \mu(\tfrac{1}{3}x^3 - x)$] en

$$dX/dt = Y - (\tfrac{1}{3}x^3 - \mu X),$$
$$dY/dt = -X.$$

$$(4)$$

Nous allons étudier (4) pour μ voisin de 0 en oubliant la restriction $\mu > 0$. Le système (4) s'écrit encore

$$\frac{d}{dt}\begin{pmatrix} X \\ Y \end{pmatrix} = \begin{pmatrix} \mu & 1 \\ -1 & 0 \end{pmatrix}\begin{pmatrix} X \\ Y \end{pmatrix} + \begin{pmatrix} -X^3/3 \\ 0 \end{pmatrix}.$$

Les valeurs propres de

$$A_\mu = \begin{pmatrix} \mu & 1 \\ -1 & 0 \end{pmatrix}$$

sont (pour $|\mu| < 2$)

$$\lambda_\mu = \frac{\mu}{2} + i\left(1 - \frac{\mu^2}{4}\right)^{1/2}, \qquad \bar{\lambda}_\mu = \frac{\mu}{2} - i\left(1 - \frac{\mu^2}{4}\right)^{1/2};$$

des vecteurs propres correspondants sont, par exemple,

$$e_\mu = \begin{pmatrix} \lambda_\mu \\ -1 \end{pmatrix}, \qquad \text{et } \bar{e}_\mu = \begin{pmatrix} \bar{\lambda}_\mu \\ -1 \end{pmatrix}.$$

Dans la base $\{e_\mu, \bar{e}_\mu\}$ les coordonnées d'un vecteur réel sont de la forme $(\xi, \eta = \bar{\xi})$; un tel vecteur est donc repéré par le seul nombre complexe ξ; dans cette identification de $\mathbb{R}^2$ à $\mathbb{C}$, l'éq. (4) s'écrit:

$$\frac{d\xi}{dt} = \lambda_\mu \xi - \frac{1}{2i(1 - \tfrac{1}{4}\mu^2)^{1/2}} \frac{(\lambda_\mu \xi + \bar{\lambda}_\mu \bar{\xi})^3}{3};$$

$$(5)$$

le changement de variables

$$z = \lambda_\mu \xi \in \mathbb{C} \tag{6}$$

transforme (5) en

$$\frac{dz}{dt} = \lambda_\mu z - \frac{\lambda_\mu}{2i(1 - \frac{1}{4}\mu^2)^{1/2}} \frac{(z + \bar{z})^3}{3}. \tag{7}$$

Si on remarque, comme l'a fait Poincaré dans sa thèse, qu'un changement de variables de la forme

$$\zeta = z + \sum_{i+j=3} \gamma_{ij} z^i \bar{z}^j \qquad (\gamma_{ij} \in \mathbb{C}) \tag{8}$$

transforme l'équation

$$\frac{dz}{dt} = \lambda z + \sum_{i+j=3} a_{ij} z^i \bar{z}^j$$

en

$$\frac{d\zeta}{dt} = \lambda \zeta + \sum_{i+j=3} [(i\lambda + j\bar{\lambda} - \lambda)\gamma_{ij} + a_{ij}]\zeta^i \bar{\zeta}^j + O(|\zeta|^4),$$

on peut ramener (7) à l'équation

$$\frac{d\zeta}{dt} = \lambda_\mu \zeta - \frac{\lambda_\mu}{2i(1 - \frac{1}{4}\mu^2)^{1/2}} \zeta |\zeta|^2 + O(|\zeta|^4). \tag{9}$$

Bien entendu, si $\mu \neq 0$, $\lambda_\mu + \bar{\lambda}_\mu \neq 0$ et on peut supprimer tous les termes d'ordre 3 dans (7); pour $\mu = 0$ ce n'est plus possible à cause de la "résonance" $2\lambda_0 + \bar{\lambda}_0 = \lambda_0$; si on veut des changements de coordonnées dépendant gentiment de μ, il faut conserver les termes en $z|z|^2$ pour tous les μ proches de 0 (voir fig. 26). Une excellente référence est le chapitre 5 de Arnold (1980): nous verrons en détail le cas des difféomorphismes un peu plus loin.

L'éq. (9) est facile à étudier si l'on oublie les termes $O(|\zeta|^4)$; elle a en effet la propriété particulièrement agréable de laisser invariant le "feuilletage" des cercles centrés à l'origine: l'image d'un tel cercle au bout du temps t est encore un cercle analogue (fig. 5). En particulier, une orbite périodique ne peut être qu'un cercle centré à l'origine et la recherche d'une telle orbite se ramène à la recherche des réels $r > 0$ tels que

$$\lambda_\mu \left(1 - \frac{r^2}{2i(1 - \frac{1}{4}\mu^2)^{1/2}} \right)$$

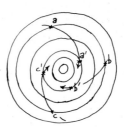

Fig. 5. Si $\zeta(0) = a, b, c \ldots$, $\zeta(t) = a', b', c' \ldots$.

soit purement imaginaire (en effet, si $|\zeta| = r$, $d\zeta/dt$ est alors orthogonal à ζ).

La condition s'écrit encore $r = \mu^{1/2}$ ce qui prouve l'existence pour $\mu > 0$, d'une orbite périodique (attractante) pour l'éq. (9) dans laquelle on oublie le terme $O(|\zeta|^4)$.

La fig. 6 représente le "portrait de phase" de cette équation pour les diverses valeurs de μ, dans l'espace produit $\mathbb{R} \times \mathbb{C}$ (coordonnées μ, ζ): cela signifie qu'on trace le portrait de phase de l'équation

$$\frac{d\zeta}{dt} = \lambda_\mu \zeta - \frac{\lambda_\mu}{2i(1 - \frac{1}{4}\mu^2)^{1/2}} \zeta \, |\zeta|^2,$$

$$d\mu/dt = 0. \tag{10}$$

Il s'agit de ce qui est classiquement appelé (au moins en Occident) une

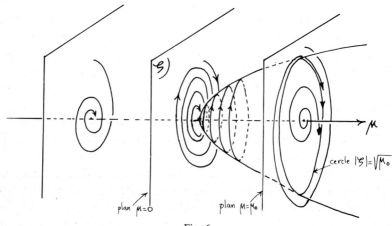

Fig. 6.

"bifurcation de Hopf" supercritique (voir Arnold (1980) chapitre 6 §.33, en particulier le haut de la page 259 où Arnold préfère l'appeler bifurcation de Poincaré-Andronov).

Lorsque $\mu > 0$ est assez proche de 0, on peut montrer (nous indiquerons comment à propos des bifurcations de Hopf des difféomorphismes, plus générales) que les termes en $O(|\zeta|^4)$ sont suffisamment petits par rapport à l'attraction de l'orbite périodique de l'équation tronquée pour que cette orbite périodique subsiste, légèrement perturbée, pour l'éq. (9), et donc pour l'équation de Van der Pol.

Remarques.

(1) Contrairement aux apparences, l'éq. (4) (ou plutôt la famille d'éqs. (4) paramétrée par μ) présente le phénomène de "bifurcation" d'une orbite périodique à partir d'une position d'équilibre dans $\mathbb{R}^2$ dans toute sa généralité (il faudrait dire "généricité"): tout d'abord, pour la valeur critique $\mu = 0$, le spectre de l'équation linéarisée rencontre l'axe imaginaire ($\lambda_0 = i$); s'il n'en était pas ainsi, l'origine serait un attracteur stable (deux valeurs propres de partie réelle négative) ou un répulseur stable (deux valeurs propres de partie réelle positive) ou un col stable (deux valeurs propres réelles non nulles de signes différents) et il ne se passerait rien de nouveau au voisinage de $\zeta = 0$ pour $\mu \neq 0$ assez petit.

Si une des valeurs propres est égale à 0 pour $\mu = 0$, une petite perturbation fait en général disparaître le point singulier et aucune orbite périodique n'apparait [voir Arnold (1980) chapitre 6 §.32]. Il reste le cas où le spectre de l'équation linéarisée est, pour $\mu = 0$, sur l'axe imaginaire (et non nul); puisqu'aucune valeur propre n'est nulle pour $\mu = 0$ une application immédiate du théorème des fonctions implicites montre qu'une petite perturbation d'une telle équation possède encore un point singulier proche de 0; par un petit changement de coordonnées on peut supposer que ce point singulier est 0: le fait que pour tout μ l'éq. (4) ait un point singulier en 0 n'est donc pas une restriction.

Enfin, dans une famille à 1 paramètre générale, le spectre de l'équation linéarisée au point singulier traverse l'axe imaginaire lorsque μ traverse la valeur critique $\mu = 0$, ce qui est bien réalisé dans l'éq. (4).

Sous ces seules conditions un théorème de nature *purement topologique*, dû initialement à Alexander et Yorke (voir Lemaire (1977) pour un exposé qui présente l'intérêt d'apprendre à ceux qui ne le savent pas ce qu'est un groupe d'homotopie et le J. homomorphisme) affirme l'existence *dans l'es-*

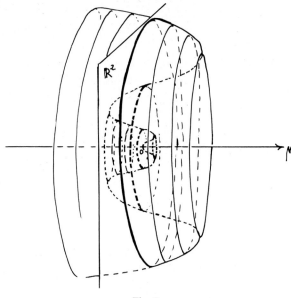

Fig. 7.

pace produit $\mathbb{R} \times \mathbb{R}^2$ d'une famille à 1 paramètre d'orbites périodiques issues de $(\mu, X, Y) = (0,0,0)$ (il s'agit d'un disque plongé dans $\mathbb{R} \times \mathbb{R}^2$, dont les intersections avec les plans $\mu = \mu_0$ petit sont des orbites périodiques de l'équation correspondante: voir fig. 7). "Génériquement", les termes non linéaires font de l'origine un attracteur ou un répulseur (faible) pour $\mu = 0$. La famille d'orbites périodiques se trouve alors du côté $\mu > 0$ ou du côté $\mu < 0$ puisqu'elle ne peut intersecter le plan $\mu = 0$ (au moins au voisinage de l'origine).

C'est ce qui se passe pour la famille (4) (pour $\mu = 0$ l'origine est un attracteur faible) mais pas pour la famille de Van der Pol, qui au voisinage de O ne se distingue pas qualitativement d'une famille d'équations linéaires (la bifurcation est alors "dégénérée", toutes les orbites périodiques voisines de l'origine apparaissant pour la seule valeur $\mu = 0$ ainsi que le montre la fig. 8).

Le mérite de notre calcul approché n'est donc pas tellement de prouver l'existence d'orbites périodiques pour μ petit, mais de les localiser assez précisément (en particulier pour l'équation de Van der Pol, on constate en suivant dans l'ordre inverse les changements de variable, que le cycle obtenu pour μ petit est voisin du cercle $x^2 + y^2 = 4$; pour une démonstration rapide utili-

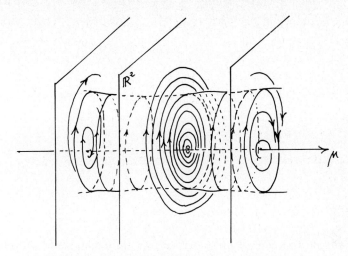

Fig. 8. L'équation de Van der Pol pour μ voisin de 0:

$$dx/dt = y - \mu(\tfrac{1}{3}x^3 - x)$$

$$dy/dt = -x.$$

sant la méthode de moyennisation, voir Arnold (1980) chapitre 4, §.17.D).

(2) Il existe beaucoup de démonstrations analytiques du théorème de bifurcation que nous venons d'évoquer; pour une méthode utilisant l'application de premier retour de Poincaré au voisinage de O, voir Ruelle and Takens (1971) §6.

(3) Les théorèmes de bifurcation sont des théorèmes locaux: dès que μ grandit, on ne sait rien dire par ces méthodes de la localisation (ou du destin) de l'orbite périodique ainsi dénichée. Un bel exemple de cette affirmation se trouve dans Benoit et al. (1980) à propos de l'équation

$$\frac{d^2 x}{dt^2} + \mu(x^2 - 1)\frac{dx}{dt} + \mu(x - a) = 0, \tag{11}$$

a voisin de 1, μ très grand,

que les auteurs étudient (sous l'impulsion de G. Reeb) par des méthodes d'analyse non standard (μ infiniment grand).

On y relève en particulier les dessins suivants (fig. 9).

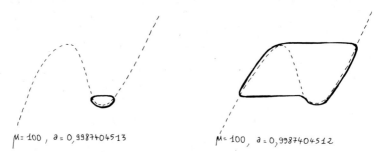

M = 100, a = 0,9987404513   M = 100, a = 0,9987404512

Fig. 9. Sur la partie gauche de la figure, une orbite péridique vient d'apparaître par bifurcation de Hopf.

0.3. Bifurcations de familles à 2 paramètres

La famille de fonctions $F_\mu(x) = \frac{1}{3}x^3 - \mu x$ est un déploiement universel de $F_0(x) = \frac{1}{3}x^3$ (pour une vue plongeante et des références sur la notion de déploiement universel, voir Arnold (1980), Chenciner (1978, 1980).

Si on remplace x^3 par x^5 et que l'on ne considère comme Van der Pol que des fonctions impaires, on est amené à étudier la famille à deux paramètres d'équations du type (2) obtenue en posant

$$F_{\mu,\nu}(x) = \frac{x^5}{5} - \nu \frac{x^3}{3} - \mu x. \tag{12}$$

La fig. 10 résume la forme du graphe de $F_{\mu,\nu}$ pour les différentes valeurs de μ et ν; les experts reconnaitront dans la réunion de la droite $\mu = 0$ et de la parabole pointillée une section de la surface appelée ''queue d'aronde''. Remarquons que, si (ν,μ) est en dessous de la courbe formée de la demi-droite $\mu = 0$, $\nu \leq 0$, et de la demi-parabole $\mu = -5\nu^2/36$, $\nu \geq 0$, on a $x F_{\mu,\nu}(x) > 0$ pour $x \neq 0$; nous avons vu au début que ceci implique que l'origine est un attracteur global (en particulier il n'y a pas d'orbite périodique). De même, si $\mu > 0$, on peut conclure à l'existence d'une unique orbite périodique (attracteur) par le théorème de Lienard (voir Lefschetz (1977) chapitre XI §.2).

Fixons maintenant une valeur $\nu = \nu_0$ et considérons la famille à un paramètre μ ainsi obtenue:

Si $\nu_0 < 0$, on assiste lorsque μ passe d'une petite valeur négative à une petite valeur positive au phénomène de bifurcation de Hopf décrit dans le paragraphe précédent (tout se passant au voisinage de l'origine, on peut pratiquement oublier le terme en x^5).

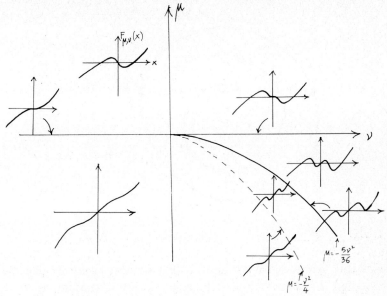

Fig. 10.

Si $v_0 > 0$, la situation est plus complexe: lorsque μ passe d'une valeur légèrement positive à une valeur légèrement négative il apparait une petite orbite périodique répulsive (phénomène analogue à la bifurcation discutée plus haut à l'exception d'un signe...); par ailleurs, l'orbite périodique attractante d'amplitude finie qui existe pour $\mu > 0$ par le théorème de Liénard continue à exister (stabilité), et un théorème de Ryckov (cité dans Lins et al. (1977)) affirme que l'éq. (2) a au plus deux orbites périodiques lorsque le degré de F est égal à 5.

Il existe donc une région bordée par la demi-droite $\mu = 0$, $v \geq 0$ et contenue entre cette demi-droite et la demi-parabole $\mu = -5v^2/36$, $v \geq 0$ correspondant à des $F_{\mu,v}$ ayant exactement deux orbites périodiques, l'une (la plus grande) attractante, l'autre répulsive (fig. 11). Lorsque $\mu < 0$, l'origine est un répulseur (local); il existe donc un voisinage de 0 dans $\mathbb{R}^2$ (d'autant plus petit que μ est proche de0) dans lequel l'éq. (2) n'a pas d'orbite périodique. Il est alors facile de montrer que la seule façon qu'ont les deux orbites périodiques ci-dessus de disparaitre lorsque μ décroît ($v = v_0 > 0$ fixé) est d'entrer en collision (fig. 12).

Exercice. (Ayant été prévenu trop tard qu'il aurait à faire ce cours, l'auteur

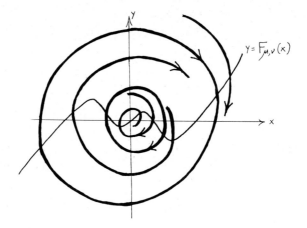

Fig. 11.

n'a pas eu le temps de faire l'exercice mais est néanmoins persuadé de la justesse de l'affirmation dont il demande la démonstration, sinon de la facilité de cette dernière.) Montrer qu'il existe une courbe Γ régulière comme sur la fig. 13, limitant le domaine du plan (μ, ν) correspondant à des équations ayant deux orbites périodiques (domaine noté 2).

Remarque. Le phénomène de collision de 2 orbites périodiques l'une attractante, l'autre répulsive se produit "génériquement" pour des valeurs

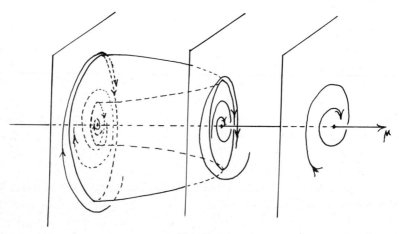

Fig. 12.

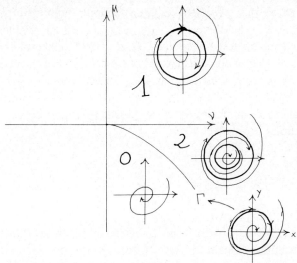

Fig. 13.

isolées du paramètre dans des familles à 1 paramètre d'équations différentielles; pour ces valeurs critiques, les deux orbites périodiques sont confon-
dues en une orbite périodique attractante d'un côté et répulsive de l'autre
(on dit qu'elle est "non normalement hyperbolique"): bien que cette orbite
soit d'amplitude finie, l'étude de ses avatars est encore un problème local
car il se ramène par la méthode des sections de Poincaré à l'étude au voisinage d'un point fixe d'un difféomorphisme local de $\mathbb{R}$ dont la dérivée au
point fixe est égale à 1 (voir Arnold (1980) chapitre 6, §.34B). Ce ne sera
plus le cas pour la situation analogue concernant des familles de difféomorphismes; cette dernière sera en fait la source des phénomènes de type
Hamiltonien qui nous intéresseront.

0.4. *Equations non autonomes et difféomorphismes*

Le problème de la démultiplication des fréquences en électricité conduit à
des équations de la forme

$$\frac{d^2x}{dt^2} + f(x)\frac{dx}{dt} + x = \varphi(t) \tag{13}$$

qui ne diffèrent de l'éq. (1) que par l'existence d'un second membre périodique en t (force électromotrice appliquée).

Une bonne partie des méthodes de la dynamique qualitative peut-être mise en oeuvre sur ce type d'équations, qui est d'ailleurs à l'origine de certaines d'entre elles [voir les souvenirs de Smale (1980) concernant les travaux classiques de Cartwright-Littlewood et Levinson; ceux-ci n'étant pas notre propos, nous renvoyons à Cartwright (1950) Levi (1978)].

Si T est la période de φ, l'éq. (13) définit un champ de vecteurs sur le tore plein $\mathbb{R}^2 \times \mathbb{R}/T\mathbb{Z}$ (coordonnées x, $y = dx/dt + F(x)$, θ défini modulo T) par les formules

$$dx/dt = y - F(x),$$

$$dy/dt = -x + \varphi(\theta), \tag{14}$$

$$d\theta/dt = 1.$$

Les solutions de (14) se projettent sur le cercle $\mathbb{R}/T\mathbb{Z}$ en les solutions de $d\theta/dt = 1$; une solution issue du plan $\theta = 0$ recoupe donc ce plan au temps $\theta = T$ (fig. 14) ce qui fournit un difféomorphisme $P : \mathbb{R}^2 \to \mathbb{R}^2$ appelé "application de Poincaré". Il est clair que l'étude qualitative de (14) equivaut à l'étude qualitative de P du point de vue de l'itération [une orbite périodique de (14) de période nT correspond à une orbite périodique de P de période n, un tore invariant de (14) correspond à une courbe fermée invariante de P, etc.). Si $\varphi = 0$, le difféomorphisme P (que nous noterons P_0) n'est autre que celui qui, au couple (x, y), associe la solution au temps

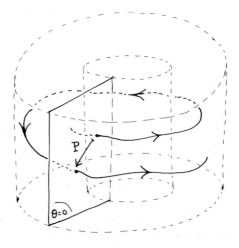

Fig. 14.

T, $(x(T), y(T))$, de l'éq. (2) qui vérifie $x(0) = x$, $y(0) = y$. Pour des fonctions φ suffisamment petites, le difféomorphisme P obtenu est une perturbation de P_0; en particulier, les points fixes de P_0 en lesquels le spectre de DP_0 ne contient pas la valeur propre 1 existent encore (légèrement déplacés) pour P, ce qui fournit des orbites périodiques de période T de (14).

Puisque le spectre de DP_0 en un point fixe (x_0, y_0) de P_0 n'est autre que l'image par l'application $z \to e^{Tz}$ du spectre de l'éq. (2) linéarisée en ce point, nous voyons apparaître des problèmes de résonance: c'est en particulier le cas lorsqu'on perturbe une famille d'éqs. (2) présentant une bifurcation de Hopf par un terme $\varphi(t)$ dont la période est égale à un multiple entier de la période de l'orbite périodique qui est en train de naître (le spectre de l'éq. (2) linéarisée pour la valeur critique du paramètre est ($\pm i\omega$), la période en question est $2\pi/\omega$; si $T = n\, 2\pi/\omega$, le spectre de DP_0 au point fixe correspondant est réduit à une valeur propre double égale à $1 = \exp[n(2\pi/\omega)(\pm i\omega)]$.

La méthode utilisée au paragraphe 2 nous ramène à l'étude d'une famille d'équations définies au voisinage de $0 \in \mathbb{R}^2 = \mathbb{C}$ par

$$\mathrm{d}z/\mathrm{d}t = (\mu + i\omega)z - \alpha z |z|^2 + \mathrm{O}(|z|^4) + \varepsilon \psi(t). \tag{15}$$

Les paramètres sont $(\mu, \omega, \varepsilon)$ voisin de $(0, \omega_0, 0)$, ψ est périodique de période $2\pi/\omega_0$, et P est le difféomorphisme local (i.e. défini au voisinage de 0) qui à $z(0)$ associe $z(T)$.

En utilisant la méthode de variations des constantes pour écrire $z(T)$, on approche la famille des difféomorphismes P par une famille de difféomorphismes $\tilde{P}$ (définis au voisinage de 0) de la forme:

$$\tilde{P}(z) = (1 + \mu)e^{i\gamma}z - a|z|^2 z + \varepsilon b. \tag{16}$$

Les paramètres sont $(\mu, \gamma, \varepsilon)$ proches de $(0, 0, 0)$, et $\mathrm{Re}\,a > 0$. La fig. 15 indique suivant les régions de l'espace $(\mu, \gamma, \varepsilon)$ le nombre de points fixes de $\tilde{P}$; signalons que les difféomorphismes locaux $\tilde{P}$ definis par (16) sont loin d'être complètement compris. [voir cependant Gambaudo (1982)].

Dans la suite de ce cours nous étudions les modifications apportées par une petite perturbation périodique aux bifurcations qui ont été illustrées dans les paragraphes 2 et 3; si on excepte les phénomènes de résonance tel celui qui vient d'être évoqué, on est ramenés au problème suivant:

Etudier, du point de vue de l'itération, une famille de difféomorphismes locaux de $\mathbb{R}^2$ perturbant un difféomorphisme local $P_0: \mathbb{R}^2, 0 \to \mathbb{R}^2, 0$ qui admette 0 comme point fixe elliptique (ce qui signifie que le spectre de $DP_0(0)$ est contenu dans le cercle unité de $\mathbb{C}$ et non réel).

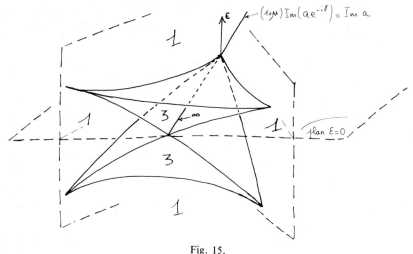

Fig. 15.

Remarque. La méthode des sections de Poincaré permet d'associer à toute orbite périodique d'une équation différentielle définie dans une variété de dimension n un difféomorphisme local $P: \mathbb{R}^{n-1}, 0 \to \mathbb{R}^{n-1}, 0$. Pour une vue générale des bifurcations d'orbites périodiques, voir Arnold (1980) chapitre 6 § 34.

1. Du théorème des contractions au théorème des fonctions implicities: persistance des points fixes

> *J'étais autrefois bien nerveux. Me voici*
> *sur une nouvelle voie:*
> *Je mets une pomme sur ma table. Puis je me*
> *mets dans cette pomme. Quelle tranquillité!*
>
> *Henri Michaux*

Soit $P: V, X_0 \to W, X_0$ un difféomorphisme de classe C^∞ (C^1 suffirait) defini sur un voisinage V du point $X_0 = (x_0, y_0) \in \mathbb{R}^2$, tel que $P(X_0) = X_0$ (fig. 16). Rappelons que ''P de classe C^r'' signifie que les dérivées partielles de P et P^{-1} jusqu'à l'ordre r existent et sont continues sur le domaine de définition de P, et que, d'après la règle de dérivation des applications

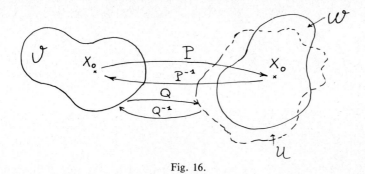

Fig. 16.

composées, la dérivée de P en X_0, $DP(X_0)$, est un isomorphisme linéaire de $\mathbb{R}^2$ dont l'inverse est $DP^{-1}(X_0)$.

Soit $Q : V \to U$ un difféomorphisme de classe C^∞ proche de P dans la topologie C^1 (Q et ses dérivées partielles sont respectivement proches de P et ses dérivées partielles). Nous allons examiner plusieurs arguments prouvant sous certaines hypthèses que Q possède un point fixe Y_0 [i.e. $Q(Y_0) = Y_0$] proche de X_0.

1.1. Contractions

Supposons que le spectre de $DP(X_0)$ soit contenu dans l'intérieur du disque unité de $\mathbb{C}$ (s'il est contenu dans l'extérieur, on applique le même argument à l'inverse P^{-1}). Il existe alors un voisinage $V' \subset V$ de X_0 d'adhérence compacte (comme un disque) tel que

(i) $P(\bar{V}') \subset \bar{V}'$,

(ii) la restriction de P à $\bar{V}'$ est une contraction.

Si Q est assez proche de P dans la topologie C^1, Q possède également ces deux propriétés et a donc un unique point fixe dans $\bar{V}'$.

Le raisonnement ci-dessus est presque correct. Le seul problème est que, si le spectre de $DP(X_0)$ n'est pas réel, pour que $\|DP(X_0)\|$ soit inférieur à 1 et donc que $P|V'$ soit vraiment une contraction, il faut bien choisir la mesure des longueurs (la norme) dans $\mathbb{R}^2$ ou, ce qui revient au même, faire un changement de coordonnées linéaires qui transforme $DP(X_0)$ en le composé d'une rotation et d'une homothétie de rapport < 1 (voir fig. 17). Ce n'est pas parce que $DP(X_0)$ envoie chaque ellipse d'un feuilletage sur une ellipse strictement plus petite que $\|DP(X_0)X\| < \|X\|$; cela devient vrai après un changement de variables transformant les ellipses en cercles.

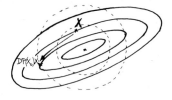 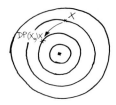

Fig. 17.

Remarquons que nous avons non seulement prouvé que Q possède un point fixe Y_0 proche de X_0, mais nous avons également une idée précise (au sens topologique) de la dynamique de Q au voisinage de Y_0 (Y_0 est un attracteur). C'est en quelque sorte cette dynamique qui force le point fixe Y_0 à exister.

1.2. Contractions déguisées: hyperbolicité

Supposons maintenant que le spectre de $DP(X_0)$ soit formé de deux valeurs propres réelles λ_s, λ_u, telles que

$$0 < |\lambda_s| < 1 < |\lambda_u|$$

(λ_s comme stable, λ_u comme instable en anglais).

Il sera commode de supposer que $X_0 = 0$ et que les vecteurs propres de $DP(0)$ sont portés par les axes de coordonnées (l'axe des x pour λ_u, l'axe des y pour λ_s).

Si Q est proche de P, on peut montrer que Q possède un point fixe Y_0 proche de $X_0 = 0$ de la manière suivante:

(i) On montre que P laisse (localement) invariantes deux courbes C^∞, $W_0^u(P)$ et $W_0^s(P)$, passant par 0 et respectivement tangentes aux axes de coordonnées, telles que la restriction $P|_{W_0^u(P)}$ (resp. $P|_{W_0^s(P)}$) soit une dilatation (resp. une contraction): ces deux courbes sont une version locale non-linéaire des directions propres de $DP(0)$ (voir la fig. 18 où quelques orbites de P sont représentées); on les appelle variété instable et variété stable de P en 0.

(ii) On montre que si Q est assez proche de P, Q laisse également (localement) invariantes deux courbes C^∞, $W^u(Q), W^s(Q)$, respectivement proches de $W_0^u(P)$, $W_0^s(P)$, et s'intersectant donc transversalement en un point Y_0 proche de $X_0 = 0$; une telle intersection vérifie forcément $Q(Y_0) = Y_0$ (fig. 19).

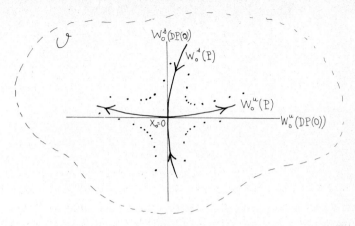

Fig. 18.

La démonstration de ces deux points utilise encore le théorème des contractions, mais cette fois dans un espace de courbes (donc de dimension infinie) et non plus dans un domaine de $\mathbb{R}^2$. Comme précédemment, le (ii) se déduit du (i) parce qu'une petite (dans le sens C^1) perturbation d'une contraction est encore une contraction. Précisons un peu cette idée (qui remonte à Hadamard) car elle sera importante dans l'étude des bifurcations: soit $\sigma : \mathbb{R} \to \mathbb{R}$ une fonction continue; l'image par $DP(0)$ du graphe de σ est encore le graphe d'une application qu'on note $F(\sigma) : \mathbb{R} \to \mathbb{R}$. Le choix des axes de coordonnées implique que $DP(0)(x, y) = (\lambda_u x, \lambda_s y)$ et donc (voir fig. 20):

$$F(\sigma)(x) = \lambda_s \sigma(x/\lambda_u). \tag{17}$$

On remarque que la fonction nulle (dont le graphe est l'axe des x, variété

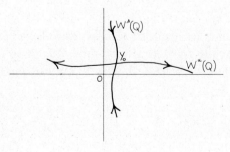

Fig. 19.

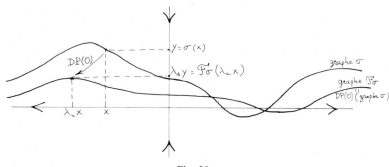

Fig. 20.

instable de $DP(0)$, est un point fixe de cette transformation; de plus, si σ est uniformément bornée, $F^n(\sigma)$ tend uniformément vers 0 lorsque n tend vers $+\infty$; Enfin, si σ est C^k avec des dérivées uniformément bornées, $F(\sigma)$ est C^k et les k dérivées de $F^n(\sigma)$ tendent uniformément vers 0 lorsque n tend vers $+\infty$. Il est donc naturel de chercher $W_0^u(P)$ comme point fixe d'une "transformée de graphes" F qui soit une contraction dans un espace fonctionnel bien choisi: pour que l'image d'un graphe soit encore un graphe, on est amenés à ne considérer que des fonctions au moins Lipshitziennes $(|\sigma(x) - \sigma(x')| \le k|x - x'|)$ et définies seulement sur un intervalle $[-r, +r]$ [$F(\sigma)$ est alors définie comme la restriction à cet intervalle de la fonction dont P (graphe σ) est le graphe, restriction qui existe car P dilate au voisinage de 0 dans la direction horizontale].

En choisissant un espace fonctionnel formé de fonctions ayant une certaine classe de dérivabilité, on montre que $W_0^u(P)$ (qui est unique, car caractérisée comme l'ensemble des points X proches de 0 tels que

$$\lim_{n \to +\infty} P^{-n}(X) = 0$$

est C^k; ceci étant vrai pour tout k, $W_0^u(P)$ est C^∞. Pour une démonstration raffinée donnant les meilleurs résultats (pas de perte de différentiabilité en différentiabilité finie) voir le chapitre 5 de Shub (1978).

Quant à l'existence de $W_0^s(P)$, elle se prouve en remarquant que $W_0^s(P) = W_0^u(P^{-1})$.

1.3. Fonctions implicites

Bien entendu, la meilleure démonstration de persistance d'un point fixe est celle qui, supposant simplement que 1 n'est pas dans le spectre de $DP(X_0)$ en déduit (c'est la définition) que $DP(X_0) - Id$ est inversible, donc (théorème des fonctions implicites) que $(P - Id)$ est un difféomorphisme d'un voisinage de X_0 sur un voisinage de 0; il en est donc de même de $Q - Id$ si Q est C^1 proche de P, ce qui montre l'existence d'un point Y_0 proche de X_0 tel que $Q(Y_0) - Y_0 = 0$.

La différence essentielle entre cette approche et la méthode des contractions est que maintenant nous n'avons aucun moyen de comparer la dynamique de Q au voisinage de Y_0 à la dynamique de P au voisinage de X_0 (dans les deux cas précédents, on peut montrer que ces dynamiques sont topologiquement les mêmes dans le sens où il existe un changement local de coordonnées continu qui transforme Q en P et même en $DP(0)$: il y a stabilité locale; voir Shub (1978) chapitre 7). A titre d'exemple, si $DP(0)$ est une rotation, Q peut présenter des dynamiques aussi différentes que celles représentées sur la fig. 21....Heureusement, puisque le but du cours est justement d'étudier la dynamique de tels difféomorphismes Q, au moins de ceux qu'on rencontre dans des familles génériques dépendant d'un nombre fini de paramètres.

Remarque. La démonstration du théorème des fonctions implicites se

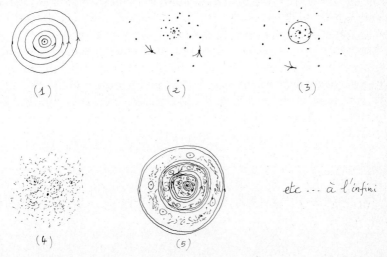

Fig. 21. (1) Rotation. (2) Point fixe attractant. (3) Point fixe répulsif entouré d'une courbe fermée invariante. (4) Point fixe entouré d'une zône ergodique (Anosov et Katok). (5) Point fixe entouré d'un Cantor de coubes fermées invariantes (situation hamiltonienne générique).

ramène elle-même au théorème des contractions; mais la contraction utilisée n'a pas de signification dynamique simple.

1.4. Courbes invariantes

Supposons qu'un difféomorphisme P de $\mathbb{R}^2$ laisse invariante une courbe
fermée C (par exemple un cercle entourant l'origine). Comme dans le cas
des points fixes, on peut se demander si un difféomorphisme Q proche de
P laisse également invariante une courbe fermée D proche de C.

On peut essayer la méthode de contraction qui vient d'être invoquée en
1.2: on identifie un voisinage Ω de C au produit de C par un intervalle
$I \subset \mathbb{R}$, ce qui permet d'associer à une application $\sigma: C \to I$ son graphe contenu dans Ω, et de définir pour des σ "assez petites" la transformée de graphe $F(\sigma)$ comme l'application dont le graphe est l'image par P du graphe
de σ (fig. 22). La différence essentielle entre le cas d'un point fixe et le cas
d'une courbe est que, dans ce dernier, ce n'est que lorsque $\sigma \mapsto F\sigma$ est une
contraction (resp. une dilatation) dans un espace fontionnel bien choisi
qu'on peut affirmer la persistance d'une courbe fermée invariante de classe
C^1 pour toute perturbation assez petite Q (pour une courbe dans $\mathbb{R}^n$, la
condition est que $\sigma \to F(\sigma)$ soit hyperbolique: il peut y avoir des directions
normales dilatantes et des directions normales contractantes). On dit alors
que la courbe C est *normalement hyperbolique* (dans $\mathbb{R}^2$, normalement
contractante ou normalement dilatante): intuitivement, cela signifie que la
contraction (resp. dilatation) de P normalement à C l'emporte sur la contraction (resp. dilatation) de P tangentiellement à C. Les définitions techniques ce trouvent dans Hirsch et al. (1977) qui n'est pas d'une lecture facile,
mais dans le cas de $\mathbb{R}^2$ il nous suffira de prendre le vocable "normalement
hyperbolique" comme synonyme de la propriété de contraction de la trans-

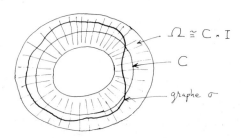

Fig. 22.

formée de graphe associée à P ou de celle associée à P^{-1}. L'exemple indiqué sur la fig. 23 devrait éclairer ceci (le nombre de flèches indique la force de la contraction ou de la dilatation; la courbe fermée invariante n'est autre que l'adhérence de la variété instable d'un point fixe hyperbolique; seul le cas (i) est normalement hyperbolique). Remarquons que non seulement l'existence mais la différentiabilité de la courbe invariante D dépendent du rapport entre contraction (resp. dilatation) de P normalement à C et contraction (resp. dilatation) de P tangentiellement à C (on suppose ici P et C de classe C^∞); examinons en effet le cas le plus simple, celui d'un point fixe attractant sur la courbe C, qui se présente dans l'exemple de la fig. 23: supposons que $P = DP(0)$ est linéaire et que ses valeurs propres sont $\lambda_1, \lambda_2, 0 < |\lambda_2| < |\lambda_1| < 1$, la direction propre relative à λ_1 (resp. λ_2) étant l'axe des x (resp. l'axe des y) (fig. 24). La transformée de graphes associée à P (remarquer qu'elle ne peut plus être définie localement, le domaine de

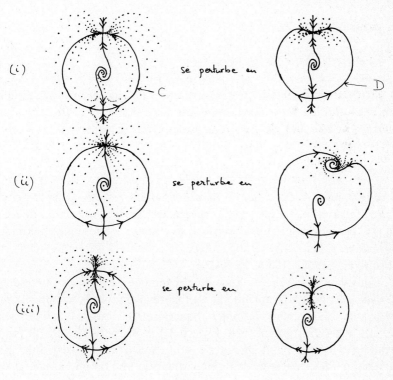

Fig. 23.

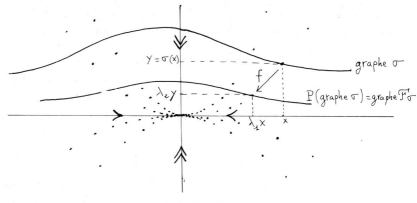

Fig. 24.

définition se rétrécissant) s'écrit

$$F(\sigma)(x) = \lambda_2 \sigma(x/\lambda_1).$$ (18)

En particulier,

$$(F(\sigma))^{(k)}(x) = \frac{\lambda_2}{\lambda_1^k} \sigma^{(k)}(x/\lambda_1).$$ (19)

Le degré k de différentiabilité des fonctions d'un espace sur lequel F puisse être une contraction est donc limité par une condition du type $|\lambda_2/\lambda_1^k| < 1$ qui ne peut être réalisée pour tous les entiers k puisque

$$\lim_{k \to +\infty} |\lambda_2/\lambda_1^k| = +\infty.$$

Si cette condition est violée, il existe des perturbations Q de P dont la courbe invariante D n'est pas C^k. En particulier, si la restriction de P à la courbe fermée invariante C possède des points périodiques génériques (attractants et répulsifs) ce phénomène de perte de différentiabilité se produira [voir Arnold (1980) fig. 141].

La raison pour laquelle un théorème général de persistance d'une courbe invariante ne peut être obtenu par un théorème de fonctions implicites est la suivante: on cherche à résoudre l'équation $(F - Id)(\sigma) = 0$ mais en général, si C n'est pas normalement hyperbolique, le spectre de $DF(0)$ [ou de ce qui tient lieu de $DF(0)$] opérant dans un Banach bien choisi de fonctions, contient tout le cercle unité de $\mathbb{C}$; en particulier, 1 est dans le spectre et l'argument du paragraphe 1.3. ne peut fonctionner.

L'exemple le plus simple d'une telle situation est fourni par un difféomorphisme P de $C \times \mathbb{R}$ (C cercle unité) défini au voisinage de $C \times 0$ par

$$P(\theta, \mathrm{x}) = (f(\theta) + \mathrm{O}(|x|),\ a(\theta)x + \mathrm{O}(|x|^2)).$$

La transformée de graphes F associée, définie sur un voisinage de 0 dans $C^k(C, \mathbb{R})$ (fonctions de classe C^k de C dans $\mathbb{R}$) n'est dérivable que si on la considère comme à valeurs dans $C^{k-1}(C, \mathbb{R})$; sa dérivée en 0 est alors la transformée de graphes linéaire G associée à

$$N(\theta, x) = (f(\theta), a(\theta)x).$$

On remarquera que G applique $C^k(C, \mathbb{R})$ dans lui-même. Si f est une rotation d'angle $2\pi\omega$, et si ω est irrationnel, le spectre de G coïncide pour tout k avec le cercle de $\mathbb{C}$ de rayon

$$\exp\left[\int_0^1 \log|a(\theta)|\,\mathrm{d}\theta \right].$$

Si f est un difféomorphisme du cercle structurellement stable, le spectre de G peut contenir toute une couronne [voir Chenciner et Iooss (1979) chapitre II].

Il est cependant possible de montrer que *certaines* perturbations de P préservent l'existence de C lorsque C n'est pas normalement hyperbolique; l'outil est un théorème raffiné de fonctions implicites fonctionnant dans des espaces de Fréchet de fonctions C^∞: le premier résultat de ce type est le célèbre théorème de Kolmogorov. Arnold. Moser sur l'existence de courbes invariantes pour des difféomorphismes préservant les aires au voisinage d'un point fixe elliptique [voir Moser (1973) chapitre 2,§4 et Herman (1980)]. Nous montrerons au chapitre 4 que ce type de méthode fournit également des résultats dans le cas dissipatif qui nous intéresse. Bien entendu, on ne contrôle plus la dynamique de Q au voisinage de la courbe invariante D dont on montre l'existence (dans le cas normalement hyperbolique il y a, comme pour les points fixes, une équivalence topologique entre la dynamique de P au voisinage de C et la dynamique de Q au voisinage de D et c'est précisément cette dynamique qui force l'existence de D).

2. Formes normales de difféomorphismes de $\mathbb{R}^2$ au voisinage d'un point fixe

Nous promenions notre visage
(Nous fûmes deux, je le maintiens)
Sur maints charmes de paysage,
O seur, y comparant les tiens.

Stéphane Mallarmé

2.1. Calculs formels

Soit $P: \mathbb{R}^2, 0 \to \mathbb{R}^2, 0$ un difféomorphisme local C^∞ de $\mathbb{R}^2$ défini au voisinage de son point fixe 0.

Dans tout ce qui suit nous supposerons que le spectre de $DP(0)$ est formé de deux valeurs propres non réelles $\lambda = \varrho e^{2\pi i \omega}$ et $\bar{\lambda}$.

Se plaçant dans une base propre sur $\mathbb{C}$ de $DP(0)$ et identifiant $\mathbb{R}^2$ à $\mathbb{C}$ comme en 1.2, on écrit P sous la forme

$$P(z) = \lambda z + O(|z|^2), \tag{20}$$

(1e terme O est fonction de z et $\bar{z}$). Par des changements de variables formels (i.e. définis par des séries dont on ne se préoccupe pas de la convergence), on cherche à simplifier le développement de Taylor en 0 de P. Le résultat obtenu, s'il converge, donne souvent une compréhension géométrique très (trop, d'où la divergence dans bien des cas) simple de P.

Plus précisément, la conjugaison de P par

$$H(z) = z + \sum_{i+j=r} \gamma_{ij} z^i \bar{z}^j \tag{21}$$

ne change pas les termes de degré inférieur à r et ajoute aux termes de degré r le commutateur

$$\sum_{i+j=r} (\lambda^i \bar{\lambda}^j - \lambda) \gamma_{ij} z^i \bar{z}^j. \tag{22}$$

Une relation de la forme $\lambda^i \bar{\lambda}^j - \lambda = 0 (i+j \geq 2)$ est appelée une *résonance*; notons

$$\mathcal{T}P(z) = \lambda + \sum_{i+j=2}^{+\infty} C_{ij} z^i \bar{z}^j \tag{23}$$

le développement de Taylor de P en 0. Un terme $c_{ij}z^i\bar{z}^j$ de ce développement est dit résonnant si $\lambda^i\bar{\lambda}^j - \lambda = 0$.

Le calcul précédent montre qu'il existe toujours une série formelle

$$H(z) = z + \sum_{i+j=2}^{+\infty} h_{ij}z^i\bar{z}^j, \tag{24}$$

obtenue en composant une infinité de polynômes du type (21) de degrés croissant, telle que la série formelle $H \circ \mathscr{T}P \circ H^{-1}(z)$ ne comporte que des termes résonnants (on appelle cette dernière une *"forme normale formelle"* de P).

S'il n'y a pas de résonance, c'est-à-dire si $|\lambda| \neq 1$, $H \circ \mathscr{T}P \circ H^{-1}(z) = \lambda z$; autrement dit, il existe un changement de coordonnées formel qui "linéarise" P. En fait, dans ce cas, la situation est bien meilleure: si P est analytique au voisinage de 0, H l'est également (i.e. son rayon de convergence est non nul), si P est C^∞, H est le développement de Taylor d'un difféomorphisme local C^∞ qui linéarise P par conjugaison au voisinage de 0. On comprend donc parfaitement la dynamique de P dans un petit voisinage de 0 (voir Arnold (1980) chapitre 5, § 25 et Chaperon (1981) pour les démonstrations de tous les théorèmes possibles de linéarisation C^∞).

Si $|\lambda| = 1$, il y a au moins les *résonances inévitables* de la forme

$$\lambda(\lambda\bar{\lambda})^i - \lambda = 0.$$

Ce sont les seules si λ n'est pas racine de l'unité; dans ce cas, on peut choisir H de façon que

$$H \circ \mathscr{T}P \circ H^{-1}(z) = z\left(\lambda + \sum_{i=1}^{+\infty} c_i|z|^{2i}\right) \tag{25}$$

mais il y a divergence en général (comparer la fig. 21 et la propriété (*) de (32), et lire la dernière remarque du chapitre 5). Si λ est racine de l'unité de nouvelles résonances apparaissent; si $\lambda^q = 1$, les nouveaux termes résonnants de degré le plus bas seront des termes en $\bar{z}^{q-1}$ (résonance $\bar{\lambda}^{q-2} - \lambda = 0$).

Remarque importante. Il résulte de l'étude ci-dessus que l'on peut, par un changement de coordonnées local H polynomial [composé d'un nombre fini de changements de variables du type (21)] supprimer les termes non résonnants du développement de Taylor de P en 0 jusqu'à un ordre fixé. Par exemple, si λ est de module 1 mais n'est pas racine q-ième de l'unité pour $q \le 2n + 3$, on peut toujours choisir des coordonnées locales au voisinage de 0 telles que P s'écrive

$$P(z) = z\left(\lambda + \sum_{i=1}^{n} c_i |z|^{2i}\right) + O(|z|^{2n+3}). \tag{26}$$

(Ecrire P est un abus de notation, il faudrait écrire $H \circ P \circ H^{-1}$.)

2.2. Familles à paramètres

Soit $P_\mu(z)$ une famille de difféomorphismes locaux de $\mathbb{R}^2$ dépendant différentiablement de paramètres $\mu \in \mathbb{R}^N$, telle que
  (i) $P_0(0) = 0$,
  (ii) spectre $DP_0(0) = \{\lambda_0, \bar{\lambda}_0\}$, $\lambda_0 = e^{2\pi i\omega_0} \ne 1$.
D'après le théorème des fonctions implicites (voir 1.3) il existe pour tout μ voisin de 0 un point $X_\mu \in \mathbb{R}^2$ proche de 0 (et dépendant différentiablement de μ) tel que $P_\mu(X_\mu) = 0$. Après une éventuelle famille de changements affines de coordonnées dépendant différentiablement de μ, on peut supposer que $X_\mu = 0$ et que

$$P_\mu(z) = \lambda_\mu z + O(|z^2|) \tag{27}$$

pour (μ, z) voisin de $(0,0)$.

Supposons de plus que, pour $\mu \ne 0$, $|\lambda_\mu| \ne 1$: pour chaque $\mu \ne 0$ voisin de 0, on peut linéariser P_μ par un changement C^∞ de coordonnées H_μ mais lorsque μ tend vers 0 les coefficients des termes en $z^i \bar{z}^j$ du développement de Taylor de H correspondant à des résonances $\lambda_0^i \bar{\lambda}_0^j - \lambda_0 = 0$ de P_0 tendent vers l'infini (ils ont $\lambda_\mu^i \bar{\lambda}_\mu^j - \lambda_\mu$ au dénominateur); ceci implique que le voisinage de 0 dans lequel est défini H_μ se rétrécit au fur et à mesure que μ tend vers 0 (voir fig. 25). Si l'on veut simplifier le développement de Taylor de P_μ par une famille de changements de coordonnées formels dépendant différentiablement de μ au voisinage de $\mu = 0$, il faut conserver, même pour $\mu \ne 0$, les termes correspondant à des résonances de P_0. Par exemple, si λ_0 n'est pas une racine de l'unité, on se ramènera suivant la situation à:

$$\mathcal{T}P_\mu(z) = z\left(\lambda_\mu + \sum_{i=1}^{+\infty} c_i(\mu)|z|^{2i}\right) \tag{28}$$

ou à

$$P_\mu(z) = z\left(\lambda_\mu + \sum_{i=1}^{n} c_i(\mu)|z|^{2i}\right) + O(|z|^{2n+3}). \tag{29}$$

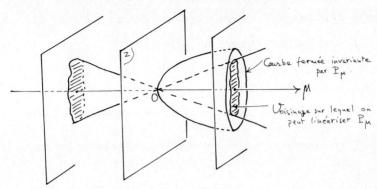

Fig. 25. La linéarisation de $P_\mu(\mu \neq 0)$ se fait obligatoirement dans un voisinage de 0 trop petit pour voir la courbe fermée invariante apparue par bifurcation en $\mu = 0$.

Remarque. Même s'il existe dans tout voisinage de $0 \in \mathbb{R}^n$ des valeurs de μ telles que $|\lambda_\mu| = 1$ (ce sera le cas en général pour $N \geq 2$) une condition du type $\lambda_0^q \neq 1$ pour $q \leq 2n + 3$ implique la même condition sur λ_μ dès que μ est assez proche de 0. On pourra donc toujours mettre la famille sous la forme (29).

2.3. Etude géométrique des familles génériques de formes normales tronquées (cas "non fortement résonant")

Soit P_0 comme précédemment $[P_0(0) = 0$, spectre $DP_0(0) = \{\lambda_0, \bar{\lambda}_0\}$, $|\lambda_0| = 1]$. On suppose que $\lambda_0 = \exp(2\pi i \omega_0)$ n'est pas une "petite" racine de l'unité ($\lambda_0^q \neq 1$ pour $q \leq 2n + 3$).

Dans un système de coordonnées convenable, une famille P_μ dépendant différentiablement de $\mu \in \mathbb{R}^N$ peut alors s'écrire sous la forme (29) pour (μ, z) voisin de $(0,0)$, ou ce qui revient au même sous la forme

$$P_\mu(z) = z[1 + f(\mu, |z|^2) + O(|z|^{2n+2})] \exp\{2\pi i [g(\mu, |z|^2) + O(|z|^{2n+2})]\}$$

$$(30)$$

où f et g sont des polynômes de degré n en la deuxième variable à coefficients réels dépendant différentiablement des paramètres μ, que l'on notera

$$f(\mu, X) = a_0(\mu) + a_1(\mu)X + \ldots + a_n(\mu)X^n, \qquad a_0(0) = 0,$$
$$g(\mu, X) = b_0(\mu) + b_1(\mu)X + \ldots b_n(\mu)X^n, \qquad b_0(0) = \omega_0. \qquad (31)$$

On appelle famille de formes normales tronquées (à l'ordre n) associée à la famille P_μ la famille de difféomorphismes locaux

$$N_\mu(z) = z[1 + f(\mu, |z|^2)] \exp[2\pi i g(\mu, |z|^2)]. \tag{32}$$

Les difféomorphismes N_μ ont les propriétés caractéristiques suivantes (comparer la fig. 5 et la fig. 26):

(*) $\begin{cases} \text{(i)} & \text{le feuilletage des cercles de centre 0 est invariant (le cercle } |z| = r \\ & \text{est envoyé sur le cercle } |z| = r[1 + f(\mu, r^2)]); \\ \text{(ii)} & \text{un cercle de centre 0 est envoyé sur un autre cercle de centre 0 par} \\ & \text{une rotation (pour le cercle de rayon } r, \text{ la rotation est d'un angle} \\ & \text{de } 2\pi g(\mu, r^2)).* \end{cases}$

Remarque. (*) traduit l'équivariance de N_μ sous l'action du groupe des rotations: Pour tout $\alpha, N_\mu(z e^{i\alpha}) = N_\mu(z) e^{i\alpha}$.

Supposons qu'il existe $N \le n$ tel que

$$a_0(0) = a_1(0) = \ldots = a_{N-1}(0) = 0,$$

$a_N(0) \neq 0$ (cette condition implique que 0 est un attracteur ou un répulseur pour N_0). $\tag{33}$

On dira alors que P_0 est formellement de codimension N. (La notion de codimension se réfère à la codimension d'une classe d'équivalence dans un espace fonctionnel (voir Arnold (1980) chapitre 6, § 29 et Chenciner (1980) Chapitre 5); il apparaîtra qu'au sens de l'équivalence topologique les dif-

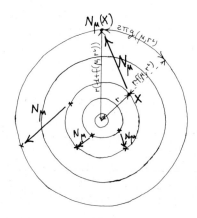

Fig. 26.

féomorphismes envisagés sont sans doute de codimension infinie. La forme normale N_0 est de codimension finie si on prend une relation d'équivalence très grossière ne tenant compte que de la variable radiale: on se sert du fait que N_0 commute avec les rotations pour passer au quotient par la variable angulaire.) Pour une famille générique dépendant de N paramètres (i.e. d'un paramètre $\mu = (\mu_0, \mu_1, \ldots, \mu_{N-1}) \in \mathbb{R}^N$), l'application de $\mathbb{R}^N$ dans $\mathbb{R}^N$

$$\mu \mapsto (a_0(\mu), \ldots, a_{N-1}(\mu)), \tag{34}$$

a une dérivée en 0 qui est un isomorphisme. D'après le théorème des fonctions implicites cette application est un difféomorphisme local d'un voisinage de $0 \in \mathbb{R}^N$ sur un voisinage de $0 \in \mathbb{R}^N$. On peut donc prendre $a_0(\mu)$, $a_1(\mu), \ldots, a_{N-1}(\mu)$, comme coordonnées locales dans $\mathbb{R}^N$, ce qui revient à écrire

$$f(\mu_0, \ldots, \mu_{N-1}, X) = \mu_0 + \mu_1 X + \ldots + \mu_{N-1} X^{N-1}$$
$$+ a_N(\mu) X^N + \ldots + a_n(\mu) X^n, \tag{35}$$

$a_N(0) \neq 0$.

Sous cette forme, il est facile de déterminer les courbes fermées invariantes (proches de 0) de N_μ: tout d'abord la propriété caractéristique (*)(i) implique qu'une telle courbe est nécessairement un cercle de centre 0; d'autre part les rayons r des cercles invariants sont les solutions de l'équation

$$f(\mu, r^2) = 0. \tag{36}$$

Les figs. 27 et 28 indiquent dans les cas $N = 1$ et $N = 2$ la forme de la surface définie par (36) dans $\mathbb{R}^N \times \mathbb{R}$ et le nombre de cercles invariants par N_μ en fonction de la position de μ dans $\mathbb{R}^N$: dans les deux cas, on a choisi $a_N(0) < 0$ (par exemple $a_N(0) = -1$); le cas $a_N(0) > 0$ est analogue avec des dynamiques inversées.

Ne pas oublier en lisant ces figures que tout est local au voisinage de $(\mu, r) = (0,0)$.

Une fois déterminées les courbes invariantes de N_μ, la dynamique de N_μ est transparente; elle est indiquée dans les cas $N = 1$ et $N = 2$ sur la fig. 29 et la fig. 30 que l'on comparera respectivement à la fig. 6 et à la fig. 13 (le sens de rotation n'est pas significatif).

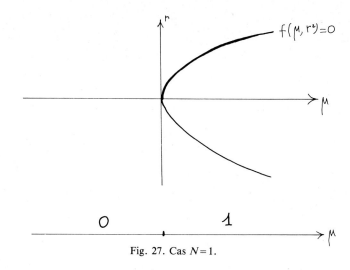

Fig. 27. Cas $N = 1$.

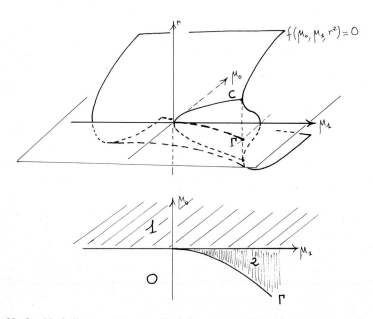

Fig. 28. Cas $N = 2$: l'axe $\mu_0 = 0$ est une ligne de points de bifurcation de Hopf, générique si $\mu_1 \neq 0$.

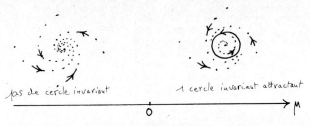

pas de cercle invariant   1 cercle invariant attractant

Fig. 29. Cas $N=1$.

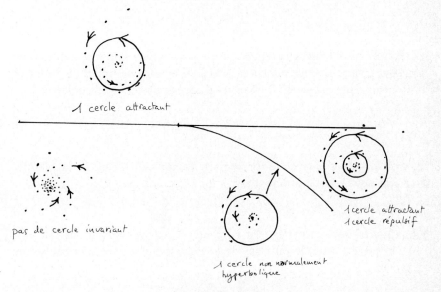

1 cercle attractant

pas de cercle invariant

1 cercle non normalement hyperbolique

1 cercle attractant
1 cercle répulsif

Fig. 30. Cas $N=2$.

2.4. *Résonances fortes*

On suppose maintenant que le spectre de $DP_0(0)$ est formé de deux valeurs propres λ_0, $\bar{\lambda}_0$ non réelles telles que $\lambda_0^q = 1$ ($q \geq 3$ est le plus petit entier ayant cette propriété).

Une résonance $\lambda_0^i \bar{\lambda}_0^j - \lambda_0 = 0 (i+j \geq 2)$ peut toujours s'écrire sous l'une des deux formes suivantes:

$$
\begin{align}
&(1) \quad \lambda_0^{kq+1}(\lambda_0\bar{\lambda}_0)^j - \lambda_0 = 0 \ (k \geq 0) \ \text{si } i \geq j, \\
&(2) \quad \bar{\lambda}_0^{lq-1}(\lambda_0\bar{\lambda}_0)^i - \lambda_0 = 0 \ (l \geq 1) \quad \text{si } i \leq j.
\end{align} \tag{37}
$$

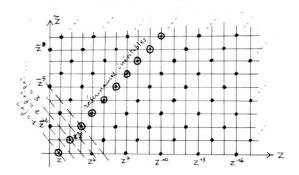

Fig. 31.

Les termes résonnants correspondant $(i-j=1 \bmod q)$ sont représentés pour $q=3$ sur la fig. 31 (les puissances de z sont en abcisse, celles de $\bar z$ sont en ordonnée). Une propriété fondamentale apparaît sur la formule

$$N_0(z)=\lambda_0 z\left(1+\sum_{k+j\geq 1}\alpha_{kj}z^{kq}(z\bar z)^j+\sum_{\substack{l\geq 1\\ i\geq 0}}\beta_{li}\bar z^{lq}(z\bar z)^{i-1}\right) \tag{38}$$

à savoir l'équivariance de la forme normale sous une rotation de $2\pi/q$ [et non plus sous tout le groupe des rotations comme en (*)]:

$$N_0(ze^{2\pi i/q})=N_0(z)e^{2\pi i/q}. \tag{39}$$

Remarquons que, parmi les termes ajoutés à ceux provenant des résonances inévitables, celui de plus bas degré correspond à $\bar z^{q-1}$. On peut donc se ramener à

$$P_0(z)=z(1+f(|z|^2))\exp[2\pi ig(|z|^2)]+c\bar z^{q-1}+O(|z|^q) \tag{40}$$

où f et g sont des polynômes de degré $\leq[(q-2)/2]$ ([a] signifie "le plus grand entier $n\leq a$").

Soit $d\geq 1$ la valuation de f (plus petit degré d'un terme non nul du polynôme f): si $1\leq d\leq[(q-3)/2]$, on peut écrire

$$P_0(z)=z(1+a_d|z|^{2d})\exp[2\pi ig(|z|^2)]+O(|z|^{2d+2}), \tag{41}$$

et l'origine est un attracteur (faible) ou un répulseur (faible) suivant que $a_d<0$ ou $a_d>0$.

Dans le cas contraire, et si $c\neq 0$, le terme $c\bar z^{q-1}$ intervient qualitativement dans la dynamique de P_0 au voisinage de 0 (voir par exemple la fig. 32): on dira alors qu'il y a *résonance forte*. C'est en particulier le cas pour

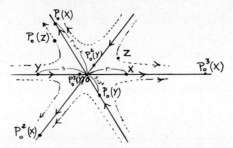

Fig. 32. $P_0(z) = \exp(2\pi i/3)\,(z + \bar{z}^2)$. Les droites $\theta = 2k\pi/3$, $k = 0, 1, 2$, sont permutées par P_0; l'origine n'est ni attracteur, ni répulseur.

$q \leq 4$ dès que $c \neq 0$, et c'est dans cette dernière circonstance qu'on emploie le terme de résonance forte dans la littérature [Arnold (1980) chapitre 6, § 34, et Iooss (1979) chapitre 4]. (Si $c = 0$ on peut donner les mêmes définitions en remplaçant $c\bar{z}^{q-1}$ par le premier terme non nul ne provenant pas des résonances inévitables.)

Nous reviendrons sur la formule (40) au dernier chapitre, lorsque nous chercherons des points homoclinique dans les bifurcations génériques de codimension 2. Nous n'étudierons pas ici les bifurcations dans le cas des résonances fortes [voir Arnold (1980) chapitre 6, § 34 E et 35 K; et Iooss (1979) chapitre 4]. Donnons simplement un exemple dans le cas où $q = 3$. Soit P_μ une famille de difféomorphismes locaux de $\mathbb{R}^2$ de la forme

$$P_\mu(z) = \lambda_\mu z + c_\mu \bar{z}^2 + O(|z|^3),$$
$$\lambda_0^3 = 1. \tag{42}$$

On calcule l'itéré troisième de P_μ:

$$P_\mu^3(z) = \lambda_\mu^3 z + c(\lambda_\mu^2 + \bar{\lambda}_\mu^4 + \lambda_\mu \bar{\lambda}_\mu^2)\bar{z}^2 + O(|z|^3). \tag{43}$$

Les points périodiques de période 3 de P_μ (proches de 0) sont les solutions de

$$(\lambda_\mu^3 - 1) + c(\lambda_\mu^2 + \bar{\lambda}_\mu^4 + \lambda_\mu \bar{\lambda}_\mu^2)\frac{\bar{z}^2}{z} + O(|z|^2) = 0. \tag{44}$$

Il n'est pas difficile de voir qu'il y en a 3 si $\lambda_\mu^3 - 1 \neq 0$.

Remarque. Dans une famille à 1 paramètre générique de difféomorphismes ou de formes normales telle que (42) il n'apparaît pas de courbe fermée invariante au voisinage de 0 pour μ voisin de 0. Cependant, de telles courbes pauvent apparaître dans des familles particulières. Ce phénomène est beaucoup mieux compris si l'on considère des familles à deux paramètres du type (42) dans lesquelles les deux paramètres ne sont autre que le module et l'argument de $\lambda_\mu^3 - 1$ [voir Arnold (1980) fig. 144]. On reviendra sur ce point au chapitre suivant (voir 3.3).

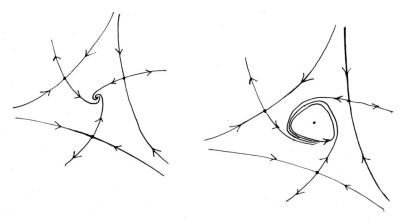

Fig. 33. Exemples de déformations du difféomorphisme P_0 de la fig. 32. On a représenté les variétés stables et instables des points périodiques de période 3 et les éventuelles courbes fermées invariantes.

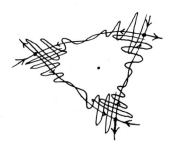

Fig. 34. Une autre déformation du même (comparer au chapitre 5).

3. Bifurcations de courbes invariantes normalement hyperboliques

> *O belle, je ne sais pas pourquoi j'aime ta trace*
> *Dans le sable durci par l'habitude des vagues,*
> *Miroir, je ne demande rien au reflet de ma face,*
> *Tout est si simple.*
>
> Georges Ribemont-Dessaignes

On a vu (chapitre 2) que la propriété pour un difféomorphisme de laisser invariante une courbe fermée normalement hyperbolique persistait pour une déformation suffisamment petite.

Nous exploitons maintenant cette idée: étant donnée une famille de difféomorphismes locaux de la forme

$$P_\mu(z) = z[1 + f(\mu,|z|^2) + O(|z|^{2n+2})]\exp\{2\pi i[g(\mu,|z|^2) + O(|z|^{2n+2})]\},$$

$$\mu = (\mu_0,\ldots,\mu_{N-1}),$$

$$f(\mu,X) = \mu_0 + \mu_1 X + \ldots \mu_{N-1}X^{N-1} + a_N(\mu)X^N + \ldots a_n(\mu)X^n, \quad (45)$$

$$a_N(0) \neq 0,$$

nous déterminons des régions (voisines de 0) de l'espace des paramètres dans lesquelles la perturbation apportée par les termes $O(|z|^{2n+2})$ est suffisamment petite par rapport à l'attraction (ou la répulsion) des cercles invariants de la forme normale N_μ pour que ceux-ci se perturbent en des courbes fermées invariantes de P_μ. Contrairement à ce qui est fait dans beaucoup de textes sur le sujet nous montrons que, pour ces valeurs des paramètres, la dynamique de P_μ ressemble (dans un sens qu'on précisera) à celle de N_μ dans un voisinage de $0 \in \mathbb{R}^2$ *uniforme* par rapport aux μ suffisamment proches de $0 \in \mathbb{R}^N$.

3.1. Domaines d'hyperbolicité normale $\mathscr{H}$

Soit p un réel, $N \leq p \leq n$; on note

$$f_p^\pm(\mu,X) = f(\mu, X) \pm \varepsilon |X|^p, \quad [0 < \varepsilon \ll |a_N(0)| \text{ si } p = N,]. \quad (46)$$

Les figs. 35 et 36 représentent $f_p^\pm(\mu, r^2) = 0$ dans les cas $N = 1$ et $N = 2$ lorsque $a_N(0) < 0$ (situation des figs. 27 et 28). Si $N = 1$, le contour apparent des surfaces $f_p^\pm(\mu, r^2) = 0$ sur l'espace des paramètres est le même que celui

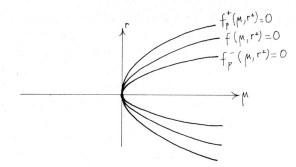

Fig. 35. Cas $N = 1$.

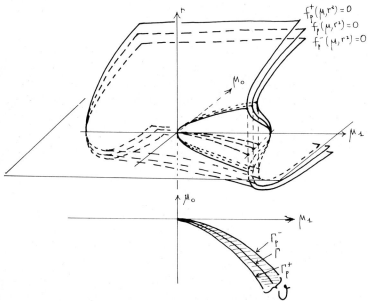

Fig. 36. Cas $N = 2$: Γ_p^- (resp. Γ_p^+) est contenu dans le contour apparent de $f_p^-(\mu, r^2) = 0$ [resp. de $f_p^+(\mu, r^2) = 0$].

de $f(\mu, r^2) = 0$. Si $N = 2$, ce contour apparent contient, en plus de l'axe $\mu_0 = 0$, une courbe $\Gamma_p^\pm$; les deux courbes Γ_p^+ et Γ_p^- définissent un voisinage effilé V de la courbe $\Gamma = \{\mu | N_\mu$ laisse invariant un cercle non normalement hyperbolique$\}$. Si $N \geq 2$, tout se passe comme pour $N = 2$: la partie non triviale du contour apparent de $f_p^+(\mu, r^2) = 0$ et $f_p^-(\mu, r^2) = 0$ définit un voisinage effilé V de l'hypersurface Γ de $\mathbb{R}^N$ correspondant aux valeurs de μ pour lesquelles N_μ possède au moins un cercle invariant non normalement hyperbolique. (On remarquera que V est d'autant plus effilé que p est grand.)

Nous appellerons *domaine d'hyperbolicité normale* (et nous noterons $\mathscr{H}$) le complémentaire de V dans un voisinage de 0 de $\mathbb{R}^N$ (fig. 37). Remarquons que dans le cas $N=1$ et dans ce cas seulement un domaine d'hyperbolicité normale est un voisinage de 0. (Tous ces voisinages de 0 devront être choisis assez petits pour que ce qui suit soit correct.) Si $\mu \in \mathscr{H}$ les inéquations $f_p^-(\mu,|z|^2) \leq 0 \leq f_p^+(\mu,|z|^2)$ définissent un anneau au voisinage de chaque cercle invariant de N_μ; l'appartenance de μ à $\mathscr{H}$ équivaut à la disjonction de tous ces anneaux (fig. 38)

3.2. *Théorèmes de bifurcation lorsque* $\mu \in \mathscr{H}$

Nous sommes maintenant en mesure d'esquisser la démonstration d'un théorème de bifurcation dans un domaine d'hyperbolicité normale:

(i) On montre que dans chacun des anneaux définis à l'instant la transformée de graphes (voir 1.4) associée à P_μ (resp. P_μ^{-1} suivant les cas) est une contraction (à condition d'avoir bien choisis p et n). P_μ possède donc dans chacun de ces anneaux une courbe C^k invariante attractante (resp. répulsive) dont le bassin d'attraction (resp. de répulsion) contient

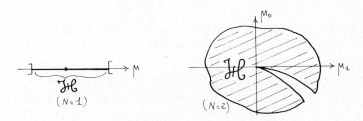

Fig. 37. Domaines d'hyperbolicité normale.

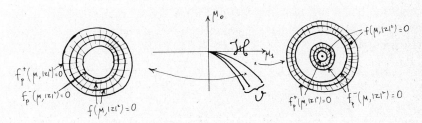

Fig. 38.

tout l'anneau. Le degré de différentiabilité k est d'autant plus grand que μ est proche de $0 \in \mathbb{R}^N$.

(ii) On montre qu'étant donné un voisinage $\mathcal{N}$, suffisamment petit, de 0 dans $\mathbb{R}^N$ et un domaine d'hyperbolicité $\mathcal{H} \subset \mathcal{N}$, il existe un voisinage Ω de 0 dans $\mathbb{R}^2$ ayant les propriétés suivantes:

(*) P_μ a autant de courbes fermées invariantes dans Ω que N_μ;
(**) les partitions de Ω en bassins d'attraction et de répulsion de l'origine et des courbes fermées invariantes sont les mêmes pour P_μ et N_μ.

Remarque. Nous verrons que (ii) n'implique pas du tout l'équivalence topologique de N_μ et P_μ dans Ω: leurs dynamiques sur les courbes fermées invariantes sont en général différentes.

Commençons par démontrer (ii) en admettant provisoirement le point (i): il suffit de remarquer que, puisque $p \leq n$, on a pour $(\mu, z) \in \mathcal{N} \times \Omega$ assez petit

$$|P_\mu(z)| \leq |z| \cdot \left(1 - \frac{\varepsilon}{2} |z|^{2p} \right) \text{ si } f_p^+ (\mu, |z|^2) \leq 0,$$

$$|P_\mu(z)| \geq |z| \cdot \left(1 + \frac{\varepsilon}{2} |z|^{2p} \right) \text{ si } f_p^- (\mu, |z|^2) \geq 0. \tag{47}$$

On en déduit, si $\mu \in \mathcal{H}$, qu'à l'extérieur des anneaux entourant les cercles invariants de N_μ, la dynamique est la même pour P_μ et pour N_μ: le seul problème pourrait être l'éventualité d'un saut de l'orbite d'un point par dessus un des anneaux, mais ceci est impossible à cause de (i).

En ce qui concerne la démonstration de (i), il suffit d'indiquer le cas $N=1$ (le cas général est analogue, la condition $\mu \in \mathcal{H}$, vide pour $N=1$, impliquant que l'attraction (resp. la répulsion) des cercles invariants de N_μ est suffisante par rapport à la perturbation par les termes $O(|z|^{2n+2})$ (à condition que p soit bien choisi et que μ soit assez proche de $0 \in \mathbb{R}^N$). On considère donc une famille à 1 paramètre de difféomorphismes locaux de la forme (45), avec

$$f(\mu, X) = \mu + a_1(\mu)X,$$

$$n = N = 1, \qquad a_1(0) < 0 \tag{48}$$

(on se ramène à ce cas dès que $\lambda_0 = \exp(2\pi i \omega_0)$, $\lambda_0^q \neq 1$ pour $q \leq 5$). En coordonnées polaires $z = r \exp(2\pi i \theta)$, P_μ s'écrit

$$\tilde{P}_\mu(r,\theta) = ((1+\mu)r + a_1(\mu)r^3 + \mathrm{O}(r^5), \theta + b_0(\mu) + b_1(\mu)r^2 + \mathrm{O}(r^4)). \tag{49}$$

Exercice. Soit $P:\mathbb{R}^2,0 \to \mathbb{R}^2,0$ un difféomorphisme local C^∞ fixant l'origine; soit $\psi:\mathbb{T}^1 \times \mathbb{R} \to \mathbb{R}^2$ le "passage en coordonnées polaires" défini par $\psi(\theta,r) = (r\cos 2\pi\theta, r\sin 2\pi\theta)$ [$\mathbb{T}^1 = \mathbb{R}/\mathbb{Z}$ est le cercle (= tore de dimension 1) de longueur 1]. Montrer qu'il existe un unique difféomorphisme C^∞ (local au voisinage de $\mathbb{T}^1 \times 0$) $\tilde{P}:\mathbb{T}^1 \times \mathbb{R}, \mathbb{P}^1 \times 0 \to \mathbb{T}^1 \times \mathbb{R}, \mathbb{T}^1 \times 0$ laissant $\mathbb{T}^1 \times 0$ invariant, tel que

$$\psi \circ \tilde{P} = P \circ \psi. \tag{50}$$

L'exercice consiste à montrer que $\tilde{P}$ est défini et C^∞ sur tout un voisinage de $\mathbb{T}^1 \times 0$ dans $\mathbb{T}^1 \times \mathbb{R}$ et pas seulement sur un voisinage de $\mathbb{T}^1 \times 0$ dans $\mathbb{T}^1 \times \mathbb{R}^+$.

On choisit des coordonnées locales dans un anneau entourant le cercle invariant

$$|z| = \left(\frac{\mu}{-a_1(\mu)}\right)^{1/2}$$

de N_μ pour $\mu > 0$ (ce serait pour $\mu < 0$ si on avait supposé $a_1(0) > 0$) en posant:

$$r = \left(\frac{\mu}{-a_1(\mu)}\right)^{1/2}(1 + y\mu^{1/2}). \tag{51}$$

On constate alors que $\tilde{P}_\mu$ s'écrit ($\mu > 0$):

$$\tilde{P}_\mu(y,\theta) = (Y,\Theta),$$

$$Y = (1-2\mu)y + \mu^{3/2}H_\mu(y,\theta), \tag{52}$$

$$\Theta = \theta + b_0(\mu) - \frac{\mu b_1(\mu)}{a_1(\mu)} + \mu^{3/2}K_\mu(\mu,\theta),$$

où les termes H_μ et K_μ sont uniformément bornés en norme C^k dans un voisinage de $y = 0$. C'est la petitesse de la perturbation (en $\mu^{3/2}$) par rapport à l'attraction du cercle invariant de N_μ (en $1-2\mu$) qui permet alors de conclure en montrant que, dans l'anneau qui nous intéresse, la transformée de graphe F associée à P_μ est une contraction (si $|\mu|$ est assez petit).

Techniquement, on définit F_k sur l'espace U_k formé des fonctions $u:\mathbb{T}^1 \to \mathbb{R}$ de classe C^k telles que

(1) $|u| \leq 1$ (le graphe reste dans un anneau du type considéré à
   condition de choisir $p = 3/2$),

(2) pour tout $i, 1 \leq i \leq k, |D^{(i)}u| \leq 1,$ (53)

(3) $D^{(k)}u$ est Lipshitzienne de rapport ≤ 1, c'est-à-dire que pour
   tous $\theta_1, \theta_2 \in \mathbb{T}^1, |D^{(k)}u(\theta_1) - D^{(k)}u(\theta_2)| \leq |\theta_1 - \theta_2|.$

(On munit U_k de la topologie de la convergence uniforme, ce qui le rend
bien complet (Merci M. Ascoli!).)

Remarque. Lorsque $\lambda_0^5 = 1$, on peut se débrouiller pour faire encore mar-
cher cette méthode [il suffit d'approcher un peu mieux la courbe fermée
invariante recherchée: voir Iooss(1979) chapitre 3 page 43].

On vient donc de "montrer" (et non "démontrer") le *théorème de bifurcation de* (disons) *Hopf pour les difféomorphismes de* $\mathbb{R}^2$, qui assure la
bifurcation (pour $\mu > 0$ ou $\mu < 0$ suivant les cas) d'une courbe fermée in-
variante C^k (k d'autant plus grand que $|\mu|$ est petit) dans toute famille à
1 paramètre *générique* P_μ de difféomorphismes locaux de $\mathbb{R}^2$ telle que
$P_0(0) = 0$ *et spectre* $DP_0(0) = \{\lambda_0, \bar{\lambda}_0\}, \lambda_0^q \neq 1$ pour $q \leq 4, |\lambda_0| = 1$.

Remarque importante. Contrairement au théorème de bifurcation de Hopf
pour les équations différentielles évoqué dans l'introduction (et qui est
d'ailleurs dans les cas génériques une conséquence de l'énoncé ci-dessus),
ce théorème n'est pas de nature topologique; autrement dit, la condition
de généricité, qui assure que l'origine est un attracteur (faible) ou un
répulseur (faible) pour P_0 n'a pas pour seul rôle de forcer une famille de
courbes fermées invariantes (ou de points périodiques) qui existerait forcément à se trouver dans le demi-espace $\mu > 0$ ou $\mu < 0$ (comparer aux figs.
6 et 7); elle est ici fondamentale pour assurer que ces courbes invariantes
existent! Pour s'en convaincre, il n'est que de considérer une famille de la
forme:

$$P_\mu(z) = (1 + \mu) P_0(z) \tag{54}$$

où P_0 est un difféomorphisme d'Anosov–Katok, c'est-à-dire un difféomorphisme préservant les aires ergodique dans le disque D^2, et tel que
$P_0(0) = 0$, spectre $DP_0(0) = \{\lambda_0, \bar{\lambda}_0\}, \lambda_0^q \neq 1$ pour tout q [voir Moser (1973)
page 56]. On voit facilement que pour tout μ assez proche de 0 (y compris,
bien sûr, $\mu = 0$) P_μ n'a pas de courbe invariante proche de 0, ni de points
périodiques bifurquant à partir de 0. Pourtant le spectre de $DP_\mu(0)$

traverse aussi gentiment que possible (i.e. "transversalement") le cercle unité lorsque μ traverse la valeur 0.

3.3. *Dynamique sur les courbes invariantes*

Lorsque $\mu \in \mathcal{H}$, ce n'est qu'au niveau de la restriction de P_μ à une courbe invariante C_μ que se révèle la différence entre P_μ et N_μ: alors que le difféomorphisme du cercle ainsi obtenu est une rotation lorsqu'il s'agit de N_μ, le difféomorphisme $P_\mu | C_\mu$ peut n'être pas même topologiquement conjugué à une rotation: c'est en général le cas lorsque son nombre de rotation $\varrho_\mu = \varrho(P_\mu | C_\mu)$ est rationnel: nous verrons qu'on obtient un difféomorphisme du cercle "structurellement stable" [au sens de l'article "Differentiable Dynamical Systems" dans Smale (1980)] ayant un nombre fini de points périodiques attractants et un nombre égal de points périodiques répulsifs.

(Rappelons [voir Herman (1979) chapitre II pour les démonstrations] qu'étant donné un homéomorphisme du cercle préservant l'orientation $f: \mathbb{T}^1 \to \mathbb{T}^1 (\mathbb{T}^1 = \mathbb{R}/\mathbb{Z})$ on peut le relever en un homéomorphisme $\tilde{f}: \mathbb{R} \to \mathbb{R}$ de la forme $\tilde{f}(\theta) = \theta + \varphi(\theta)$, $\varphi(\theta + 1) = \varphi(\theta)$ et que la suite de fonctions $(\tilde{f}^n - Id)/n$ ($\tilde{f}^n = $ itéré $n^{\text{ème}}$ de $\tilde{f}$) converge uniformément vers une fonction constante dont la valeur $\varrho(\tilde{f}) \in \mathbb{R}$ est appelée nombre de rotation de $\tilde{f}$ (H. Poincaré). Deux relèvements $\tilde{f}_1$ et $\tilde{f}_2$ de f diffèrent d'un entier, et il en est de même de $\varrho(\tilde{f}_1)$ et $P(\tilde{f}_2)$, ce qui permet d'associer à f son nombre du rotation $\varrho(f) \in \mathbb{T}^1 = \mathbb{R}/\mathbb{Z}$ [classe de $\varrho(\tilde{f})$ modulo $\mathbb{Z}$]; $\varrho(f)$ est une "rotation moyenne" de f, en particulier égal à α si $f = R_\alpha$ est la rotation de α définie par $R_\alpha(\theta) = \theta + \alpha$. Les propriétés essentielles de $f \to \varrho(f)$ sont les suivantes:

(1) $f \to \varrho(f)$ est continu pour la topologie de la convergence uniforme sur f;

(2) ϱ est un invariant de conjugaison C°: $\varrho(f) = \varrho(h \circ f \circ h^{-1})$ si $h: \mathbb{T}^1 \to \mathbb{T}^1$ est un homéomorphisme;

(3) $\varrho(f)$ est rationnel si et seulement si f possède au moins un point périodique (de période q si $\varrho(f) = p/q$);

(4) Lorsque f est un difféomorphisme de classe C^2, $\varrho(f)$ est irrationnel si et seulement si $f = h \circ R_{\varrho(f)} \circ h^{-1}$, $h: \mathbb{T}^1 \to \mathbb{T}^1$ homéomorphisme (A. Denjoy);

(5) il existe $A \subset \mathbb{T}^1 - \mathbb{Q}/\mathbb{Z}$ de mesure de Lebesgue égale à 1 tel que si f est un difféomorphisme C^∞ de nombre de rotation $\varrho(f) \in A$, $f = h \circ R_{\varrho(f)} \circ h^{-1}$, $h: \mathbb{T}^1 \to \mathbb{T}^1$ difféomorphisme C^∞ (M. Herman).)

Pour comprendre la façon dont apparaissent des points périodiques sur les courbes invariantes données par les théorèmes de bifurcation du paragraphe précédent, nous considérons maintenant certaines familles génériques à deux paramètres $P_{\mu,t}$ (on peut étudier de la même manière des familles à $(N+1)$ paramètres $\mu_0,\ldots,\mu_{N-1},t$; voir le chapitre 5 pour le cas $N=2$) telles que

(1) spectre $DP_{0,0}(0) = \{\lambda_{0,0}, \bar{\lambda}_{0,0}\}$, $\lambda_{0,0} = e^{2\pi i p/q}$, $q \geq 5$;
(2) l'application $(\mu,t) \mapsto \lambda_{\mu,t}$ est un difféomorphisme local de $\mathbb{R}^2$ (55) dans $\mathbb{C}$ au voisinage de $(0,0)$.
$\{\lambda_{\mu,t}, \bar{\lambda}_{\mu,t}\}$ est le spectre de $(DP_{\mu,t}(0))$

[comparer à Arnold (1980) chapitre 6, § 35 L].

On peut supposer après changement éventuel de coordonnées [comparer à (40)] que

$$P_{\mu,t}(z) = N_{\mu,t}(z) + \exp(2\pi i p/q)c(\mu,t)\bar{z}^{q-1} + O(|z|^q),$$

$c(0,0)$ réel $\neq 0$,

$$N_{\mu,t}(z) = z[1 + f(\mu,t,|z|^2)\exp[2\pi i g(\mu,t,|z|^2)]], \qquad (56)$$

$$f(\mu,X) = \mu + a_1(\mu,t)X + \ldots, a_1(0,0) \neq 0 \text{ (par ex. } = -1),$$

$$g(\mu,t,X) = \frac{p}{q} + t + b_1(\mu,t)X + \ldots, b_1(0,0) \neq 0 \text{ (par ex. } = 1).$$

Si on note l'itéré $k$ème de $N_{\mu,t}$ sous la forme

$$N_{\mu,t}^k(z) = e^{2\pi i k p/q} z \alpha_k(|z|^2)\exp[2\pi i \beta_k(|z|^2)] \qquad (57)$$

on constate sans peine que

$$P_{\mu,t}^k(z) = N_{\mu,t}^k(z) + \exp(2\pi i k p/q)c_k(\mu,t)\bar{z}^{q-1} + O(|z|^q). \qquad (58)$$

Soit $r_q = r_q(\mu,t)$ l'unique (si elle existe) petite racine positive de l'équation

$$\beta_q(r_q^2) = 0. \qquad (59)$$

r_q est le rayon de l'unique cercle (s'il existe) transformé "radialement" par $N_{\mu,t}^q$ (fig. 39): si $b_1(0,0) > 0$ un tel cercle existe pour $t < 0$. Si de plus $\alpha_q(r_q^2) - 1 = 0$, le cercle $|z| = r_q$ est fixé par $N_{\mu,t}^q$. C'est clairement au voisinage d'un tel cercle qu'il faut chercher d'éventuels points périodiques de période q de $P_{\mu,t}$ qui bifurquent à partir de l'origine [cas $(\mu,t) = (0,0)$]. Le changement de variables (pour $t < 0$)

$$r = r_q \varrho, \qquad z = r e^{2\pi i \theta} \qquad (60)$$

Fig. 39. Le cercle $|z| = r_q$ est transformé "radialement" par $N_{\mu,\,t}^q$.

conduit à l'expression suivante de $P_{\mu,\,t}^q$:

$$P_{\mu,\,t}^q(\theta, \varrho) = (\theta_q, \varrho_q),$$

$$\theta_q = p + \theta + \tilde{\beta}_q(r_q^2 \varrho^2) r_q^2 \hat{f}_1(\mu, t, \varrho, \theta), \tag{61}$$

$$\varrho_q = \varrho + \delta_q r_q^{q-1} \hat{f}_2(\mu, t, \varrho, \theta),$$

où

$$\hat{f}_1(\mu, t, \varrho, \theta) = (\varrho^2 - 1) - \frac{\delta_q r_q^{q-4} \varrho^{q-2}}{2\pi \tilde{\beta}_q(r_q^2 \varrho^2) \alpha_q(r_q^2 \varrho^2)} \sin 2\pi U_q + O(r_q^{q-3}),$$

$$\hat{f}_2(\mu, t, \varrho, \theta) = \varrho^{q-1} \cos 2\pi U_q + \frac{\tilde{\alpha}_q(r_q^2 \varrho^2) \varrho (\varrho^2 - 1)}{\delta_q r_q^{q-4}} \tag{62}$$

$$+ \frac{\varrho}{\delta_q} \frac{(\alpha_q(r_q^2) - 1)}{r_q^{q-2}} + O(r_q),$$

et

$$c_q(\mu, t) = \delta_q\, e^{2\pi i u_q},$$

$$\tilde{\alpha}_q(r^2) = \frac{\alpha_q(r^2) - \alpha_q(r_q^2)}{r^2 - r_q^2},$$

$$\tilde{\beta}_q(r^2) = \frac{\beta_q(r^2) - \beta_q(r_q^2)}{r^2 - r_q^2} = \frac{\beta_q(r^2)}{r^2 - r_q^2}, \tag{63}$$

$$U_q = q\theta - u_q + \beta_q(r_q^2 \varrho^2).$$

Les points fixes de $P_{\mu,\,t}^q$ proches de l'origine sont donnés par les équations

$$\hat{f}_1(\mu, t, \varrho, \theta) = 0,$$

$$\hat{f}_2(\mu, t, \varrho, \theta) = 0. \tag{64}$$

La première se résoud sans problème par le théorème des fonctions implicites et fournit pour (μ, t) proche de $(0,0)$ une courbe de la forme $\varrho = \varrho_{\mu, t}(\theta)$ proche du cercle $|z| = r_q$ (i.e. $\varrho = 1$) transformée "radialement" par $P^q_{\mu, t}$ (fig. 40). Remarquons que, si $\hat{f}_1(\mu, t, \varrho, \theta) = 0$, la deuxième équation s'écrit

$$\cos 2\pi U_q + \frac{\alpha_q(r_q^2\varrho^2)}{2\pi\tilde{\beta}_q(r_q^2\varrho^2)\alpha_q(r_q^2\varrho^2)}\sin 2\pi U_q$$

$$+ \frac{1}{\delta_q\varrho^{q-2}}\frac{\alpha_q(r_q^2) - 1}{r_q^{q-2}} + O(r_q) = 0. \tag{65}$$

C'est donc la taille de $[\alpha_q(r_q^2) - 1]/r_q^{q-2}$ qui détermine l'existence ou non de solutions de l'eq. (64) ou, ce qui revient au même, l'existence ou non d'intersection entre la courbe $\varrho = \varrho_{\mu, t}(\theta)$ et son image par $P^q_{\mu, t}$.

Il y aura en particulier $2q$ solutions si

$$\frac{\alpha_q(r_q^2) - 1}{r_q^{q-2}} < \gamma \qquad \text{assez petit.}$$

Il reste à remarquer que $\alpha_q(r_q^2) - 1$ est de l'ordre de $q(\mu + a_1(0, 0)r_q^2)$ et r_q de l'ordre de $[-t/b_1(0, 0)]^{1/2}$ pour obtenir le domaine $\mathscr{P}_{p/q}$ de l'espace des paramètres dans lequel $P_{\mu, t}$ possède deux orbites de points périodiques de période q (fig. 41).

Remarques.

(1) A t fixé, μ est dans un domaine d'hyperbolicité normale et les points périodiques obtenus ne peuvent qu'appartenir à une courbe invariante [voir eq. (46)]. On comparera à la fig. 145 de Arnold (1980).

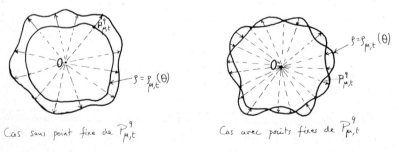

Fig. 40. Suivant que la courbe transformée "radialement" par $P^q_{\mu, t}$ intersecte ou non son image, on obtient ou non des points périodiques de période q de $P_{\mu, t}$.

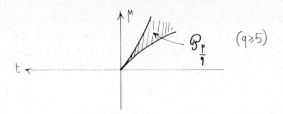

Fig. 41. Le domaine $\mathscr{P}p/q$: approximativement

$$\left| \mu - \frac{a_1(0,0)}{b_1(0,0)} t \right| < \gamma |t|^{(q-2)/2}.$$

On a représenté le cas où $a_1(0,0) = -1$, $b_1(0,0) = 1$. L'orientation bizarre de l'axe des t vient de ce qu'on le pense comme un morceau de cercle orienté dans le sens trigonométrique.

(2) L'idée de chercher une courbe $\varrho = \varrho_{\mu,t}(\theta)$ transformée "radialement" par $P^q_{\mu,t}$ remonte à Birkhoff et Lewis dans le cas conservatif (où alors cette courbe recoupe forcément son image par $P^q_{\mu,t}$).

(3) Nous verrons au chapitre 5 que $c(0,0)$ est simplement relié à l'angle des variétés stables et instables des points fixes de $P^q_{\mu,t}$ qui sont des cols.

Comme le remarque Arnold, les considérations précédentes permettent de comprendre en termes locaux l'infinité de bifurcations que subit la restriction du difféomorphisme P_μ à une courbe fermée invariante C_μ apparue par bifurcation de Hopf générique (la généricité est ici au sens de la catégorie de Baire comme dans le chapitre 5): il suffit de plonger cette famille dans une famille à deux paramètres dans laquelle le spectre de la dérivée en 0 des difféomorphismes de la famille parcourt tout un voisinage *dans* $\mathbb{C}$ du spectre de $DP_0(0)$; au voisinage de chaque racine de l'unité, le phénomène décrit ci-dessus d'apparition de points périodiques isolés sur une courbe fermée invariante se produit, la période tendant vers l'infini lorsque le spectre du difféomorphisme se rapproche du cercle unité (fig. 42: remarquer que ceci reste vrai même si $\lambda^q_0 = 1$, $q \geq 5$). Il y a donc, pour la famille $P_\mu | C_\mu$ de difféomorphismes du cercle, une infinité de bifurcations du type collision et apparition de points périodiques dans tout voisinage de $\mu = 0$.

On a représenté sur la fig. 43 la fonction $\mu \to \varrho\,(P_\mu | C_\mu)$ pour une famille à 1 paramètre générique présentant une bifurcation de Hopf [formules (48)] et pour la famille de formes normales N_μ qui lui correspond.

D'une fonction C^∞ (formes normales), cette fonction devient un

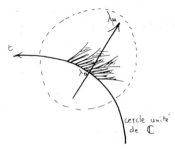

Fig. 42. On a représenté une partie des valeurs des paramètres pour lesquelles le difféomorphisme $z \mapsto \exp(2\pi i t)\, P_\mu(z)$ possède des points périodiques proches de 0: la famille P_μ est supposée générique, et telle que Spectre $DP_\mu(0) = \{\lambda_\mu, \bar{\lambda}_\mu\}$, $\lambda_\mu = (1 + \mu)\lambda_0, |\lambda_0| = 1$, $\lambda_0^q \neq 1$ pour $q < 5$.

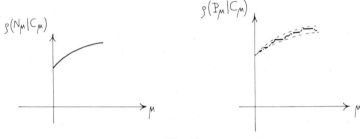

Fig. 43.

"escalier du diable" ayant des paliers à chaque valeur rationnelle. Au voisinage de $\mu = 0$, le graphe est contenu entre les graphes de deux fonctions polynomiales ayant un contact d'ordre arbitrairement grand pour $\mu = 0$ (voir Iooss (1979), III.3 théorème 3). Génériquement, l'ensemble des valeurs de μ pour lesquelles $\varrho(P_\mu|C_\mu)$ est rationnel est un ouvert dense dans $\mathbb{R}$ au voisinage de 0; mais on déduit de Herman (1977) que la mesure relative de l'ensemble des μ pour lesquels $\varrho(P_\mu|C_\mu)$ est irrationnel tend vers 1 lorsque μ tend vers 0 [voir également Chenciner (1978)].

4. Bifurcations de courbes invariantes non normalement hyperboliques

Que saisir sinon qui s'échappe,
Que voir sinon qui s'obscurcit,
Que désirer sinon qui meurt,
Sinon qui parle et se déchire?

         Yves Bonnefoy

4.1. Introduction

Les phénomènes étudiés dans ce chapitre et le suivant apparaissent dans leur généralité dès la codimension 2. Nous supposons donc que $N = 2$, et étudions des familles génériques à deux paramètres de difféomorphismes locaux de $(\mathbb{R}^2, 0)$ de la forme (45) (au changement près du nom des paramètres, qui a plus à voir avec le sentiment qu'avec la logique, et du nom de l'argument du difféomorphisme, qui aura, lui, son utilité):

$$P_{\mu, a}(z) = e^{2\pi i \omega} z[1 + f(\mu, a, |z|^2) + O(|z|^{2n+2})] \tag{66}$$

$$\times \exp\{2\pi i[g_\omega(\mu, a, |z|^2) + O(|z|^{2n+2})]\},$$

$$f(\mu, a, X) = \mu + aX + a_2(\mu, a)X^2 + \ldots + a_n(\mu, a)X^n, \quad a_2(0, 0) \neq 0,$$

$$g_\omega(\mu, a, X) = \omega(\mu, a) - \omega + b_1(\mu, a)X + \ldots + b_n(\mu, a)X^n,$$

$$b_1(0, 0) \neq 0.$$

Remarquons la nouvelle condition (générique) $b_1(0, 0) \neq 0$. Pour fixer les idées, nous supposerons

$$a_2(0, 0) = -1, \qquad b_1(0, 0) = 1, \qquad \omega(0, 0) = \omega_0.$$

Rappelons qu'une telle famille est générique si $\omega_0 \neq p/q$ pour $q \leq 2n + 3$ ce que nous supposerons. Remarquons enfin que l'indice ω est muet ($P_{\mu, a}$ ne dépend pas de ω). Nous supposerons que (μ, a) est dans le voisinage effilé V de la courbe Γ représenté sur la fig. 36 (où l'on se contentera de remplacer μ_0 par μ et μ_1 par a). Lorsque $(\mu, a) \in \Gamma$, la forme normale $N_{\mu, a}$ de $P_{\mu, a}$ possède, au voisinage de 0, un unique cercle invariant non normalement hyperbolique. Un exemple typique de difféomorphisme ayant des courbes invariantes non normalement hyperboliques est un difféomorphisme qui conserve les aires, au voisinage d'un point fixe elliptique [voir Moser (1973)]. L'expérience de ce genre de difféomorphismes (qui est, pour nous, hautement non générique) nous apprend deux choses:

(1) il ne faut pas s'attendre à une dynamique trop simple de $P_{\mu,a}$ lorsque $(\mu, a) \in V$,

(2) le paramètre important (qui va remplacer l'attraction ou la répulsion normale d'un cercle invariant de $N_{\mu,a}$) est le cisaillement tangentiel (twist) de $N_{\mu,a}$ au voisinage d'un cercle invariant, c'est-à-dire le taux de variation par rapport au rayon du cercle de l'angle de rotation de la restriction de $N_{\mu,a}$ à un cercle proche d'un cercle invariant); sa non-nullité est assurée par $b_1(0,0) \neq 0$.

Enfin, heuristiquement, la famille à un paramètre de cercles invariants de $N_{\mu,a}$ obtenue lorsque (μ, a) parcourt la courbe Γ, va remplacer la famille des cercles centrés à l'origine qui sont tous laissés invariants par la forme normale de Birkhoff d'un difféomorphisme P conservant les aires. On est donc amenés à faire, en plus de l'hypothèse $b_1(0,0) \neq 0$ l'hypothèse (génériquement vérifiée) que le nombre de rotation de la restriction de $N_{\mu,a}$ a son unique cercle invariant *varie effectivement* lorsque (μ, a) parcourt Γ. (C'est l'analogue de l'hypothèse de "twist" dans [Moser (1973)]; si on ne la fait pas, on peut s'attendre à des pathologies analogues à celles décrites par Anosov et Katok, dont on a parlé à la fin de 3.2 [voir Moser (1973) p. 56].

Si l'on pose

$$\eta(\mu, a) = 1 - \frac{2a_2(\mu, a)}{b_1(\mu, a)} \frac{\partial \omega}{\partial a}(\mu, a) \tag{67}$$

cette dernière condition est

$$\eta(0, 0) = 1 + 2\frac{\partial \omega}{\partial a}(0, 0) \neq 0. \tag{68}$$

En effet, en un point $(\mu, a) \in \Gamma$ le rayon r de l'unique cercle invariant de $N_{\mu,a}$ est de la forme

$$r^2 = \tfrac{1}{2}a + O(|a|^2) \tag{69}$$

et la rotation induite par $N_{\mu,a}$ sur ce cercle a pour angle

$$2\pi\omega = 2\pi[\omega_0 + \tfrac{1}{2}\eta(0, 0)a + O(|a|^2)]. \tag{70}$$

Comparons maintenant les stratégies de recherche de courbes fermées invariantes de $P_{\mu,a}$ suivant que $(\mu, a) \in V$ ou $(\mu, a) \in \mathcal{H}$ (on fera le parallèle avec 1.1 et 1.3):

(1) Lorsque $(\mu, a) \in \mathcal{H}$, on cherche une courbe fermée invariante par $P_{\mu,a}$ pour une valeur donnée (μ_0, a_0) de (μ, a). Lorsque $(\mu, a) \in V$, on pourra seule-

ment affirmer l'existence de valeurs de (μ, a) proches de $(\mu_0, a_0) \in \Gamma$ pour lesquelles $P_{\mu, a}$ aura une courbe fermée invariante.

(2) Lorsque $(\mu, a) \in \mathcal{H}$ les courbes obtenues sont de classe C^k, k fini. Lorsque $(\mu, a) \in V$, les courbes obtenues seront de classe C^∞. Plus précisément, les courbes sont obtenues dans le premier cas comme conséquence du théorème des contractions dans un Banach de fonctions C^k et sont donc automatiquement normalement hyperboliques; dans le deuxième, elles viendront de l'application d'un théorème raffiné de fonctions implicites dans un Fréchet de fonctions C^∞, et pourront ne pas être normalement hyperboliques.

(3) Lorsque $(\mu, a) \in \mathcal{H}$ on ne contrôle pas la dynamique de $P_{\mu, a}$ sur la courbe invariante cherchée. Lorsque $(\mu, a) \in V$, on trouvera uniquement des courbes fermées invariantes sur lesquelles $P_{\mu, a}$ induise un difféomorphisme conjugué à une rotation d'un angle irrationnel fixé.

Remarquons l'opposition entre (2) et (3): dans un cas on contrôle bien la dynamique au voisinage de la courbe mais par sur la courbe; dans l'autre on contrôle la dynamique sur la courbe mais pas au voisinage.

Dans la suite du chapitre, nous donnons une idée précise des résultats concernant l'existence pour $P_{\mu, a}$ de courbes fermées invariantes lorsque $(\mu, a) \in V$, ainsi que des étapes de leur démonstration. Dans le chapitre suivant, nous étudions, toujours lorsque $(\mu, a) \in V$, les problèmes liés à l'existence et au comportement des points périodiques de $P_{\mu, a}$.

4.2. Résultats

Les résultats énoncés ci-dessous sont démontrés dans Chenciner (1982d) sous l'hypothèse que ω_0 ne s'écrive pas p/q avec p, q entiers et $q \le 33$. En particulier, ces résultats sont de nature générique (dans tous les sens du terme, ce qui n'était pas le cas pour la première version annoncée dans Chenciner (1981).

Une telle hypothèse de non-résonnance est à rapprocher de celle qu'on doit faire pour démontrer la stabilité des points fixes elliptiques dans le cas hamiltonien.

Nous allons montrer, sous les hypothèses précédentes, que:

(i) Lorsqu'on passe de la famille $N_{\mu, a}$ de formes normales à une famille $P_{\mu, a}$ générique de la forme (66), la courbe Γ de la fig. 36 est remplacée par un ensemble de Cantor $\tilde{\Gamma}$ (adhérant à 0) de valeurs de (μ, a) pour lesquelles $P_{\mu, a}$ possède une courbe fermée invariante C^∞ non normalement hyperbo-

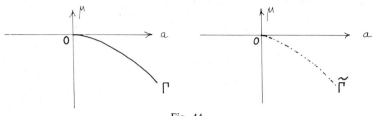

Fig. 44.

lique $\tilde{\mathscr{C}}_{\mu,a}$, telle que $P_{\mu,a}|\tilde{\mathscr{C}}_{\mu,a}$ soit C^∞-conjuguée à une rotation irration-
nelle $R_{\omega(\mu,a)}$ (voir fig. 44).

(ii) Chaque point $(\mu_0,a_0)\in\tilde{\Gamma}$ est centre d'un petit morceau de courbe
$\tilde{C}_{\omega(\mu_0,a_0)}$ dont les extrémités sont au-dessus de V, formé de valeurs de
(μ,a) pour lesquelles $P_{\mu,a}$ possède, si $(\mu,a)\ne(\mu_0,a_0)$, une courbe fermée
invariante C^∞ normalement hyperbolique $\tilde{\mathscr{C}}_{\mu,a}$, telle que $P_{\mu,a}|\tilde{\mathscr{C}}_{\mu,a}$ soit C^∞
conjuguée à la rotation $R_{\omega(\mu_0,a_0)}$; cette courbe est un attracteur si (μ,a)
est d'un côté de (μ_0,a_0), un répulseur de l'autre [de quel côté dépend du
signe de $\eta(0,0)$] (voir fig. 45).

(iii) On déduit de (ii), et de la stabilité des courbes invariantes normalement hyperboliques, que chaque point $(\mu_0,a_0)\in\tilde{\Gamma}$ est en fait centre d'un
double cône effilé formé de valeurs de (μ,a) pour lesquelles $P_{\mu,a}$ possède
une courbe fermée invariante C^k normalement hyperbolique si $(\mu,a)\ne$
(μ_0,a_0), attracteur si (μ,a) est d'un côté de (μ_0,a_0), répulseur s'il est de
l'autre (fig. 46).

(iv) En un point d'intersection (μ,a) de deux des demi-cônes ouverts
dont il est question en (iii), $P_{\mu,a}$ possède deux courbes fermées invariantes
C^k l'une attractante, l'autre répulsive.
(Il ne faudrait cependant pas en déduire que $P_{\mu,a}$ "ressemble" à $N_{\mu,a}$.)

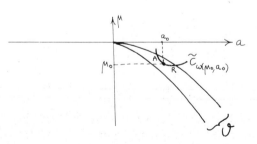

Fig. 45. Cas $\eta(0,0)>0$; $A=$ attracteur, $R=$ répulseur.

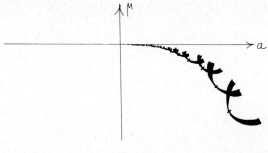

Fig. 46.

4.3. *Idée des demonstrations*

Les démonstrations de ces résultats reposent sur une analyse des structures obtenues dans l'espace des paramètres (μ, a) lorsqu'on s'intéresse aux $N_{\mu, a}$ ayant un cercle invariant $\mathscr{C}$ tel que $N_{\mu, a} | \mathscr{C}$ soit une rotation de nombre de rotation fixé. Plus précisément, on montre facilement que

$$C_\omega = \{(\mu, a)| \text{ il existe un cercle } \mathscr{C}_{\omega, \mu, a} \text{ invariant par } N_{\mu, a}$$

$$\text{tel que } N_{\mu, a} | \mathscr{C}_{\omega, \mu, a} = R_\omega\} \tag{71}$$

est une courbe C^∞ si $\eta(0, 0) \neq 0$. (On identifie un cercle $|z| = \varrho$ avec $\mathbb{R}/\mathbb{Z}$ par $\theta \mapsto \varrho e^{2\pi i\theta}$; R_ω correspond donc à une rotation d'angle $2\pi\omega$.) De plus, lorsque $\eta(0,0) > 0$ (resp. < 0), C_ω est tangente en un unique point γ_ω à Γ si $\omega \gtrless \omega_0$ (resp. $\omega \lesseqgtr \omega_0$) (la fig. 47 représente respectivement les quatre cas génériques; on a toujours $\omega_2 < \omega_0 < \omega_1$).

Nous pouvons introduire maintenant ce qui, dans Chenciner (1981) est appelé "système de coordonnées adapté" à un $\omega(\gtrless \omega_0$ si $\eta(0,0) \gtrless 0)$: Pour (μ, a) proche de γ_ω, soit $r(\omega, \mu, a)$ l'unique solution positive proche de 0 de l'équation

$$g_\omega(\mu, a, r(\omega, \mu, a)^2) = 0 \tag{72}$$

$N_{\mu, a}$ applique le cercle $|z| = r(\omega, \mu, a)$ sur le cercle $|z| = r(\omega, \mu, a)[1 + f(\mu, a, r(\omega, \mu, a)^2)]$ par la rotation R_ω. Posons, pour α donné,

$$v = f(\mu, a, r(\omega, \mu, a)^2) \text{ (la courbe } C_\omega \text{ est définie par } v = 0),$$

$$\varepsilon = f(\mu, a, r(\omega, \mu, a)^2) + 2\frac{\partial f}{\partial x}(\mu, a, r(\omega, \mu, a)^2) \cdot r(\omega, \mu, a)^2,$$

$$\tag{73}$$

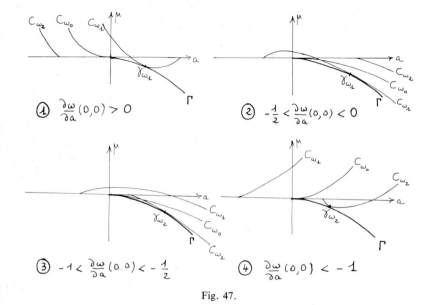

Fig. 47.

$$\tau = 2 \frac{\partial g_\omega}{\partial X}(\mu, a, r(\omega, \mu, a)^2) \cdot r(\omega, \mu, a)^2 \cdot \alpha,$$

$$r = |z| = r(\omega, \mu, a)\ (1 + \alpha\sigma).\ ^*$$

On montre:

(1) L'application $\Phi_\omega : (\mu, a) \mapsto (\nu, \varepsilon)$ est un difféomorphisme local au voisinage de γ_ω, pourvu que ω soit assez proche de ω_0; son déterminant jacobien au point γ_ω est en effet égal à

$$2\eta(\mu_\omega, a_\omega)\varrho_\omega^2 + O(\varrho_\omega^4) \quad \text{où } \gamma_\omega = (\mu_\omega, a_\omega),\ \varrho_\omega = r(\omega, \mu_\omega, a_\omega).$$

(2) Soit $(\nu, \varepsilon) = \Phi_\omega(\mu, a)$; si (μ, a) est voisin de $(0,0)$, et si α est assez petit, $N_{\mu, a}$ définit un difféomorphisme de $\mathbb{T}^1 \times [-1, +1]$ sur un voisinage de $\mathbb{T}^1 \times \{0\}$ dans $\mathbb{T}^1 \times \mathbb{R}$ de la forme suivante:

$$N_{\mu, a}(\theta, \sigma) = \Theta, \Sigma,$$

$$\Theta = \theta + \omega + \tau\sigma + O(\tau\alpha)\sigma^2 + O(\tau^2\alpha)\sigma^3 + \ldots + O(\tau^n\alpha^n)\sigma^{2n}, \tag{74}$$

$$\Sigma = \frac{\nu}{\alpha} + (1 + \varepsilon)\sigma + O[(\varepsilon - \nu)\alpha + \tau r(\omega, \mu, a)^2]\sigma^2 + \ldots + O(\tau^n\alpha^n)\sigma^{2n+1},$$

$^*r = r(\omega, \mu, a)(1 + \alpha\sigma)^{1/2}$ est un meilleur changement de variables; il permet en particulier de choisir $\alpha = 1$ dans le lemme (75) [voir Chenciner (1982d)].

où les termes notés $O(x)$ sont de la forme $x\,\alpha^k\varphi(\omega,\mu,a)$, k entier ≥ 0, φ fonction C^∞ définie au voisinage de $(\omega_0,0,0)$.

La signification de $\tau,\nu/\alpha,\varepsilon$ apparaît clairement sur cette formule: ce sont respectivement le twist (cisaillement tangentiel), la "translation" (différence des rayons) et la contraction normale, associées à l'application $N_{\mu,a}$ au voisinage du cercle $|z| = r(\omega,\mu,a)$.

Remarquons que lorsqu'on parcourt $C_\omega(\nu = 0)$, le passage par $\gamma_\omega(\varepsilon = 0)$ correspond à un changement de nature (attractant ou répulsif) du cercle invariant $|z| = r(\omega,\mu,a)$ de $N_{\mu,a}$. C'est ce fait qui va nous permettre de prouver que, pour un Cantor de valeurs de (μ,a) proches de Γ, le difféomorphisme $P_{\mu,a}$ possède une courbe fermée C^∞ invariante non normalement hyperbolique.

Le lemme fondamental est le suivant:

Si $\alpha = \varrho_\omega^2, |\varepsilon| \leq \varrho_\omega^6, |\nu| \leq \varrho_\omega^8, n \geq 3, (\nu,\varepsilon) = \Phi_\omega(\mu,a)$, on a

$$P_{\mu,a}(\theta,\sigma) = (\theta + \omega + \tau_\omega\sigma + \tau_\omega^{3/2}A_{\omega,\mu,a}(\theta,\sigma), \sigma + \tau_\omega^{3/2}B_{\omega,\mu,a}(\theta,\sigma)), \quad (75)$$

où $\tau_\omega = \tau(\omega,\mu_\omega,a_\omega) = 2\varrho_\omega^4 + O(\varrho_\omega^6)$, et où les fonctions $A_{\omega,\mu,a}, B_{\omega,\mu,a}$, sont bornées en norme C^k (pour tout k) sur $\mathbb{T}^1 \times [-1,+1]$ *uniformément* par rapport à ω,μ,a vérifiant les conditions ci-dessus.

Pa contraste, dès qu'on s'éloigne de γ_ω, on a des estimations du type "hyperbolicité normale", où le rôle joué par τ dans le lemme (75) est maintenant tenu par ε:

Si $\alpha = \varepsilon^{3/2}, 0 < r(\omega,\mu,a)^{(2n+2)/3} \leq |\varepsilon|, |\nu| \leq \varepsilon^3, (\nu,\varepsilon) = \Phi_\omega(\mu,a)$, on a

$$P_{\mu,a}(\theta,\sigma) = (\theta + \omega + |\varepsilon|^{3/2}C_{\omega,\mu,a}(\theta,\sigma), (1+\varepsilon)\sigma + |\varepsilon|^{3/2}D_{\omega,\mu,a}(\theta,\sigma)),$$

$$(76)$$

où les fonctions $C_{\omega,\mu,a}$ et $D_{\omega,\mu,a}$ sont bornées en norme C^k (pour tout k) sur $\mathbb{T}^1 \times [-1,+1]$, *uniformément* par rapport à ω,μ,a vérifiant les conditions ci-dessus.

La fig. 48 représente le système de coordonnées (ν,ε) au voisinage de γ_ω, la fig. 49 indique les zônes respectives de validité des lemmes (75) et (76) (les deux figures sont faites dans le cas où $(\partial\omega/\partial a)(0,0) > 0$).

Nous supposerons dorénavant satisfaites les hypothèses du lemme (75); on peut voir que $A_{\omega,\mu,a}(\theta,\sigma)$ et $B_{\omega,\mu,a}(\theta,\sigma)$ sont chacun la somme d'un

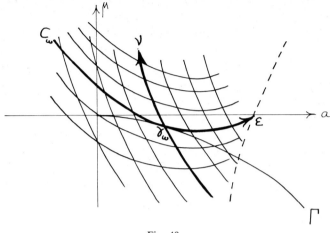

Fig. 48.

polynôme en σ (indépendant de θ) et d'un terme de la forme $\tau_\omega^{(n-3)/2} \cdot U_{\omega,\mu,a}(\theta,\sigma)$; si ce terme supplémentaire est nul (cas de la forme normale $N_{\mu,a}$), le cercle $\sigma=0$ [c'est à dire $|z|=r(\omega,\mu,a)$] est appliqué sur le cercle "translaté" (au sens des coordonnées θ,σ) $\sigma = \nu/\varrho_\omega^2$. Un théorème de Rüssmann [voir Herman (1980)] adapté à notre situation [voir Chenciner (1982d) nous permet d'affirmer qu'une situation analogue se produit dans le cas général à condition que ω soit mal approché par les rationnels (i.e. $\exists C, \beta > 0$ tels que pour tout rationnel p/q,

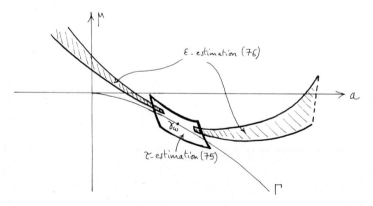

Fig. 49.

$$|\omega - p/q| \geq \frac{C\tau_\omega}{|q|^{2+\beta}}.$$

Plus précisément, dans le domaine de validité du lemme (75), il existe, pour de tels ω, une unique courbe fermée C^∞, $\tilde{\mathscr{C}}_{\omega,\mu,a}$ d'équation $\sigma = \psi_{\omega,\mu,a}(\theta)$, que $P_{\mu,a}$ applique sur la courbe "translatée" d'équation $\sigma = \psi_{\omega,\mu,a}(\theta) + \lambda_{\omega,\mu,a}$ par un difféomorphisme C^∞ – conjugué à R_ω (on paramètre les deux courbes par θ).

Le théorème de Rüssmann est déduit dans Herman (1980) du théorème de fonctions implicites de Hamilton; ceci a pour nous une conséquence fondamentale: la "translation" $\lambda_{\omega,\mu,a}$ dépend de manière C^∞ (C^1 suffirait), dans un sens raisonnable, des perturbations $\tau_\omega^{3/2} \cdot A_{\omega,\mu,a}$ et $\tau_\omega^{3/2} \cdot B_{\omega,\mu,a}$, et est donc proche de ν/ϱ_ω^2 si ω est proche ω_0 (il faut bien entendu des estimations précises). On en déduit que, si ω est assez proche de ω_0 (et n assez grand), $\lambda_{\omega,\mu,a}$ s'annule sur une courbe $\tilde{C}_\omega$ d'équation $\nu = t(\varepsilon)$, où $t: [-\varrho_\omega^6, \varrho_\omega^6] \to [-\varrho_\omega^8, \varrho_\omega^8]$ est C^∞ et proche de o (fig. 50). Si $\Phi_\omega(\mu, a) = (t(\varepsilon), \varepsilon)$ (c'est-à-dire si $(\mu, a) \in \tilde{C}_\omega$), la courbe $\tilde{\mathscr{C}}_{\omega,\mu,a}$ est donc laissée invariante par $P_{\mu,a}$, et $P_{\mu,a} | \tilde{\mathscr{C}}_{\omega,\mu,a}$ est C^∞ – conjugué à R_ω. Nous venons donc de trouver pour $P_{\mu,a}$ l'analogue de C_ω, et en fait l'analogue des courbes $\nu = $ constante (les courbes $\lambda_{\omega,\mu,a} = $ cste) lorsque ω est mal approché par les rationnels. Nous allons maintenant trouver l'analogue des courbes $\varepsilon = $ constante: par un changement de coordonnées $(\theta, \sigma) \mapsto (\xi, x)$ tel que $\tilde{\mathscr{C}}_{\omega,\mu,a}$ s'écrive $x = 0$, $P_{\mu,a}$ devient

$$P_{\mu,a}(\xi, x) = [\xi + \omega + O(|x|), \lambda_{\omega,\mu,a} + X_{\omega,\mu,a}(\xi)x + O(|x|^2)]. \tag{77}$$

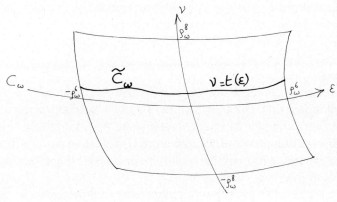

Fig. 50.

Au voisinage de y_ω, $X_{\omega,\mu,a}(\xi)$ est proche de 1, donc positif; puisque ω est mal approché par les rationnels, l'équation aux différences

$$\log Y_{\omega,\mu,a}(\xi+\omega) - \log Y_{\omega,\mu,a}(\xi) = \log X_{\omega,\mu,a}(\xi) - \int_0^1 \log X_{\omega,\mu,a}(\xi)\mathrm{d}\xi$$

(78)

a une unique solution $Y_{\omega,\mu,a}(\xi)C^\infty$ et proche de 1 (démonstration par séries de Fourier). Le changement de variables

$$(\xi,x) \mapsto \left(\xi, y = \frac{x}{Y_{\omega,\mu,a}(\xi)}\right)$$

(79)

transforme alors $P_{\mu,a}$ en

$$P_{\mu,a}(\xi,y) = \left(\xi+\omega+\mathrm{O}(|y|), \ \exp\left(\int_0^1 \log X_{\omega,\mu,a}(\xi)\mathrm{d}\xi\right)y+\mathrm{O}(y^2)\right)$$ (80)

pourvu que $(\mu,a) \in \tilde{C}_\omega$ (i.e. que $\tilde{\mathscr{C}}_{\omega,\mu,a}$ soit une courbe fermée *invariante* de $P_{\mu,a}$).

Si ω est assez proche de ω_0, et n assez grand, l'application ψ_ω définie par

$$\psi_\omega(v,\varepsilon) = \left(\lambda_{\omega,\Phi_\omega^{-1}(v,\varepsilon)}, \ \exp\left(\int_0^1 \log X_{\omega,\Phi_\omega^{-1}(v,\varepsilon)}(\xi)\mathrm{d}\xi\right)\right)$$

(81)

est un difféomorphisme de $[-\varrho_\omega^8,\varrho_\omega^8] \times [-\varrho_\omega^6,\varrho_\omega^6]$ sur son image et cette dernière contient $(0,1)$ (fig. 51 sous les mêmes hypothèses que les figs. 48, 49, 50). Soit $\tilde{\gamma}_\omega = (\psi_\omega \circ \Phi_\omega)^{-1}(0,1)$: on déduit de l'éq. (80) que $\tilde{\gamma}_\omega$ est l'unique point de la courbe $\tilde{C}_\omega$ tel que la courbe invariante $\tilde{\mathscr{C}}_{\omega,\tilde{\gamma}_\omega}$ de $P_{\tilde{\gamma}_\omega}$ soit non normalement hyperbolique, ce qui démontre toutes les assertions du paragraphe 2.

Remarques.

(1) Même pour les valeurs de (μ,a) telles que $P_{\mu,a}$ ait deux courbes fermées invariantes C^∞, l'une attractante, l'autre répulsive, on est loin de pouvoir décrire complètement la dynamique de $P_{\mu,a}$: les bassins d'attraction ou de répulsion des courbes sont à priori très étroits et il est vraisemblable qu'existent entre eux des points périodiques homoclines du type de ceux décrits dans le chapitre 5.

(2) Le théorème de Rüssmann était initialement destiné à prouver le théorème de Kolmogorov–Arnold–Moser: dans ce dernier, la condition de

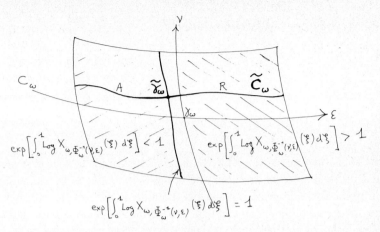

Fig. 51. En tous les points de $\hat{C}_\omega$, $P_{\mu,a}$ possède une courbe fermée C^∞ invariante $\tilde{\mathscr{C}}_{\omega,\mu,a}$, attractante ($A$) si (μ, a) est à gauche de $\tilde{y}_\omega$, répulsive (R) s'il est à droite.

conservation de l'aire intervient en effet comme une condition de codimension 1: chaque courbe fermée entourant l'origine doit recouper son image. En particulier, si la dite image est une courbe "translatée", la translation ne peut être que nulle. Dans la situation dissipative que nous avons considérée, c'est la présence de paramètres qui fournit des valeurs de (μ, a) pour lesquelles la "translation" s'annule.

On remarquera que, contrairement à ce que semblait penser son auteur à l'époque, le théorème de Rüssmann n'est nullement artificiel: il suffit de le comparer aux propriétés caractéristiques (*) des difféomorphismes $N_{\mu,a}$ énoncées en 2.3.

(3) On pourrait montrer qu'on obtient un Cantor Γ de mesure positive (en projection sur l'axe des a).

5. Bifurcations de points homoclines

Vous vous êtes abordés
Comme coquelicot et blé
Ma fille, ma fille, je tremble

René Char

Dans le chapitre précédent, les techniques fournissant la structure générique d'un difféomorphisme local de $\mathbb{R}^2$ préservant les aires au voisinage d'un point fixe elliptique ont permis d'établir l'existence de courbes fermées invariantes non normalement hyperboliques de $P_{\mu, a}$ pour certaines valeurs de (μ, a) voisines de Γ: le Cantor de courbes invariantes de la situation conservative est devenu un Cantor de valeurs de (μ, a); en quelque sorte, la situation conservative générique a été "déployée" dans la direction de Γ [de même que, dans la bifurcation de Hopf, la situation linéaire elliptique est "déployée" dans la direction du paramètre (fig. 6)]. Nous poursuivons maintenant l'analogie au niveau des points homoclines[*] relatifs à des orbites périodiques. Alors que dans le cas conservatif générique, tout voisinage de l'origine contient un point homocline transversal relatif à une orbite périodique dont la période tend vers l'infini lorsque le voisinage se rétrécit [Zehuder (1973)], nous allons montrer que pour une famille $P_{\mu, a}$ générique, étant donné un voisinage Ω de 0 dans $\mathbb{R}^2$, il existe une valeur de (μ, a) dans V, proche de $(0,0)$, telle que $P_{\mu, a}$ possède dans Ω un point homocline transversal [voir Chenciner (1982a)].

Remarque. Jusqu'ici, le mot "générique" signifiait le plus souvent "pourvu que certaines expressions algébriques ne s'annulent pas". Dans ce chapitre, la signification est [comme dans Zehnder (1973)] plus restreinte: c'est la généricité au sens de la catégorie de Baire (propriété vraie pour tous les éléments d'une intersection dénombrable d'ouverts denses) telle qu'elle intervient dans la théorie de la stabilité structurelle [voir Smale (1980)].

Comme dans Zehnder (1973), on commence par considérer des familles de difféomorphismes analytiques, que l'on rend génériques par une suite infinie de modifications élémentaires. Nous nous bornerons ici à la description de ces modifications élémentaires.

---

[*] Rappelons qu'un point est homocline s'il est dans l'intersection des variétés stable et instable de deux points périodiques appartenant à la même orbite.

L'idée principale est de trouver les points périodiques comme traces des bifurcations qui se produisent lorsque ω_0 prend des valeurs rationnelles (souvenir des résonances). Pour cela, on plonge comme dans 3.3 la famille $P_{\mu, a}$ étudiée dans une famille à trois paramètres qu'on peut supposer de la forme suivante (à ceci près qu'on a supposé c constant):

$$P_{\mu, a, t}(z) = N_{\mu, a, t}(z) + \exp(2\pi i p/q) c\bar{z}^{q-1} + O(|z|^q), c \text{ réel} \neq 0,$$

$$N_{\mu, a, t}(z) = \exp\left[2\pi i\left(\frac{p}{q} + t\right)\right] z(1 + f(\mu, a, |z|^2))$$

$$\times \exp[2\pi i g_{p/q}(\mu, a, |z|^2)],$$

$$f(\mu, a, X) = \mu + aX + a_2(\mu, a)X^2 + \ldots, \tag{82}$$

$$g_{p/q}(\mu, a, X) = \omega(\mu, a) - \frac{p}{p} + b_1(\mu, a)X + \ldots,$$

$$a_2(0, 0) = -1, \; b_1(0, 0) = 1, \; \omega(0, 0) = p/q$$

que l'on comparera à eq. (56). Dans la suite, on supposera $t < 0$ assez petit. Tous les calculs faits en 3.3 restent valables; on garde les mêmes notations $\alpha_k, \beta_k, r_q, \ldots$ [voir eqs. (57), (59), (63)].

Exercice. Montrer qu'on a les évaluations grossières

$$\alpha_q(r_q^2) - 1 \sim qf(\mu, a, r_q^2), \; \tilde{\alpha}_q(r_q^2) \sim q\frac{\partial f}{\partial x}(\mu, a, r_q^2),$$

$$\tilde{\beta}_q(r_q^2) \sim q\frac{\partial g}{\partial X}(\mu, a, r_q^2), \; \delta_q \sim q \cdot c$$

Le changement de variables (60) conduit aux formules (61) dont on déduit comme précédemment l'existence de $2q$ points fixés par $P_{\mu, a, t}^q$ pourvu que

$$|\alpha_q(r_q^2) - 1| < \gamma r_q^{q-2}, \qquad \gamma \text{ assez petit.} \tag{83}$$

Remarquons que cette condition signifie exactement que (μ, a) est assez proche de la courbe [comparer à eq. (71)]

$$C_{p/q}(t) = \{(\mu, a) \in \mathbb{R}^2 | \text{ il existe un cercle } \mathscr{C}_{p/q, \mu, a, t} \text{ invariant par}$$

$$N_{\mu, a, t}, \text{ tel que } N_{\mu, a, t} | \mathscr{C}_{p/q, \mu, a, t} = R_{p/q}\} \tag{84}$$

dont une équation est justement $\alpha_q(r_q^2) - 1 = 0$, puisque la dernière condition équivaut, si t est proche de 0, à $(N_{\mu,a,t} \,|\, \mathscr{C}_{p/q,\mu,a,t})^q = $ Identité (on sait a priori que c'est une rotation car $N_{\mu,a,t}$ est une forme normale, et si p'/q est assez proche de p/q, on a $p = p'$).

Notons $(\theta_\nu(t), \varrho_\nu(t))$, $\nu = 1, 2, \ldots, 2q$ les points fixes de $P^q_{\mu,a,t}$ ainsi obtenus, et faisons le changement de variables

$$
\begin{aligned}
\theta &= \theta_\nu(t) + \psi \\
\varrho &= \varrho_\nu(t) + r_q^{q/2-1} \cdot x
\end{aligned}
\tag{85}
$$

dans un anneau entourant le cercle $\varrho = 1$.

Si, en plus de (83), on impose que

$$
\tilde{\alpha}_q(r_q^2) = \mathrm{O}(r_q) \qquad \text{[notations de (63)]}
\tag{86}
$$

c'est-à-dire que (μ, a) n'est pas trop loin du point $\gamma_{p/q}(t) \in C_{p/q}(t)$, on obtient,

$$
\begin{aligned}
P^q_{\mu,a,t}(\psi, x) &= (\psi_q, x_q), \\
\psi_q &= p + \psi + 2\tilde{\beta}_q(r_q^2) r_q^{q/2} x + \mathrm{O}(r_q^{q-2}), \\
x_q &= (-1)^\nu \delta_q r_q^{q/2} \sin 2\pi q \psi + [1 + 2\tilde{\alpha}_q(r_q^2) r_q^2] x + \mathrm{O}(r_q^{q/2+1})
\end{aligned}
\tag{87}
$$

(ν pair correspond aux cols, ν impair aux centres, puits, ou sources).

Remarquons l'analogie avec les formules de Zehnder (1973), la seule différence provenant du terme $2\tilde{\alpha}_q(r_q^2) r_q^2 x$ qui s'annule en particulier au point $\gamma_{p/q}(t)$.

Très près de $\gamma_{p/q}(t)$, c'est-à-dire de la courbe Γ, plus précisément dès que

$$
\tilde{\alpha}_q(r_q^2) = \mathrm{O}(r_q^{q/2-1}),
\tag{88}
$$

ce terme disparaît et on retrouve les formules de Zehnder (1973) dans le cas conservatif.

Lorsque ν est pair, $P^q_{\mu,a,t}$ est proche, au voisinage de 0, de l'application "solution au temps 1" de l'équation différentielle

$$
\begin{aligned}
\mathrm{d}\psi/\mathrm{d}t &= 2\tilde{\beta}_q(r_q^2) r_q^{q/2} x, \\
\mathrm{d}x/\mathrm{d}t &= \delta_q r_q^{q/2} \sin 2\pi q \psi + 2\tilde{\alpha}_q(r_q^2) r_q^2 x,
\end{aligned}
\tag{89}
$$

c'est-à-dire [avec les notations de eq. (63)]

$$
\frac{\mathrm{d}^2 \psi}{\mathrm{d}t^2} = 2\delta_q \tilde{\beta}_q(r_q^2) r_q^q \sin 2\pi q \psi + 2\tilde{\alpha}_q(r_q^2) r_q^2 \frac{\mathrm{d}\psi}{\mathrm{d}t}
\tag{90}
$$

qui admet la fonction de Liapounov [de Hamilton si $\tilde{\alpha}_q(r_q^2)=0$]

$$F(x, \psi) = 2\tilde{\beta}_q(r_q^2)r_q^{q/2}\frac{x^2}{2} + \frac{\delta_q}{2\pi q}\cos 2\pi q\psi \qquad (91)$$

(équation du pendule avec frottement).

Le portrait de phase de cette équation est donné sur la fig. 52 pour les diverses valeurs du frottement $2\tilde{\alpha}_q(r_q^2)r_q^2$; on a indiqué sur la fig. 53 les régions correspondantes de l'espace des paramètres.

Exercice. Calculer l'angle des variétés stables et instables des cols en fonction de δ_q, et donc de c (on trouvera une expression qui s'annule avec c). Remarquons que les variétés stables et instables de deux cols consécutifs de l'équation différentielle se traversent lorsqu'on passe de 1 à 5. On peut montrer que la perturbation qui fait passer de la solution au temps 1 des diverses équations (89) (correspondant aux q cols) à l'application $P_{\mu, a, t}^q$ est suffisamment petite pour que cette propriété soit préservée. Un argument standard permet alors d'affirmer que l'intersection se fera transversalement en général, CQFD. (Figure 54 où le mouvement d'ensemble est indiqué.)

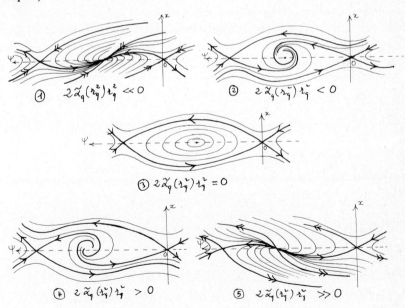

Fig. 52. L'orientation de ψ correspond à celle du cercle.

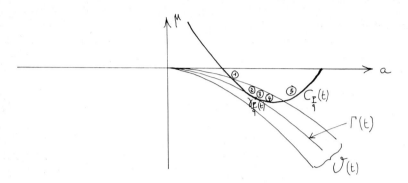

Fig. 53. La figure correspond à $N_{\mu, a, t}$ pour $t < 0$ petit fixé. Les points 1, 2, 3, 4, 5 sont voisins de $C_{p/q}(t)$ et peuvent en particulier être choisis sur cette courbe.

Remarques.

(1) Les cas 1 et 5 correspondent à des valeurs de (μ, a) en dehors de $V(t)$; les points périodiques appartiennent donc à une courbe différentiable invariante de $P_{\mu, a, t}$. Cette courbe est un attracteur en 1, un répulseur en 5.

(2) Un simple argument de continuité montre qu'il existe des valeurs de (μ, a) proches de $\gamma_{p/q}(t)$ pour lesquelles les variétés stables et instables de deux cols consécutifs de période q de $P_{\mu, a, t}$ sont tangentes; il est alors possible que des phénomènes du type de ceux décrits par S. Newhouse ("infinitely many sinks") apparaissent.

(3) Lorsqu'on s'éloigne de la courbe $C_{p/q}(t)$, les points périodiques s'éliminent par paires; près de $\gamma_{p/q}(t)$ on obtient ainsi des valeurs de (μ, a)

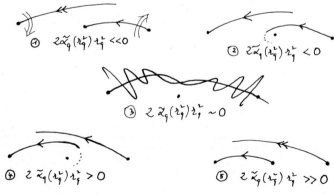

Fig. 54.

au voisinage desquelles se produit (en plus complexe) l'analogue pour les difféomorphismes de ce que Bogdanov a décrit pour les champs de vecteurs [voir Arnold (1980) fig. 135]: on peut donc s'attendre à de nouvelles bifurcations de Hopf pour $P_{\mu, a, t}^q$, etc.

(4) Pour une famille *donnée* $P_{\mu, a}$ l'existence de points périodiques se montre directement par un argument du type Birkhoff. Lewis (fig. 40) à condition que la période q ne soit pas trop grande par rapport à la distance $|\omega_0 - p/q|^{-1}$ (par exemple si $P_{\mu, a}$ est de la forme $P_{\mu, a, t}$ [voir eq. (82)] avec $|t|$ assez petit): en suivant un chemin transverse à $C_{p/q}$ dans le plan (μ, a), la courbe transformée ''radialement'' par $P_{\mu, a}^q$ est à l'intérieur puis à l'extérieur de son image; il existe donc un moment où elle rencontre cette image (en général transversalement); la continuité de la dépendance par rapport à (μ, a) vient de ce qu'une telle courbe s'obtient à partir d'un banal théorème de fonctions implicites. Si l'intersection est transverse, il y a un nombre fini de points périodiques de période q, et la courbe $C_{p/q}$ définie pour $N_{\mu, a}$ ''explose'' en une courbe épaisse (voir fig. 55a) $\tilde{C}_{p/q}$ de valeurs de (μ, a) pour lesquelles $P_{\mu, a}^q$ possède une orbite q-periodique de nombre de rotation p/q.

(5) Dans le cas conservatif générique, chaque voisinage de l'origine contient un point homoclinie transversal. Si ω_0 est irrationnel, ceci prouve *géométriquement* la divergence des séries transformant un tel difféomorphisme (supposé analytique) en une forme normale: cette dernière devrait en effet laisser invariant chaque cercle de centre l'origine.

(6) Lorsque q est grand par rapport à $|\omega_0 - p/q|^{-1}$, on prouve l'existence d'orbites périodiques de période q pour des valeurs de (μ, a) proches de $\gamma_{p/q}$ en utilisant une version dissipative du théorème géométrique de Poincaré–Birkhoff [voir Chenciner (1982b)]; on en déduit que l'intersection de V avec l'ensemble $\tilde{C}_{p/q}$ contient au-moins un ensemble de Cantor [voir Chenciner (1982c) et la fig. 55b].

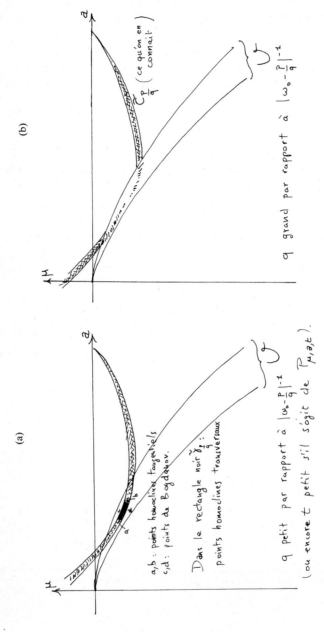

Fig. 55. La région $\tilde{C}_{p/q}(t)$ est l'intersection avec un plan "t = constante" de l'analogue à 3 paramètres de la région $\mathcal{P}_{p/q}$ représentée sur la fig. 41.

6. Conclusion

> *Nuit putride et glaciale, épouvantable nuit,*
> *Nuit du fantôme infirme et des plantes pourries,*
> *Incandescente nuit, flamme et feu dans les puits,*
> *Ténèbres sans éclairs, mensonges et roueries!*
>
> *Robert Desnos*

Les deux voies évoquées par Lefschetz se sont rejointes: dès la codimension deux, des phénomènes de type conservatif se présentent ''déployés'' au voisinage des hypersurfaces de l'espace des paramètres qui correspondent à des formes normales ayant au moins un cercle invariant non normalement hyperbolique.

Ainsi qu'on a commencé de le faire au début du chapitre 5, on compare dans la fig. 56 les couples

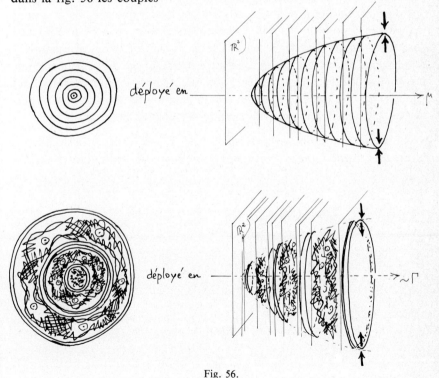

Fig. 56.

(difféomorphisme linéaire elliptique/Bifurcation de Hopf générique), et (difféomorphisme conservatif générique/Bifurcation de Hopf de codimension deux au voisinage de Γ).

En fait, la fig. 56 n'est qu'un lointain reflet de la réalité; en particulier, des points homoclines transversaux peuvent a priori coexister avec une courbe invariante.

Un exemple intéressant est obtenu en composant un difféomorphisme conservatif générique $\mathscr{A}$ avec une famille à deux paramètres $N_{\mu, a}$ de formes normales comme celles que nous avons considérées [voir les éqs. (32) et (35)]: On peut s'attendre à voir apparaitre des "attracteurs de Birkhoff" [voir Birkhoff (1932)] construits en composant un difféomorphisme conservatif générique $\mathscr{A}$ avec un difféomorphisme ayant une courbe fermée invariante très légèrement attractante séparant en deux une zône d'instabilité de $\mathscr{A}$.

Une façon de montrer l'existence de tels "attracteurs" dans une famille générique $P_{\mu, a}$ à deux paramètres serait de trouver une des courbes $\tilde{C}_\omega$ et une des "courbes épaisses" $\tilde{C}_{p/q}$ s'intersectant dans une zône de $\tilde{C}_{p/q}$ pour laquelle $P_{\mu, a}$ possède des points homoclines transversaux, et dans une zône de C_ω pour laquelle $P_{\mu, a}$ possède une courbe fermée invariante C^∞ repulsive. Il existerait alors un anneau A invariant par $P_{\mu, a}$ contenant une orbite homocline et l'"attracteur" serait l'intersection des itérés de cet anneau (fig. 57); malheureusement, les estimations dont on dispose ne permettent pas d'assurer que cette situation se produit en général.

On peut retirer de ce qui précède le "slogan" suivant: pour étudier les bifurcations génériques, au voisinage d'un point fixe, de familles de difféo-

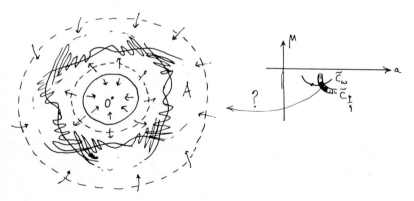

Fig. 57.

morphismes locaux de $\mathbb{R}^2$, il est utile de comprendre la structure de ceux qui préservent les aires.

Ces derniers concentrent en effet dans leur structure les éléments (points périodiques, courbes invariantes, etc...) qui, dans des familles génériques de difféomorphismes ne préservant pas les aires, apparaissent un à un pour différentes valeurs des paramètres (voir fig. 56); la fig. 58 donne des exemples de champs de vecteurs et de difféomorphismes conservatifs pour lesquels le lecteur pourra chercher des analogues en théorie de bifurcations.

Je finirai par deux commentaires:

Le premier sur le fait, qu'en général, l'ouvert de l'espace des difféomorphismes d'une variété formé par ceux qui possèdent un attracteur d'un certain type (autre qu'un point fixe, par exemple une courbe fermée) est

Fig. 58. Ces figures ne sont bien sûr que qualitatives.

"bordé" par une zône complexe dont la structure est au moins aussi riche que celle du produit d'une variété de dimension infinie par un ensemble de Cantor; le problème de la bifurcation de cet attracteur en fonction de paramètres n'est donc pas un problème local dans l'espace de ces paramètres. Cependant, dans le cas où une courbe fermée attractante d'un difféomorphisme local de $\mathbb{R}^2$ disparait par "collision" avec une courbe fermée répulsive, les deux courbes venant de naître d'un point fixe elliptique (chemin transverse à Γ), il y a une probabilité non nulle (à condition d'avoir effectivement montré que la mesure de $\tilde{\Gamma}$ est positive) de rencontrer une valeur du paramètre pour laquelle le difféomorphisme considéré laisse invariante une courbe fermée C^∞ non normalement hyperbolique. Ceci donne un intérêt au problème *local* de bifurcation consistant à comprendre les déformations possibles d'un tel difféomorphisme.

Le second commentaire concerne la notion d'attractivité (répulsivité) "vague", c'est à dire "non linéaire" (ou faible) telle qu'elle est postulée en (33). Si P est un difféomorphisme local de $(\mathbb{R}^2,0)$ dont la dérivée en 0, $DP(0)$, est elliptique, les termes non linéaires du développement de Taylor de P en 0 font génériquement de l'origine un attracteur ou un répulseur sauf dans les cas de résonance forte (dont on comprend cependant la dynamique locale, voir chapitre 2.4). Il n'en est plus du tout de même lorsqu'on remplace le point fixe 0 par une variété invariante V (une courbe fermée par exemple) d'un difféomorphisme de $\mathbb{R}^n, n \geq 2$. La dérivée $DP(0)$ est remplacée par une application "skew-produit" $S: V \times \mathbb{R}^{n-k} \to V \times \mathbb{R}^{n-k}$ ($k = \dim. V$) de la forme $S(\theta, x) = [f(\theta), A(\theta)x]$, où f est un difféomorphisme de V, et $A(\theta)$ une famille de matrices inversibles $n \times n$ paramétrées par $\theta \in V$ (on a supposé le fibré normal de V dans $\mathbb{R}^n$ trivial). Le spectre de $DP(0)$ peut être remplacé par le spectre de la transformée de graphes $\mathscr{S}$ de S, définie sur l'espace $C^k(V, \mathbb{R}^{n-k})$ (par exemple) par

$$S \text{ (graphe } \sigma) = \text{graphe } (\mathscr{S}\sigma).$$

Mais, même si le spectre de $\mathscr{S}$ est contenu dans le cercle unité de $\mathbb{C}$, on ne peut en général rien dire de la dynamique de S; par exemple, si V est le cercle et f une rotation irrationnelle, les orbites peuvent être localement denses et le slogan qui précède devient vide de sens: c'est comme si la géométrie d'un difféomorphisme local générique de $\mathbb{R}^2$ préservant les aires était remplacée par celle d'un des exemples d'Anosov et Katok cités à la fin de 3.2).

Ceci explique que des hypothèses très fortes doivent être faites pour assurer la persistance d'une courbe invariante pour de petites perturbations du difféomorphisme (comparer à la fin du chapitre 1) ou la bifurcation de variétés invariantes telles le tore (voir par exemple Chenciner et Iooss (1979), en particulier 5.3).

La seule façon d'éviter partiellement ce type de problèmes est de considérer des familles dépendant de suffisamment de paramètres pour contenir la naissance de la variété V à partir d'un point fixe. Mais de toute façon, les problèmes liés aux "skew-produits" persistent.

Bibliographie

Arnold, V.I. (1980) Chapitres supplémentaires de la théorie des equations différentielles ordinaires (Ed. de Moscou).

Birkhoff, G.D. (1932) Sur quelques courbes fermées remarquables, Bull. Soc. Math. France, 60, 1–26.

Benoit, E.B., J.L. Callot, F. Diener et M. Diener (1980) Chasse au Canard (Publ. Mme I.R.M.A., Strasbourg).

Cartwright, M.L. (1950) Forced oscillations in nonlinear systems, in: Lefschetz, Contribution to the Theory of Nonlinear Oscillations (Princeton).

Chaperon, M. (1981) Thèse et cours à L'ENS, à paraître.

Chenciner, A. (1978) Stabilité structurelle et ergodicité (ou catégorie et mesure), J. de Physique, Suppl. au no. 8, 39, C5.69–77.

Chenciner, A. (1980) Singularités des fonctions différentiables, Encyclopédia Universalis, 2e Ed.

Chenciner, A. (1981) Courbes fermées invariantes non normalement hyperboliques au voisinage d'une bifurcation de Hopf dégénérée de difféomorphismes de $\mathbb{R}^2$, Note C.R.Acad. Sci. 292, Série I, 507–510.

Chenciner, A. (1982a) Points homoclines au voisinage d'une bifurcation de Hopf dégénérée de difféomorphismes de $\mathbb{R}^2$, Note C.R.Acad Sci 294, Série I, 269–272.

Chenciner, A. (1982b) Sur un énoncé dissipatif du théorème géométrique de Poincaré–Birkhoff, Note C.R.Acad. Sci. 294, Série I, 243–245.

Chenciner, A. (1982c) Points périodiques de longues périodes au voisinage d'une bifurcation de Hopf dégénérée de difféomorphismes de $\mathbb{R}^2$, Note C.R.Acad. Sci. 294, Série I, 661–663.

Chenciner, A. (1982d) Bifurcations de points fixes elliptiques, I. Courbes invariantes; Preprint Université Paris VII, à paraître.

Chenciner, A. et G. Iooss (1979) Bifurcation de tores invariants, Archives for Rat. Mech. and Anal. 69, 2, 109–198.

Gambando, J.M. (1982) Thèse de $3^{\text{ème}}$ cycle, Université de Nice.

Herman, M. (1977) Mesure de Lebesgue et nombre de rotation, in: Geometry and Topology, Lecture Notes in Math., Vol. 597 (Springer, Berlin).

Herman, M. (1979) Sur la conjugation differentiable des diffeomorphismes du cercle à des rotations, Pure Math. IHES, 49.

Herman, M. (1980) Cours à l'ENS, à paraitre dans Astérisque.

Hirsch, M.W. C.C. Pugh et M. Shub (1977) Invariant Manifolds, Lecture Notes in Math., Vol. 583 (Springer, Berlin).

Hirsch, M.W. et S. Smale (1974) Differential Equations, Dynamical Systems, and Linear Algebra (Academic Press, New York).

Iooss, G. (1979) Bifurcation of Maps and Applications (North-Holland, Amsterdam).

Lefschetz, S. (1977) Differential Equations: Geometric Theory (Dover Publ.).

Lemaire, J.M. (1977) Bifurcation de Hopf, Preprint, Université de Nice.

Levi, M. (1978) Thèse, Courant Institute, New York.

Lins, A., W. De Melo et C.C. Pugh (1977) On Lienard's Equation, in: Geometry and Topology, Lecture Notes in Math., Vol. 597 (Springer, Berlin).

Moser, J. (1973) Stable and random motion in dynamical systems, Ann. Math. Study 77.

Ruelle, D. et F. Takens (1971) On the nature of turbulence, Commun. Math. Phys. 20, 167–192; 23, 343–344.

Shub, M. (1978) Stabilité globale des systèmes dynamiques, Astérisque 56.

Smale, S. (1980) The Dynamics of Time (Springer, Berlin).

Van der Pol, B. (1934) Nonlinear theory of electric oscillations, Proc. I.R.E. 22, 1051–1086.

Zehnder, E. (1973) Homoclinic points near elliptic fixed points, Commun. Pure Appl. Math. XXVI, 131–182.

COURSE 5

STABILITY AND BIFURCATION THEORY

Daniel D. JOSEPH

Department of Aerospace Engineering and Mechanics, The University of Minnesota, Minneapolis, MN 55455, U.S.A.

G. Iooss, R.H.G. Helleman and R. Stora, eds.
Les Houches, Session XXXVI, 1981 – Comportement Chaotique des Systèmes Déterministes/ Chaotic Behaviour of Deterministic Systems
© *North-Holland Publishing Company, 1983*

Contents

1. Bifurcation in $\mathbb{R}'$

In this lecture we consider the theory of singular points of plane curves. And to these considerations we add the study of stability. To make a point, not to be taken literally, I will say that sixty per cent of the ideas of bifurcation theory can be most readily understood from this elementary study in $\mathbb{R}'$.

We study the evolution equation

$$\mathrm{d}u/\mathrm{d}t = F(\mu, u); \qquad \mu, u \in \mathbb{R}', \tag{1.1}$$

where $F(\cdot, \cdot)$ has two continuous derivatives with respect to μ and u. Equilibrium of eq. (1.1) satisfy $u = \varepsilon$, independent and

$$F(\mu, \varepsilon) = 0. \tag{1.2}$$

The study of bifurcation of equilibrium solutions of the automous problem (1.1) is equivalent to the study of singular points of the curve (1.2) in the (μ, ε) plane.

It is desirable to classify points of the curves (1.2):

(i) *A regular point* of $F(\mu, \varepsilon) = 0$ is one for which the implicit function theorem works

$$F_\mu \neq 0 \text{ or } F_\varepsilon \neq 0. \tag{1.3}$$

If eq. (1.3) holds, then we can find a unique curve $\mu = \mu(\varepsilon)$ or $\varepsilon = \varepsilon(\mu)$ through the point.

(ii) *A regular turning point* is a point at which $\mu_\varepsilon(\varepsilon)$ changes sign and $F_\mu(\mu, \varepsilon) \neq 0$.

(iii) *A singular point* of the curve $F(\mu, \varepsilon) = 0$ is a point at which

$$F_\mu = F_\varepsilon = 0. \tag{1.4}$$

(iv) *A double point* of the curve $F(\mu, \varepsilon) = 0$ is a singular point through which pass two and only two branches of $F(\mu, \varepsilon) = 0$ possessing distinct tangents. We shall assume that all second derivatives of F do not simultaneously vanish at a double point.

(v) *A singular turning (double) point* of the curve $F(\mu, \varepsilon) = 0$ is a double point at which μ_ε changes sign on one branch.

(vi) *A cusp point* of the curve $F(\mu, \varepsilon) = 0$ is a point of second order contract between two branches of the curve. The two branches have the same tangent at a cusp point.

(vii) *A conjugate point* is an isolated singular point solution of $F(\mu, \varepsilon) = 0$.

(viii) *A higher-order singular point* of the curve $F(\mu, \varepsilon) = 0$ is a singular point at which all three second derivatives of $F(\mu, \varepsilon)$ are null.

Double points are most important for bifurcation. Suppose (μ_0, ε_0) is a singular point. Then equilibrium curves passing through the singular points satisfy

$$2F(\mu, \varepsilon) = F_{\mu\mu}\, \delta\mu^2 + 2F_{\varepsilon\mu}\, \delta\varepsilon\, \delta\mu + F_{\varepsilon\varepsilon}\, \delta\varepsilon^2 + \mathrm{O}[(|\delta\mu| + |\delta\varepsilon|)^2] = 0 \quad (1.5)$$

where $\delta\mu = \mu - \mu_0$, $\delta\varepsilon = \varepsilon - \varepsilon_0$ and $F_{\mu\mu} = F_{\mu\mu}\,(\mu_0, \varepsilon_0)$, etc. In the limit, as $(\mu, \varepsilon) \to (\mu_0, \varepsilon_0)$ the eq. (1.5) for the curves reduces to the quadratic equation

$$F_{\mu\mu}\, \mathrm{d}\mu^2 + 2F_{\varepsilon\mu}\, \mathrm{d}\varepsilon\, \mathrm{d}\mu + F_{\varepsilon\varepsilon}\, \mathrm{d}\varepsilon^2 = 0 \quad (1.6)$$

for the tangents to the curve. We find that

$$\begin{bmatrix} \mu_\varepsilon^{(1)}(\varepsilon_0) \\ \mu_\varepsilon^{(2)}(\varepsilon_0) \end{bmatrix} = -\frac{F_{\varepsilon\mu}}{F_{\mu\mu}}\begin{bmatrix} 1 \\ 1 \end{bmatrix} + \left(\frac{D}{F_{\mu\mu}^2}\right)^{1/2}\begin{bmatrix} 1 \\ -1 \end{bmatrix}, \quad (1.7)$$

$$\begin{bmatrix} \mu_\varepsilon^{(1)}(\varepsilon_0) \\ \varepsilon_\mu^{(2)}(\mu_0) \end{bmatrix} = -\frac{F_{\varepsilon\mu}}{F_{\varepsilon\varepsilon}}\begin{bmatrix} 1 \\ 1 \end{bmatrix} - \left(\frac{D}{F_{\varepsilon\varepsilon}^2}\right)^{1/2}\begin{bmatrix} 1 \\ -1 \end{bmatrix}, \quad (1.8)$$

where

$$D = F_{\varepsilon\mu}^2 - F_{\mu\mu}F_{\varepsilon\varepsilon}.$$

If $D < 0$ there are not real tangents through (μ_0, ε_0) and the point (μ_0, ε_0) is an isolated (conjugate) point solution of $F(\mu, \varepsilon) = 0$.

We shall consider the case when (μ_0, ε_0) is not a higher-order singular point. Then (μ_0, ε_0) is a double point if and only if $D > 0$. If the two curves pass through the singular point and $D = 0$ then the slope at the singular point of higher contact is given by eqs. (1.7) or (1.9). If $D > 0$ and $F_{\mu\mu} \neq 0$ then there are two tangents with slopes $\mu_\varepsilon^{(1)}(\varepsilon_0)$ and $\mu_\varepsilon^{(2)}(\varepsilon_0)$ given by eq. (1.7). If $D > 0$ and $F_{\mu\mu} = 0$, then $F_{\varepsilon\mu} \neq 0$ and

$$\mathrm{d}\varepsilon\,(2\mathrm{d}\mu\, F_{\varepsilon\mu} + \mathrm{d}\varepsilon\, F_{\varepsilon\varepsilon}) = 0$$

and there are two tangents $\varepsilon_\mu(\mu_0) = 0$ and $\mu_\varepsilon(\varepsilon_0) = -F_{\varepsilon\varepsilon}/2F_{\varepsilon\mu}$. If $\varepsilon_\mu(\mu_0) = 0$ then $F_{\mu\mu}(\mu_0, \varepsilon_0) = 0$. So all the possibilities are covered in the following two cases:

(A) $D>0$, $F_{\mu\mu}\neq 0$ with tangents $\mu_\varepsilon^{(1)}(\varepsilon_0)$ and $\mu_\varepsilon^{(2)}(\varepsilon_0)$.

(B) $D>0$, $F_{\mu\mu}=0$ with $\varepsilon_\mu(\mu_0)=0$ and $\mu_\varepsilon(\varepsilon_0)=-F_{\varepsilon\varepsilon}/2F_{\varepsilon\mu}$.

The existence of two branches passing through the point (μ_0,ε_0) is guaranteed by the implicit function theorem when $D>0$ (see ESBT*, section II.4).

When $D=0$ and all second derivatives are not zero there is a cusp at the origin (ESBT, section II.5). There are two typical situations:

(i) Bifurcation with two curves having common tangents and different curvatures at $(\mu_0,\varepsilon_0)=(0,0)$, an example is given by fig. 1.

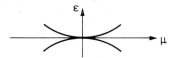

Fig. 1. Bifurcation at a cusp point.

(ii) A cusp point of a single curve. This is degenerate form of a turning point. An example is given at fig. 2.

Fig. 2. A degenerate turning point.

When $D=0$ because all second derivatives are null it is necessary to consider cubic equation to determine the number real tangents. If there are three real, distinct roots then three bifurcating solutions pass through the singular point (μ_0,ε_0). If two roots are complex, then there is no bifurcation.

We next consider the stability of bifurcating solutions using the linearized theory of stability. The linearized equation is

$$Z_t=F_\varepsilon(\mu,\varepsilon)Z. \tag{1.10}$$

The general solution of eq. (1.10) is

$$Z=e^{\sigma t}Z_0, \qquad \sigma=F_\varepsilon(\mu,\varepsilon). \tag{1.11}$$

---

* We will use the abbreviation ESBT to denote ref. [4].

Since all solutions of eq. (1.10) are in the form eq. (1.11) we find that disturbances Z of ε grow when $\sigma > 0$ and decay when $\sigma < 0$. The linearized theory implies that $[\mu(\varepsilon), \varepsilon]$ satisfying $F(\mu, \varepsilon) = 0$ is stable when $\sigma < 0$ and is unstable when $\sigma > 0$. This criterion applies even to the nonlinear problem when the initial disturbance is sufficiently small (cf. section II.7 in ESBT).

A very general and important result is easy to deduce from the second part of eq. (1.11) under the hypothesis that eq. (1.2) may be solved for $\mu = \mu(\varepsilon)$. Then, differentiating $F[\mu(\varepsilon), \varepsilon] = 0$ with respect to ε we find that

$$\sigma(\varepsilon) = F_\varepsilon[\mu(\varepsilon), \varepsilon] = -\mu_\varepsilon F_\mu[\mu(\varepsilon), \varepsilon]. \tag{1.12}$$

It follows easily from eq. (1.12) that $\sigma(\varepsilon)$ must change sign as ε is varied across a regular turning point. This implies that the $u = \varepsilon$, $\mu = \mu(\varepsilon)$ is stable on one side of regular turning point and is unstable on the other side (see fig. 3).

The study of stability may be tied to the study of bifurcation by the hypothesis of strict loss of stability which was introduced by Hopf. This hypothesis is a non-degeneracy condition which guarantees double-point bifurcation. More precisely, we have the following theorem: *Suppose that* (μ_0, ε_0) *is a singular point* (A) $\sigma_\varepsilon(\varepsilon_0) \neq 0$ *or* (B) $\sigma_\mu(\mu_0) \neq 0$. *Then* (μ_0, ε_0) *is a double point.* For the proof under hypothesis (A) see ESBT, section II.9. For case B we must solve $F(\mu, \varepsilon)$ for $\varepsilon(\mu)$. At the singular point (μ_0, ε_0) we have strict loss of stability because $\sigma_\mu = F_{\varepsilon\mu} + F_{\varepsilon\varepsilon}\varepsilon_\mu = F_{\varepsilon\mu} = D^{1/2} \operatorname{sgn} F_{\varepsilon\mu}$.

It is easy to derive formulas which show that there is an exchange of stability at a double point (ESBT, section II.10). These formulas can be used to prove the following theorem. *Assume that all singular points of solutions of* $F(\mu, \varepsilon) = 0$ *are double points. The stability of such solutions must change at each regular turning point and at each singular point (which is not a turning point), and only at such points.*

We shall prove this theorem for the case in which $u = 0$ is a solution of the evolution problem

$$F(\mu, 0) = 0 \qquad \forall \mu \in \mathbb{R}. \tag{1.13}$$

Fig. 3. Exchange of stability at a regular turning point. The same type of exchange of stability can be demonstrated for degenerate case shown in fig. 2.

Then, differentiating the second part of eq. (1.11) with respect to μ on the solution $\varepsilon = 0$, we get

$$\sigma_\mu^{(1)}(0) = F_{\mu\varepsilon}(0,0) \neq 0, \qquad \text{say} > 0. \tag{1.14}$$

On the bifurcating branch $F[\mu(\varepsilon), \varepsilon] = 0$ and

$$\sigma^{(2)} = F_\varepsilon[\mu(\varepsilon), \varepsilon] = -\mu_\varepsilon F_\mu[\mu(\varepsilon), \varepsilon],$$
$$= -\mu_\varepsilon[F_{\mu\varepsilon}(0,0)\varepsilon + O(\varepsilon)],$$
$$= -\mu_\varepsilon \sigma_\mu^{(0)}(0)\{\varepsilon + O(\varepsilon)\}. \tag{1.15}$$

The following bifurcation diagrams are implied by eqs. (1.14) and (1.15) (fig. 4).

To bring the ideas developed so far we give a demonstration here of the stability and bifurcation of the bent wire arch described in fig. II.5 of ESBT. We replace u by σ, the angle of deflection, and μ by l, the length. We imagine that the equation of motion of the bent arch is

$$d\theta/dt = \theta[l - l(\theta)], \tag{1.16}$$

where

$$l(-\theta) = l(\theta)$$

is even. The upright position is $\theta = 0$ and the bifurcating solution is $l = l(\theta)$, shown in fig. 5.

The ideas developed so far have a much wider range of applicability than might at first be supposed. The local analysis near turning point and singular point applies even to partial differential equations under rather common conditions (called bifurcation at a simple eigenvalue) under which the important part of the problem is a part which can be projected into one dimension.

There is a very important global result which holds strictly in $\mathbb{R}'$ and not

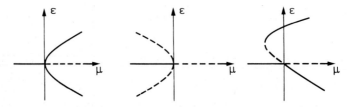

Fig. 4. Bifurcation and stability at a double point.

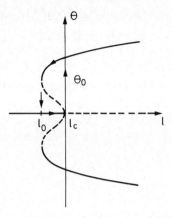

Fig. 5. Bifurcation diagram for the bent arch. The arch bifurcates subcritically and exhibit hysteresis. A demonstration of the actual bifurcation is given here and is described in ESBT.

necessarily in $\mathbb{R}'$ in projection. In the one-dimensional case it is possible to prove that the stability of solutions which pierce the line $\mu = $ constant is of alternating sign, as in fig. 6. In higher dimensions curves of solutions which appear to intersect when projected onto the plane of the bifurcation diagram actually do not intersect in the higher-dimensional space. We may write the evolution eq. (1.1) in following factored form

$$\mathrm{d}u/\mathrm{d}t = F_1 F_2 F_3 F_4 F_5 F_6 F_7, \tag{1.17}$$

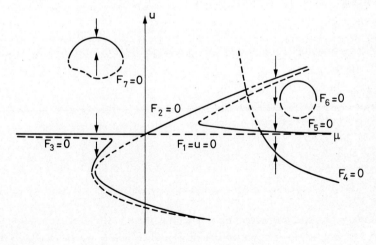

Fig. 6. Bifurcation, stability and domains of attraction of equilibrium solutions of eq. (1.17).

where each $F_i = F_i(\mu, u) = 0$ gives an equilibrium solution. An example of seven equilibrium solutions is shown in fig. 6.

In this simple example we see bifurcating solutions, solutions which perturb bifurcation and isolated solutions. I call the intersecting solutions F_1, F_2 and F_4 "bifurcating" (from one another). All the remaining solutions can be isolated F_3 and F_5 perturb bifurcation. F_6 is an "isola" which can be treated as a perturbation of a conjugate singular point. The stability of solutions on the line $\mu = $ constant alternates. You see that nonuniqueness is endemic, even in $\mathbb{R}'$.

2. Bifurcation in $\mathbb{R}^2$

For the moment we will use a general notation for our fundamental (autonomous) problem

$$du/dt = f(\mu, u) = f_u(\mu \mid u) + N(\mu, u). \tag{2.1}$$

Here $u \in \mathbb{R}^n$ or, say, u is an element in a normed space and $u = 0$ is a solution for all μ

$$f(\mu, 0) = 0,$$

$f_u(\mu \mid u)$ is the derivative of f with respect to u at $u = 0$, a linear operator $f_u(\mu \mid \cdot)$, and $N(\mu, u) = O(|u|^2)$. The linearized stability problem for the stability of the solution $u = 0$ is

$$dv/dt = f_u(\mu \mid v). \tag{2.2}$$

A spectral problem for stability can be obtained from solutions of eq. (2.2) in the form

$$v = e^{\sigma t} \zeta, \tag{2.3}$$

where ζ is independent of t and

$$\sigma = \zeta + i\eta$$

is an eigenvalue of

$$\sigma\zeta = f_u(\mu \mid \zeta). \tag{2.4}$$

We say $u = 0$ is stable (according to spectral theory) if $\zeta(\mu) < 0$ for all eigenvalues of eq. (2.4). The problem (2.1) arises when we have a problem governed by differential equations which is forced by steady data. Then there is a forced steady solution and u is the difference between the forced solution and any other solution of the same forced problem. Many very general problems may be represented by eq. (2.1).

Consider eq. (2.4) in $\mathbb{R}^2$. Then σ is an eigenvalue of

$$f_u(\mu \mid \cdot) = \begin{bmatrix} a(\mu) & b(\mu) \\ c(\mu) & d(\mu) \end{bmatrix} \overset{\text{def}}{=} A, \tag{2.5}$$

a root of

$$\sigma^2 - \sigma(a+d) + ad - bc = 0. \tag{2.6}$$

There are two roots

$$\sigma_1 = \frac{a+d}{c} + \varDelta^{1/2},$$

and

$$\sigma_2 = \frac{a+d}{2} - \varDelta^{1/2},$$

where

$$\varDelta = \frac{(a-d)^2}{4} + bc = \frac{(a+d)^2}{4} - ad + bc.$$

The adjoint matrix, the transpose

$$A^T = \begin{bmatrix} a & b \\ c & d \end{bmatrix}$$

has the same two eigenvalues but, if $c \neq b$, different eigenvectors.

There are four cases in two categories to consider. Category one are the algebraically simple eigenvalues.

Case 1: $\varDelta > 0$. $\sigma_1 \neq \sigma_2$ are real. There are two adjoint eigenvectors.

Case 2: $\varDelta < 0$. σ_1 and σ_2 are complex and $\sigma_2 = \bar{\sigma}_1$. There are two eigenvectors and they are conjugate.

Category two are the algebraically double eigenvalues $\sigma_1 = \sigma_2$; i.e., $\varDelta = 0$.

Case 3: $\sigma_1 = \sigma_2$ is a semi-simple double eigenvalue, $(a-d)^2 = b = c = 0$. Then $A = a\mathbf{I} = A^T$ and every vector ζ is an eigenvector belonging to $\sigma_1 = \sigma_2 = a$. We can select two orthonormal ones $\zeta_1 \cdot \zeta_2 = 0$. The eigenvalue a is said to have a Riesz index 1.

Case 4: Riesz index two. There is only one eigenvector, one vector satisfying $(A - \sigma I)\zeta = 0$ and one generalized eigenvector satisfying $(A - \sigma I)^2 X = 0$ (section IV.4 of ESBT).

We shall not consider bifurcation for cases 3 and 4 (double eigenvalues) in these lectures. Case 1 can be formulated as a problem in $\mathbb{R}'$ by the method of projection. I will show this in the next lecture. The remaining case 2 in which $\sigma = \zeta + i\eta$ is complex and

$$v = e^{\xi t} e^{i\eta t} \zeta \tag{2.7}$$

is time periodic, leads to a time-periodic bifurcating solution as I now shall show.

The evolution eq. (2.1) may be written in component form

$$\dot{u}_j = A_{ij}(\mu)u_j + B_{ijk}u_ju_k + \text{higher order terms.} \tag{2.8}$$

We suppose that near $\mu = 0$ the discriminant Δ is negative, so that

$$\sigma(\mu)\zeta = A\zeta,$$
$$\sigma(\mu)\bar{\zeta}^* = A^T\bar{\zeta}^*, \tag{2.9}$$

where $\zeta^*(\mu)$ is the adjoint vector with eigenvalue $\bar{\sigma}(\mu) = \xi(\mu) - i\eta(\mu)$ in the scalar product

$$\langle X \cdot Y \rangle \overset{\text{def}}{=} X \cdot \bar{Y}. \tag{2.10}$$

We may normalize so that

$$\langle \zeta, \zeta^* \rangle = \zeta \cdot \bar{\zeta}^* = \zeta_k \bar{\zeta}_k^* = 1. \tag{2.11}$$

It is easy to deduce

$$\langle \zeta, \bar{\zeta}^* \rangle = \zeta_k \zeta_k^* = 0. \tag{2.12}$$

We suppose that the loss of stability of $u = 0$ occurs at $\mu = 0$ so that $\xi(0) = 0$. We will get bifurcation into periodic solutions if

$$\eta(0) = \omega_0 \neq 0 \quad \text{and} \quad d\xi(0)/d\mu = \xi_\mu(0) \neq 0 \tag{2.13}$$

[say $\xi_\mu(0) > 0$].

Since ζ and $\bar{\zeta}$ are linearly independent any real-valued two-dimensional vector $u = (u_1, u_2)$ may be represented as

$$u_i = a(t)\zeta_i + \bar{a}(t)\bar{\zeta}_i. \tag{2.14}$$

Combining eqs. (2.14) and (2.8), using eq. (2.9) we get

$$\dot{a}\zeta_i + \dot{\bar{a}}\bar{\zeta}_i = \sigma(\mu)\zeta_i + \bar{\sigma}(\mu)\bar{\zeta}_i + a^2 B_{ijk}\zeta_j\zeta_k + 2|a|^2 B_{ijk}\zeta_i\bar{\zeta}_k$$

$$+ \bar{a}^2 B_{ijk}\bar{\zeta}_i\bar{\zeta}_k + O(|a|^3).$$

The orthogonality properties, [eqs. (2.11) and (2.12)], are now employed to reduce the preceding into a single, complex-valued, amplitude equation

$$\dot{a} = f(\mu, a) = \sigma(\mu)a + \alpha(\mu)a^2 + 2\beta(\mu)|a|^2 + \gamma(\mu)\bar{a}^2 + O(|a|^3), \qquad (2.15)$$

where, for example, $\alpha(\mu) = B_{ijk}(\mu)\zeta_j\zeta_k\zeta_i^*$. (For simplicity we shall suppress cubic terms of $f(\mu, a)$ here. These terms come into the bifurcating solution at second order but do not introduce new features.) The linearized stability of the solution $a = 0$ of eq. (2.15) is determined by $\dot{a} = \sigma(\mu)a$, $a = $ constant $\times e^{\sigma(\mu)t}$. At criticality ($\mu = 0$), $a = $ constant $\times e^{i\omega_0 t}$ is 2π-periodic in $s = \omega_0 t$.

We shall show that a bifurcating time-periodic solution may be constructed from the solution of the linearized problem at criticality. This bifurcating solution is in the form

$$a(t) = b(s, \varepsilon), \quad s = \omega(\varepsilon)t, \quad \omega(0) = \omega_0, \quad \mu = \mu(\varepsilon), \qquad (2.16)_1$$

where ε is the amplitude of a defined by

$$\varepsilon = \frac{1}{2\pi}\int_0^{2\pi} e^{-is} b(s, \varepsilon)\, ds = [b]. \qquad (2.16)_2$$

The solution, eq. (2.16), of eq. (2.15) is unique to within an arbitrary translation of the time origin. This means that under translation $t \to t + c$ the solution $b(s + c\omega(\varepsilon), \varepsilon)$ shifts its phase. This unique solution is analytic in ε when $f(\mu, a)$ is analytic in the variables $(\mu, a, \bar{a})$ and it may be expressed as a series:

$$\begin{bmatrix} b(s, \varepsilon) \\ \omega(\varepsilon) - \omega_0 \\ \mu(\varepsilon) \end{bmatrix} = \sum_{n=1}^{\infty} \varepsilon^n \begin{bmatrix} b_n(s) \\ \omega_n \\ \mu_n \end{bmatrix}. \qquad (2.17)$$

The perturbation problems which govern $b_n(s)$, ω_n and μ_n can be obtained by identifying the coefficient of ε^n which arises when eq. (2.17) is substituted into the two equations: $\omega b = f(\mu, b)$ and $\varepsilon = [b]$. We find that at order one

$$\omega_0 \dot{b}_1 - i\omega_0 b_1 = 0, \quad [b_1] = 1, \quad b_1(s) = e^{is}.$$

At order two we find that $[b_2] = 0$ and

$$\omega_0[\dot{b}_2 - ib_2] + \omega_1\dot{b}_1 = \mu_1\sigma_\mu b_1 + \alpha_0 b_1^2 + 2\beta_0|b_1|^2 + \gamma_0\bar{b}_1^2,$$

where $\sigma_\mu = d\sigma(0)/d\mu$ and, for example, $\alpha_0 = \alpha(0)$.

Equations of the form $\dot{b}(s) - ib(s) = f(s) = f(s + 2\pi)$ are solvable for $b(s) = b(s + 2\pi)$ if and only if the Fourier expansion of $f(s)$ has no term proportional to e^{is}. Hence, because $\xi_\mu \neq 0$ we obtain

$$\mu_1 = \omega_1 = 0$$

in eq. (2.17) and

$$\dot{b}_2 - ib_2 = (\alpha_0 e^{2is} + 2\beta_0 + \gamma_0 e^{-2is})/\omega_0.$$

We find that

$$b_2(s) = [\alpha_0 e^{2is} - 2\beta_0 - (\gamma_0 e^{-2is}/3)]/i\omega_0.$$

The problem which governs at order three, with cubic terms in b neglected, is

$$\dot{b}_3 - ib_3 = [-\omega_2\dot{b}_1 + \mu_2\sigma_\mu b_1 + 2\alpha_0 b_1 b_2 + 2\beta_0(b_1\bar{b}_2 + \bar{b}_1 b_2) + 2\gamma_0\bar{b}_1\bar{b}_2]/\omega_0. \tag{2.18}$$

To solve eq. (2.18), we must eliminate terms proportional to e^{is} from the right-hand side of eq. (2.18). This is done if $[b_3] = 0$; that is, if

$$i\omega_2 - \mu_2\sigma_\mu = -[4\alpha_0\beta_0 - 4|\beta_0|^2 - 2\alpha_0\beta_0 - (2|\gamma_0|^2/3)]/i\omega_0. \tag{2.19}$$

The real part of eq. (2.19) is solvable for μ_2 provided that $\xi_\mu \neq 0$. The imaginary part of eq. (2.19) is always solvable for ω_2.

Proceeding to higher orders, it is easy to verify that all of the perturbation problems are solvable when eq. (2.13) holds and, in fact $\omega(\varepsilon) = \omega(-\varepsilon)$ are even functions. It follows that periodic solutions which bifurcate from steady solutions bifurcate to one or the other side of criticality and never to both sides; periodic bifurcating solutions cannot undergo two-sided or transcritical bifurcation.

We now search for the conditions under which the bifurcating periodic solutions are stable. We consider a small disturbance $z(t)$ of $b(s, \varepsilon)$. Setting $a(t) = b(s, \varepsilon) + z(t)$ in eq. (2.15), we find the linearized equation $\dot{z}(t) = f_a[\mu(\varepsilon), b(s, \varepsilon)]z(t)$ where $f_a = \partial f/\partial a$ and $s = \omega(\varepsilon)t$. Then, using Floquet theory, we set $z(t) = e^{\gamma t}y(s)$ where $y(s) = y(s + 2\pi)$ and find that

$$\gamma y(s) = -\omega\dot{y}(s) + f_a(\mu, b)y(s) \stackrel{\text{def}}{=} [J(s, \varepsilon)y](s) \tag{2.20}$$

where $\dot{y}(s) = dy(s)/ds$.

The stability result we need may be stated as a factorization theorem. To prove this theorem we use the fact that $\gamma = 0$ is always an eigenvalue of J with eigenfunction $b(s, \varepsilon)$

$$J\dot{b} = 0 \tag{2.21}$$

and the relation

$$\omega_\varepsilon(\varepsilon)\dot{b}(s, \varepsilon) = \mu_\varepsilon(\varepsilon)f_\mu(\mu(\varepsilon), b(s, \varepsilon)) + Jb_\varepsilon \tag{2.22}$$

which arises from differentiating $\omega\dot{b} = f(\mu, b)$ with respect to ε at any ε.

Factorization theorem. The eigenfunction y of eq. (2.20) and the Floquet exponent γ are given by the following formulas:

$$y(s, \varepsilon) = c(\varepsilon)\left(\frac{\tau}{\gamma}\dot{b}(s, \varepsilon) + b_\varepsilon(s, \varepsilon) + \mu_\varepsilon(\varepsilon)\varepsilon q(s, \varepsilon)\right),$$

$$\tau(\varepsilon) = \omega_\varepsilon(\varepsilon) + \mu_\varepsilon(\varepsilon)\hat{\tau}(\varepsilon),$$

$$\gamma(\varepsilon) = \mu_\varepsilon(\varepsilon)\hat{\gamma}(\varepsilon), \tag{2.23}$$

where $c(\varepsilon)$ is an arbitrary constant and $q(s, \varepsilon) = q(s + 2\pi, \varepsilon)$, $\hat{\tau}(\varepsilon)$ and $\hat{\gamma}(\varepsilon)$ satisfy the equation

$$\hat{\tau}\dot{b} + \hat{\gamma}b_\varepsilon + f_\mu(\mu, b) + \varepsilon(\gamma q - Jq) = 0 \tag{2.24}$$

and are smooth functions in a neighborhood of $\varepsilon = 0$. Moreover $\hat{\tau}(\varepsilon)$ and $\hat{\gamma}(\varepsilon)/\varepsilon$ are even functions and such that

$$\hat{\gamma}_\varepsilon(0) = -\xi_\mu(0), \qquad \hat{\tau}(0) = -\eta_\mu(0). \tag{2.25}$$

Remark. If $\omega_\varepsilon(0) \neq 0$, $c(\varepsilon)$ may be chosen so that $y(s, \varepsilon) \to b(s, \varepsilon)$ when $\varepsilon \to 0$.

Proof. Substitute the representations (2.23) into (2.20) utilizing eq. (2.21) to eliminate $J\dot{b}$ and eq. (2.22) to eliminate Jb_ε. This leads to eq. (2.23) which may be solved by series

$$\begin{bmatrix} q(s, \varepsilon) \\ \hat{\gamma}(\varepsilon)/\varepsilon \\ \hat{\tau}(\varepsilon) \end{bmatrix} = \sum_{l=0}^{\infty} \begin{bmatrix} q_l(s) \\ \hat{\gamma}_l \\ \hat{\tau}_l \end{bmatrix} \varepsilon^l, \tag{2.26}$$

where $\gamma_0 = \hat{\gamma}_\varepsilon(0)$ and $\hat{\tau}_0 = \hat{\tau}(0)$. Using the fact that to the lowest order $b = \varepsilon e^{is}$, $\gamma = O(\varepsilon^2)$ and $f_\mu(\mu, b) = \sigma_\mu(0)e^{is}\varepsilon$ we find that

$$e^{is}[i\hat{\tau}(0) + \hat{\gamma}_\varepsilon(0) + \sigma_\mu] - J_0 q_0 = 0, \qquad J_0 \overset{\text{def}}{=} J(\cdot, 0). \tag{2.27}$$

Eq. (2.27) is solvable for $q_0(s) = q_0(s + 2\pi)$ if and only if the term in the bracket vanishes; that is if eq. (2.25) holds. The remaining properties asserted in the theorem may be obtained by mathematical induction using the power series (2.26).

The linearized stability of the periodic solution for small values of ε may now be obtained from the spectral problem: $u(s, \varepsilon) = u(s + 2\pi, \varepsilon)$ is stable when $\gamma(\varepsilon) < 0$ [$\gamma(\varepsilon)$ is real] and is unstable when $\gamma(\varepsilon) > 0$ where

$$\gamma(\varepsilon) = \mu_\varepsilon(\varepsilon)\hat{\gamma}(\varepsilon) = -\mu_\varepsilon(\varepsilon)[\xi_\mu(0)\varepsilon + O(\varepsilon^3)].$$

Two examples are given in fig. 7.

3. Projections into $\mathbb{R}^2$

In this section we shall show that the analysis of bifurcation of periodic solutions from steady ones in $\mathbb{R}^2$, also applies in $\mathbb{R}^n$ and in infinite dimensions; say, for partial differential equations and for functional differential equations, when the steady solution loses stability at a simple, complex-valued eigenvalue.

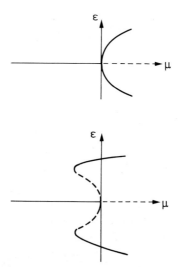

Fig. 7. (a) Supercritical (stable) Hopf bifurcation. (b) Subcritical (unstable) Hopf bifurcation with a turning point. In (b), if zero loses stability strictly as μ is increased past zero, then $\xi_\mu > 0$ and zero is unstable for $\mu > 0$ (as shown); the double eigenvalue of J_0 splits into two simple eigenvalues of $J(\cdot, \varepsilon)$: one eigenvalue is 0 and the other, $\gamma(\varepsilon)$, controls stability.

Our basic problem is again

$$\dot{u} = f(\mu, u) = f_u(\mu \mid u) + N(\mu, u), \tag{3.1}$$

where $N(\mu, u) = O(|u|^2)$. A small disturbance $v = e^{\sigma t}\zeta$ of $u = 0$ satisfies

$$\sigma\zeta = f_u(\mu \mid \zeta). \tag{3.2}$$

The adjoint problem is

$$\sigma\bar{\zeta}^* = f_u^*(\mu \mid \bar{\zeta}^*) \tag{3.3}$$

and very often in applications, there are a countably infinite number of eigenvalues $\{\sigma_n\}$ which are arranged in a sequence corresponding to the size of their real parts

$$\xi_1 \geq \xi_2 \geq \cdots \geq \xi_n \geq \cdots,$$

clustering at $-\infty$. To each eigenvalue there corresponds, at most, a finite number of eigenvectors ζ_n and adjoint eigenvectors ζ_n^*. In the case of a semi-simple eigenvalue σ_n we may choose the eigenvectors of $f_u(\mu \mid \cdot)$ and $f_u^*(\mu \mid \cdot)$ such that they form biorthonormal families

$$\langle \zeta_{nk}, \zeta_{nj}^* \rangle = \delta_{kj}, \qquad k, j = 1, \ldots, m_n, \tag{3.4}$$

m_n being the multiplicity of the eigenvalue σ_n (assumed to be semi-simple). Taking now the scalar product of eq. (3.1) with ζ_n^* we obtain

$$\frac{d}{dt} \langle u, \zeta_n^* \rangle = \langle f_u(\mu \mid u), \zeta_n^* \rangle + \langle N(\mu, u), \zeta_n^* \rangle$$

$$= \langle u, f_u^*(\mu \mid \zeta_n^*) \rangle + \langle N(\mu, u), \zeta_n^* \rangle$$

$$= \sigma_n \langle u, \zeta_n^* \rangle + \langle N(\mu, u), \zeta_n^* \rangle. \tag{3.5}$$

When u is small the linearized equations lead to

$$\langle u(t), \zeta_n^* \rangle \simeq \langle u(0), \zeta_n^* \rangle e^{\xi_n(\mu)t} e^{i\eta_n(\mu)t},$$

so that if $\xi_n(\mu) < 0$, the projection $\langle u(t), \zeta_n^* \rangle$ decays to zero. In fact, for the full nonlinear problem there is a coupling between different projections, and if some of these do not decay, this last result is no longer true. Nevertheless, the important part of the evolution problem (3.1) is related to the part of the spectrum of $f_u(\mu \mid \cdot)$ for which $\xi_n(\mu) \geq 0$.

In the problem of bifurcation studied in this section we shall assume that the real part of two complex-conjugate simple eigenvalues $\sigma(\mu)$, $\bar{\sigma}(\mu)$

changes sign when μ crosses 0 and the remainder of the spectrum stays on the left hand side of the complex plane. Suppose ζ and $\bar{\zeta}^*$ are the eigenvectors of $f_u(\mu \mid \cdot), f_u^*(\mu \mid \cdot)$ belonging to the eigenvalue $\sigma(\mu)$. Then, the equation governing the evolution of the projection

$$\frac{d}{dt} \langle u, \zeta^* \rangle = \sigma(\mu) \langle u, \zeta^* \rangle + \langle N(\mu, u), \zeta^* \rangle, \tag{3.6}$$

is complex-valued, that is, two-dimensional. So our problem is essentially two-dimensional whenever

$$u - \langle u, \zeta^* \rangle \zeta - \langle u, \bar{\zeta}^* \rangle \bar{\zeta}$$

is an "extra little part".

Now we shall delineate the sense in which the essentially two-dimensional problem is strictly two-dimensional. We first decompose the bifurcating solution u into a real-valued sum

$$u(t) = a(t)\zeta + \bar{a}(t)\bar{\zeta} + w(t), \tag{3.7}$$

where

$$\langle w, \zeta^* \rangle = \langle \bar{\zeta}, \zeta^* \rangle = \langle \zeta, \zeta^* \rangle - 1 = 0. \tag{3.8}$$

Substituting eq. (3.7) into eq. (3.1), we find, using eq. (3.2) that

$$[\dot{a} - \sigma(\mu)]\zeta + [\dot{\bar{a}} - \bar{\sigma}(\mu)\bar{a}]\bar{\zeta} + \frac{dw}{dt} = f_u(\mu \mid w) + N(\mu, u). \tag{3.9}$$

Projecting eq. (3.9) with ζ^* leads us to an evolution problem for the "little part" w on a supplementary space of the space spanned by ζ and $\bar{\zeta}$:

$$\frac{dw}{dt} = f_u(\mu \mid w) + [N(\mu, u) - \langle N(\mu, u), \zeta^* \rangle \zeta - \langle N(\mu, u), \bar{\zeta}^* \rangle \bar{\zeta}], \tag{3.10}$$

and to an evolution equation for the projected part

$$\dot{a} - \sigma(\mu)a = \langle N(\mu, u), \zeta^* \rangle. \tag{3.11}$$

In deriving eq. (3.11) we made use of the relations

$$\left\langle \frac{dw}{dt}, \zeta^* \right\rangle = \frac{d}{dt} \langle w, \zeta^* \rangle = 0$$

and

$$\langle f_u(\mu \mid w), \zeta^* \rangle = \langle w, f_u^*(\mu \mid \zeta^*) \rangle = \sigma \langle w, \zeta^* \rangle = 0.$$

Eq. (3.10) now follows easily from eqs. (3.9) and (3.11).

In sum, eq. (3.11) governs the evolution of the projection of the solution u into the eigensubspace belonging to the eigenvalue $\sigma_1(\mu) = \sigma(\mu)$ and eq. (3.10) governs the evolution of the part of the solution which is orthogonal to the subspace spanned by ζ^* and $\bar{\zeta}^*$.

In bifurcation problems the complementary projection w plays a minor role; it arises only as a response generated by nonlinear coupling to the component of the solution spanned by ζ and $\bar{\zeta}$. To see this we note that

$$\langle N(\mu, u), \zeta^* \rangle = \tfrac{1}{2}\langle (f_{uu}(\mu \,|\, u \,|\, u) + \mathrm{O}(\|u\|^3)), \zeta^* \rangle$$

$$\tfrac{1}{2}\langle f_{uu}(\mu \,|\, u \,|\, u), \zeta^* \rangle = \alpha(\mu)a^2 + 2\beta(\mu)|a|^2 + \gamma(\mu)\bar{a}^2 + 2a\langle f_{uu}(\mu \,|\, \zeta \,|\, w), \zeta^* \rangle$$

$$+ 2\bar{a}\langle f_{uu}(\mu \,|\, \bar{\zeta} \,|\, w), \zeta^* \rangle + \langle f_{uu}(\mu \,|\, w \,|\, w), \zeta^* \rangle,$$

$$\alpha(\mu) = \tfrac{1}{2}\langle f_{uu}(\mu \,|\, \zeta \,|\, \zeta), \zeta^* \rangle,$$

$$\beta(\mu) = \tfrac{1}{2}\langle f_{uu}(\mu \,|\, \zeta \,|\, \bar{\zeta}) \bar{\zeta}^* \rangle,$$

$$\gamma(\mu) = \tfrac{1}{2}\langle f_{uu}(\mu \,|\, \bar{\zeta} \,|\, \bar{\zeta}) \bar{\zeta}^* \rangle. \tag{3.12}$$

It follows that amplitude equation (3.11) may be written as

$$\dot{a} - \sigma(\mu)a = \alpha(\mu)a^2 + 2\beta(\mu)|a|^2 + \gamma(\mu)\bar{a}^2 + \mathrm{O}(|a|^3 + |a|\,\|w\| + \|w\|^2). \tag{3.13}$$

Returning now to eq. (3.10) with eq. (3.12) we find that after a long time $w = \mathrm{O}(|a|^2)$ and dramatize the two-dimensional structure of Hopf bifurcation in the general case by comparing eq. (3.13) with eq. (2.5) which governs the stability of the strictly two-dimensional problem.

4. Bifurcation from periodic orbits. Normal forms

We consider the equation

$$\mathrm{d}V/\mathrm{d}t = F(t, \mu, V). \tag{4.1}$$

Here $V(t, \mu)$ lies in a real Hilbert space $(H, \langle \, \rangle)$, μ is a real bifurcation parameter, and F is T-periodic i.e., $F(t, \mu, V) = F(t + T, \mu, V)$. Assume that there is a T-periodic solution

$$V = U(t, \mu) = U(t + T, \mu). \tag{4.2}$$

We rewrite eq. (4.1) in local form about U. If $u = V - U$, then

$$\mathrm{d}u/\mathrm{d}t = f(t, \mu, u)$$

where

$$f(t, \mu, u) = F(t, \mu, U + u) - F(t, \mu, U).\qquad(4.3)$$

We shall study eq. (4.3) with

$$f(t, \mu, \cdot) = f_u(t, \mu \mid \cdot) + N(t, \mu, \cdot),$$

where $f_u(t, \mu \mid \cdot)$ is linear and $N(t, \mu, v) = O(\|v\|^2)$.

We assume that the periodic orbit U, that is the orbit $u = 0$ of eq. (4.3) is stable if $\mu < 0$, and loses stability for $\mu > 0$. To express this precisely consider the linearisation of eq. (4.3)

$$\mathrm{d}v/\mathrm{d}t = f_u(t, \mu \mid v).\qquad(4.4)$$

This is to be thought of as a complex linear equation (with real coefficients) on $H^{\mathbb{C}}$, the complexification of H. Associated with eq. (4.4) is a linear operator on the space $\mathbb{P}_T^{\mathbb{C}}$ of T-periodic vector fields on $H^{\mathbb{C}}$,

$$J_\mu = -\mathrm{d}/\mathrm{d}t + f_u(t, \mu \mid \cdot)\qquad(4.5)$$

Eigenvalues of J_μ are called *Floquet exponents*. The orbit $u = 0$ is stable if all Floquet exponents have negative real part, and unstable if any has positive real part. The loss of stability at $\mu = 0$ is assumed to occur in the simplest way.

Bifurcation assumptions:
   There is a Floquet exponent $\sigma(\mu) = \xi(\mu) + i\eta(\mu)$ such that
   (i) $\sigma(0) = i\omega_0 = 2\pi r/T$, $0 \le r < 1$.
   (ii) $\sigma(\mu)$ and $\bar{\sigma}(\mu)$ are isolated algebraically simple eigenvalues of J_μ.
   (iii) $\mathrm{d}\xi/\mathrm{d}\mu(0) > 0$.
   (iv) All eigenvalues of J_0 other than $\sigma(0)$ and $\bar{\sigma}(0)$ have negative real part.

   The type of bifurcation that occurs depends on the value of r.
   (i) Strong resonance: if $r = m/n$ and $n = 1, 2, 3$, or $n = 4$ and a certain inequality holds then nT-periodic solutions bifurcate.
   (ii) Wan [6] has shown that there is an invariant torus when $n = 4$ and the inequality does not hold.
   (iii) Weak resonance: If $r = m/n$, $n \ge 5$, and certain exceptional conditions hold then nT-periodic solutions bifurcate.
   (iv) If $r \ne m/n$, $n = 1, 2, 3, 4$ there is a Hopf bifurcation to an invariant torus.

The next section describes how to approximate the original problem (4.3) with an autonomous equation in $\mathbb{R}^2$. It should be mentioned that the asymptotic representations can be constructed directly, without normal forms, by methods of applied analysis (see appendices to chapter X in ref. [4]).

4.1. Derivation of the autonomous equation

We assume that $r \neq 0, \frac{1}{2}$ (see refs. [3–5] for a study of these cases). This means that the periodic orbit $u = 0$ loses stability in two real dimensions instead of just one. The first step is to decompose eq. (4.3) into a part in this plane and a complementary part.

There is an inner product on $\mathbb{P}_T^C$,

$$[\xi_1, \xi_2] = \frac{1}{T} \int_0^T \langle \xi_1(t), \xi_2(t) \rangle \, dt.$$

Let J_μ^* be the adjoint of J_μ with respect to $[\cdot, \cdot]$. It can be verified that

$$J_\mu^* = d/dt + f_u^*(t, \mu \mid \cdot), \tag{4.6}$$

where $f^*(t, \mu \mid \cdot)$ is the adjoint of $f(t, \mu \mid \cdot)$ with respect to $\langle , \rangle$. Now $\sigma(\mu), \bar{\sigma}(\mu)$ are eigenvalues of J_μ, J_μ^* respectively; let ξ_μ, ξ_μ^* be corresponding eigenfunctions. Using eq. (4.6) and the assumption that $r \neq 0, \frac{1}{2}$, one can show that

$$\langle \xi_\mu(t), \xi_\mu^*(t) \rangle \equiv \langle \xi_\mu(0), \xi_\mu^*(0) \rangle$$

$$\langle \bar{\xi}_\mu(t), \xi_\mu^*(t) \rangle \equiv 0.$$

Normalise ξ_μ, ξ_μ^* so $\langle \xi_\mu, \xi_\mu^* \rangle = 1$. Now we can write

$$u = z \xi_\mu + \bar{z} \bar{\xi}_\mu + W,$$

where $z = \langle u, \xi_\mu^* \rangle$ and W is real. Eq. (4.3) becomes

$$dz/dt = \sigma(\mu) z + b, \tag{4.7a}$$

$$dW/dt = f_u(t, \mu \mid W) + B, \tag{4.7b}$$

where

$$b(t, \mu, z, \bar{z}, W) = \langle N(t, \mu, u), \xi_\mu^*(t) \rangle,$$

$$B(t, \mu, z, \bar{z}, W) = N(t, \mu, u) - \langle N(t, \mu, u), \xi_\mu^* \rangle \xi_\mu - \langle N(t, \mu, u), \bar{\xi}_\mu^* \rangle \bar{\xi}_\mu.$$

We have $b = b_0 + b_1$, $\boldsymbol{B} = \boldsymbol{B}_0 + \boldsymbol{B}_1$, where $b_0 = b(t, \mu, z, \bar{z}, 0)$, $b_1 = \mathrm{O}(|z| \, \|\boldsymbol{W}\| + \|\boldsymbol{W}\|^2)$, $\boldsymbol{B}_0 = \boldsymbol{B}(t, \mu, z, \bar{z}, 0)$, $\boldsymbol{B}_1 = \mathrm{O}(|z| \, \|\boldsymbol{W}\| + \|\boldsymbol{W}\|^2)$.

Roughly speaking eq. (4.7b) will be eliminated and eq. (4.7a) made autonomous up to $\mathrm{O}(|z|^{N+1})$. To do this we change variables

$$y = z + \gamma(t, \mu, z, \bar{z}) = z + \sum_{p+q \geq 2}^{N} z^p \bar{z}^q \gamma_{pq}(t, \mu),$$

$$\boldsymbol{Y} = \boldsymbol{W} + \boldsymbol{\Gamma}(t, \mu, z, \bar{z}) = \boldsymbol{W} + \sum_{p+q \geq 2}^{N} z^p \bar{z}^q \boldsymbol{\Gamma}_{pq}(t, \mu), \tag{4.8}$$

where N is arbitrary, γ_{pq} and $\boldsymbol{\Gamma}_{pq}$ are T-periodic, and $\boldsymbol{\Gamma}_{pq} \perp \xi_\mu^*, \bar{\xi}_\mu^*$. We choose $\gamma_{pq}, \boldsymbol{\Gamma}_{pq}$ later, after eq. (4.7) has been rewritten in terms of $y, \boldsymbol{Y}$. Now

$$\frac{\mathrm{d}y}{\mathrm{d}t} = \sigma z + b + \frac{\partial \gamma}{\partial t} + \frac{\partial \gamma}{\partial z}(\sigma z + b) + \frac{\partial \gamma}{\partial \bar{z}}(\bar{\sigma}\bar{z} + \bar{b})$$

$$= \sigma y + \left(\frac{\partial \gamma}{\partial t} + \sigma z \frac{\partial \gamma}{\partial z} + \bar{\sigma}\bar{z}\frac{\partial \gamma}{\partial \bar{z}} - \sigma\gamma + \tilde{b}\right) + b_1\left(1 + \frac{\partial \gamma}{\partial z}\right) + \bar{b}_1 \frac{\partial \gamma}{\partial \bar{z}},$$

where

$$\tilde{b}(t, \mu, z, \bar{z}) = b_0\left(1 + \frac{\partial \gamma}{\partial z}\right) + \bar{b}_0 \frac{\partial \gamma}{\partial \bar{z}};$$

$$\frac{\mathrm{d}\boldsymbol{Y}}{\mathrm{d}t} = f_u(t, \mu \mid \boldsymbol{Y}) + \left(\frac{\partial \boldsymbol{\Gamma}}{\partial t} + \sigma z \frac{\partial \boldsymbol{\Gamma}}{\partial z} + \bar{\sigma}\bar{z}\frac{\partial \boldsymbol{\Gamma}}{\partial \bar{z}} - f_u(t, \mu \mid \boldsymbol{\Gamma}) + \tilde{\boldsymbol{B}}\right)$$

$$+ \boldsymbol{B}_1\left(1 + \frac{\partial \boldsymbol{\Gamma}}{\partial z}\right) + \tilde{\boldsymbol{B}}_1 \frac{\partial \boldsymbol{\Gamma}}{\partial \bar{z}},$$

where

$$\tilde{\boldsymbol{B}}(t, \mu, z, \bar{z}) = \boldsymbol{B}_0\left(1 + \frac{\partial \boldsymbol{\Gamma}}{\partial z}\right) + \bar{\boldsymbol{B}}_0 \frac{\partial \boldsymbol{\Gamma}}{\partial \bar{z}}.$$

Expand

$$\begin{matrix} \tilde{b} \\ \tilde{\boldsymbol{B}} \end{matrix} = \sum_{p+q \geq 2}^{N} \begin{matrix} \tilde{b}_{pq} \\ \tilde{\boldsymbol{B}}_{pq} \end{matrix}(t, \mu)z^p \bar{z}^q + \mathrm{O}(|z|^{N+1}),$$

where $\tilde{b}_{pq}, \tilde{\boldsymbol{B}}_{pq}$ are T periodic and $\tilde{\boldsymbol{B}}_{pq} \perp \xi^*, \bar{\xi}^*$. Then

$$\frac{\mathrm{d}y}{\mathrm{d}t} = \sigma y + \sum_{p+q \geq 2}^{N} \left(\frac{\partial \gamma_{pq}}{\partial t} + [\sigma(p-1) + \bar{\sigma}q]\gamma_{pq} + \tilde{b}_{pq}\right)z^p \bar{z}^q$$

$$+ \mathrm{O}(|z| \, \|w\| + \|w\|^2 + |z|^{N+1}),$$

$$\frac{dY}{dt} = f_u(t, \mu \mid Y) + \sum_{p+q\geq 2} \left(\frac{\partial \Gamma_{pq}}{\partial t} - f_u(t, \mu \mid \Gamma_{pq}) + [\sigma p + \bar{\sigma} q]\Gamma_{pq} + \tilde{B}_{pq} \right) z^p \bar{z}^q$$

$$+ O(|z| \|w\| + \|w\|^2 + |z|^{N+1}).$$

Finally use eq. (4.8) on the right hand side to get

$$\frac{dy}{dt} = \sigma y + \sum_{p+q\geq 2} \left(\frac{\partial \gamma_{pq}}{\partial t} + [\sigma(p-1) + \bar{\sigma} q]\gamma + \bar{b}_{pq} \right) y^p \bar{y}^q$$

$$+ O(|y| \|Y\| + \|Y\|^2 + |y|^{N+1}), \tag{4.9a}$$

$$\frac{dY}{dt} = f_u(t, \mu \mid Y) + \sum_{p+q\geq 2}^{N} \{ -J_\mu(\Gamma_{pq}) + [\sigma p + \bar{\sigma} q]\Gamma_{pq} + \bar{\bar{B}}_{pq} \}$$

$$+ O(|y| \|Y\| + \|Y\|^2 + |y|^{N+1}), \tag{4.9b}$$

where $\bar{b}_{pq}$ and $\bar{\bar{B}}_{pq}$ are functions of γ_{ij}, Γ_{ij} with $i+j < p+q$ with T-periodic coefficients and such that all terms in eq. (4.9b) are orthogonal to $\xi_\mu^*, \bar{\xi}_\mu^*$.

Now γ_{pq}, Γ_{pq} are chosen successively for $p+q = 2, 3, \ldots, N$ so as to simplify eq. (4.9). This is the key step. We choose Γ_{pq} to make $-J_\mu(\Gamma_{pq}) + [\sigma p + \bar{\sigma} q] + \bar{\bar{B}}_{pq} \equiv 0$ for small μ. This is always possible since $\Gamma_{pq}, \bar{\bar{B}}_{pq} \in \{\xi \in \mathbb{P}_T : \xi \perp \xi_\mu^*, \bar{\xi}_\mu^*\}$ and the bifurcation assumptions mean that for small μ none of the eigenvalues of J_μ on this space has real part as small as $\text{Re}(\sigma p + \bar{\sigma} q)$. This reduces eq. (4.9b) to

$$\frac{dY}{dt} = f_u(t, \mu \mid Y) + O(\|Y\| |y| + \|Y\|^2 + |y|^{N+1}). \tag{4.10a}$$

In order to choose γ_{pq}, write

$$\bar{b}_{pq}(t, \mu) = \sum_{l \in \mathbb{Z}} b_{pql}(\mu) \exp(2\pi i l t / T),$$

$$\gamma_{pq}(t, \mu) = \sum_{l \in \mathbb{Z}} \gamma_{pql}(\mu) \exp(2\pi i l t / T).$$

Then

$$\frac{\partial \gamma_{pq}}{\partial t} + [\sigma(p-1) + \bar{\sigma} q]\gamma_{pq} + \bar{b}_{pq} = \sum_{l \in \mathbb{Z}} \alpha_{pql}(\mu) \exp(2\pi i l t / T),$$

where

$$\alpha_{pql}(\mu) = \left(\frac{2\pi i l}{T} + [\sigma(p-1) + \bar{\sigma} q] \right) \gamma_{pql}(\mu) + b_{pql}(\mu),$$

$$\alpha_{pql}(0) = \frac{2\pi i}{T} \{l + r[p-1-q]\} \gamma_{pql}(0) + b_{pql}(0).$$

We see that we can always choose γ_{pql} to make $\alpha_{pql}(\mu)\equiv0$ for small μ unless $l+r[p-1-q]=0$. We call $\{(p,q,l,r):l+r[p-l-q]=0\}$ the *Exceptional Set*. It is the union of two disjoint subsets:

I   the mean set: $(p,q,l,r)=(q+1,q,0,r)$ $2\le 2q+1\le N$,
II   the resonant set: $(p,q,l,r)=(q+1+nk,q,-km,m/n)$
$$0\le m<n,\ k\ge1,\ 2\le2q+1+nk\le N.$$

The mean set is present for any r, but the resonant set arises only when r is rational.

When (p,q,l,r) is in the exceptional set choose $\gamma_{pql}(\mu)\equiv0$; otherwise choose $\gamma_{pql}(\mu)$ to make $\alpha_{pql}(\mu)\equiv0$. This reduces eq. (4.9a) to

$$\frac{dy}{dt}=\sigma(\mu)y+\sum_{q\ge1}^{2q+1\le N}y^{q+1}\bar{y}^q b_{q+1,q,0}(\mu)$$

$$+\sum_{k>0}\sum_{q\ge0}^{2q-1+nk\le N}[y^{q+1+nk}\bar{y}^q b_{q+1+nk,q,-mk}(\mu)e^{-2\pi imkt/T}$$

$$+y^q\bar{y}^{q-1+nk}b_{q,q-1+nk,mk}(\mu)e^{2\pi imkt/T}]$$

$$+O(|y|\,\|Y\|+\|Y\|^2+|y|^{N+1}).\tag{4.10b}$$

The asymptotic representation is obtained by neglecting the order terms in eqs. (10a, b). The truncation number N in eq. (4.10b) is arbitrary. The justification of this approximation will not be attempted here; see refs. [3–5]. We proceed to study the approximate problem.

It is clear that eq. (4.10a) gives $Y(t,\mu)\equiv0$. To study eq. (4.10b) set

$$y=xe^{i\omega_0 t}.\tag{4.11}$$

Substitution in eq. (4.10b) gives an autonomous equation of the form

$$\frac{dx}{dt}=\mu\hat{\sigma}(\mu)x+\sum_{q\ge1}^{2q+1<N}+x|x|^{2q}a_q(\mu)$$

$$+\sum_{k>0}\sum_{q\ge0}|x|^{2q}\{x^{1+nk}a_{qk}+\bar{x}^{nk-1}a_{q,-k}\},\tag{4.12}$$

where $\mu\hat{\sigma}(\mu)=\sigma(\mu)-\sigma(0)$ and $a_{q,k}(\mu)\equiv0$ if r is irrational.

We shall look for the equilibrium solutions of eq. (4.12). We expect to find fixed points and closed curves. These will be cross sections of sub-harmonic trajectories and invariant tori for the original problem. The type of solution will depend on which terms on the right hand side of eq. (4.12) have lowest order in x after $\mu\hat{\sigma}(\mu)x$. If $n=3$ the term from the resonant set

$a_{0,-1}\bar{x}^{n-1}$ is the only term of order 2, and we shall find fixed points for eq. (4.12). If $n = 4$, $a_{0,-1}\bar{x}^{n-1}$ from the resonant set and $a_1 x|x|^2$ from the mean set both have order 3, and either fixed points or an invariant circle can occur. If $n \geq 5$ then terms from the mean set have lower order, and we expect a closed orbit of eq. (4.12). Normally it is traversed at a speed $O(\varepsilon^2)$, but if enough exceptional conditions hold this speed can be so low that the terms from the resonant set break up the closed orbit into fixed points. This is weak resonance.

All of the above remarks assume that various terms are $\neq 0$. The exceptional cases where this is not true are ignored here. Also it will be assumed for simplicity that $\hat{\sigma}, a_q, a_{q,k}, a_{q,-k} \cdots$ are independent of μ. This does not change the essence of the arguments.

5. Bifurcation from periodic solutions. Hopf bifurcation into a torus of subharmonic and asymptotically quasiperiodic solutions

This section outlines how to compute the trajectories on the torus when $n \geq 5$. We introduce an amplitude which is the mean radius of the invariant circle,

$$\varepsilon = \frac{1}{2\pi} \int_0^{2\pi} x(s) e^{-is} \, ds.$$

We assume the orbit can be written in the form

$$x(t, \mu) = \varepsilon \, e^{is} \chi(s, \varepsilon),$$

$$\mu = \varepsilon \bar{\mu}(\varepsilon),$$

$$s = \varepsilon \Omega(\varepsilon) t, \tag{5.1}$$

where χ is 2π periodic in θ. Note that $2\pi/\varepsilon\Omega(\varepsilon)$ is the period of the closed orbit of eq. (4.12).

Substitution in eq. (4.12) gives

$$(i\Omega - \bar{\mu}\hat{\sigma})\chi + \Omega \frac{d\chi}{ds} = \sum \chi|\chi|^{2q} a_q \varepsilon^{2q-1}$$

$$+ \sum_{k>0} \sum_{q\geq 0} |\chi|^{2q} [a_{q,k} e^{ink\theta} \chi^{1+nk} \varepsilon^{2q+nk-1}$$

$$+ a_{q,-k} e^{-ink\theta} \chi^{nk-1} \varepsilon^{2q+nk-3}]. \tag{5.2}$$

Expand in powers of ε:

$$\chi(s, \varepsilon) = \sum_{j=0}^{\infty} \chi_j(s)\varepsilon^j,$$

$$\tilde{\mu}(\varepsilon) = \sum_{j=0}^{\infty} \tilde{\mu}_j \varepsilon^j,$$

$$\Omega(\varepsilon) = \sum_{j=0}^{\infty} \Omega_j \varepsilon^j. \tag{5.3}$$

The functions $\chi_j(\cdot)$ are 2π-periodic; and

$$1 = \frac{1}{2\pi} \int_0^{2\pi} \chi(s, \varepsilon) \, ds,$$

so

$$\frac{1}{2\pi} \int_0^{2\pi} \chi_j(s, \varepsilon) \, ds = 1 \qquad j = 0,$$
$$= 0 \qquad j \geq 1. \tag{5.4}$$

We now solve by evaluating coefficients of successive powers of ε. From the terms of order 0,

$$(i\Omega_0 - \tilde{\mu}_0 \hat{\sigma})\chi_0 + \Omega_0 \frac{d\chi_0}{ds} = 0.$$

Taking the mean over $(0, 2\pi)$ gives

$$i\Omega_0 = \tilde{\mu}_0 \hat{\sigma}.$$

Now it follows from the bifurcation assumptions that $\hat{\sigma}$ has positive real part. Hence, since Ω_0 and $\tilde{\mu}_0$ are both real,

$$\Omega_0 = \tilde{\mu}_0 = 0.$$

The terms of order 1 in ε now give

$$(i\Omega_1 - \tilde{\mu}_1 \hat{\sigma})\chi_0 + \Omega_1 \frac{d\chi_0}{ds} = |\chi_0|^2 \chi_0 a_1. \tag{5.5}$$

Taking the mean over $(0, 2\pi)$ gives

$$i\Omega_1 - \tilde{\mu}_1 \hat{\sigma} = a_1 \frac{1}{2\pi} \int_0^{2\pi} |\chi_0|^2 \chi_0 \, ds. \tag{5.6}$$

It can be shown from eqs. (5.4), (5.5) and (5.6) that $\chi_0(s) \equiv 1$. From eq.

(5.13) we now obtain

$$i\Omega_1 - \tilde{\mu}_1\hat{\sigma} = a_1.$$

Taking real and imaginary parts gives

$$\tilde{\mu}_1\hat{\xi} + \alpha_1 = 0$$

$$\Omega_1 - \tilde{\mu}_1\hat{\eta} = \beta_1.$$

To continue we have to assume that $\Omega_1 \neq 0$. It will be seen in the next section that $\Omega_1 = 0$ is the first of the special conditions leading to weak resonance.

The terms in ε^2 give

$$\Omega_1 \frac{d\chi_1}{ds} - a_1(\chi_1 + \bar{\chi}_1) = g_1(s) - (i\Omega_2 - \tilde{\mu}_2\hat{\sigma})$$

where

$$g_1(s) = a_{0,-1}e^{-5is} \qquad n = 5$$

$$= 0 \qquad n \geq 5.$$

We see from eq. (5.4) that we must have

$$\int_0^{2\pi} [g_1(s) - (i\Omega_2 - \tilde{\mu}_2\hat{\sigma})]\, ds = 0.$$

This is true if and only if

$$\Omega_2 = \tilde{\mu}_2 = 0.$$

It is easily shown using Fourier series that the equation

$$\Omega_1 \frac{dy}{ds} - a_1(y + \bar{y}) = \hat{g}(s)$$

where $\hat{g}$ is 2π-periodic and $\int_0^{2\pi} \hat{g}(s)\, ds = 0$ has a unique 2π-periodic solution. We see that

$$\chi_1(s) = Ae^{5is} + Be^{-5is} \qquad n = 5$$

$$= 0 \qquad n \geq 5.$$

The analysis continues along these lines. It is found that $\tilde{u}(\cdot)$ and $\Omega(\cdot)$ are both odd functions, and that $\chi(\cdot, \varepsilon)$ is $2\pi/n$ periodic (constant if r is irrational). This is to be expected since eq. (4.12) is invariant under rotation through $2\pi/n$. By tracing back through the derivation in section 2, we see that our approximate solution is quasi-periodic with the two frequencies $2\pi/T$ and $\omega_0 + \varepsilon^2 \omega(\varepsilon^2) = \omega_0 + \Omega_1 + \varepsilon^2 \Omega_3 + \cdots$.

5.1. Subharmonic bifurcation

Suppose $x = \delta e^{i\varphi(\delta)}$ is a steady solution of eq. (4.12). Note that $\delta \exp[i\varphi(\delta)] \exp(2\pi ik/n)$, $0 \le k \le n-1$, are all steady solutions of eq. (4.12). They are the n-piercing points of a single nT-periodic trajectory. We have

$$0 = \mu\hat{\sigma} + \delta^2 a_1 + \delta^4 a_2 + \cdots + \delta^{n-2} e^{-in\varphi} a_{0,-1} + \cdots.$$

Assume

$$\varphi(\delta) = \varphi_0 + \varphi_1\delta + \varphi_2\delta^2 + \cdots$$

$$\mu = \mu^{(1)}\delta + \mu^{(2)}\delta^2 + \cdots.$$

We evaluate the coefficients of increasing powers of δ.

For $n = 3$: the terms in δ give

$$\mu^{(1)}\hat{\sigma} + a_{0,-1}e^{-3i\varphi_0} = 0.$$

Hence

$$\mu^{(1)} = |a_{0,-1}/\hat{\sigma}|,$$

$$\varphi_0 = \tfrac{1}{3}\arg(a_{0,-1}/\hat{\sigma}) + \frac{2k-1}{3} \qquad k = 0, 1, 2$$

(taking $\mu^{(1)} = -|a_{0,-1}/\hat{\sigma}|$ will give the same solution). The higher order terms can now be calculated. We obtain a single $3T$-periodic trajectory. The bifurcation is two sided since $\mu(\delta) = 0(\delta)$.

If $n \ge 4$: the terms in δ give

$$\mu^{(1)} = 0.$$

For $n = 4$ the terms in δ^2 give

$$\mu^{(2)}\hat{\sigma} + a_1 + e^{-4i\varphi_0} a_{0,-1} = 0,$$

so

$$|\mu^{(2)}\hat{\sigma} + a_1|^2 = |a_{0,-1}|^2.$$

This gives a quadratic equation for $\mu^{(2)}$. If the discriminant is positive we have two different values of $\mu^{(2)}$ which lead to two different $4T$-periodic trajectories.

If $n \geq 5$: the terms in δ^2 give

$$\mu^{(2)}\hat{\sigma} + a_1 = 0.$$

This is the first special condition for weak resonance; the requirement that $\mu^{(2)}$ be real restricts $\hat{\sigma}$ and a_1. It can be verified that this restriction is equivalent to the requirement that $\Omega_1 = 0$ which was used in section 3.

For $n = 5$: the terms in δ^3 give

$$\mu^{(3)}\hat{\sigma} + a_{0,-1}e^{-5i\varphi_0} = 0.$$

This determines $\mu^{(3)}$ and φ_0. Higher order terms can then be calculated. Since $\mu(\delta) = O(\delta^2)$ the bifurcation is one sided. Since $\mu^{(3)} \neq 0$, $\mu(\delta)$ is not even, and we obtain two $5T$-periodic trajectories.

If $n \geq 6$: the terms in δ^3 give

$$\mu^{(3)} = 0.$$

For $n = 6$: the terms in δ^4 give

$$\mu^{(4)}\hat{\sigma} + a_2 + a_{0,-1}e^{-6i\varphi_0} = 0.$$

This gives a quadratic equation for $\mu^{(4)}$; if the discriminant is positive two $6T$-periodic trajectories bifurcate.

If $n \geq 7$: the terms in δ^4 give

$$\mu^{(4)}\hat{\sigma}_0 + a_2 = 0.$$

This is the second special condition for weak resonance.

The results continue along these lines. As n increases subharmonic trajectories are possible only if more and more special conditions hold.

5.2. *Rotation number and lock-ins*

We conclude with a few remarks about the phenomenon of frequency locking when there is an invariant torus. This occurs when all the trajectories on the torus are captured by a single (subharmonic) trajectory.

Consider the Poincaré (first return) map. This is a map from the invariant circle to itself, this map takes a point on the circle to where the trajectory passing through it meets the circle again after going round the torus once (i.e. after time T). Consider its rotation number, ϱ (defined for example in ref. [4]; the reader may think of ϱ as a frequency ratio). If ϱ

is irrational there is a change of coordinates which makes the Poincaré map a rotation, and the flow on the torus is quadiperiodic. The Poincaré map has no periodic points. If $\varrho = p/q$ is rational, the Poincaré map must have periodic points of order q, to which correspond subharmonic trajectories. Generally there will be two such trajectories one attracting, the other repelling.

It is important to distinguish between the rotation number $\hat{\varrho}(\varepsilon)$ for the asymptotic representation and the rotation number $\varrho(\varepsilon)$ for the real flow. It is known that $\varrho(\varepsilon)$ is continuous but it is generally not differentiable. What happens is that if $\varrho(\varepsilon_0) = p/q$ then $\varrho(\varepsilon) \equiv p/q$ on an interval about ε_0. The rotation number locks on to each rational value. This happens because if θ_0 is a periodic point of order q of the Poincaré map, f_{ε_0}, then generically $\partial/\partial\theta(f_\varepsilon^q)\big|_{\varepsilon = \varepsilon_0, \theta = \theta_0} \neq 0$. This enables us to solve for a fixed point of f_ε^q when ε is near ε_0, so $\varrho(\varepsilon)$ cannot change near ε_0.

In particular the set of values of ε for which $\varrho(\varepsilon)$ is rational has positive measure. It is an important result of Herman [2] that the set on which $\varrho(\varepsilon)$ is irrational also has positive measure.

The results show that the approximate rotation number is of the form

$$\hat{\varrho}(\varepsilon) = \omega_0 + \varepsilon^2 \omega(\varepsilon^2).$$

It can be concluded from this that the true rotation number lies between two polynomials

$$\varrho(\varepsilon) = \hat{\varrho}(\varepsilon) \pm K\varepsilon^N,$$

where N is arbitrary. It follows that the lengths of the flat line segments on which lock-ins occur must tend to zero faster than any power of N as $\varepsilon \to 0$.

5.3. Experiments

The type of dynamics that I have discussed here is characteristic of the observed dynamics in some mechanical systems involving fluid motions. The fact that an analysis of the kind given here does seem to fit well the observations of motion in small boxes of liquid heated from below, and in flow systems like the Taylor problem may surprise some readers. The surprise is that an analysis in two dimensions, and low dimensions greater than 2 give results in agreement with observations of continuum systems with "infinitely" many dimensions. In fact, this kind of agreement is associated with the fact the spectrum of eigenmodes in the small scale systems for which agreements is sought is widely separated and the dimension of active eigenvalues is actually small.

I do not want to give a too cryptic explanation of the relevance to real fluid mechanics of the kind of analysis sketched in these lectures. In fact this kind of analysis is recommended for actual computation of bifurcated objects in fluid mechanics near the point of bifurcation [4]. A not too cryptic explanation of relevance can be found in my two review papers (D.D. Joseph, Hydrodynamic Instability and Turbulence, eds. H. Swinney and J. Gollub, Topics in Physics (Springer, 1980)) or in Bifurcation in Fluid Mechanics, in the translation of the XIIIth Int. Congr. of Theoretical and Applied Mechanics (IUTAM), Toronto (1980).

References

[1] M. Herman, Sur la conjugaison différentiable des difféomorphismes du cercle à des rotations, Publ. I.H.E.S. 49 (1979) 5–234.
[2] M. Herman, Mesure de Lebesgue et nombre de rotation, Lecture Notes in Math., Vol. 597 (Springer, Berlin, 1977) pp. 271–293.
[3] G. Iooss, Bifurcation of maps and applications, Mathematics Studies No. 36 (North-Holland, Amsterdam, 1979).
[4] G. Iooss and D.D. Joseph, Elementary Stability and Bifurcation Theory, Undergraduate Textbook in Mathematics (Springer, New York, 1980).
[5] G. Iooss and D.D. Joseph, Bifurcation and stability of nT-periodic solutions at a point of resonance, Arch. Rat. Mech. Anal. 66 (1977) 135–172.
[6] Y.H. Wan, Bifurcation into invariant tori at points of resonance, Arch. Rat. Mech. Anal. 68 (1978) 343–357.

COURSE 6

THE CREATION OF NON-TRIVIAL RECURRENCE IN THE DYNAMICS OF DIFFEOMORPHISMS

Sheldon E. NEWHOUSE

*Mathematics Dapartment, University of North Carolina,
Chapel Hill, NC 27514, U.S.A.*

G. Iooss, R.H.G. Helleman and R. Stora, eds.
Les Houches, Session XXXVI, 1981 – Comportement Chaotique des Systèmes Déterministes/
Chaotic Behaviour of Deterministic Systems
© North-Holland Publishing Company, 1983

Contents

0. Introduction

In recent years there has been increasing interest in the study of differential equations whose solutions exhibit a complicated or chaotic behavior. This has been stimulated by several developments. On the one hand, it has been observed that the motions in many physical systems exhibit structural changes which at times lead to complicated or erratic time dependence. On the other hand, new mathematical methods have been developed which enable one to begin to understand the mechanisms underlying such motions. Our purpose in these lectures will be to describe some mathematical results connected with the way in which systems which have a very simple recurrence structure can be modified to obtain complicated motions. We will deal mostly with discrete dynamical systems (i.e. diffeomorphisms). The results apply to differential equations having cross-sections and to behavior of periodic solutions away from equilibria in a well-known manner. One takes a cross-section H which is a hypersurface transverse to the flow, and one considers the first return map on H which is a local diffeomorphism.

Lecture 1 deals with the behavior of periodic orbits and their associated stable and unstable manifolds in generic or typical one-parameter families of diffeomorphisms. Lecture 2 deals with the structure associated to homoclinic motions and some simple ways in which such motions are created. Also, Morse–Smale diffeomorphisms are defined and their basic properties are discussed. Lecture 3 describes a general phenomenon arising in two-dimensional dynamics. This is the frequent occurrence of infinitely many attracting periodic orbits when a homoclinic tangency is created in a curve of diffeomorphisms. In lectures 4 and 5, we consider generic curves of diffeomorphisms which begin at a Morse–Smale system and which cease to be Morse–Smale at some parameter value b. Under certain conditions, the structure of the diffeomorphisms for parameters near b is described. Lecture 6 contains suggestions for future research concerning curves of diffeomorphisms beginning at more general structurally stable elements.

384

1. Periodic orbits, their stable and unstable manifolds, under diffeomorphisms

In this lecture, we shall describe the typical behavior of smooth one-parameter families of diffeomorphisms at least regarding periodic orbits and intersections of stable and unstable manifolds. We shall state our results in terms of a compact smooth manifold M, but in most cases the topology plays no crucial role. Thus, not much is lost if the reader thinks of M as a closed ball D^m in Euclidean space $\mathbb{R}^m$ or a sphere S^m (the boundary of a ball in $\mathbb{R}^{m+1}$).

Let M be a compact C^∞ manifold. A C^r diffeomorphism f from M to itself is a one-to-one C^r differentiable map from M to itself such that f^{-1} is also C^r where $r \geq 1$. When M has a boundary we will allow diffeomorphisms from M into itself which are not necessarily onto. Thus f^{-1} will be defined on the image of f. Let $\mathbb{D}^r(M)$ be the set of C^r diffeomorphisms of M. One defines the C^r topology in $\mathbb{D}^r(M)$ as follows.

Let $(U_1, \phi_1), \ldots, (U_s, \phi_s)$ be a finite family of C^r coordinate charts on M such that

$$M = \bigcup_{i=1}^{s} U_i.$$

Thus, each U_i is an open subset of M and $\phi_i : U_i \to \mathbb{R}^m$ is a homeomorphism (where m is the dimension of M) and for each i and j with $U_i \cap U_j \neq \emptyset$, the map $\phi_i \circ \phi_j^{-1} : \phi_j(U_i \cap U_j) \to \phi_i(U_i \cap U_j)$ is C^r. For $f \in \mathbb{D}^r(M)$, one has a local coordinate representative $\phi_i \circ f \circ \phi_j^{-1}$ defined on $\phi_j(U_j \cap f^{-1} U_i)$ provided $U_j \cap f^{-1} U_i \neq \emptyset$. This is a C^r map between the open sets $\phi_j(U_j \cap f^{-1} U_i)$ and $\phi_i(f U_j \cap U_i)$ in $\mathbb{R}^m$. For $f \in \mathbb{D}^r(M)$, set $c(f) = \{(i, j) : U_j \cap f^{-1} U_i \neq \emptyset\}$. One says that a sequence $(f_k)_{k \geq 1}$ in $\mathbb{D}^r(M)$ converges to f in $\mathbb{D}^r(M)$ iff
   (1) for large k, $c(f) = c(f_k)$
   (2) the first r derivatives of $\phi_i \circ f_k \circ \phi_j^{-1}$ converge to those of $\phi_i \circ f \circ \phi_j^{-1}$ as $k \to \infty$ for all $(i, j) \in c(f)$.

This notion of convergence defines the C^r topology on $\mathbb{D}^r(M)$. The statement g is C^r near f with mean there is a finite covering of M by coordinate charts such that the local representatives of g and f have their first r derivatives close to each other. It is easy to check that each of the inclusions $i : \mathbb{D}^{r+1}(M) \subset \mathbb{D}^r(M)$ is continuous. One defines the C^∞ topology on $\mathbb{D}^\infty(M)$ to be the weakest topology making all of the inclusions

$\mathbb{D}^\infty(M) \subset \mathbb{D}^r(M)$ continuous for $r \geq 0$. Note $\mathbb{D}^0(M)$ is the space of homeomorphisms of M (i.e., one-to-one continuous maps with continuous inverses). The topology in $\mathbb{D}^0(M)$ is that of uniform convergence.

We first recall the structure of typical elements in $\mathbb{D}^r(M)$.

Let $P \in \mathbb{R}^m$ be a fixed point for a diffeomorphism f defined in a neighborhood of p. That is, $f(p) = p$. One says that p is *hyperbolic* if the derivative $T_p f : \mathbb{R}^m \to \mathbb{R}^m$ has all of its eigenvalues off the unit circle. In this case, by linear algebra, there is a direct sum decomposition $\mathbb{R}^n = E^s \oplus E^u$ such that $T_p f(E^s) = E^s$, $T_p f(E^u) = E^u$, the eigenvalues of $T_p f / E^s$ have norm less than one, and the eigenvalues of $T_p f / E^u$ have norm greater than one.

For $f \in \mathbb{D}^r(M)$, a fixed point P of f is hyperbolic if it is a hyperbolic fixed point of some (or any) local coordinate representative of f. One gets a corresponding splitting of the tangent space $T_p M = E_p^s \oplus E_p^u$. A periodic point p of f is a point such that $f^n(p) = p$ for some $n \geq 1$. The least such n is called the period of p. One says that p is *hyperbolic* if it is hyperbolic as a fixed point for f^n. Suppose p is a hyperbolic periodic point of $f \in \mathbb{D}^r(M)$. Let d be the distance function on M. Set $W^s(p) = W^s(p,f) = \{y \in M : d(f^n x, f^n y)$ $\to 0$ as $n \to \infty\}$, and set $W^u(p) = W^u(p,f) = W^s(p, f^{-1})$. One calls $W^s(p)[W^u(p)]$ the stable (unstable) manifold of p. They have parametrizations as C^r injectively immersed copies of Euclidean spaces, and $W^s(p)[W^u(p)]$ is tangent at p to $E_p^s(E_p^u)$. Typical pictures in two and three dimensions are given in fig. 1.1.

If N_1 and N_2 are submanifolds of M and $x \in N_1 \cap N_2$, one says that N_1 is transverse to N_2 at x if $T_x N_1 + T_x N_2 = T_x M$. In this case, one writes $N_1 \pitchfork_x N_2$. If N_1 and N_2 are transverse at any point of intersection, then one says that N_1 is transverse to N_2, and one writes $N_1 \pitchfork N_2$. Note that if $N_1 \cap N_2 = \emptyset$, they are automatically said to be transverse.

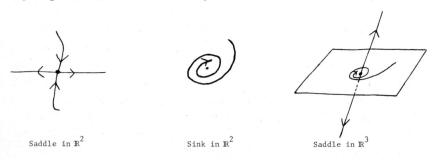

Saddle in $\mathbb{R}^2$ Sink in $\mathbb{R}^2$ Saddle in $\mathbb{R}^3$

Fig. 1.1.

Recall that a subset $\mathbb{B} \subset \mathbb{D}^r(M)$ is residual if it contains a countable intersection of dense open sets. Every residual set is dense and a countable intersection of residual sets is again residual. Elements of residual sets are called typical or generic elements.

Theorem (1.1) (Kupka–Smale). There is a residual set $\mathbb{B} \subset \mathbb{D}^r(M)$ for $r \geq 1$ such that for $f \in \mathbb{B}$,
   (1) all periodic points of f are hyperbolic.
   (2) if p and q are hyperbolic periodic points of f, then $W^s(p)$ is transverse to $W^u(q)$.

A diffeomorphism satisfying (1) and (2) in theorem (1.1) is called *Kupka–Smale*.

Now we proceed to state the one-parameter analog of theorem (1.1). We begin with the *non-hyperbolic* periodic points which appear in typical one-parameter families of diffeomorphisms.

(1) *Saddle-node*
A fixed point p of a C^2 diffeomorphism f is called a saddle-node if
   (a) 1 is an eigenvalue of $T_p f$ with multiplicity one, and the other eigenvalues of $T_p f$ have norm different from 1;
   (b) if v is a non-zero eigenvector of $T_p f$ associated with 1, then $T_p^2 f(v, v) \neq 0$ where $T_p^2 f(v, v)$ is the second derivative of f at p on the pair (v, v).
If p is a saddle-node, then there is a locally f-invariant curve γ through p such that in some C^2 coordinates (u, v) near p with $u \in \mathbb{R}$, $v \in \mathbb{R}^{m-1}$, γ has the form $(v = 0)$ and f on γ has the form $u \mapsto u + u^2 +$ higher order terms in u.

The origin $p = (0,0,0)$ is a saddle-node in $\mathbb{R}^3$ for the map $(u, v, w) \mapsto (u + u^2, \frac{1}{2} v, 2w)$ [here the map is not one-to-one on all of $\mathbb{R}^3$, so one only considers it near $(0,0,0)$]. The local picture is given in fig. 1.2.

Double arrows indicate sharper rates of attraction or repulsion than single arrows. Note that the local stable set of p is the set $w = 0$, $u \leq 0$. The local unstable set of p is the set $v = 0$, $u \geq 0$. Any saddle-node p for f is locally topologically conjugate to a product $\phi_1 \times \phi_2$ where $\phi_1 : \mathbb{R} \to \mathbb{R}$ has a saddle node at 0 and $\phi_2 : \mathbb{R}^{m-1} \to \mathbb{R}^{m-1}$ has a hyperbolic fixed point at 0. That is, there is a homeomorphism h from a neighborhood U of p in M to a neighborhood V of $(0,0)$ in $\mathbb{R} \times \mathbb{R}^{m-1}$ such that $hfh^{-1} = \phi_1 \times \phi_2$ on V.

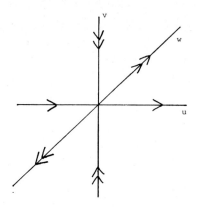

Fig. 1.2.

Now suppose $\{f_\mu\}_{\mu \in [0,1]}$ is a one-parameter family of diffeomorphisms of M having a saddle node p at $\mu = \bar{\mu}$.

We say that p unfolds generically with $\{f_\mu\}$ at $\bar{\mu}$ if there is an $\varepsilon > 0$ such that for $\nu \in (0, \varepsilon)$,

(c) $f_{\bar{\mu} - \nu}$ has no fixed point near p, and

(d) $f_{\bar{\mu} + \nu}$ has two distinct hyperbolic fixed points near p, or (c) and (d) hold for $\nu \in (-\varepsilon, 0)$.

The typical picture in the case $\nu \in (0, \varepsilon)$ is given in fig. 1.3.

A saddle node unfolding generically with $\{f_\mu\}$ at $\bar{\mu}$ is locally topologically equivalent to the family $\{\phi_{1\mu} \times \phi_2\}$ where $\phi_{1\mu}(u) = u \pm \mu + u^2$ on $\mathbb{R}^1$ and $\phi_2 : \mathscr{R}^{m-1} \to \mathbb{R}^{m-1}$ is a linear hyperbolic automorphism. The equivalence can be chosen to depend continuously on μ if $\{f_\mu\}$ is C^∞ [6].

(2) *Flip*

A fixed point p of a C^3 diffeomorphism f is called a flip if

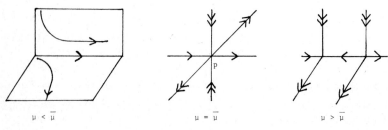

$\mu < \bar{\mu}$ $\mu = \bar{\mu}$ $\mu > \bar{\mu}$

Fig. 1.3.

(a) -1 is an eigenvalue of $T_p f$ with multiplicity one and all other eigen-values of $T_p f$ with multiplicity one and all other eigenvalues of $T_p f$ have norm different from one;

(b) if v is a non-zero eigenvector of $T_p f$ associated to -1, then the third derivative at p of f^2 in the direction v is not zero; i.e., $T_p^3 f^2(v, v, v) \neq 0$.

A typical example of a flip on $\mathbb{R}^1$ is 0 for the map $f(x) = -x + \beta x^3$ where $\beta \neq 0$. Any flip is locally topologically conjugate to the product $\phi_1 \times \phi_2$ where $\phi_1 : \mathcal{R} \to \mathbb{R}$ has a flip at 0 and $\phi_2 : \mathbb{R}^{m-1} \to \mathbb{R}^{m-1}$ has a hyperbolic fix-ed point at 0.

We say that the flip p unfolds generically with $\{f_\mu\}_{\mu \in [0,1]}$ at $\bar{\mu}$ if there is an $\varepsilon > 0$ such that for $v \in (0, \varepsilon)$.

(c) the largest $f_{\bar{\mu}-v}$-invariant set in a neighborhood of p consists of one hyperbolic fixed point near p; and

(d) the largest $f_{\bar{\mu}+v}$-invariant set in a neighborhood of p consists of one hyperbolic fixed point near p and a pair of hyperbolic periodic points of period two near p which are permuted by $f_{\bar{\mu}+v}$;

or (c) and (d) hold for $v \in (-\varepsilon, 0)$.

A picture of a flip unfolding generically is given by the following family of diffeomorphisms in $\mathbb{R}^2$ shown in fig. 1.4. In fig. 1.4, for $\mu < \bar{\mu}$, f_μ has a hyperbolic fixed point with one positive eigenvalue $\lambda_{2\mu} > 1$ and a second eigenvalue $\lambda_{1\mu} \in (-1, 0)$. At $\mu = \bar{\mu}$, the eigenvalues satisfy $\lambda_{2\bar{\mu}} > 1$, $\lambda_{1\bar{\mu}} = -1$. For $\mu > \bar{\mu}$, the eigenvalues of the fixed point satisfy $\lambda_{2\mu} > 1$, $\lambda_{1\mu} < -1$.

Remark: A flip unfolding generically at $\bar{\mu}$ is locally topologically equivalent to $\{\phi_{1\mu} \times \phi_2\}$ where $\phi_{1\mu} = (-1 \pm \mu)x \pm x^3$ on $\mathbb{R}^1$ and ϕ_2 is a hyperbolic linear automorphism of $\mathbb{R}^{m-1}$.

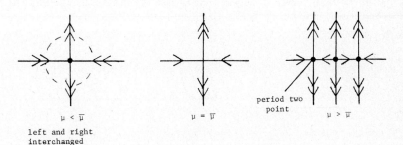

$\mu < \bar{\mu}$ $\mu = \bar{\mu}$ period two point $\mu > \bar{\mu}$

left and right
interchanged

Fig. 1.4.

(3) *Hopf point*

A fixed point p of a C^3 diffeomorphism f is called a Hopf point if

- (a) $T_p f$ has a pair of complex conjugate non-real eigenvalues $\lambda, \bar{\lambda}$ of norm 1 and multiplicity one with $\lambda^3 \neq 1 \neq \lambda^4$, and all other eigenvalues of $T_p f$ has norm different from 1;
- (b) there is an f-invariant two-dimensional manifold H through p tangent at p to the eigenspace corresponding to λ and $\bar{\lambda}$ such that in complex coordinates $z = u + iv$ centered at p, f has the form $z \mapsto z(\lambda + a_1|z|^2) + 0(|z|^3)$ where the complex number a_1 is such that the dot product $a_1 \cdot \lambda \neq 0$.

In (b), the point p is a weak attractor in H if $a_1 \cdot \lambda < 0$ and a weak repeller in H if $a_1 \cdot \lambda > 0$. As in the case of saddle nodes and flips any Hopf point is locally topologically conjugate to the product of a Hopf point in $\mathbb{R}^2$ and a hyperbolic fixed point in $\mathbb{R}^{m-2}$.

We say the Hopf point p *unfolds generically* with $\{f_\mu\}$ at $\bar{\mu}$ if there is an $\varepsilon > 0$ such that for $v \in (0, \varepsilon)$,

- (c) the largest $f_{\bar{\mu}-v}$-invariant set in a neighborhood of p consists of a single hyperbolic fixed point near p *and*
- (d) the largest $f_{\bar{\mu}+v}$-invariant set in a neighborhood of p consists of a single hyperbolic fixed point near p and an invariant circle near p,

or (c) and (d) hold for $v \in (-\varepsilon, 0)$.

One typical picture of a Hopf point unfolding generically in $\mathbb{R}^2$ is that a strongly attracting fixed point becomes weakly attracting as its derivative has a pair of complex conjugate eigenvalues cross the unit circle from inside to outside, and then it becomes repelling and an invariant attracting circle is created (see fig. 1.5).

In this case, one has $a_1 \cdot \lambda < 0$ at the Hopf point. Another possibility in which the pair of eigenvalues crosses the unit circle from inside to outside is that at the Hopf point $a_1 \cdot \lambda > 0$. In this case, an invariant repelling circle

Strong Attraction   Weak Attraction   Repulsion with
                                                                attracting circle

Fig. 1.5.

collapses down onto the Hopf point p with increasing μ and then for $\mu > \bar{\mu}$ there is only an expanding hyperbolic fixed point near p. This is called an inverse or reverse Hopf bifurcation.

One defines saddle node, flip, and Hopf periodic points and their generic unfoldings in the obvious way by passing to the least period of the points. It is not difficult to see that the stable and unstable sets of flips and Hopf points are smoothly immersed submanifolds without boundary, and the stable and unstable sets of saddle nodes are smoothly immersed submanifolds with boundary.

Now we define the quasi-transversal intersection of two submanifolds N_1 and N_2 of M. A point $x \in N_1 \cap N_2$ is such an intersection if

(a) $\dim(T_x N_1 + T_x N_2) = \dim M - 1$ or

(b) if $\dim T_x N_1 + \dim T_x N_2 \geq \dim T_x M$, then there are coordinates $x_1, \ldots, x_m$ on a neighborhood of x such that N_1 and N_2 have the following representations

$$N_1 = \{(x_1, \ldots, x_m) : x_1 = x_2 = \ldots = x_{m-s} = 0\}$$

$$N_2 = \{(x_1, \ldots, x_m) : x_1 = \Psi(x_{m-s+1}, \ldots, x_{u+1})$$

$$\text{and } x_{u+2} = x_{u+3} = \ldots = x_m = 0\}$$

where $m = \dim M$, $s = \dim N_1$, $u = \dim N_2$ and Ψ is a homogeneous quadratic function (non-degenerate).

If $\dim M = 2$, a quasi-transversal intersection of two curves is a tangency of order 1. We will also call such intersections non-degenerate tangencies. If $\dim M = 3$, there are a number of possibilities for quasi-transversal intersections.

Example 1: The intersection of any two distinct straight lines in $\mathbb{R}^3$ is a quasi-transverse intersection.

Example 2: A line and a surface in $\mathbb{R}^3$. Let (x_1, x_2, x_3) denote coordinates in $\mathbb{R}^3$, and let N_1 be the line $x_1 = x_2 = 0$ in $\mathbb{R}^3$. Then, if $N_2 = \{(x_1, x_2, x_3) : x_1 - x_3^2\}$, one has fig. 1.6. N_2 is a parabolic cylinder. This is the general quasi-transversal intersection of a line and surface in $\mathbb{R}^3$ (up to a smooth coordinate change).

Example 3: Two surfaces in $\mathbb{R}^3$. N_1 is the plane $x_1 = 0$ and N_2 is the quadratic surface $x_1 = x_2^2 \pm x_3^2$. We draw the saddle case $x_1 = x_2^2 - x_3^2$ (see

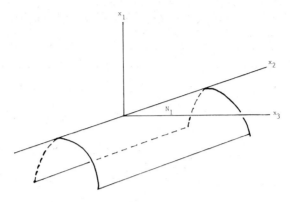

Fig. 1.6.

fig. 1.7).

There is a notion of generic unfolding of quasi-transversal intersections, but we will only need it in special cases below so we will omit the general definition.

Let us say that a diffeomorphism f (of class C^3) is *elementary* if exactly one of the following two possibilities holds

(1) all periodic orbits are hyperbolic and all stable and unstable manifolds of the periodic points meet transversely except at one orbit which consists of quasi-transversal intersections)

(2) there is exactly one non-hyperbolic periodic orbit which is a saddle-node, flip, or Hopf orbit, and all stable and unstable manifolds meet transversely.

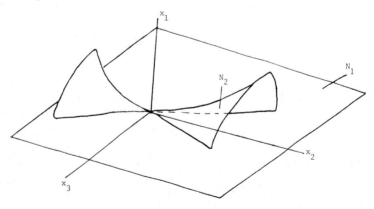

Fig. 1.7.

We can now state the one-parameter analog of theorem (1.1). Let $I = [0,1]$ and for $r \geq 1$, let $\mathbb{P}^r(M)$ be the set of C^r maps ξ from $M \times I$ to M of the form $(m, \mu) \to \xi_\mu(m)$ such that for each $\mu \in I, \xi_\mu : M \to M$ is a diffeomorphism. The topology on $\mathbb{P}^r(M)$ is that induced by the C^r maps from $M \times I$ to M. We will also write $\{\xi_\mu\}$ for the family ξ.

Theorem (1.2). For $r \geq 5$, there is a residual set $\mathbb{B} \subset \mathbb{P}^r(M)$ such that if $\xi \in \mathbb{B}$, then
  (1) ξ_μ is Kupka–Smale except for at most countably many μ's;
  (2) if ξ_μ fails to be Kupka–Smale, then it is an elementary diffeomorphism;
  (3) any saddle-node, flip, or Hopf orbit unfolds generically with $\{\xi_\mu\}$;
  (4) the periodic orbits of the mappings $\{\xi_\mu\}$ lie on one-dimensional submanifolds of $M \times I$; these submanifolds are transverse to the submanifold $M \times \{\mu\}$ except at a saddle node p for ξ_μ.

Statement (4) needs some clarification. If p is a saddle node for $\xi_{\bar\mu}$ where $\{\xi_\mu\}$ is an element of $\mathbb{B}$, then there is a curve C in $M \times I$ containing $(p, \bar\mu)$ such that C has a quadratic contact with the submanifold $M \times \{\bar\mu\}$ as in fig. 1.8.

The curve C is C^r and shows why ξ_μ has two periodic points of the same period as p near p either for $\mu > \bar\mu$ [in fig. (1.8a)] or for $\mu < \bar\mu$ [in fig. (1.8b)].

The picture for a flip is given in fig. 1.8c. The periodic orbit of double the period of p lies on a curve C_2 which comes into p, but at $\mu = \bar\mu$, this curve ceases to be defined (since the periodic orbit of double the period ceases to exist). The smooth curve C_1 containing $(p, \bar\mu)$ crosses $M \times \{\bar\mu\}$ transversely at $(p, \bar\mu)$.

In the case of the Hopf point there is a curve C_1 through p such that for

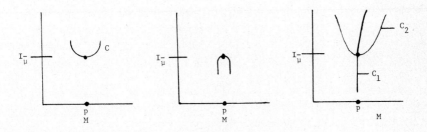

Fig. 1.8.

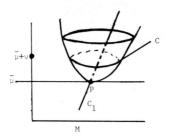

Fig. 1.9.

$v \in (0,\varepsilon)$ [or $(-\varepsilon,0)$], $M \times \{\bar{\mu}+v\} \cap S$ is the $\xi_{\bar{\mu}+v}$-invariant circle C near p (see fig. 1.9).

The dynamics of $\xi_{\bar{\mu}+v}$ on C is given by a numerical invariant $\zeta(\xi_{\bar{\mu}+v})$ called the rotation number (see ref. [7] for definition). When $\zeta(\xi_{\bar{\mu}+v})$ is irrational, $\xi_{\bar{\mu}+v}|C$ is topologically conjugate to a geometric rotation. When $\zeta(\xi_{\bar{\mu}+v})$ is rational, $\xi_{\bar{\mu}+v}$ has finitely many periodic orbits of the same period. They are either all hyperbolic or there is one saddle node orbit, and all points are forward and backward asymptotic to periodic orbits.

Of course, the global structure of the curves in $M \times I$ on which the periodic orbits of ξ lie is not well-understood. The structure is complicated even for the family of one-dimensional maps $\xi(x,\mu)=\mu-x^2$. The Feigenbaum doubling phenomenon [3] gives rise to curves in $\mathbb{R} \times \mathbb{R}$ with infinitely many branches as in fig. 1.10.

Notes. Various parts of Theorem (1.2) have been proved by many people. The local structure of saddle nodes (for diffeomorphisms and vector fields) has been considered by Sotomayor [10], Brunovsky [1], and

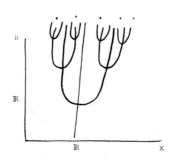

Fig. 1.10.

Chochitaichvili [2]. Flips were studied by Sotomayor [10] and Brunovsky [1], and Hopf points were studied by Sotomayor [10], Sacker [10], Ruelle and Takens [8], and Naimark. Various versions of quasi-transversal intersections were known to Sotomayor and systematically used in refs. [4], [5], and [6]. The spirit of Theorem (1.2) is essentially due to Sotomayor who studied related questions for one-parameter families of vector fields on surfaces [10].

References

[1] P. Brunovsky, On one-parameter families of diffeomorphisms, Comm. Math. Univ. Carolinae 11 (1970) 559, 582.

[2] A. Chochitaichvili, see refs. on p. 253 of V.I. Arnold, Chapitres Supplémentaires de la Theorie des Equations Différentielles Ordinaires, Traduction francaise (Editions Mir, Moscow, 1980).

[3] M. Feigenbaum, Quantitive universality for a class of non linear transformations, J. Stat. Phys. 19 (1978) 25–52; 21 (1979) 669–706.

[4] S. Newhouse and J. Palis, Bifurcations of Morse–Smale dynamical systems, in: Dynamical Systems, ed., Peixoto (Academic Press, New York, 1973) pp. 303–366.

[5] S. Newhouse and J. Palis, Cycles and bifurcation theory, Astérisque 31 (1976) 43–141.

[6] S. Newhouse, J. Palis and F. Takens, Stable families of diffeomorphisms, preprint.

[7] Z. Nitecki, Differentiable Dynamics (MIT Press, Cambridge, MA., 1971).

[8] D. Ruelle and F. Takens, On the nature of turbulence, Commun. Math. Phys. 20 (1971) 167–192.

[9] R. Sacker, On invariant surfaces and bifurcation of periodic solutions of ordinary differential equations, New York University IMM-NYU, 333 (1964).

[10] J. Sotomayor, Generic one parameter families of vector fields on two dimensional manifolds, Publ. Math. IHES 43 (1973); Generic bifurcations of dynamical systems, in: Dynamical Systems, ed., Peixoto (Academic Press, New York, 1973).

2. Homoclinic motions. Morse–Smale diffeomorphisms

The results of Lecture 1 are essentially of a local nature. They do not describe how orbits in one area of space interact with those in another area, and they do not concern non-trivial recurrence. In this lecture, we consider some simple ways in which non-trivial recurrence occurs and we will describe some results which give a good understanding of many motions involved.

Almost all complicated motion in smooth dynamics is connected in one way or another with homoclinic orbits, so we begin by describing these objects.

Let p be a hyperbolic periodic point of a C^r diffeomorphism f. Write $O(p)$ for the orbit of p,

$$W^u[O(p)] = \bigcup_{x \in O(p)} W^u(x),$$

and

$$W^s[O(p)] = \bigcup_{x \in O(p)} W^s(x).$$

A point $q \in W^u[O(p)] \cap W^s[O(p)] - O(p)$ is called a homoclinic point. If the intersection of $W^u[O(p)]$ and $W^s[O(p)]$ at q is transverse then q is called a transverse homoclinic point. Consider the example in fig. 2.1 in which p is a saddle fixed point of a diffeomorphism f on $\mathbb{R}^2$.

The positive orbit of q accumulates on p and, hence, $W^u(p)$ accumulates on itself oscillating wildly. A similar statement holds for $W^s(p)$. To begin to understand the orbit structure near q, take a small rectangular neighborhood Q of an interval in $W^s(p)$ containing p and q. If the vertical height of Q is small, its iterates will begin to approach $W^u(p)$ as in fig. 2.2.

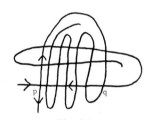

Fig. 2.1.

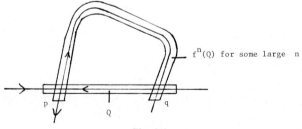

$f^n(Q)$ for some large n

Fig. 2.2.

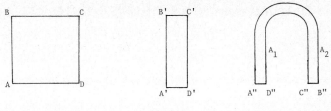

Fig. 2.3.

Thus, following Smale, we can idealize the mapping f^n on Q as follows. Consider the rectangle Q with vertices A, B, C, D mapped into $\mathbb{R}^2$ as indicated (see fig. 2.3). The images of A, B, etc. are denoted by A', B', etc. Then, $Q \cap f(Q) = A_1 \cup A_2$ where A_1 and A_2 are two rectangles. In our idealization, we suppose that $f|f^{-1}A_1$ is affine with derivative

$$\begin{pmatrix} \alpha & 0 \\ 0 & \alpha^{-1} \end{pmatrix}$$

and $f|f^{-1}A_2$ is affine with derivitive

$$\begin{pmatrix} -\alpha & 0 \\ 0 & -\alpha^{-1} \end{pmatrix}$$

where $0 < \alpha < \frac{1}{2}$. Consider $Q \cap f(Q) \cap [Q \cap f(Q)] = Q \cap f(Q) \cap f^2(Q)$. This consists of four rectangles as indicated in fig. 2.4. Continuing, one sees that

$$\bigcap_{0 \le j \le n} f^j Q$$

consists of 2^n rectangles, each with a horizontal width of α^n. Similarly,

$$\bigcap_{-n \le j \le 0} f^j Q$$

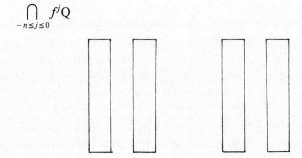

Fig. 2.4.

consists of 2^n rectangles each with vertical width α^n. The set

$$\Lambda = \bigcap_{j \in \mathbb{Z}} f^j Q$$

is a compact uncountable set without interior and every point in Λ is a limit of other points.

Λ is what is called a Cantor set. It is zero dimensional; i.e., each point in Λ has a neighborhood whose boundary misses Λ. Its orbit structure can be conveniently coded as follows. Let $\Sigma_2 = \{1,2\}^{\mathbb{Z}}$ be the set of double infinite sequences of 1's and 2's. Denote the elements of Σ_2 by $\underline{a}$, and let the ith entry of the sequence $\underline{a}$ be written $\underline{a}(i)$.

Give Σ_2 a metric d by setting

$$d(a, b) = \sum_{i = -\infty}^{\infty} \frac{|a(i) - b(i)|}{2^{|i|}}.$$

The shift automorphism $\sigma : \Sigma_2 \to \Sigma_2$ is defined by $\sigma(\underline{a})(i) = \underline{a}(i+1)$. One calls (σ, Σ_2) (or sometimes just Σ_2) the full 2-shift. For any $\underline{a} \in \Sigma_2$, the set

$$\bigcap_{-n \leq j \leq n} f^{-j}(A_{\underline{a}(j)})$$

is a square whose sides have length α^n, so

$$\bigcap_{j \in \mathbb{Z}} f^{-j}(A_{\underline{a}(j)})$$

is a unique point. Setting

$$h(\underline{a}) = \bigcap_{j \in \mathbb{Z}} f^{-j}(A_{\underline{a}(j)})$$

defines a homeomorphism $h : \Sigma_2 \to \Lambda$ such that $h\sigma = fh$ or $h\sigma h^{-1} = f$.

In general, if $f_1 : X_1 \to X_1$ and $f_2 : X_2 \to X_2$ are homeomorphisms (or even continuous maps) of compact metric spaces and there is a homeomorphism $h : X_1 \to X_2$ such that $hf_1 h^{-1} = f_2$, then one says that f_1 and f_2 are topologically conjugate. The map h (called a topological conjugacy between f_1 and f_2) is a continuous change of coordinates carrying the orbits of f_1 to those of f_2.

In the above example, f/Λ is topologically conjugate to σ on Σ_2.

The following theorem of Smale shows that the example is quite general. For an intuitive proof along the above lines, see ref. [7].

Theorem (2.1) (Smale). Let p be a hyperbolic fixed point of a C^1 diffeomorphism f, and let q be a transverse homoclinic point. Then there are a compact neighborhood U of $\{p, q\}$ and an integer $n > 0$ so that f^n restricted to the largest f^n-invariant subset of U is topologically conjugate to (σ, Σ_2).

Let Λ be the largest f^n-invariant set in U. Then Λ naturally determines the set

$$\Lambda_1 = \bigcup_{0 \le j < n} f^j \Lambda$$

which is f-invariant. To describe f on Λ_1 requires a slight generalization of the 2-shift as follows.

Let $\Sigma_N = \{1, \dots, N\}^{\mathbb{Z}}$ be the set of doubly infinite sequences of the letters 1 through N where N is an integer greater than 1. The shift $\sigma : \Sigma_N \to \Sigma_N$ and metric on Σ_N are defined as above. The pair (σ, Σ_N) is called the full N-shift. Let A be an $N \times N$ matrix of 0's and 1's, and let $\Sigma_A = \{\underline{a} \in \Sigma_N : A_{a(i), a(i+1)} = 1 \text{ for all } i\}$. One sees that Σ_A is a closed σ-invariant set. The set Σ_A or the pair (σ, Σ_A) is called a subshift of finite type. The map f on Λ_1 is topologically conjugate to a certain subshift of finite type.

Exercise. Let (σ, Σ_A) be a subshift of finite type. Let N_n be the number of fixed points of σ^n for $n \ge 1$. Show that $N_n = \text{trace } A^n$.

A fundamental problem has been to describe the changes in orbit structure which a one-parameter family $\{\xi_\mu\}$ of diffeomorphisms experiences as it acquires new homoclinic orbits. At present no complete analysis of this process has been given. We first give some examples where this occurs and make some remarks.

Example 1. Consider the horseshoe map described above, but with the image of Q translated down below Q. Call this map ϕ_0 (see fig. 2.5). Letting $\phi_\mu(x, y) = \phi_0(x, y) + (0, \mu)$ raises Q up as μ increases so that eventually one has a horseshoe. Assume ϕ_0 has Jacobian determinant less than one at each point of Q and that the image of each vertical line in Q consists of two vertical segments connected by a curve with strictly positive curvature. Then, the first change in orbit structure is the creation of a saddle node which splits into a saddle and a sink (see fig. 2.6).

As the parameter increases, the sink moves from left to right, its eigen-

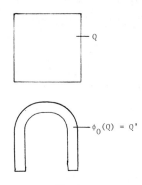

Fig. 2.5.

Fig. 2.6.

values become complex, then negative, and one of them passes through -1 giving a flip bifurcation. From this a period two sink splits off, moves to the left in Q (apparently) repeating the same phenomenon for ϕ_μ^2. It seems that one has period doubling very much like the Feigenbaum phenomenon for maps of an interval. It is possible that for certain μ's, ϕ_μ has strange attractors (i.e. complicated invariant attracting sets with dense orbits), but this has never been proved. It is known that at certain μ's, ϕ_μ will have infinitely many periodic sinks.

Example 2. This example shows that the behavior of Example 1 occurs in simple models of physical systems. Consider the motion of a Newtonian particle of unit mass on the line without friction and with potential energy $U(x) = x^3 - x$. The equation of motion is

$$\ddot{x} + x^2 - 1 = 0 \tag{1}$$

and the solutions in the phase space, i.e., $(x, \dot{x})$-plane are shown in fig. 2.7. The corresponding system is

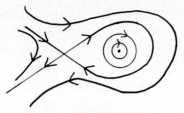

Fig. 2.7.

$$\dot{x} = v$$

$$\dot{v} = 1 - x^2.$$

There is a saddle point at $x = -1$, $v = 0$ and a center at $x = 1$, $v = 0$. The saddle point also has a saddle connection. Adding a positive linear damping term gives

$$\ddot{x} + b\dot{x} + x^2 - 1 = 0, \qquad b > 0. \tag{2}$$

The point $(1,0)$ becomes a sink, the saddle connection breaks, and one solution negatively asymptotic to $(-1,0)$ spirals toward $(1,0)$. See fig. 2.8. Now add a periodic forcing term $A \cos 2\pi\omega t$ with amplitude A and frequence ω. This gives

$$\ddot{x} + b\dot{x} + x^2 - 1 = A \cos 2\pi\omega t \tag{3}$$

or the system

$$\dot{x} = v$$

$$\dot{v} = -bv + 1 - x^2 + A \cos 2\pi\omega t \tag{4}$$

$$\dot{t} = 1.$$

We have added the last equation to remove the time-dependence and get a flow on $\mathbb{R}^3$. Consider the associated time $- 1/\omega$ map (of course, $1/\omega$ is

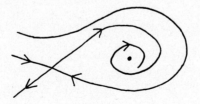

Fig. 2.8.

the period of the forcing term). Identifying the planes $t = 0$ and $t = 1/\omega$ gives us a map, say ϕ_A, of the (x, v)-plane to itself. For A small, it is a perturbation of the time $- 1/\omega$ map of the planar system associated to eq. (2). The map ϕ_A has a saddle fixed point P_A near $(-1, 0)$, a sink P'_A near $(1, 0)$, and a component of $W^u(P_A)$ spirals to P'_A. Also, the determinant of $T_{P_A}\phi_A$ is less than 1. For b small, as A increases, the results of Chow et al. [3] or Holmes [4] show that the bounded component of $W^u(P_A)$ becomes tangent to $W^s(P_A)$ as in fig. 2.9. If we analyze the creation of this tangency closely, we see that something like the translated horseshoe of example 1 occurs. If x denotes the tangency, then there is a rectangle Q near x on the inside of $W^s(P_A)$ whose image under some power ϕ_A^n of ϕ_A looks like the map in fig. 2.10.

Thus, there already are horseshoes and homoclinic points at the time the tangency occurs. At previous parameters A' the map $\phi_{A'}^n$ on Q behaves like the translated horseshoe above.

Similar phenomena occur in any periodically forced oscillation

$$x + bx + g(x) = A \cos 2\pi \omega t \tag{5}$$

whose potential energy $U(x) = \int_0^x g(u)du$ has only finitely many non-degenerate critical points, at least two such points, and satisfies

$$\lim_{|x| \to \infty} U(x) = \infty .$$

Therefore, it would not be surprising to find chaotic motions, strange attractors, etc. in such problems.

Remark. In recent years, a number of ordinary and partial differential equations arising in physics and chemistry have been shown to have homoclinic motions. Holmes and Marsden [5] have proved that homoclinic motions exist in partial differential equations for forced beam deflections.

Fig. 2.9.

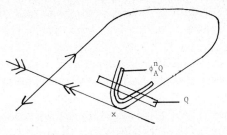

Fig. 2.10.

They also seem to find strange attractors, but this has not been proved. Zehnder [15] has shown that transverse homoclinic points accumulate on elliptic periodic orbits generically (in an analytic topology) in area preserving mappings of the plane. Pixton [11] has shown that C^r generically for area preserving diffeomorphisms of the 2-disk or 2-sphere every hyperbolic periodic point has transverse homoclinic points. Kopell [6] has found transverse homoclinic points in model reaction diffusion equations. Takens [14] has proved that homoclinic points, tangencies, etc., occur in certain open sets of two parameter families of diffeomorphisms when an eigenvalue of the derivative at a fixed point crosses the unit circle at a non-real root of unity. Recently, Chenciner has found similar behavior in generic two parameter families of diffeomorphisms in which an eigenvalue of the derivative at a fixed point crosses the unit circle at a non-root of unity.

Consider an equation of type (5) above with the assumption that

$$\lim_{x \to \pm\infty} g(x) = \pm\infty.$$

This guarantees that for fixed A and b there will be a large disk D in the (x, v)-plane mapped into itself by ϕ_A. For A fixed and b small, it can be proved that the map ϕ_A has a simple dynamical structure. There are finitely many hyperbolic fixed points. Also there are no saddle connections (orbits which are forward and backward asymptotic to saddles). If one thinks of ∞ as a source, these maps are examples of Morse–Smale diffeomorphisms of the two-dimensional sphere.

Let us define these diffeomorphisms. For $f \in \mathscr{D}^r(M)$, a point $x \in M$ is an ω-limit point of f if there is a sequence of integers $n_1 < n_2 < \dots$ and a point $y \in M$ such that $f^{n_i}(y) \to x$ as $i \to \infty$. Similarly, x is an α-limit point of f if there is a sequence $n_1 > n_2 > \dots$ and a point $y \in M$ such that $f^{n_i}(y) \to x$ as $i \to \infty$. Let $L_\alpha(f) = $ closure $[L_\alpha(f) \cup L_\omega(f)]$. The diffeomorphism f is Morse–Smale if

(1) $L(f)$ consists of finitely many hyperbolic periodic points, and
(2) if x and y are in $L(f)$, then $W^u(x)$ is transverse to $W^s(y)$.

Remark: Conditions (1) and (2) are equivalent to the conditions that $\Omega(f)$ be hyperbolic and $W^s(x) \pitchfork W^s(y)$ for $x, \; y \in \Omega(f)$.

Morse–Smale diffeomorphisms are very simple. Their recurrent motions reduce to finitely many periodic orbits. They also enjoy strong stability properties.

Let us make this more precise. A C^1 diffeomorphism $f: M \to M$ of a compact manifold M is called *structurally stable* if whenever g is C^1 near f it follows that g is topologically conjugate to f. Thus, a structually stable diffeomorphism has its orbit structure unchanged by small C^1 perturbations.

Theorem (2.2). (i) The set of Morse–Smale diffeomorphisms is open in $\mathscr{D}^r(M)$ for $r \geq 1$. (ii) Morse–Smale diffeomorphisms are structurally stable.

Theorem (2.2) (i) and the 2-dimensional case of Theorem (2.2) (ii) were proved by Palis [9]. The general case of Theorem (2.2) (ii) was proved by Palis and Smale [10]. The theorem is surprisingly non-trivial to prove even in the two-dimensional case. More general theorems involving stability for curves of diffeomorphisms are contained in ref. [8].

Let MS = MS(M) denote the set of Morse–Smale diffeomorphisms. The above examples make it natural to consider curves $\{\xi_\mu\}_{\mu \in [0,1]}$ of diffeomorphisms with $\xi_0 \in$ MS and to study the bifurcations of ξ_μ as μ varies. At least near the first point (in [0,1]) where ξ_μ ceases to be in MS, one has a good understanding in many cases. This will be discussed in Lecture 4. We will see that one frequently encounters structurally stable diffeomorphisms with infinitely many periodic orbits. Let us give some typical examples of these.

Example 3. In this example, we complete the horseshoe diffeomorphism to a diffeomorphism of the two-dimensional sphere S^2. We do this by first

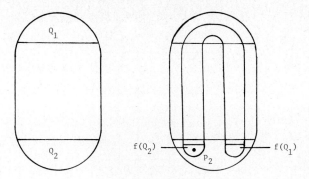

Fig. 2.11.

adding half disks to the top and bottom of the square Q, say Q_1 and Q_2 as in the fig. 2.11. Then map $Q \cup Q_1 \cup Q_2$ inside itself as pictured. This can be done so that there is a fixed sink P_2 in Q_2. Finally, we put an expanding fixed point at ∞. All of this is done so that any point in S^2 has its α or ω-limit point in $Q \cup \{P_1\} \cup \{P_2\}$. The diffeomorphism so constructed is structurally stable.

Example 4. Let

$$A = \begin{pmatrix} 2 & 1 \\ 1 & 1 \end{pmatrix}$$

and consider the induced linear automorphism $\bar{A} : \mathbb{R}^2/\mathbb{Z}^2 \to \mathbb{R}^2/\mathbb{Z}^2$ of the two-dimensional torus $T^2 = \mathbb{R}^2/\mathbb{Z}^2$. Then $\bar{A}$ is structurally stable. This is an example of an *Anosov* diffeomorphism.

Example 5. This example is like the horseshoe diffeomorphism of Example 3, but its limit set is topologically conjugate to a different subshift.

Keep the same map f as in the horseshoe on $Q \cup Q_1$. Let $B_1 = f^{-1}A_1$ and $B_2 = f^{-1}A_2$. Modify f on Q_2 to produce a diffeomorphism f_1 mapping part of Q_2 around through q as in next fig. 2.12. The dashed parts of the figure indicate part of Q_2 and its image under f_1.

We now formulate general conditions for structural stability. A compact f-invariant set Λ is *hyperbolic* if there is a splitting $T_xM = E_x^u \oplus E_x^u$ for each $x \in \Lambda$, a Riemann norm $|\cdot|$ on TM, and a constant $0 < \lambda < 1$ such that
(1) $T_xf(E_x^s) = E_{f(x)}^s$ and $T_xf(E_x^u) = E_{f(x)}^u$.
(2) $|T_xf|E_x^s| < \lambda$ and $|T_xf^{-1}|E_x^u| < \lambda$.

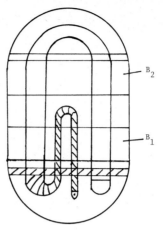

Fig. 2.12.

We note that if Λ is *hyperbolic*, it follows that the subspaces E_x^s and E_x^u are unique subject to conditions (1) and (2) and vary continuously with x in Λ. Also, by a Riemann norm $|\cdot|$ we mean the norm induced by a smooth Riemann metric on M. Let d denote the distance function on M induced by $|\cdot|$.

Suppose Λ is a hyperbolic set for f. For $x \in \Lambda$, let

$$W^s(x) = \{y \in M : d(f^n x, f^n y) \to 0 \text{ as } n \to \infty\},$$

and let

$$W^u(x) = \{y \in M : d(f^n x, f^n y) \to 0 \text{ as } n \to -\infty\}.$$

It is a fact that for each $x \in \Lambda$, $W^s(x)$ and $W^u(x)$ are smooth immersed submanifolds of M. If $L(f)$ is hyperbolic, then each $y \in M$ is in some $W^s(x)$ and some $W^u(x')$ for $x, x' \in L(f)$. One says that $L(f)$ satisfies the transversality condition if $W^u(x) \pitchfork W^s(y)$ for each $x, y \in L(f)$ (or, equivalently, each $x, y \in M$).

Theorem (2.3). If f is a C^1 diffeomorphism of a compact manifold M such that $L(f)$ is hyperbolic and $L(f)$ satisfies the transversality condition, then f is structurally stable.

This theorem is due to Robbin [12] and Robinson [13] who assumed f satisfied Axiom A and the transversality condition. That these assumptions are equivalent to those of Theorem (2.3) follows from Proposition (4.1) and Theorem (4.6) in ref. [7].

It is not difficult to verify that Examples 3, 4, and 5 satisfy the hypotheses of Theorem (2.3).

It should be pointed out that many physical systems will not give rise to structurally stable diffeomorphisms. Nevertheless, there usually are "parts" of the system which are structurally stable, and it is important to understand these parts. Theorem (2.3) and its related theorems (e.g. the ω-stability theorem [7] and Theorem (2.4) below) provide good theoretical conditions for the preservation of many qualitative orbit features under small perturbations of a system.

We will call an f-invariant set Λ a *hyperbolic basic set* for f if

(1) Λ is hyperbolic,

(2) there is a neighborhood U of Λ such that Λ is the largest invariant set in U; i.e.

$$\bigcap_{n \in \mathbb{Z}} f^n(U) = \Lambda.$$

(3) f/Λ has a dense orbit.

Two important facts about hyperbolic basic sets are contained in the next results.

Theorem (2.4). Suppose Λ is a hyperbolic basic set for f and U is a neighborhood of Λ such that

$$\bigcap_{n \in \mathbb{Z}} f^n(U) = \Lambda.$$

If g is C^1 close to f, then g restricted to

$$\bigcap_{n \in \mathbb{Z}} g^n(U)$$

is topologically conjugate to f restricted to Λ.

Theorem (2.5). Suppose Λ is a hyperbolic basic set for f. Then there is a subshift of finite type (σ, Σ_A) and a boundedly finite-to-one continuous surjection $\pi : \Sigma_A \to \Lambda$ such that $f\pi = \pi\sigma$. If Λ is zero-dimensional, then π can be chosen to be a homeomorphism.

A simple proof of Theorem (2.4) based on an idea of Bowen's is in ref. [7]. Theorem (2.5) is due to Bowen [1, 2], extending earlier work of Sinai.

Since we will often refer to zero-dimensional hyperbolic basic sets, it is convenient to have a spacial name for them. We will call them *generalized horseshoes*.

References

[1] R. Bowen, Topological entropy and Axiom A, Proc. Symp. Pure Math. 14 (1970).

[2] R. Bowen, Equilibrium states and the ergodic theory of Anosov diffeomorphisms, Lecture Notes in Math., Vol. 470 (Springer, Berlin, 1975).

[3] S. Chow, J. Hale and J. Mallet-Paret, An example of bifurcation to homoclinic orbits, to appear in J. Diff. Eq.

[4] P. Holmes, A non linear oscillator with a strange attractor, Phil. Trans. R. Soc. London Ser. A292 (1979) 419–448.

[5] P. Holmes and J. Marsden, A horseshoe in the dynamics of a forced beam, Non-linear Dynamics, Ann. NY Acad. Sci. 357 (1980) 313–21.

[6] N. Kopell, Time periodic but spacially irregular solutions to a model reaction diffusion equation, Non-linear Dynamics, Ann. NY Acad. Sci. 357 (1980) 397–410.

[7] S. Newhouse (1980) Lectures on dynamical systems, Prog. Math. 8 (Birkhauser, Boston).

[8] S. Newhouse, J. Palis and F. Takens, Stable families of diffeomorphisms, preprint.

[9] J. Palis, On Morse–Smale dynamical systems, Topology 8 (1969) 385–405.

[10] J. Palis and S. Smale, Structural stability theorems, Proc. Symp. Pure Math. 14 (1970) 223–231.

[11] D. Pixton, Planar homoclinic points, J. Diff. Eq., to appear.

[12] J. Robbin, A structural stability theorem, Ann. Math. 94 (1971) 447–493.

[13] R.C. Robinson, Structural stability of C^1 diffeomorphisms, J. Diff. Eq. 22 (1976) 28–73.

[14] F. Takens, Forced oscillations and bifurcations, Applications of Global Analysis, Commun. Math. Inst. Rijksuniversiteit Utrecht 3 (1974) 1–59.

[15] E. Zehnder, Homoclinic points near elliptic fixed points, Commun. Pure Appl. Math. 26 (1973) 131–182.

3. Infinitely many attracting periodic orbits in two dimensional dynamics

The preceding lecture dealt with special ways in which transverse homoclinic points appear and with the structure associated to these points. There are some general phenomena which occur in the dynamics of two-dimensional diffeomorphisms and we shall discuss one of them here. This is the fact that in a typical one-parameter family $\{\xi_\mu\}$ of diffeomorphisms on a surface in which homoclinic tangencies are created one finds parameter values μ' for which $\xi_{\mu'}$ has infinitely many periodic sinks or sources. A second general phenomenon – that homoclinic points occur if and only if a numerical invariant called the topological entropy is positive – will not be discussed here.

Let M be a surface, and let $\{\xi_\mu\}_{\mu \in [0,1]}$ be a family of C^2 diffeomorphisms depending continuously on a parameter μ. Suppose each ξ_μ has a hyperbolic saddle point p_μ depending continuously on μ for all μ. We say that $\{\xi_\mu\}$ creates a non-degenerate tangency of $W^u(p_\mu)$ and $W^s(p_\mu)$ at $(\bar{\mu}, x)$ if

(1) $W^u(p_{\bar{\mu}}, \xi_{\bar{\mu}})$ has a non-degenerate tangency with $W^s(p_{\bar{\mu}}, \xi_{\bar{\mu}})$ at x,
(2) there are an $\varepsilon > 0$ and two open intervals $I_\mu \subset W^s(p_\mu, \xi_\mu)$ and $J_\mu \subset W^u(p_\mu, \xi_\mu)$ such that
  (a) $x \in J_{\bar{\mu}} \cap I_{\bar{\mu}}$
  (b) J_μ and I_μ vary continuously with μ in the C^2 sense,
  (c) for μ $[\bar{\mu} - \varepsilon, \bar{\mu})$, $J_\mu \cap I_\mu = \emptyset$
  (d) for μ $(\bar{\mu}, \bar{\mu} + \varepsilon]$, $J_\mu \cap I_\mu$ consists of two points q_μ and $q_{\mu'}$ of transverse intersections and the distance between q_μ and $q_{\mu'}$ is a monotone increasing function of μ.

If one chooses (μ-dependent) coordinates in which I_μ is horizontal and fixed and J_μ is below I_μ for μ $(\bar{\mu} - \varepsilon, \bar{\mu})$, then one has fig. 3.1 near x.

Of course, generically, if $\{\xi_\mu\}$ creates a non-degenerate tangency of $W^u(p_\mu)$ and $W^s(p_\mu)$ at $(\bar{\mu}, x)$, then $\det T_{p_{\bar{\mu}}} \xi_{\bar{\mu}} \neq 1$. The following theorem assumes $\det T_{p_{\bar{\mu}}} \xi_{\bar{\mu}} < 1$. It applies when $\det T_{p_{\bar{\mu}}} \xi_{\bar{\mu}} > 1$ by replacing ξ_μ by ξ_μ^{-1}.

Theorem (3.1). Suppose $\{\xi_\mu\}$ creates a non-degenerate tangency of $W^s(p_\mu)$ and $W^u(p_\mu)$ at $(\bar{\mu}, x)$ and $\det T_{p_{\bar{\mu}}} \xi_{\bar{\mu}} < 1$. Then, given $\varepsilon > 0$, there is a residual subset A of an interval $A_\varepsilon \subset [\bar{\mu} - \varepsilon, \bar{\mu} + \varepsilon]$ such that for each $\mu \in A$

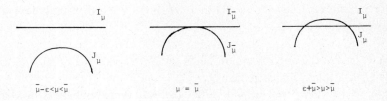

$$\bar{\mu} - \varepsilon < \mu < \bar{\mu} \qquad \mu = \bar{\mu} \qquad \varepsilon + \bar{\mu} > \mu > \bar{\mu}$$

Fig. 3.1.

(a) P_μ has transverse homoclinic points and
(b) each transverse homoclinic point of p_μ is in the closure of the set of sinks of ξ_μ.

Remark. Bowen and Ruelle have proved that a generalized horseshoe for a C^2 diffeomorphism has Lebesgue measure zero. Return to examples 3 and 5 of Lecture 2. If

$$\Lambda = \bigcap_{n \in \mathbb{Z}} f^n Q,$$

then almost all points for f and f_1 will approach the fixed sink in Q_2. Suppose we change f slightly so that it has a fixed point in Q whose derivative has determinant less than one. Consider a typical curve $\{\xi_\mu\}_{\mu \in [0,1]}$ of C^2 diffeomorphisms such that
(a) $\xi_\mu(x) = f(x)$ for $x \in Q$ and all μ,
(b) $\xi_0 = f$,
(c) $\xi_1 = f_1$.
Then, it can be proved that the hypotheses of Theorem (3.1) will hold for some $\bar\mu \in (0,1)$. It follows that for certain parameters μ near $\bar\mu$, ξ_μ will have open sets of points whose orbits frequently pass near Λ. Thus, one sees some chaotic behavior for *open sets* of initial conditions in the transition from f to f_1. On the other hand, to achieve more and more chaotic motion (at least that given by Λ) the periods of the sinks must get larger, and their basins smaller. So one has the feeling that to keep the deterministic model, one needs the presence of other sets to account for complicated asymptotic motion from initial conditions in sets of fixed positive Lebesgue measure. Such sets could be strange attractors or supports of Ruelle measures with positive entropy in the sense of Lecture 6.

Theorem (3.1) is proved (under some slightly more restrictive hypotheses) combining methods in refs. [1] and [3]. The proof without these hypotheses has not yet been published. We will sketch how the theorem is proved.

Step 1. Under the hypotheses of Theorem (3.1), there is a μ_1 near $\bar\mu$ such that ξ_{μ_1} has a periodic sink near x.
The idea of the proof is to consider fig. 3.2. For definiteness, we have supposed that the interval $J_{\bar\mu}$ lies below $I_{\bar\mu}$.

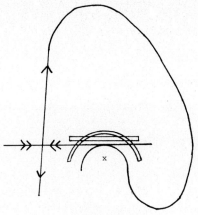

Fig. 3.2.

Very near x, one can choose a rectangle D near $I_{\bar{\mu}}$ and an integer $n > 0$ so that $\xi_{\bar{\mu}}^n D$ is a curved rectangle near $J_{\bar{\mu}}$. Raising or lowering μ produces a $\mu \in$ and a type of horseshoe diffeomorphism $\xi_{\mu'}$ from D to $\xi_{\mu'}^n D$. In the process of passing from $\bar{\mu}$ to μ', there is a saddle node created in D for some μ''. Since $\xi_{\mu}^j D$ is near p for most j's in $[0,n]$, the Jacobian of ξ_{μ}^n on D is less than 1. Then, a calculation shows that the saddle node splits into a saddle and a sink for μ_1 near μ''.

We shall indicate the idea of this calculation. We shall make several simplifying assumptions to avoid cumbersome technical details. Let $Cl[O(x)]$ denote the closure of the orbit of x and suppose there are coordinates (u, v) on a neighborhood U of $Cl[O(x)]$ such that $\xi_{\bar{\mu}}^{-1}(x) = (0, v_0)$ and $x = (u_0, 0)$. We also suppose there is a small neighborhood N of $(0, v_0)$ such that for $(u, v) \in U - N$ and all μ near $\bar{\mu}$, ξ_{μ} is the linear map $(u,v) \mapsto (\lambda_1 u, \lambda_2 v)$ where $0 < \lambda_1 < 1 < \lambda_2$ and $\lambda_1 \lambda_2 < 1$. Thus, the only non-linearity near $O(x)$ comes from ξ_{μ} on N. We also assume that for μ near $\bar{\mu}$ and z near $\xi_{\bar{\mu}}^{-1}(x)$, we have $\xi_{\mu}(z) - \xi_{\bar{\mu}}(z) + (0, \mu - \bar{\mu})$. This means that $\xi_{\mu}(z)$ merely translates $\xi_{\bar{\mu}}(z)$ vertically a distance $\mu - \bar{\mu}$. From the fact that the tangency at x of $W^u(p_{\bar{\mu}})$ and $W^s(p_{\bar{\mu}})$ is non-degenerate, it follows that near $(0, v_0)$, $\xi_{\bar{\mu}}$ has the form $\xi_{\bar{\mu}}(0, v_0) = [g_1(u, v), g_2(u, v)]$ where $g_1(0, v_0) = u_0$, $g_2(0, v_0) = 0$, $g_{1v}(0, v_0) \neq 0$, $g_{2u}(0, v_0) \neq 0$, $g_{2v}(0, v_0) = 0$, and $g_{2vv}(0, v_0) \neq 0$. Here subscripts indicate partial derivatives. For definiteness, suppose that $g_{2vv}(0, v_0) < 0$. Now, ξ_{μ}^n restricted to D (for suitably chosen D depending on n) has the form

$$\xi_{\mu}^n(u, v) = [g_1(\lambda_1^n u, \lambda_2^n v), \, g_2(\lambda_1^n u, \lambda_2^n v) + \mu - \mu]$$

for $(u, v) \in D$.

The equations for the fixed points of ξ_μ^n in D are

$$u = g_1(\lambda_1^n u, \lambda_2^n v)$$
$$v = g_2(\lambda_1^n u, \lambda_2^n v). \tag{3.1}$$

We may solve for $u = u(v)$ from the first equation since

$$\frac{\partial}{\partial u}[g_1(\lambda_1^n u, \lambda_2^n v)] = O(\lambda_1^n)$$

is very small for $(u, v) \in D$.

Thus, we get $v = g_2[\lambda_1^n u(v), \lambda_2^n v] + \mu - \bar{\mu}$ as the equation of the v-value of the fixed points. Write this as $\phi(v) = v$ where
$\phi(v) = g_2[\lambda_1^n u(v), \lambda_2^n v] + \mu - \bar{\mu}$.

Let $a = g_{2vv}(0, v_0) < 0$ and suppose $D = D_n$ is the rectangle

$$D_n = \left\{ (u, v) : |u - u_0| \le \frac{1}{|a|} \lambda_2^{-n}, \ |v - v_0| \le \frac{1}{|a|} \lambda_2^{-2n} \right\}.$$

We consider the graph of $\phi(v)$ for

$$|v - \lambda_2^{-n} v_0| \le \frac{1}{|a|} \lambda_2^{-2n}$$

with n large. All statements refer to $(u, v) \in D_n$. First, since $g_2(0, \lambda_2^n v) = a(\lambda_2^n v - v_0)^2 + O[(\lambda_2^n v - v_0)^2]$ we have for large n, that $|g_{2vv}(\lambda_1^n u, \lambda_2^n v) - a| < |a|/2$, so $g_{2vv}(\lambda_1^n u, \lambda_2^n v) < a/2 < 0$. Now,

$$\phi'(v) = g_{2u}(\lambda_1^n u, \lambda_2^n v)\lambda_2^n u'(v) + g_{2v}(\lambda_1^n u, \lambda_2^n v)\lambda_2^n$$

and

$$\phi''(v) = g_{2uu}(\lambda_1^n u, \lambda_2^n v)\lambda_1^{2n}[u'(v)]^2 + g_{2uv}(\lambda_1^n u, \lambda_2^n v)\lambda_1^n \lambda_2^n u'(v)$$
$$+ g_{2u}(\lambda_1^n u, \lambda_2^n v)\lambda_1^n u''(v) + g_{2vu}(\lambda_1^n u, \lambda_2^n v)\lambda_1^n \lambda_2^n u'(v)$$
$$+ g_{2vv}(\lambda_1^n u, \lambda_2^n v)\lambda_2^{2n}.$$

But, $u'(v) = O(\lambda_2^n)$, $u''(v) = O(\lambda_2^{2n})$ and $\lambda_1 \lambda_2 < 1$, so $\phi''(v)$ behaves as $g_{2vv}(\lambda_1^n u, \lambda_2^n v)\lambda_2^{2n}$ for n large. Hence, $\phi''(v) < \frac{1}{2}a\lambda_2^{2n}$ for large n. We claim there is a fixed point $[u(v), v]$ (for some μ near $\bar{\mu}$)

for which $v \in D_n$, and $g_{2v}(\lambda_1^n u, \lambda_2^n v)\lambda_2^n = \frac{1}{2}$. $\qquad (3.2)$

Assuming (3.2) for the moment, let us show that $[u(v), v]$ is a sink for ξ_μ. The derivative of ξ_μ^n at (u, v) has the form

$$T_{(u, v)}\xi_\mu^n = \begin{pmatrix} g_{1u} \cdot \lambda_1^n & g_{1v}\lambda_2^n \\ g_{2u}\lambda_1^n & g_{2v}\lambda_2^n \end{pmatrix},$$

where the partial derivatives of g_i are taken at $(\lambda_1^n u, \lambda_2^n v)$. Therefore, the eigenvalues of $T_{(u, v)}\xi_\mu^n$ satisfy

$$\lambda^2 - b\lambda + c = 0 \quad \text{with } b = g_{1u}(\lambda_1^n u, \lambda_2^n v)\lambda_1^n + g_{2v}(\lambda_1^n u, \lambda_2^n v)\lambda_2^n$$

and $c = \lambda_1^n \lambda_2^n c_1$ where $c_1 = g_{1u} \cdot g_{2v} - g_{1v}g_{2u}$. So the eigenvalues are, simply,

$$\lambda = \frac{b \pm (b^2 - 4c)^{1/2}}{2}.$$

If $g_{2v} \cdot \lambda_2^n = \frac{1}{2}$, then b is nearly $\frac{1}{2}$ and c is near 0, so $b^2 - 4c > 0$. Thus, the largest eigenvalue is

$$\frac{b + (b^2 - 4c)^{1/2}}{2} < b < 1$$

and $[u(v), v]$ is a sink. Thus, we must prove (3.2). Consider the values of $\phi'(v)$ for v at the top and bottom values in D_n. That is, consider $\phi'[\lambda_2^n v_0 \pm (1/|a|)\lambda_2^{-2n}]$. For $v = \lambda_2^{-n}v_0 \pm (1/|a|)\lambda_2^{-2n}$, we have

$$\phi'(v) \approx g_{2v}[\lambda_1^n u(v), \lambda_2^n v]\lambda_2^n \approx g_{2v}(0, \lambda_2^n v)\lambda_2^n$$

$$\approx 2a(\lambda_2^n v - v_0)\lambda_2^n \approx 2a[\pm (1/|a|)\lambda_2^{-n}]\lambda_2^n \approx \pm 2.$$

The wavy lines mean, of course, that other terms may be neglected. The graph of $\phi(v)$ is shown in fig. 3.3. Pick v_1 such that $g_{2v}[\lambda_1^n u(v_1), \lambda_2^n v_1] \times \lambda_2^n = \frac{1}{2}$. Clearly, translating by some $\mu - \bar\mu$ small makes v_1 satisfy $v_1 = \phi(v_1)$. Then, $[u(v_1), v_1]$ is the required sink. Before proceeding, it is

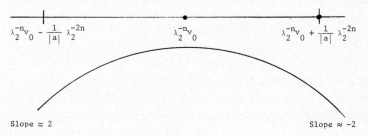

Fig. 3.3.

convenient to generalize the creation of non-degenerate tangencies between $W^s(p_\mu)$ and $W^u(p_\mu)$ to $W^s(\Lambda_\mu)$ and $W^u(\Lambda_\mu)$ where Λ_μ is a generalized horseshoe for ξ_μ depending continuously on μ. The definition is obtained by formally replacing p_μ by Λ_μ. We note that to say that $I_\mu \subset W^s(\Lambda_\mu)$ and $J_\mu \subset W^u(\Lambda_\mu)$ means there are points $z_\mu \in \Lambda_\mu$ and $w_\mu \in \Lambda_\mu$ such that $I_\mu \subset W^s(z_\mu)$ and $J_\mu \subset W^u(w_\mu)$.

Remark. The fact the tangencies of stable and unstable manifolds give rise to sinks after perturbation has been observed by several people. See, for instance, Gavrilov and Silnikov [4].

Step 2. Under the hypotheses of Theorem (3.1), there is a μ_1 near $\bar{\mu}$ and an interval A_{μ_1} about μ_1 so that for each $\tilde{\mu} \in A_{\mu_1}$, $\xi_{\tilde{\mu}}$ has a generalized horseshoe $\Lambda_{\tilde{\mu}}$ which contains p_μ and $\{\xi_\mu\}$ creates a non-degenerate tangency of $W^u(\Lambda_\mu)$ and $W^s(\Lambda_\mu)$ at $(\tilde{\mu}, \tilde{x})$ for some $\tilde{x}$ near x.

This step involves some complicated estimates and is proved (under some more restrictive hypotheses) in ref. [1]. Let us assume Step 2 holds.

It follows from general facts about hyperbolic sets (see ref. [2]) that Cl $W^u(p_\mu) \supset W^u(z_\mu)$ and Cl $W^s(p_\mu) \supset W^s(w_\mu)$. This and Step 2 imply that there is a countable dense set B in A_{μ_1} such that $\tilde{\mu} \in B$ implies $\{\xi_\mu\}$ creates a non-degenerate tangency of $W^u(p_\mu)$ and $W^s(p_\mu)$ at $(\tilde{\mu}, \tilde{x})$ for some $\tilde{x}$ near x.

Now, let us say that two hyperbolic periodic points p, p' for a diffeomorphism f are homoclinically related (or h-related) if

$$W^u[O(p)] - O(p) \text{ has a point } z \text{ of transverse intersection}$$
$$\text{with } W^s[O(p')] - O(p'), \tag{3.3}$$

and

$$W^u[O(p')] - O(p') \text{ has a point } z' \text{ of transverse}$$
$$\text{intersection with } W^s[O(p)] - O(p). \tag{3.4}$$

Fact. If p is a hyperbolic periodic point which has a p' homoclinically related to p, then p has transverse homoclinic points. In this case, let $H(p)$ be the set of transverse homoclinic points of p, and let $\tilde{H}(p)$ be the set of hyperbolic periodic points which are homoclinically related to p. Then, Closure $H(p) =$ Closure $\tilde{H}(p)$ and this set is f-invariant with a dense orbit. (For a proof, see ref. [2].)

Step 3. Let $\mu \in A_{\mu_1}$, q_μ be a transverse homoclinic point of p_μ and $\varepsilon > 0$. There is a μ' ε-close to μ such that $\xi_{\mu'}$ has a sink p' ε-close to $q_{\mu'}$.

The idea of the proof is the following. Arbitrarily near q_μ, one can find a $p''_\mu \in \tilde{H}(p_\mu)$. Then, parts of $W^u[O(p''_\mu)]$ and $W^s[O(p''_\mu)]$ accumulate well on $W^u(p_\mu)$ and $W^s(p_\mu)$. Choose $\bar{\mu} \in B$ near μ. Then, there is a μ_2 near $\bar{\mu}$ such that $\{\xi_\mu\}$ creates a non-degenerate tangency of $W^u[O(p''_\mu)]$ and $W^s[O(p''_\mu)]$ at μ_2. It follows from Step 1 that for some μ' near μ_2, $\xi_{\mu'}$ has a sink close to $p''_{\mu'}$ which is near $q_{\mu'}$.

Now the proof of Theorem (3.1) requires a standard topological semi-continuity argument.

Consider a set valued function S defined on an interval $J \subset \mathbb{R}$ such that S_μ is a compact subset of M for each $\mu \in J$. One says that S is lower-semi-continuous at μ_0 if for any finite open cover $U_1, \ldots, U_\alpha$ of S_{μ_0} with $U_i \cap S_{\mu_0} \neq \emptyset$ for $1 \leq i \leq \alpha$, there is an $\varepsilon > 0$ such that $|\mu - \mu_0| < \varepsilon$ and $\mu \in J$, imply that $S_\mu \cap U_i \neq \emptyset$ for all i. The function S is called upper-semi-continuous at μ_0 if for any open set U containing S_{μ_0}, there is an $\varepsilon > 0$ such that $|\mu - \mu_0| < \varepsilon$ and $\mu \in J$ imply that $S_\mu \subset U$. It is proved in topology (see ref. [5]), that if S is lower-semi-continuous at each point μ in J, then $\{\mu \in J : S$ is upper-semi-continuous at $\mu\}$ is residual in J. The same statement holds if J is replaced by a dense open subset of J.

It follows from Step 3 that there is a dense open set $\bar{A}_{\mu_1}$ in A_{μ_1} such that for each $\mu \in \bar{A}_{\mu_1}$, ξ_μ has a periodic sink. Then, it can be shown that the function $\mu \to S_\mu = \text{Closure (set of sinks of } \xi_\mu)$ is lower semi-continuous at each $\mu \in \bar{A}_{\mu_1}$.

Let $A \subset \bar{A}_{\mu_1}$ be the residual subset on which S is upper-semi-continuous. We claim that $\mu \in A$ implies $S_\mu \supset \text{Closure } H(p_\mu)$. This follows from Step 3. More precisely, if S_μ did not contain $\text{Cl } H(p_\mu)$ for some $\mu \in \bar{A}_{\mu_1}$, then pick a point $q \in \text{Cl } H(p_\mu)$ not in S_μ. Let $\varepsilon > 0$ be small enough so that the closed ε-ball about $q, \bar{B}_\varepsilon(q)$ is disjoint from S_μ. Then, $S_\mu \subset M - \bar{B}_\varepsilon(q)$. By Step 3, there is a $\mu' \in A$ near μ such that $S_{\mu'} \cap B_\mu(q) \neq \emptyset$. Thus, S is not upper-semi-continuous at μ, so μ cannot be in A.

References

[1] S. Newhouse, The abundance of wild hyperbolic sets and non-smooth stable sets for dif-feomorphisms, Publ. Math. IHES 50 (1979) 101–151.

[2] S. Newhouse, Lectures on dynamical systems, Progr. Math., Vol. 8 (Birkhauser, Boston, 1980).
[3] S. Newhouse, Diffeomorphisms with infinitely many sinks, Topology 13 (1974) 9–18.
[4] N.K. Gavrilov and L.P. Silnikov, On three-dimensional systems close to systems with a structurally unstable homoclinic curve, Math. USSR Sbor. 17 (1972) 467–485; 19 (1973) 139–156, 1973.
[5] C. Kuratowski, Topologie, Vol. II (Warsaw, 1950).

4. Curves of diffeomorphisms starting at Morse–Smale systems

In this lecture, we consider curves $\{\xi_\mu\}, 0 \leq \mu \leq 1$, of diffeomorphisms with ξ_0 Morse–Smale, and we discuss what is known about the ξ_μ's for μ near the first point at which ξ_μ ceases to be Morse–Smale. First, it is convenient to have some definitions. Let $\Sigma = \Sigma(M)$ denote the set of structurally stable diffeomorphisms on M. It is easy to see that Σ is open in $\mathbb{D}^r(M)$. Theorem (2.2) (ii) says that $MS \subset \Sigma$. For a curve $\xi = \{\xi_\mu\}_{\xi \in [0,1]}$ of diffeomorphisms, define $B(\xi) = \{\mu \in [0,1] : \xi_\mu \notin \Sigma\}$; i.e., $B(\xi)$ is the set of parameter values μ where ξ_μ fails to be structurally stable. One calls $B(\xi)$ the bifurcation set of ξ. Since $\mu \mapsto \xi_\mu$ is continuous, it follows that $B(\xi)$ is a closed subset of the interval $[0,1]$. Assuming $\xi_0 \in MS$ and $B(\xi) \neq \emptyset$, let $b = b(\xi) = \inf\{\mu : \mu \in B(\xi)\}$. One would like to describe the structure of ξ_μ for μ near b for a generic curve $\{\xi_\mu\}$. This has not yet been done completely, but we believe we do have a general idea of what the answer should be. What kind of recurrence can ξ_b have? Since, for generic curves ξ, Theorem (1.2) says the periodic orbits lie in curves in $M \times I$ and at most one orbit can be non-hyperbolic at a given value of μ, we see that ξ_b can have at most finitely many periodic orbits. In fact, the number of such orbits differs from that for ξ_0 by at most 1. Since non-trivially recurrent orbits usually are accompanied by periodic orbits, it is natural to suspect that $L(\xi_b)$ has only finitely many orbits. We can state this as a precise conjecture: For $r \geq 5$, there is a dense open set $\mathcal{B} \subset \mathcal{P}^r(M)$ of curves of diffeomorphisms such that if $\xi \in \mathcal{B}$ and $\xi_0 \in MS$, then $L(\xi_b)$ consists of only finitely many orbits. This conjecture (first stated in ref. [6]) remains unproved. Nevertheless, we believe it is true and we shall explore its consequences.

In the sequel, we shall write $\mathcal{P}(M)$ for $\mathcal{P}^r(M)$ where $r \geq 5$ is some fixed integer. Let $\mathcal{A} = \{\xi : \xi_0 \in MS \text{ and } L(\xi_b) \text{ has only finitely many orbits}\}$.

Proposition (4.1) [6]. There is a residual set $\mathscr{B} \subset \mathscr{P}(M)$ such that if $\xi \in \mathscr{B} \cap \mathscr{A}$, then either $L_\alpha(\xi_b)$ or $L_\omega(\xi_b)$ consists of finitely many periodic points.

There is a condition which reduces the analysis of ξ_μ for μ near b to the local bifurcation theory of Lecture 1. A *cycle* is a sequence $O(p_0)$, $O(p_1), \ldots, O(p_j)$ of periodic orbits such that $O(p_0) = O(p_j)$ and

$$W^u[O(p_i)] - O(p_i) \cap W^s[O(p_{i+1})] - O(p_{i+1}) \neq \emptyset$$

for $0 \le i < j$.

A cycle is *simple* if it has only one periodic orbit $O(p)$, that orbit is a saddle node, and for each $q \in O(p)$, $W^u(q) \cap W^s(q)$ consists of a single smooth circle.

Theorem (4.2). There is a residual set $\mathscr{B} \subset \mathscr{P}(M)$ such that if $\xi \in \mathscr{B} \cap \mathbb{A}$ and ξ_b has no cycles or a simple cycle then there is an $\varepsilon > 0$ such that $B(\xi) \cap [b, b + \varepsilon)$ is nowhere dense and ξ_μ is Morse–Smale for each $\mu \in [b, b + \varepsilon) - B(\xi)$. Furthermore, $B(\xi) \cap [b, b + \varepsilon)$ is uncountable if and only if ξ_b has a Hopf point or a simple cycle.

Let us describe some examples.

(1) We begin with the North Pole, South Pole diffeomorphism ξ_0 on S^2. There is an expanding fixed point p_1 at the North Pole and a contracting fixed point p_2 at the South Pole (see fig. 4.1).

We assume we have some box Q which is mapped as in the picture. It is contracted horizontally and translated down. Now we translate $\xi_0(Q)$

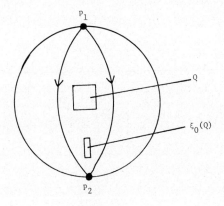

Fig. 4.1.

upward smoothly. Under typical conditions on ξ_μ, at the first bifurcation point one has a saddle node fixed point q which then splits into a saddle q_1 and a sink q_2. The first bifurcation point b_1 is isolated and ξ_μ is again Morse–Smale for $\mu > b_1$. Now as μ moves the eigenvalues at p_2 could become complex and then p_2 could pass through a Hopf bifurcation, say at $\mu = b_2$. In this case generically, the map $\xi_\mu, \mu > b_2$ on the invariant circle C_μ near p_2 is topologically conjugate to an irrational rotation for uncountably many values of μ near b_2 (even positive linear measure [3]) and each such μ is in $B(\xi)$. For a dense open set of μ's in $[b_2, b_2 + \varepsilon)$, however, ξ_μ will be Morse–Smale, having high period periodic orbits on C_μ. Thus although there are many bifurcations, the structure is completely described.

(2) Consider a Morse–Smale diffeomorphism ξ_0 on S^2 having an invariant smooth circle C_0 on which there are a saddle p_1 and a sink p_2 and two fixed sources q_1 and q_2 outside of C_0. This can be seen in fig. 4.2. We can bring p_1 and p_2 together to form a saddle node p in a simple cycle so that $W^u(p) \cap W^s(p) = C_0$. If $\{\xi_\mu\}$ is the family which does this with saddle node at $\mu = b$, then for $\mu > b$ and near b, ξ_μ has a unique invariant circle C_μ near C_0 and the dynamics of ξ_μ on C_μ is again determined by its rotation number.

The proof of Theorem (4.2) when ξ_b has no cycles is contained in ref. [5]. The case of a simple cycle requires techniques in ref. [6] together with techniques involving normally hyperbolic invariant manifolds as in ref. [4].

More interesting bifurcations occur when there are non-simple cycles. For diffeomorphisms there are always non-trivial bifurcations which occur.

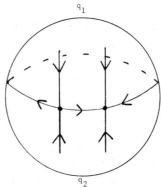

Fig. 4.2.

To avoid technicalities, we shall consider the two-dimensional case in the next theorem.

Theorem (4.3). There is a residual set $\mathscr{B} \subset \mathscr{P}(M)$ of curves of diffeomorphisms on a compact 2-manifold M such that if $\xi \in \mathscr{B} \cap \mathscr{A}$ and ξ_b has a non-simple cycle, then

(1) for each $\varepsilon > 0$, $B(\xi) \cap [b, b + \varepsilon)$ contains open intervals,
(2) there is a sequence $\mu_i \to b$ such that ξ_{μ_i} has infinitely many sources or sinks,
(3) if $L_\alpha(\xi_b)$ or $L_\omega(\xi_b)$ is hyperbolic, then
$$\lim_{\varepsilon \to 0} \frac{\text{Lebesgue Measure } (B(\xi) \cap [b, b + \varepsilon))}{\varepsilon} = 0.$$

The proof of Theorem (4.3) combines results in refs. [6], [9], [10], and [11]. We shall sketch the proof of a certain case of Theorem (4.3) (3) in the next Lecture.

Theorem (4.3) (3) says that all but a small fraction of μ's in $[b, b + \varepsilon)$ are such that ξ_μ is structurally stable. These structurally stable diffeomorphisms have infinite, zero-dimensional limit sets. The limit sets are like the generalized horseshoes we described in the last lecture. It follows from a theorem of Bowen and Ruelle [2], that for these ξ_μ's, almost all points (with respect to Lebesgue measure on M) have their limit sets equal to the finite set of sinks and sources near those for ξ_b. Thus, the probability of introducing even one new sink or source is small relative to ε if ε is small. This is not obvious. The proof involves some elementary number theory of the type encountered in small denominator problems in celestial mechanics.

We shall describe several examples in which Theorem (4.3) applies. Begin with the diffeomorphism f on S^2 pictured in fig. 4.3.

The circles represent sources and sinks and there are two saddles p_1 and p_2. Pick points q_1 and q_2 in $W^u(p_1)$ and $W^s(p_2)$ so that there is a curve γ joining q_1 to q_2 so that the forward orbit of γ never meets γ. Let V be a small neighborhood of γ so that $f^n(V) \cap V = \emptyset$ for $n > 0$. Let I be a small interval in $W^u(p_1)$ containing q_1. Let $\{\zeta_\mu\}$ be a curve of diffeomorphisms such that $\zeta_0 = \text{id}$, $\zeta_\mu(x) = x$ for $x \notin V$ and $\zeta_1(I)$ has precisely two transversal intersections with $W^s(q_2)$ as shown in fig. 4.4.

Consider $\xi_\mu = \zeta_\mu \circ f$. Then, $f_1 = \xi_1$ is such that $W^u(p_1)$ meets $W^s(p_2)$ transversely.

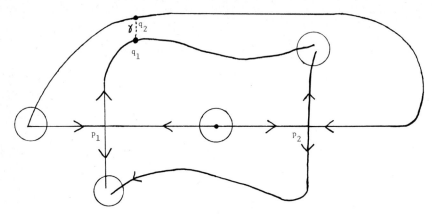

Fig. 4.3.

Now we create a non-degenerate tangency of $W^u(p_2)$ and $W^s(p_1)$ *at a point x* as indicated in fig. 4.5. Call the final diffeomorphism ξ_2. Note that this forces the right component of $W^s(p_1) - \{p_1\}$ to oscillate wildly near p_2, but only on the left of p_2. One can check that $L(\xi_2)$ is still the set of fixed points and ξ_2 has a 2-cycle. Note that for each μ slightly larger than 2, and near 2, p_1 will have transverse homoclinic points. So each such ξ_μ has non-trivial recurrence. Theorem (4.3) (3) applies to say that for most $\mu \in [2, 2 + \varepsilon)$ with ε small, the recurrence occurs in a describable way.

Notice that the most complicated kind of recurrence in ξ_b is that some points near x above $W^s(p_1)$, return in future time near x below $W^u(p_2)$ and then they approach a sink.

The next example is such that $L_\omega(\xi_b)$ is infinite even though $L_\alpha(\xi_b)$ is finite.

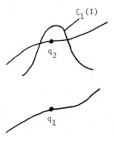

Fig. 4.4.

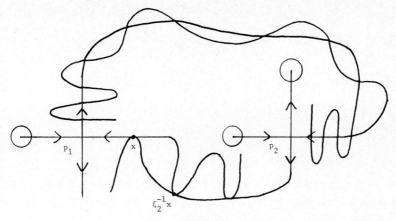

Fig. 4.5.

We begin with the equation we have studied

$$\ddot{x} + x^2 - 1 = 0$$

and the associated system

$$\dot{x} = v,$$

$$\dot{v} = 1 - x^2.$$

If we add negative friction with magnitude c, we get the system

$$\dot{x} = v,$$

$$\dot{v} = 1 - x^2 + cv, \tag{4.1}$$

whose orbit structure is shown in fig. 4.6. Now, instead suppose the $\dot{v}$ term has the form $1 - x^2 + c(x, v)v$ where $c(x, v)$ is a function which is negative in a small neighborhood of the saddle point $(-1, 0)$ and strongly positive off a slightly larger neighborhood. The new system has the form

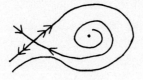

Fig. 4.6.

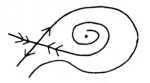

Fig. 4.7.

$$\dot{x} = v,$$
$$\dot{v} = 1 - x^2 + c(x, v)v, \tag{4.2}$$

and its orbit structure is essentially the same as that of eq. (1) except that the divergence is negative at $(-1,0)$, so we have fig. 4.7. Next add the forcing term $\mu \cos 2\pi t$ to obtain

$$\dot{x} = v,$$
$$\dot{v} = 1 - x^2 + c(x, v)v + \mu \cos 2\pi t, \tag{4.3}$$
$$\dot{t} = 1.$$

Under certain conditions on $c(x, v)$, one would expect that as μ increases, the saddle point for the time-one map varies continuously with μ and its stable and unstable manifolds become tangent as in fig. 4.8.

If before putting in the external force, we add to the vector field of eq. (2) a vector field which has small magnitude for $(x-2)^2 + v^2 > \frac{1}{2}$ and has a strong sink at $(-2,0)$, we obtain the orbit structure shown in fig. 4.9.

After forcing with $\mu \cos 2\pi t$, the time-one map ξ_μ is a diffeomorphism with a saddle fixed point p in which a tangency between $W^s(p_\mu, \xi_\mu)$ and $W^u(p_\mu, \xi_\mu)$ is created. We suppose b is the parameter at which the tangency is created. We also assume the tangency is non-degenerate and that $\det T_p \xi_b < 1$.

This is our required curve $\xi = \{\xi_\mu\}$. Again we regard each ξ_μ as a diffeomorphism of the 2-sphere with a source at ∞. We claim that $L_\alpha(\xi_b)$ is

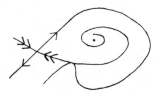

Fig. 4.8.

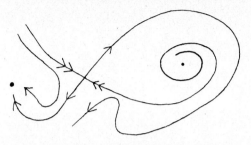

Fig. 4.9.

finite but $L_\omega(\xi_b)$ is infinite. To see why this is so, first choose a neighborhood U of $p = p_b$ in which linear estimates given by $T_p\xi_b$ dominate. Take a point x of tangency of $W^u(p, \xi_b)$ and $W^s(p, \xi_b)$ so that $\xi_b^n(x) \in U$ for $n \geq 0$. Because b is the first bifurcation point of ξ, it follows that for $\mu < b$ and near b, all points are in the stable and unstable manifolds of the fixed points of ξ_μ. From this one can show that if V is a neighborhood of x, then any point y whose ω-limit set (α-limit set) is not a fixed point of ξ_b must have its forward (backward) orbit meet V infinitely many times. Let R be a small rectangle centered at x. Because $\det T_p\xi_b < 1$, any point in R which returns to R in positive time must do so much closer to x. Thus, if $\xi_\beta^{n_1}(y)$, $\xi_b^{n_2}(y), \ldots$ is an infinite sequence of iterates of y with $n_1 < n_2 < \ldots$ and each $\xi_b^{n_i}(y)$ is in R, then $\mathrm{dist}[\xi_b^{n_i}(y), x] \to 0$ as $i \to \infty$. This shows that if y is any point such that $\omega(y)$ is not a fixed point of ξ_b, then $\omega(y) \cap R = \{x\}$. It follows that $\omega(y) = 0(x) \cup \{p\}$. One can also easily show that there are such points y. Thus, if F is the set of fixed points of ξ_b, then $L_\omega(\xi_b) = 0(x) \cup F$. On the other hand, any point $y \in R$ for which $\xi_b^{-n}y \in R$ for some $n > 0$ has $\mathrm{dist}(\xi_b^{-n}y, x) > c$ $\mathrm{dist}(y, x)$ where $c > 1$ and c is independent of y and n. This implies that no point y has infinitely many backward iterates in R. Hence $L_\alpha(\xi_b) = F$.

For our next example, consider a saddle node p of a diffeomorphism f on a surface for which $W^u(p) \subset W^s(p)$. Then $W^s(p)$ is two-dimensional and there is a unique continuous f-invariant family of curves in $W^s(p)$ which contains the boundary of $W^s(p)$. This is called the strong stable foliation of $W^s(p)$ and is denoted $\mathscr{F}^{ss}$. If $W^u(p)$ is transverse to each curve in $\mathscr{F}^{ss}$, one can show that $\{p\}$ is a simple cycle, and we have already discussed this. If $W^u(p)$ is not transverse to some curve in $\mathscr{F}^{ss}$, then one can show that $W^u(p)$ is a non-smooth circle. A picture of this situation is

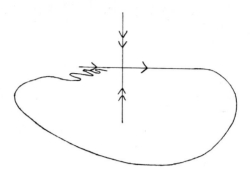

Fig. 4.10.

depicted in fig. 4.10. Suppose now that $f = \xi_b$ where $\{\xi_\mu\}$ is a smooth curve of diffeomorphisms, and for $\mu > b$ and near b, there is no fixed point near p, i.e., the saddle disappears for $\mu > b$. Under certain restrictive conditions, Afraimovic and Silnikov [1] proved that there are μ's near b for which ξ_μ has hyperbolic periodic points with transverse homoclinic points. This has been generalized to the generic case in the next theorem.

Theorem (4.4) [11]. There is a residual set $\mathcal{B} \subset \mathcal{P}(M^2)$ such that if $\xi \in \mathcal{B}$ and ξ_b has a saddle node p such that $W^u(p) \subset W^s(p)$ and $W^u(p)$ is not a smooth circle, then there are $\bar{\mu}$'s near b such that ξ_μ has a hyperbolic saddle periodic point q_μ and $\{\xi_\mu\}$ creates a non-degenerate tangency of $W^u(q_\mu, \xi_\mu)$ and $W^s(q_\mu, \xi_\mu)$ at $\bar{\mu}$.

Observe that saddle nodes can appear in cycles involving some hyperbolic periodic orbits as in fig. 4.11. In dimension two, if p is a saddle node for ξ_b such that $\dim W^s(p) = 2$, $W^u(p)$ has a non-tranverse intersection with some leaf of $\mathcal{F}^{ss}$, there is a hyperbolic saddle point q such that $W^u(q) \cap W^s(p) \neq \emptyset$, and $W^u(p)$ has a transverse intersection with $W^s(q)$, then generically it is easy to show that non-degenerate tangencies of $W^u(q)$ and $W^s(q)$ are created for μ's near b.

Certain parts of Theorem (4.3) have been proved in higher dimension. Suppose $\{\xi_\mu\}$ is a generic curve in $\mathcal{A}$ (where $\dim M > 2$) and $L_\alpha(\xi_b)$ or $L_\omega(\xi_b)$ is a finite hyperbolic set with a cycle. If all of the stable manifolds of the periodic points in the cycles have the same dimension, then the conclusion of part (3) of Theorem (4.3) holds [6]. Such cycles are called equidimensional. An example of a diffeomorphism ξ_b with a non-

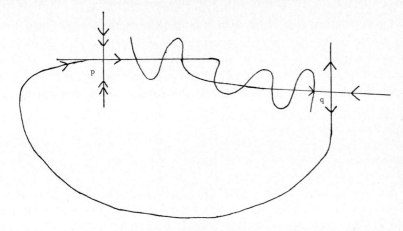

Fig. 4.11.

equidimensional cycle is on p. 121 of ref. [6].

Except in very particular cases, if ξ_b has a non-equidimensional cycle, it is not known if there are μ's near b with ξ_μ structurally stable and $L(\xi_\mu)$ infinite.

We remark that Theorem (4.4) holds in higher dimension for saddle nodes $p_{\bar\mu}$ such that $W^u(p_{\bar\mu}, \xi_{\bar\mu}) \subset W^s(p_{\bar\mu}, \xi_{\bar\mu})$ [or $W^s(p_{\bar\mu}, \xi_{\bar\mu}) \subset W^u(p_{\bar\mu}, \xi_{\bar\mu})$]. That is, it holds if the saddle node unfolds to a sink (or source) and a saddle point. In the case where the saddle node unfolds to two saddle points and is the only periodic point in the cycle at $\bar\mu$, the theorem is not known.

Miscellaneous remarks. One can define various notions of stability for arcs of diffeomorphisms. One such notion is that ξ is *stable* if whenever η is near ξ there is a heomorphism $\zeta_t : M \to M$, depending continuously on $t \in [0,1]$ and a homeomorphism $\phi : [0,1] \to [0,1]$ such that $\zeta_t \xi_t = \eta_{\phi(t)} \zeta_t$. In ref. [11], necessary and sufficient conditions are given for an arc ξ to be stable provided that $L(\xi_t)$ consists of finitely many orbits for each t. While the definition is natural, it turns out that there are relatively few stable arcs. One might try to enlarge the class of stable arcs by dropping the requirement that the conjugacies ζ_t depend continuously on t. This is called *mild* stability. Again, there are relatively few such arcs. In particular, no arc containing a diffeomorphism with a non-transverse intersection of stable and unstable manifolds (of periodic orbits) is mildly stable. It is probably the case that any mildly stable arc of diffeomorphisms has its topological entropy constant along the arc.

Flows (vector fields) offer slightly more flexible bifurcation possibilities as can be seen in ref. [7]. In particular, there are arcs of flows joining a Morse–Smale flow with a structurally stable flow with transverse homoclinic points having a single saddle node bifurcation. Such arcs were first considered by Silnikov [12] (and independently by Sotomayor). It is quite possible that such arcs are stable in the above sense.

References

[1] V. Afraimovic and L. Silnikov, On some global bifurcations connected with the disappearance of a fixed point of saddle node type, Sov. Math. Dokl. 15 (1974) 1761–1765.
[2] R. Bowen and D. Ruelle, The ergodic theory of Axiom A flows, Invent. Math. 29 (1975) 181–202.
[3] M. Herman, Mesure de Lebesgue et nombre de rotation, Lecture Notes in Math., Vol. 597 (Springer, Berlin, 1976) pp. 271–294.
[4] M. Hirsch, C. Pugh and M. Shub, Invariant manifolds, Lecture Notes in Math., Vol. 583 (Springer, Berlin, 1977).
[5] S. Newhouse and J. Palis, Bifurcations of Morse–Smale dynamical systems, in: Dynamical Systems (Academic Press, New York, 1973) pp. 303–366.
[6] S. Newhouse and J. Palis, Cycles and bifurcation theory, Asterisque 31 (1976) 43–140.
[7] S. Newhouse and M. Peixoto, There is a simple arc joining any two Morse–Smale flows, Asterisque 31 (1976) 15–43.
[8] S. Newhouse, On simple arcs between structurally stable flows, Dynamical Systems (Warwick, 1975), Lecture Notes in Math., Vol. 468 (Springer, Berlin, 1975) pp. 209–234.
[9] S. Newhouse, Diffeomorphisms with infinitely many sinks, Topology 13 (1974) 9–18.
[10] S. Newhouse, The abundance of wild hyperbolic sets and non smooth stable sets for diffeomorphisms, Publ. Math. IHES 50 (1979) 101–151.
[11] S. Newhouse, J. Palis and F. Takens, Stable families of diffeomorphisms, to appear.
[12] L. Silnikov, On a new type of bifurcation of multidimensional dynamical systems, Sov. Math. Dokl. 10 (1969) 1368–1371.

5. Elements of proofs of Theorem (4.3)(3)

In this lecture, we wish to sketch parts of the proof of Theorem (4.3) (3) in one of the examples which was discussed in Lecture 4. Our goal is to present the main ideas, so we will make various simplifying assumptions. The reader interested in a more complete treatment is referred to ref. [6] of Lecture 4.

Let us return to the example in which $L_\alpha(\xi_b)$ is finite and equal to the set F of fixed points of ξ_b, but $L_\omega(\xi_b) = F \cup O(x)$ where x is a non-degenerate tangency of $W^u(p, \xi_b)$ and $W^s(p, \xi_b)$ and p is a fixed saddle point of ξ_b.

We assume we have the situation shown in fig. 5.1.

Letting $f = \xi_b$, we will assume we have coordinates (u, v) on a neighborhood U of p in which f is the linear map $(u, v) \mapsto (\lambda_1, u, \lambda_2 v)$ where $0 < \lambda_1 < 1 < \lambda_2$, and $\lambda_1 \lambda_2 < 1$. We also assume $(u = 0) \subset W^u(p)$ and $(v = 0) \subset W^s(p)$. In general, there will be finitely many points on the orbit of x outside U. If we assume for simplicity, that $f^n(x) \in U$ for all n, then near the orbit of x the only non-linearity of f occurs in a neighborhood of $f^{-1}x$. We will assume there is a fixed source at ∞, a finite source q_1, and a finite sink q_2. The set of fixed points F of f equals $\{q_1, q_2, \infty, p\}$. Any point which has only finitely many iterates near x is in the stable or unstable manifold of a sink or source, and, hence, will not contribute to complicated recurrence. To prove structural stability of ξ_μ for certain μ's *near b* and greater than b, one has to verify that $L(\xi_\mu)$ is hyperbolic and ξ_μ satisfies the strong transversality condition. Then, one applies Theorem (2.3). The hardest part of this is to show that the non-trivial part of $L(\xi_\mu)$ is a generalized horseshoe, and here we will deal exclusively with this part. We begin with a more detailed description of the orbit structure of f.

Any point near x below $W^s(p)$ is in the stable manifold of the sink q_2 and any point near $f^{-1}x$ to the left of $W^s(p)$ is in the stable manifold of the sink q_2 and any point near $f^{-1}x$ to the left of $W^u(p)$ is in the unstable

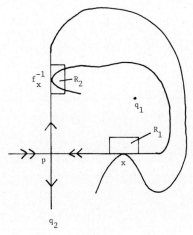

Fig. 5.1.

manifold of ∞. Let R_1 be a small rectangle whose sides are parallel to the coordinate axes and whose bottom side is an interval in $W^s(p)$ centered at x, and let R_2 be a small rectangle whose sides are parallel to the coordinate axes and whose left side is an interval in $W^u(p)$ centered at $f^{-1}x$ as in fig. 5.1.

Let us denote by $h(R_1)$ and $h(R_2)$ the vertical heights of R_1 and R_2, respectively, and by $w(R_1)$ and $w(R_2)$ the horizontal widths of R_1 and R_2, respectively. Adjust $h(R_1)$ and $w(R_2)$ so that the forward orbit of the top of R_1 never meets R_2 and the backward orbit of the right side of R_2 never meets R_1.

Let $D(R_1, R_2) = \{y \in R_1 : \text{there is an } n \geq 1 \text{ such that } f^n y \in R_2\}$. Then, for $y \in D(R_1, R_2)$, let $T(y) = \inf\{n \geq 1 : f^n y \in R_2\}$. For $n \geq \inf\{T(y) : y \in D(R_1, R_2)\}$, let $D_n = \{y \in D(R_1, R_2) : T(y) = n\}$. We have set things up so that each D_n is a rectangle in $R_1, f^n D_n$ is a rectangle in R_2, and the map $f^n : D_n \to f^n D_n$ is affine with derivative

$$\begin{pmatrix} \lambda_1^n & 0 \\ 0 & \lambda_2^n \end{pmatrix}.$$

Typical examples of D_n and $f^n D_n$ are shown in fig. 5.2.

Let v_0 be the v-coordinate of $f^{-1}x$ and let u_0 be the u-coordinate of x. Then the distance from D_n to x is $[v_0 - \frac{1}{2}h(R_2)]\lambda_2^{-n}$ and the distance from $f^n(D_n)$ to $f^{-1}x$ is $[u_0 - \frac{1}{2}w(R_1)]\lambda_1^n$. Since $\lambda_1\lambda_2 < 1$, $f^n(D_n)$ is much closer to $(u = 0)$ than D_n is to $(v = 0)$. This means that $f^{n+1}D_n$ is near $W^u(p)$ below D_n as shown in fig. 5.3.

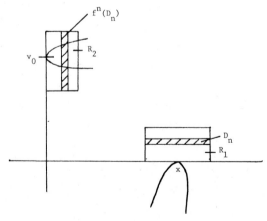

Fig. 5.2.

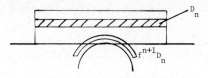

Fig. 5.3.

The main effect of ξ_μ for μ near b is to move a small interval about x in $W^u(p, \xi_b)$ up, so for simplicity, we will assume that ξ_μ only differs from ξ_b in the following special way. Suppose there is an $\varepsilon > 0$ such that

(a) $\xi_\mu(y) = \xi_b(y)$ for $y \notin \xi_b^{-1}[B_\varepsilon(x)]$,

(b) for $y \in \xi_b^{-1}[B_{\varepsilon/2}(x)]$, $\xi_\mu(y) = A_\mu[\xi_b(y)]$ where $A_\mu(u, v) = (u, \mu - b + v)$,

(c) $B_\varepsilon(x) \subset U$, $R_1 \subset B_{\varepsilon/2}(x)$, and $\xi_b R_2 \subset B_{\varepsilon/2}(x)$.

We wish to prove that for most $\mu > b$ and near b, the limit points of ξ_μ which meet R_1 are contained in a generalized horseshoe. The idea of the proof is the following. For each $\mu > b$ near b, we will construct two sets $D_{1,\mu}$ and $D_{2,\mu}$ consisting of finitely many rectangles as shown in fig. 5.4. The rectangles in $D_{2,\mu}$ all have their vertical lines running from $v = 0$ to $v = v_0 + c_1(\mu)$, and the rectangles $D_{1,\mu}$ all have their horizontal lines running from $u = 0$ to $u = u_0 + c_2(\mu)$ where $c_1(\mu)$ and $c_2(\mu)$ are certain functions of μ. $L(\xi_\mu)$ will be contained in the union of F and finitely many ξ_μ iterates of $D_{1,\mu} \cup D_{2,\mu}$. Assuming all this is done, notice that $\xi_\mu(D_{2,\mu}) \cap R_1$ consists of finitely many curved rectangles near x as shown in fig. 5.5. We will choose the rectangles $D_{1,\mu}$ and $D_{2,\mu}$ fairly thin, and then the good μ's [i.e., those for which $L(\xi_\mu) \cap R_1$ is contained in a generalized horseshoe] will be those μ's for which the angles between lines in $D_{1,\mu}$ are not too

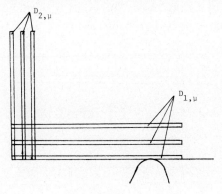

Fig. 5.4.

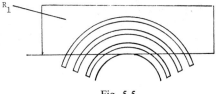

R_1

Fig. 5.5.

small as shown in fig. 5.6. The tangents to the ξ_μ-images of the vertical lines in $D_{2,\mu}$ will then be near the unstable subbundle E^u on $L(\xi_\mu) \cap R_1$ which is needed for hyperbolicity. The tangents to the horizontal lines will be near the subbundle E^s on $L(\xi_\mu) \cap R_1$. To accomplish this, we will have to choose the number of rectangles in $D_{1,\mu}$ and $D_{2,\mu}$ to be $C_1|\log(\mu - b)|$ and the maximum horizontal width of the rectangles in $D_{2,\mu}$ and the maximum vertical height of the rectangles in $D_{1,\mu}$ will be $C_2|\mu - b|^{1/2}$ where C_1 and C_2 are constants.

We need some more information about ξ_b before we construct $D_{1,\mu}$ and $D_{2,\mu}$. Since the tangency of $W^u(p, \xi_b)$ and $W^s(p, \xi_b)$ at x is non-degenerate, we can write $W^u(p, \xi_b)$ near x as $v = -a_1(u - u_0)^2 + o[(u - u_0)^2]$, and we can write a small interval in $W^s(p, \xi_b)$ containing $f^{-1}x$ as $u = a_2(v - v_0)^2 + o[(v - v_0)^2]$ where a_1 and a_2 are positive constants, and, as usual, $o(t)$ is a function of t such that

$$\lim_{t \to 0+} \frac{o(t)}{t} = 0.$$

For $\mu - b > 0$ small, any point y in R_1 with $\mathrm{dist}[y, (v = 0)] \geq 4(\mu - b)$ is such that if $\xi_\mu^{-n}(y) \in R_1$ for some $n > 0$ and n is the least such integer, then $\mathrm{dist}[\xi_\mu^{-n}(y), (v = 0)] > 2\,\mathrm{dist}[y, (v = 0)]$. Thus, the only points in R_1 which return to R_1 infinitely often in backward time must be closer to $(v = 0)$

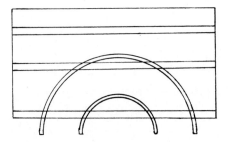

Fig. 5.6.

than $4(\mu - b)$. Similarly, points in R_2 which return to R_2 infinitely often in backward time must be closer to $(u = 0)$ than $4(\mu - b)$.

Consider the set G of points in R_2 whose distance to $(u = 0)$ is less than or equal to $4(\mu - b)$. Then $\xi_\mu(G)$ is contained in a rectangle $R_{1,\mu}$ of the form

$$\left[u_0 - 2\left(\frac{8b_1(\mu - b)^{1/2}}{a_1} \right), \ u_0 + 2\left(\frac{8b_1(\mu - b)^{1/2}}{a_1} \right) \right] \times [0, \ 8b_1(\mu - b)],$$

where b_1 is the Lipschitz constant of ξ_b on M. For this, we need R_2 and $\mu - b$ to be small. So, $L(\xi_\mu) \cap R_1 \subset R_{1,\mu}$. Similarly, if we set

$$R_{2,\mu} = [0, \ 8b_2(\mu - b)]$$
$$\times \left[v_0 - 2\left(\frac{8b_2(\mu - b)^{1/2}}{a_2} \right), \ v_0 + 2\left(\frac{8b_2(\mu - b)^{1/2}}{a_2} \right) \right]$$

where b_2 is the Lipschitz constant of ξ_b^{-1}, then $L(\xi_\mu) \cap R_2 \subset R_{2,\mu}$. Increasing b_1 and b_2 slightly, we may assume that the negative ξ_μ orbit of the right side of $R_{2,\mu}$ does not hit $R_{1,\mu}$ without first leaving U, and the positive ξ_μ orbit of the top of $R_{1,\mu}$ does not hit $R_{2,\mu}$ without first leaving U.

Let $0 < \zeta < 1$ be such that $\lambda_2 < \lambda_1^{-1} < \zeta$. Consider μ of the form $\mu = b + a\zeta^n$ where $a \in [\zeta, 1]$ and $n \geq n_0$ with n_0 suitably large.

Let i_0 be the least positive integer for which $\lambda_2^{-i_0} v_0 \leq 8b_1(\mu - b)$, and let j_0 be the least positive integer for which $\lambda_1^{j_0} u_0 \leq 8b_2(\mu - b)$. We now define the rectangles in $D_{1,\mu}$ and $D_{2,\mu}$. For $i_0 \leq i < n$, let

$$D_{1,\mu}^i = \left[0, u_0 + \frac{16b_1(\mu - b)^{1/2}}{a_1} \right]$$
$$\times \left[\lambda_2^{-i}\left(v_0 - \frac{16b_2(\mu - b)^{1/2}}{a_2} \right), \ \lambda_2^{-i}\left(v_0 + \frac{16b_2(\mu - b)^{1/2}}{a_2} \right) \right].$$

Set

$$D_{1,\mu}^n = \left[0, u_0 + \frac{16b_1(\mu - b)^{1/2}}{a_1} \right] \times [0, \lambda_2^{-n} v_0].$$

For $j_0 \leq j < n$, let

$$D_{2,\mu}^j = \left[\lambda_1^j\left(u_0 - \frac{16b_1(\mu - b)^{1/2}}{a_1} \right), \ \lambda_1^j\left(u_0 + \frac{16b_1(\mu - b)^{1/2}}{a_1} \right) \right]$$

$$\times\left[0,\,v_0+\frac{16b_2(\mu-b)^{1/2}}{a_2}\right]$$

and let

$$D_{2,\mu}^n=[0,\lambda_1^n u_0]\times\left[0,\,v_0+\frac{16b_2(\mu-b)^{1/2}}{a_2}\right].$$

Then let

$$D_{1,\mu}=\bigcup_{i_0\le i\le n}D_{1,\mu}^i\qquad\text{and}\qquad D_{2,\mu}=\bigcup_{j_0\le j\le n}D_{2,\mu}^j.$$

Notice that the vertical heights of the rectangles in $D_{1,\mu}$ and the horizontal widths of the rectangles in $D_{2,\mu}$ get smaller as $\mu\to b$.

For particular $i\in[i_0,n]$ and $j\in[j_0,n]$ such that $\xi_\mu(D_{2,\mu}^j)\cap D_{1,\mu}^i\ne\emptyset$, and $z\in\xi_\mu(D_{2,\mu}^j)\cap D_{1,\mu}^i$, let l_z be the horizontal line through z, and let h_z be the ξ_μ image of the vertical line throuth $\xi_\mu^{-1}(z)$. It can be proved that the set of μ's for which $L(\xi_\mu)$ is hyperbolic contains those μ's for which the angle between l_z and h_z is greater than $c\alpha(z)$ where c is a constant independent of n and $\alpha(z)$ satisfies

(a) $\alpha(z)=\dfrac{\zeta^{n/2}}{n^2}\qquad$ if $i=n$ or $j=n$,

(b) $\alpha(z)=\max\left\{\left(\dfrac{\lambda_2^{-i}}{n^4}\right)^{1/2},\left(\dfrac{\lambda_1^j}{n^4}\right)^{1/2}\right\}\qquad$ if $i<n$ and $j<n$.

To give a very brief glimpse of why this is true, let us prove that a unit vector $v=(v_1,v_2)$ tangent at z to h_z is expanded exponentially by $T_z\xi_\mu^i$ if condition (b) above holds. For this, we use the norm $|(v_1,v_2)|=\max\{|v_1|,|v_2|\}$. If $(v_1',v_2')=T_z\xi_\mu^i(v_1,v_2)$, then (assuming $\mu-b$ is small), $|v_2|<|v_1|$, so $1=|(v_1,v_2)|=|v_1|$. Also

$$\frac{|v_2'|}{|v_1'|}\ge\left(\frac{\lambda_2^i}{\lambda_1^i}\right)c\alpha(z)\ge\lambda_2^i\lambda_1^{-i}\lambda_2^{-i/2}\frac{c}{n^2}.$$

If v is such that $\zeta^v<\lambda_1<\lambda_2^{-1}$, then

$$\zeta^{vi_0}v_0\le\lambda_2^{-i_0}v_0\le 8b_1\zeta^n,\qquad\text{or }vi_0\ge n,\qquad\text{or }i_0\ge n/v.$$

Thus, $i\ge n/v$, so $|v_2'|/|v_1'|>1$ for n large. Thus,

$$\frac{|(v_1',v_2')|}{|(v_1,v_2)|}=|v_2'|=\lambda_2^i|v_2|\ge\lambda_2^i c\alpha(z)$$

$$\geq \lambda_2^i c \lambda_2^{-i/2} n^{1/2} = c \lambda_2^{i/2} n^{1/2} > \frac{c}{2} \lambda_2^{i/2}$$

for n large since $i \geq n/v$.

Similarly, one can verify that conditions (a) and (b) imply that if $z \in L(\xi_\mu) \cap D_{1,\mu}$, then any unit vector tangent to h_z is exponentially expanded in the future and any unit vector tangent to l_z is exponentially expanded in the past. This makes it plausible that $L(\xi_\mu)$ is hyperbolic, and then this can be proved.

Next we wish to indicate how conditions (a) and (b) are obtained for most μ near b. For $i_0 \leq i < n$, let $x_i = \lambda_2^{-i} v_0$ and for $j_0 \leq j < n$, let y_j be the maximum v-coordinate of the curve

$$\xi_\mu \left[\{\lambda_1^j u_0\} \times \left[0, v_0 + \frac{16 b_2 (\mu - b)^{1/2}}{a_2} \right] \right].$$

Let $x_n = 0$ and $y_n = \mu - b$. If $z \in \xi_\mu D_{2,\mu}^i \cap D_{1,\mu}^i$, then the angle between l_z and h_z is a constant times the square root of the distance from l_z to the point on h_z with maximum v-coordinate. Because the rectangles in $D_{1,\mu}$ and $D_{2,\mu}$ are appropriately thin, this distance is a constant times $|\mu - b + y_j - x_i|$.

Since $x_i = \text{constant} \times \lambda_2^{-i}$ and $y_j = \text{constant} \times \lambda_1^j$, we see that conditions (a) and (b) will be satisfied if

(a)′ $\quad |\mu - b + y_j - x_i| \geq \dfrac{\zeta^n}{n^4} \qquad$ for i or $j = n$,

and

(b)′ $\quad |\mu - b + y_j - x_i| \geq \max \left(\dfrac{x_i}{n^4}, \dfrac{y_j}{n^4} \right) \qquad$ for $i_0 \leq i < n$ and $j_0 \leq j < n$.

To compute the μ's which are to be excluded, we compute first those a's in $[\zeta, 1]$ for which $\mu = b + a\zeta^n$ is to be excluded. Let

$$A_n' = \{ a \in [\zeta, 1] : |a\zeta^n + y_j - x_i| < x_i/n^4$$

for some $i_0 \leq i < n$ and some $j_0 \leq j < n$ or $|a\zeta^n + y_j - x_n| < \zeta^n/n^4$ for some $j_0 \leq j \leq n\}$, and let

$$A_n'' = \{ a \in [\zeta, 1] : |a\zeta^n + y_j - x_i| < y_j/n^4 \text{ for some } j_0 \leq j < n$$

$$\text{and } i_0 \leq i \leq n \text{ or } |a\zeta^n + y_n - x_i| < \zeta^n/n^4$$

$$\text{for some } i_0 \leq i \leq n \}.$$

If $a \in A'_n$, then

$$\left| a + \frac{y_j - x_i}{\zeta^n} \right| < \frac{x_i}{\zeta^n n^4} = \frac{\lambda_2^{-i} v_0}{\zeta^n n^4} \leq \frac{8b_1}{n^4} \quad \text{or} \quad \left| a + \frac{y_j - x_i}{\zeta^n} \right| < \frac{1}{n^4}$$

for appropriate i and j. In the first case, a is in an interval about $-(y_j - x_i)/\zeta^n$ of length $16b_1/n^4$, and in the second case, a is in an interval about $-(y_j - x_i)/\zeta^n$ of length $2/n^4$. The total number of points $(y_j - x_i)/\zeta^n$ is less than n^2, so meas$(A'_n) <$ constant$/n^2$. Now, $A'_n \subset [\zeta, 1]$ and

$$\text{meas}\{\mu : \mu - b = a\zeta^n \text{ for some } a \in A'_n\} < (\text{constant}/n^2)\zeta^n.$$

Thus, if,

$$\mathscr{B}'_{n_0} = \{\mu \in [b, b + \zeta^{n_0}] : \mu = b + a\zeta^n \text{ for some } a \in A'_n \text{ and some } n \geq n_0\},$$

then

$$\text{meas } \mathscr{B}'_{n_0} \leq \sum_{n \geq n_0} \frac{\text{constant}}{n^2} \zeta^n \leq \zeta^{n_0} \text{ constant} \times \sum_{n \geq n_0} \frac{1}{n^2}.$$

Thus, for $\varepsilon \in [\zeta^{n_0+1}, \zeta^{n_0}]$,

$$\frac{\text{meas } \mathscr{B}'_{n_0}}{\text{meas}[b, b + \varepsilon]} \leq \frac{\text{constant}}{\zeta} \sum_{n \geq n_0} \frac{1}{n^2} \to 0 \quad \text{as } n_0 \to \infty.$$

Similar statements hold for $\mathscr{B}''_{n_0}$ which is defined replacing A'_n by A''_n in the definition of $\mathscr{B}'_{n_0}$. The μ's to be excluded lie in $\mathscr{B}_{n_0} = \mathscr{B}'_{n_0} \cup \mathscr{B}''_{n_0}$, and we now have meas $\mathscr{B}_{n_0}/$meas$[b, b + \varepsilon] \to \infty$ as $n_0 \to \infty$ (or $\varepsilon \to 0$).

6. Open problems on curves of diffeomorphisms starting at structurally stable elements

Up to now in these lectures, we have described the typical way in which arcs of diffeomorphisms pass from Morse–Smale diffeomorphisms to other kinds of systems, and we have described some general bifurcation phenomena in two dimensions. In this lecture, we wish to be a bit more adventurous. We wish to sketch a possible description of the orbit structure of diffeomorphisms and their dependence on parameters. Most of this will be speculative, and the serious reader should take it as a series of suggestions and problems for future development.

First, let us say a few words about our meaning of general diffeomorphism. Much has been said about the usefulness or lack of usefulness of generic properties of diffeomorphisms. It is clear that differential equations which are to be seriously used as models for physical systems must have a certain robustness (or lack of sensitivity) with respect to perturbations. Whether this robustness is appropriately described by the mathematical notion of genericity or some other notion depends ultimately of course on how successful the notion is in predicting or describing the behavior of the system. Our feeling is that the elements found in generic k-parameter families of diffeomorphisms ($1 \leq k < \infty$) and vector fields will be the most useful for applications. Hence, by a general diffeomorphism, we mean one which can occur as an element ξ_λ where $\xi = \{\xi_\lambda\}_{\xi \in L}$ is a generic family of diffeomorphism parametrized by a finite dimensional manifold Λ.

To begin the description of general diffeomorphisms, it is convenient to review the structure of diffeomorphisms with hyperbolic limit sets.

We first need some definitions. A hyperbolic attractor for f is a hyperbolic basic set Λ for which there is a neighborhood U of Λ so that for $x \in U$, $\omega(x) \subset \Lambda$.

If $f \in \mathcal{D}^r(M)$ and m is a probability measure on the Borel sets of M, then m is f-invariant if $m(f^{-1}E) = m(E)$ for every Borel set E. One says that m is ergodic if $f^{-1}E = E$ implies $m(E) = 0$ or $m(E) = 1$. The metric entropy $h_m(f)$ is defined as follows. If $\alpha = \{A_1, ..., A_s\}$ is a finite Borel partition of M, one sets

$$H_m(\alpha) = -\sum_{A \in \alpha} m(A) \log m(A).$$

If α and β are partitions, one sets $\alpha \vee \beta = \{A \cap B : A \in B \in \beta\}$. One then defines the mean entropy of the partition α to be

$$h_m(\alpha, f) = \lim_{n \to \infty} \frac{1}{n} H_m(\alpha \vee f_\alpha^{-1} \vee \cdots \vee f_\alpha^{-n+1})$$

where is is proved that the limit exists. Finally,

$$h_m(f) = \sup_\alpha h_m(\alpha, f).$$

The quantity $h_m(f)$ is a measure-theoretic isomorphism invariant which indicates the complexity of f on sets of m-measure 1.

The number

$$\sup_{m} \{h_m(f)\} = h(f)$$

is called the topological entropy of f. It is a topological conjugacy invariant and gives a crude measure of the orbit complexity of f itself. A measure m for which $h_m(f) = h(f)$ is called a maximal measure for f. Such measures do not always exist, but it seems that frequently they do, and that they have a special structure. As a simple example of these concepts consider the full shift (σ, Σ_N) on N symbols $\{1, \dots, N\}$. Think of $\{1, \dots, N\}$ as the set of possible outcomes of an experiment where the outcome i has probability $P_i > 0$. Thus, $m(\{i\}) = P_i$ defines a probability measure on $\{1, \dots N\}$. The product measure $\bar{m}$ on $\{1, \dots, N\}^{\mathbb{Z}} = \Sigma_N$ is invariant under the shift and is called the Bernoulli shift $B(P_1, \dots, P_N)$. It is the sample space for the experiment

$$\begin{pmatrix} 1 & \cdots & N \\ P_1 & \cdots & P_N \end{pmatrix}$$

repeated infinitely often in the future and past with each trial independent of the others. One can prove that

$$h_m(\sigma) = -\prod_{i=1}^{N} P_i \log P_i \quad \text{and} \quad h(\sigma) = \log N.$$

Also, the unique maximal measure is $B(1/N, \dots, 1/N)$. If (σ, Σ_A) is a subshift of finite type with a dense orbit, then $h(\sigma/\Sigma_A) = \log \lambda$ where λ is the real eigenvalue of largest modulus, and σ/Σ_A has a unique measure of maximal entropy, say m_σ. The pair $(\sigma/\Sigma_A, m_\sigma)$ is measure-theoretically isomorphic to the product of a periodic transformation and a Bernoulli shift. For more information on metric entropy and topological entropy, see ref. [3].

For future reference, we will say that a measure m (invariant for f) is essentially Bernoulli of the pair (f, m) is measure theoretically isomorphic to the product of a periodic transformation and some Bernoulli shift.

General references for the following theorem are refs. [1] and [4].

Theorem (6.1). Suppose $L(f)$ is hyperbolic. Then there is a disjoint decomposition $L(f) = \Lambda_1 \cup \cdots \cup \Lambda_n$ into isolated invariant sets such that the following facts are true.

(1) The periodic points in Λ_i are dense in Λ_i.

(2) f restricted to Λ_i has a dense orbit and a unique invariant measure of maximal entropy which is essentially Bernoulli.

(3) If we set

$$W^u(\Lambda_i) = \bigcup_{x \in \Lambda_i} W^u(x, f),$$

then

$$M = \bigcup_{i=1}^{n} W^s(\Lambda_i) = \bigcup_{i=1}^{n} W^u(\Lambda_i).$$

(4) Suppose $\Lambda_{i_1}, \ldots, \Lambda_{i_r}$ are the attractors of f. Then there is a set $A \subset M$ such that the Lebesgue measure of $M - A$ is zero and

$$A \subset \bigcup_{j=1}^{r} W^s(\Lambda_{i_j}).$$

Moreover, for each $j \in [1, r]$, there is an essentially Bernoulli f-invariant measure m_j supported in Λ_{i_j} such that for any $x \in A \cap W^s(\Lambda_{i_j})$ and any continuous real-valued function ϕ on M, one has

$$\lim_{n \to \infty} \frac{1}{n} \sum_{k=0}^{n-1} \phi(f_x^k) = \int_{\Lambda_{i_j}} \phi \, \mathrm{d}m_j.$$

In general, given a diffeomorphism f, let us say that a *Ruelle* measure m for f is an invariant probability measure m such that there is a set $A \subset M$ of positive Lebesgue measure such that for each $x \in A$ and any continuous real-valued function $\phi : M \to \mathbb{R}$, one has

$$\lim_{n \to \infty} \frac{1}{n} \sum_{k=0}^{n-1} \phi(f_x^k) = \int_{M} \phi \, \mathrm{d}m.$$

Let us call a maximal such set A the *basin* of the Ruelle measure m. One may summarize Theorem (6.1) (4) by saying that if $L(f)$ is hyperbolic, then f has finitely many Ruelle measures the union of whose basins has full Lebesgue measure in M. It turns out that a hyperbolic attractor always has a unique Ruelle measure and its basin is a set of full Lebesgue measure in the open set of points whose ω-limit sets are in the attractor.

A natural question is what is the relationship between the Ruelle measure and the measure of maximal entropy on a hyperbolic attractor. Although they are usually singular with respect to each other, they can each be obtained as limits of measures supported on the periodic points. To describe this, suppose first that Λ is a hyperbolic attractor which contains a fixed point p. Then it turns out that f/Λ is topologically mixing. This means that for any two relatively open sets U and V in Λ, there is an integer $T>0$ such that $n \geq T$ implies $f^n(U) \cap V \neq \emptyset$. Consider the set P_n of fixed points of f^n in Λ. Let δ_x be the point mass at x. Then we obtain a measure

$$m_n = \frac{1}{\operatorname{card} P_n} \sum_{x \in P_n} \delta_x$$

where $\operatorname{card} P_n$ denotes the number of elements of P_n. The measures m_n converge in the space of probability measures (with the weak $*$ topology) to the measure of maximal entropy m. Now given $x \in P_n$, let $S_n(x)$ denote the product of the eigenvalues of $T_x f^n | E_x^{\mathrm{u}}$ with multiplicities. Let

$$\bar{m}_n = \frac{1}{\sum_{n \in P_n} S_n^{-1}(x)} \cdot \frac{1}{S_n(x)} \delta_x.$$

Then, $\bar{m}_n$ converges as $n \to \infty$ to the Ruelle measure $\bar{m}$. Thus, a good knowledge of the distribution of the periodic points of f on Λ leads one to both m and $\bar{m}$. If Λ does not contain a fixed point, one chooses a periodic point p in Λ, say of period τ, and obtains m and $\bar{m}$ on $\Lambda_1 = \text{closure } [W^{\mathrm{u}}(p) \cap \Lambda]$ for f^τ. Then one obtains m and $\bar{m}$ on

$$\Lambda = \bigcup_{0 \leq j < \tau} f^j(\Lambda_1)$$

Explicitly, one writes

$$m(E) = \tau^{-1} \sum_{0 \leq j < \tau} m(\Lambda_1 \cap f^{-j}E)$$

and

$$\bar{m}(E) = \tau^{-1} \sum_{0 \leq j < \tau} \bar{m}(\Lambda_1 \cap f^{-j}E) \qquad \text{where } E \subset \bigcup_{0 \leq j < \tau} f^j \Lambda_1.$$

Now let $\mathrm{AS} = \{f \in \mathscr{D}^\infty(M) : L(f) \text{ is hyperbolic and for any } x \in L(f),$ $y \in L(f), \ W^{\mathrm{s}}(x, f) \pitchfork W^{\mathrm{u}}(y, f)\}$. It is known that AS is an open set in $\mathscr{D}^\infty(M)$ and Theorem (2.3) says each f in AS is structurally stable. The case when $L(f)$ has only finitely many orbits is precisely the Morse–Smale case. We now consider a curve $\xi = [0, 1] \to \mathscr{D}^\infty(M)$, let $B(\xi) = \{\mu \in [0, 1] : \xi_\mu \notin \mathrm{AS}\}$ and let $b(\xi) = \inf B(\xi)$. We ask for the structure of ξ_b for μ near b and most ξ. This is not yet known except in the simplest cases, but we shall try to imagine the answer.

Before giving examples, we state the general kind of structure we imagine for generic k-parameter families of diffeomorphisms. We assume $\xi: \Lambda \to \mathcal{D}^{\infty}(M)$ is a C^{∞} map (which just means that $\xi: \Lambda \times M \to M$ is C^{∞}) and is generic.

(1) For each $\lambda \in \Lambda$, ξ_λ has finitely many or countably many ergodic maximal measures and finitely many or countably many ergodic Ruelle measures.

(2) The union of the basins of the ergodic Ruelle measures has full Lebesgue measure in M.

(3) The topological entropy $h(\xi_\lambda)$ is a continuous function of λ.

(4) Each ergodic maximal measure or ergodic Ruelle measure is essentially Bernoulli.

In many cases ergodic maximal measures and ergodic Ruelle measures for ξ_λ may depend continuously on λ as well. If the above structure holds, one would have a beautiful and fairly simple description of the orbit structue of the diffeomorphisms involved. They would look more or less like AS diffeomorphisms except with countably many generalized basic sets containing most of the recurrent points. An indication that such general structure theorems are possible is given by a beautiful recent theorem of Pesin [6]. He proves that a C^2 diffeomorphism f on a compact 2-manifold M which preserves a smooth measure m (i.e., a measure given by a smooth density) has the following property. There is a countable decomposition of $M = \Lambda_0 \cup \Lambda_1 \cup \cdots$ such that

(1) $h_m(\Lambda_0) = 0$,

(2) (f, m) is essentially Bernoulli on each Λ_i for $i \geq 1$.

Thus, Pesin tells us that the positive entropy part of a smooth measure preserving diffeomorphism of a surface is metrically very simple.

Now, let us describe some ways in which non-trivial AS diffeomorphisms undergo bifurcations.

(1) *A local periodic point bifurcation in a non-trivial basic set.* Consider a hyperbolic set Λ for a diffeomorphism containing a fixed point p whose unstable manifold $W^u(p)$ is one-dimensional. First suppose the eigenvalue of norm greater than one is positive. The local picture is shown in fig. 6.1. Now take a family $\{\xi_\mu\}$ with $\xi_0 = f$ which creates a saddle node p_1 on $W^u(p)$ to the right of p at $\mu = b_0$. At the time $\mu = b_0$, and for $\mu > b_0$ and near b_0 the right component of $W^u(p) - \{p\}$ is completely contained in $W^s(p_1)$. Under appropriate conditions, ξ_μ will again be in AS for $\mu > b_0$

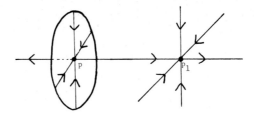

Fig. 6.1.

and near b_0. If the eigenvalue of a norm greater than one is negative, then one can use a flip bifurcation at p and again have $\xi_\mu \in AS$ for $\mu > b_0$ and μ near b_0. Such bifurcations do not change the topological entropy. On the other hand, if Λ happened to be an attractor for f, then the Ruelle measure changes from a measure supported on all of Λ to one on the fixed point p_1 at $\mu = b_0$. An explicit example of this in dimension 3 is worked out in ref. [5]. The detailed analysis of what occurs when a Hopf point is created in a non-trivial basic set has not yet been carried out.

(2) *Non-transversal intersections involving non-trivial basic sets.* There are two ways this can happen.

(a) The non-transversal intersection is a limit of periodic points and the basic set ceases to be hyperbolic. A typical example here is obtained by downward translation of the standard horseshoe f on the rectangle Q given in fig. 6.2. One may assume there is a hyperbolic saddle point p whose stable and unstable manifolds become tangent near the top of Q. If the tangency is non-degenerate, then at the first bifurcation point b, the structure of ξ_b is known. There is a unique maximal measure on the closure Λ of the homoclinic points of p_b, and ξ_b/Λ can be described as a simple quotient of the 2-shift in which two orbits become identified.

(b) The second way this can happen is for

$$\Lambda_\mu = \bigcap_n \xi_\mu^n(Q)$$

to remain hyperbolic and have $W^u(\Lambda_\mu)$ and $W^s(\Lambda_\mu)$ become tangent outside of Λ_μ. This bifurcation has much in common with the creation of cycles we described in Lecture 4. Although even if dim $\Lambda_0 = 0$, ξ_μ may fail to be in AS for μ in half-open interval $(b, b + \varepsilon)$ (see for instance section 8 in ref. [4]). In higher dimensions, one can easily imagine similar phenomena occurring.

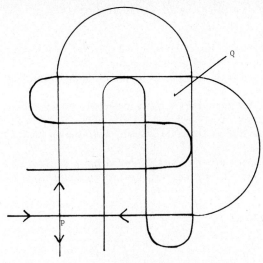

Fig. 6.2.

The above examples have one feature in common. There is a small neighborhood of a single orbit (either the periodic bifurcation or the nontransverse intersection) where non-hyperbolic phenomena occur and the orbit structure is reduced to a hyperbolic analysis or an analysis of how points pass through this neighborhood.

Now suppose we have systems depending on several parameters. Then some interplay between periodic bifurcations and non-transversal intersections will occur.

In generic systems depending on finitely many parameters, one would expect to have finitely many small neighborhoods in which non-hyperbolic phenomena occur, and again the orbit structure analysis should reduce to understanding the effects produced by orbits passing from one such neighborhood to another. The theory of singularities of mappings (e.g., as in ref. [2]) gives one a good understanding of the generic way in which submanifolds depending on finitely many parameters become non-transverse. the analysis of Lecture 5 for the proof of Theorem (4.3) (3) is only the beginning of possible applications of this theory to dynamics. Thus, one would expect some combination of methods of singularity theory, ergodic theory, and smooth dynamical systems to come together in the study of systems depending on several parameters.

References

[1] R. Bowen, Equilibrium states and the ergodic theory of Anosov diffeomorphisms, Lecture Notes in Math., Vol. 470 (Springer, Berlin, 1975).
[2] M. Golubitsky and V. Guillemin, Stable Mappings of their Singularities (Springer, Berlin, 1973).
[3] M. Denker, C. Grillenberger and K. Sigmund, Ergodic Theory on Compact Spaces, Lecture Notes in Math., Vol. 527 (Springer, Berlin, 1976).
[4] S. Newhouse, Lectures on Dynamical Systems, in: Progress in Math., Vol. 8 (Birkhäuser, Boston, 1980).
[5] S. Newhouse, Structural stability and bifurcation theory, in: Bifurcation Theory and Applications in Scientific Disciplines, eds., O. Gurel, O. Rössler, Ann. N.Y. Acad. Sci. 316 (1979) pp. 121–126.
[2] Ya. Pesin, Characteristic Lyapunov exponents and smooth ergodic theory, Russ. Math. Surveys 32 (1977) 55.

SEMINAR

GENERIC PROPERTIES OF CONSERVATIVE SYSTEMS

S.E. NEWHOUSE

Mathematics Department, University of North Carolina
Chapel Hill, NC 27514, U.S.A.

Our goal here is to describe some of the known generic (or typical) properties of symplectic diffeomorphisms. These mappings arise naturally as the first return maps to transverse hypersurfaces of the restrictions of Hamiltonian systems to constant energy submanifolds. For simplicity, we will mainly deal with the two dimensional case (i.e. with area preserving mappings).

We begin with a description of the local structure near periodic points. Let M be a C^∞ two-dimensional manifold with area element ω. Thus, ω is a nowhere vanishing differential 2-form. let $f: M \to M$ be a C^r diffeomorphism preserving ω. If M has a boundary, ∂M, we suppose $f(\partial M) = \partial M$ and all of our considerations of periodic points are assumed to occur in $M - \partial M$. A point $p \in M$ is a *fixed point* of f if $f(p) = p$. Such a point is hyperbolic if the derivative $T_p f$ has eigenvalues of absolute value different from one. The fixed point is *elliptic* if $T_p f$ has two non-

G. Iooss, R.H.G. Helleman and R. Stora, eds.
Les Houches, Session XXXVI, 1981 – Comportement Chaotique des Systèmes Déterministes/
Chaotic Behaviour of Deterministic Systems
© *North-Holland Publishing Company, 1983*

real eigenvalues of absolute value equal to one. A periodic point p of period $\tau > 1$ is a fixed point for f^τ with τ being the least such positive integer. The periodic point p is called *hyperbolic* or *elliptic* according to the eigenvalues of $T_p f^\tau$.

The local structure of f near a hyperbolic fixed point can be completely described thanks to the Grobman–Hartman theorem [7]. Introduce local coordinates x near p so that

$$f(x) = f(p) + L(x - p) + R(x),$$

where $L = T_p f$ and

$$\lim_{x \to p} |R(x)| / |x - p| = 0.$$

Then the Grobman–Hartman theorem gives a homeomorphism (1–1 continuous map with continuous inverse) h from a neighborhood U of p to a neighborhood V of 0 in $\mathbb{R}^2$ such that $hfh^{-1}(u) = L(u)$ for u in $h(U \cap f^{-1}U)$. The map h is a local continuous coordinate change transforming f to the linear map L.

The local structure near an elliptic fixed point p is much more complex. In the generic C^4 case, each elliptic fixed point is surrounded and accumulated upon by invariant circles on each of which f is C^1 conjugate to a geometric irrational rotation. Between any two such circles there are elliptic and hyperbolic periodic points of high period. At least some of the hyperbolic periodic points will have transverse homoclinic points. This is depicted in fig. 1.

The existence of the invariant circles is a consequence of the familiar

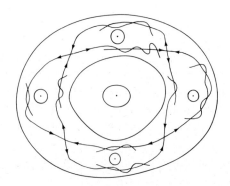

Fig. 1.

Kolmogorov–Arnold–Moser theory [1] (for refinements concerning differentiability assumptions see refs [12] and [3]). The fact that generic elliptic points are limits of hyperbolic and elliptic points was proved by Moser [4], and the existence of homoclinic points was proved by Zehnder [14]. Recently, Mather has proved the existence of minimal Cantor sets similar to Denjoy type minimal sets in C^1 circle diffeomorphisms. The regions in between the invariant circles are known as zones of instability. A detailed description of most orbits in these regions has still not yet been given. For example, it is not known whether there are sets of positive Lebesgue measure with non-zero Lyapunov characteristic exponents (see ref. [8] for definitions) or whether the periodic points (even elliptic periodic points) are dense for typical C^4 maps f. In the C^1 case, considerably more is known about generic f, but much of the detailed structure (for instance, the existence of invariant circles) disappears. It is an important and difficult problem to find out which of the C^1 theorems extend to higher differentiability.

Let us be more precise about the notion of genericity. Let $D^r_\omega(M)$ be the set of C^r diffeomorphisms of M preserving ω where $r \geq 1$. We give $D^r_\omega(M)$ the uniform C^r topology. A subset $B \subset D^r_\omega M$) is called *residual* if it contains a countable intersection of open dense sets. Residual sets are dense and a countable intersection of residual sets is again residual. A property of elements in $D^r_\omega(M)$ is called *generic* if it holds for all elements of some residual set. The term "generic f" refers to f in some residual set.

If $f \in D^r_\omega(M)$, d is a distance function on M, and $x \in M$, one defines

$$W^s(x) = \{ y \in M : d(f^n x, f^n y) \to 0 \text{ as } n \to \infty \}$$

and

$$W^u(x) = \{ y \in M : d(f^{-n} x, f^{-n} y) \to 0 \text{ as } n \to \infty \}.$$

The set $W^s(x)$ is the *stable set* of x, and $W^u(x)$ is the *unstable set* of x. If p is a hyperbolic periodic point of f, then $W^u(p)$ and $W^s(p)$ are C^r injectively immersed curves in M. A point $y \in W^s(p) \cap W^u(p) - \{p\}$ is called a homoclinic point. If $W^u(p)$ is transverse to $W^u(p)$ at y, then y is called a transverse homoclinic point. Otherwise, y is called a homoclinic tangency.

Let us now give a list of known C^1 generic properties of area preserving f's, i.e. generic properties in $D^1_\omega(M)$. Let m be Lebesgue measure on M.

(1) Every periodic point is hyperbolic or elliptic.

(2) Stable and unstable manifolds of hyperbolic periodic points meet transversely.

(3) The hyperbolic periodic points of f are dense in M.

(4) Every hyperbolic periodic point has transverse homoclinic points.

(5) The transverse homeoclinic points are dense in the stable and unstable manifolds of every hyperbolic periodic point.

(6) Either f is Anosov or the elliptic periodic points are dense in M.

(7) If $h_m(f)$ is the measure theoretic entropy of f relative to m, then either f is Anosov or $h_m(f) = 0$.

Properties (1) and (2) were proved by Robinson [11]. Property (3) is a consequence of the closing lemma of Pugh and Robinson [10] and simple arguments with the Birkhoff normal form as in Moser [4]. Properties (4) and (5) were proved by Takens [13]. Property (6) is proved in ref. [6], and property (7) was recently proved by Mãne (not yet published). Properties (1)–(4) are C^1 generic for symplectic diffeomorphisms in any dimension. Properties (1) and (2) are C^r generic for $r \geq 1$, and property (4) was proved C^r generic on the 2-sphere by Pixton [9]. At present, it is not known whether properties (3), (4), (5), (7), and the obvious analog of (6) (1-ellipticity in ref [6]) are C^r generic in all dimensions. Our feeling is that property (7) is not C^r generic for large r, but that the other properties are C^r generic, $r \geq 1$, in all dimensions.

We now proceed to sketch the proof that property (6) is C^1 generic. For more details, and the extension to higher dimensions, see ref. [6].

Let us first recall the definition and basic properties of Anosov diffeomorphisms. A diffeomorphism $f: M \to M$ is called Anosov if there is a splitting $T_x M = E_x^s \oplus E_x^u$ of each tangent space and constants $C > 0$, $0 < \lambda < 1$, such that

(a) $T_x f(E_x^s) = E_{f(x)}^s$, $T_x f(E_x^u) = E_{f(x)}^u$,

(b) for $n \geq 0$ and $v \in E_x^s$, $|T_x f^n(v)| \leq C\lambda^n |v|$,

(c) for $n \geq 0$ and $v \in E_x^u$, $|T_x f^n(v)| \geq C^{-1}\lambda^{-n} |v|$.

The norms $|\cdot|$ used in (a), (b), and (c) are those induced by a fixed Riemann metric on TM.

It can be shown that E_x^s and E_x^u are unique subject to conditions (a), (b), and (c) and the mappings $x \mapsto E_x^u, x \mapsto E_x^s$ are continuous.

Examples of Anosov diffeomorphisms on the two-torus $T^2 = \mathbb{R}^2/\mathbb{Z}^2$ are obtained as follows. Let

$$A = \begin{pmatrix} a & b \\ c & d \end{pmatrix}$$

be a matrix of integers with $ad - bc = \pm 1$ and eigenvalues of absolute value different from 1. Then $A(\mathbb{Z}^2) \subset \mathbb{Z}^2$, so A induces a map $\bar{A} : T^2 \to T^2$. The map $\bar{A}$ is called a linear hyperbolic toral automorphism. The subspaces E_x^s and E_x^u are obtained by translating the eigenspaces of A to all points of $\mathbb{R}^2$ and then projecting these spaces to $T_x T^2$. It can be proved that any Anosov diffeomorphism of a compact 2-manifold is topologically conjugate to a linear hyperbolic toral automorphism [2, 5]. In particular, *no* surface different from T^2 admits an Anosov diffeomorphism. It is noteworthy that Anosov diffeomorphisms form an open set in $D_\omega^r(M)$. An Anosov diffeomorphism has no elliptic periodic points.

The next theorem states that in $D_\omega^1(M)$, the set of Anosov diffeomorphisms is the largest open set whose elements have no elliptic periodic points. In fact, it says that any non-Anosov diffeomorphism can be C^1 perturbed to create an elliptic periodic point in any preassigned open set.

Theorem. Suppose $f \in D_\omega^1(M)$ is not Anosov. Let N be any neighborhood of f in $D_\omega^1(M)$ and let U be any open set in M. There is a g in N such that g has an elliptic periodic point in U.

The C^1 genericity of property (7) is an easy consequence of this theorem. First, it follows from the implicit function theorem that for any open set U, the set $E(U) = \{g \in D_\omega^1(M): g$ has an elliptic periodic point in $U\}$ is open in $D_\omega^1(M)$. Let $U_1, U_2, \ldots$ be a countable basis for the topology of M. Let A be the possibly empty set of Anosov diffeomorphisms in $D_\omega^1(M)$.

It follows from the theorem that each $A \cup E(U_i)$ is dense in $D_\omega^1(M)$. It is also open in $D_\omega^1(M)$. Thus,

$$B = A \cup \bigcap_{i=1}^{\infty} E(U_i) = \bigcap_{i=1}^{\infty} [A \cup E(U_i)]$$

is residual in $D_\omega^1(M)$. Any $f \in B - A$ has its elliptic periodic points dense in M.

The first part of the proof of the above theorem is the following lemma.

Lemma 1. Suppose $f \in D_\omega^1(M)$ has a hyperbolic periodic point p whose stable and unstable manifolds are tangent at some point x. Let U be any neighborhood of x and let N be any neighborhood of f in $D_\omega^1(M)$. There is a g in N which has an elliptic periodic point in U.

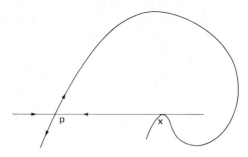

Fig. 2.

Let us sketch the proof of lemma 1. With a preliminary perturbation, we may assume the tangency at x of the manifolds $W^s(p)$ and $W^u(p)$ is quadratic as in fig. 2. Then, there is a rectangle D near x and a positive integer $n > 0$ so that $f^n D$ is near D as in fig. 3. We do not know on which side of $W^u(p)$ $f^n(D)$ is, but that does not matter. Let V be a neighborhood of x such that $V \cap f(V) = \emptyset$. Let $Q_t, t \in [0, 1]$, be a C^1 family of C^1 symplectic diffeomorphisms such that

(1) $Q_t(y) = y$ for $y \notin V$ and each $t \in [0, 1]$,

(2) $Q_0(x)$ is below x and $Q_1(x)$ is above x,

(3) $Q_{1/2}(y) = y$ for all $y \in M$,

(4) $Q_t \circ f \in N$ for each $t \in [0, 1]$.

One easily constructs such a family using degenerating functions (see ref. [6]). Let $g_t = Q_t \circ f$. For suitable D and n, the images $(g_t)^n(D)$ move from below D up through D as t rungs in some interval $[\alpha, \beta] \subset [0, 1]$. Figure 4 show typical $g_\alpha^n(D)$ and $g_\beta^n(D)$. Thus, we have the creation of a horseshoe for g_β. It is easily seen with an index argument that g_β^n has two fixed points in D. But g_α^n has no fixed points in D and, for each $t \in [\alpha, \beta]$, g_t^n has no fixed points in the boundary of D. If $t_1 = \inf\{t \in (\alpha, \beta): g_t^n \text{ has a fixed}$

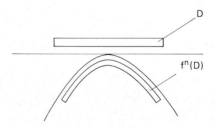

Fig. 3.

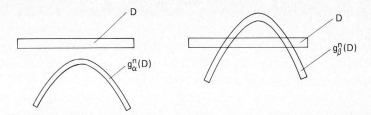

Fig. 4.

point in D}, and x_0 is a fixed point of $g_{t_1}^n$ in D, it follows that 1 is an eigenvalue of $T_{x_0} g_{t_1}^n$. Otherwise, all g_t^n with t near t_1 will have a fixed point in D by the implicit function theorem. Since $g_{t_1}^n$ preserves ω, 1 must be an eigenvalue of multiplicity two of $T_{x_0} g_{t_1}^n$. Then, one easily perturbs g_{t_1} to be g in N so that x_0 is an elliptic fixed point of g^n, proving lemma 1.

We remark that lemma 1 remains valid in the C^r topology.

Now to use lemma 1 to prove the theorem, we must show that if f is not Anosov, U is any open set in M, and N is any neighborhood of f if $D_\omega^1(M)$, then there is a g in N having a hyperbolic periodic point whose stable and unstable manifolds have a tangency in U. Applying properties (3) and (4) we may find a sequence $g_1, g_2, \ldots$ of elements of $D_\omega^1(M)$ such that $g_i \to f$ as $i \to \infty$, and for each i, the set of transverse homoclinic points of g_i is dense in M. Let $V(g_i, U)$ be the set of transverse homoclinic points of g_i which lie in U, and let $H(g_i, U)$ be the set of points on the orbits of elements of $V(g_i, U)$. The set $V(g_i, U)$ is a countable dense set in U, and the set $H(g_i, U)$ is a countable g_i invariant set. Also, for each $y \in H(g_i, U)$, there is the splitting $T_y M = E_y^s \oplus E_y^u$, where E_y^s is the tangent space to $W^s(y)$ at y and E_y^u is the tangent space to $W^u(y)$ at y. Let $\alpha(g_i, U)$ be the infimum of the angles between the subspaces E_y^s and E_y^u as y varies in $H(g_i, U)$. If

$$\liminf_{i \to \infty} \alpha(g_i, U) = 0,$$

then there are large i's for which g_i has some transverse homoclinic point y_i whose orbit meets U such that $E_{y_i}^s$ is nearly tangent to $E_{y_i}^u$. It is then possible to find $g \in C^1$ near such a g_i (and hence C^1 near f) with a homoclinic tangency in U. The proof of the theorem is obtained by showing that if f is not Anosov, then g_i may be perturbed to $\tilde{g}_i$ such that $\tilde{g}_i \to f$ and $\alpha(\tilde{g}_i, U) \to 0$ as $i \to \infty$.

The main idea of the construction of $\tilde{g}_i$ is as follows. Since f is not Anosov, for large i, the set $H(g_i, U)$ is a countable g_i-invariant set which is hyperbolic in a certain sense, but not uniformly. That is, it either takes larger and larger iterates of Tg_i to expand some E_y^u or contract some E_y^s for some $y \in H(g_i, U)$. Let us suppose for definiteness that it takes longer and longer to expand some E_y^u. The precise statement is that for each i there is a $y_i \in H(g_i, U)$ and an integer $m_i > 0$ such that $|T_{y_i} g_i^j | E_{y_i}^u| \leq 2$ for $0 \leq j \leq m_i$ and $m_i \to \infty$ as $i \to \infty$. Suppose y_i is a transverse homoclinic point of the hyperbolic periodic point p_i of g_i. Let W_i be a small neighborhood of the set $\{y_i, g_i(y_i), \ldots, g_i^{m_i}(y_i)\}$ so that $g_i^j(y_i) \notin W_i$ for $j \notin \{0, 1, \ldots, m_i\}$. Given $\varepsilon > 0$ fixed and small, let $\bar{g}_i$ be C^1 close to g_i such that

(a) p_i is a hyperbolic periodic point of $\bar{g}_i$,
(b) y_i is a homoclinic point of p_i for $\bar{g}_i$,
(c) $\bar{g}_i(y) = g_i(y)$ for $y \notin W_i$,
(d) $|T_{y_i} \bar{g}_i^j(v)| \leq 2(1 - \varepsilon)^j |v|$ for $0 \leq j \leq m_i$ and $v \in T_{y_i} W^u(y_i, \bar{g}_i)$.

Thus, we make $T\bar{g}_i$ contract its unstable space along the first m_i iterates of y_i. Again we may suppose that the angles between E_z^u and E_z^s for $z = y_i$ or $\bar{g}_i^{m_i} y_i$ are bounded away from zero. Since $\bar{g}_i$ preserves the area form ω, this implies that

$$|T_y \bar{g}_i^{m_i}(v)| \geq C(1 - \varepsilon)^{-m_i} |v| \qquad \text{for } v \in E_{y_i}^s \text{ and } C > 0,$$

independent of i. Thus, we have made $T\bar{g}^{m_i}$ contract $E_{y_i}^u$ and expand $E_{y_i}^s$.

Let $\tilde{W}_i$ be a neighborhood of $\bar{g}_i^{-1}(y_i) = z_i$ not meeting $\bar{g}_i^j(y_i)$ for $j \neq -1$. Perturb $\bar{g}_i$ to $\tilde{g}_i$ so that $\bar{g}_i = \tilde{g}_i$ off $\tilde{W}_i$, y_i remains a transverse homoclinic point for $\tilde{g}_i$, and $T_{z_i} \tilde{g}_i(E_{z_i}^u)$ makes a fixed non-zero angle with $E_{y_i}^u$. That is, a unit vector in $T_{z_i} \tilde{g}_i(E_{z_i}^u)$ has a component along $E_{y_i}^u$ of length larger than a positive constant independent of i.

Since $T_{y_i} \bar{g}_i^{m_i}$ contracts $E_{y_i}^u$ and expands $E_{y_i}^s$, we have that the $T_{y_i} \tilde{g}_i^{m_i}$ image of $T_{z_i} \tilde{g}_i(E_{z_i}^u)$ makes a small angle with $E_{\bar{g}_i^{m_i}(y_i)}^s$. This image is $E_{\tilde{g}_i^{m_i}(y_i)}^u$ for $\tilde{g}_i$, so we have made the angle between $E_{\tilde{g}_i^{m_i}(y_i)}^s$ and $E_{\tilde{g}_i^{m_i}(y_i)}^u$ small as required.

References

[1] V. Arnold and A. Avez, Ergodic Problems of Classical Mechanics (Benjamin, New York, 1968).
[2] J. Franks, Anosov diffeomorphisms, in: Global Analysis, Proc. Symp. Pure Math. XIV (AMS, Providence, RI, 1970) pp. 61–93.
[3] M. Herman, A simple proof of the existence of invariant circles for twist maps, preprint Ecole Polytechnique, Palaiseau, France.

[4] J. Moser, Non-existence of integrals for canonical systems of differential equations, Commun. Pure Appl. Math. 8 (1955) 409–436.

[5] S. Newhouse, On codimension one Anosov diffeomorphisms, Am. J. Math. 92 (1970) 761–770.

[6] S. Newhouse, Quasi-elliptic periodic points in conservative dynamical systems, Am. J. Math. 99 (1977) 1061–1087.

[7] J. Palis, Local structure of hyperbolic fixed points in Banach spaces, Ann. Acad. Bras. Cienc. (1968) 263–266.

[8] Ya. Pesin, Lyapunov characteristic exponents and smooth ergodic theory, Russ. Math. Surveys 32, No. 4 (1977) 55–114.

[9] D. Pixton, Planar homoclinic points, J. Diff. Eq. 44 (1982) 365–383.

[10] C. Pugh and R.C. Robinson, The C^1 closing lemma, J. Ergodic Theory Dynam. Systems, to appear.

[11] R.C. Robinson, Generic properties of conservative systems I, II, Am. J. Math. 92 (1970) 562–603; 897–906; Lectures on Hamiltonian systems (IMPA, Rio de Janeiro, Brazil, 1971).

[12] H. Russmann, Kleine Nenner I: Uber invariante kurven differenzierbarer Abbildungen eines Kreisrings, Nachr. Akad. Wiss. Gött. Math. Phys. Kl. II (1970) 67–105.

[13] F. Takens, Homoclinic points in conservative systems, Invent. Math. 18 (1972) 267–292.

[14] E. Zehnder, Homoclinic points near elliptic fixed points, Commun. Pure Appl. Math. 26 (1973) 131–182.

COURSE 7

ROUTES TO CHAOS WITH SPECIAL EMPHASIS ON PERIOD DOUBLING

Jean-Pierre ECKMANN

Départment de Physique Théorique,
Université de Genève, 1211 Genève 4, Switzerland

Notes written by

Robert S. MACKAY

G. Iooss, R.H.G. Helleman and R. Stora, eds.
Les Houches, Session XXXVI, 1981 – Comportement Chaotique des Systèmes Déterministes/
Chaotic Behaviour of Deterministic Systems
© *North-Holland Publishing Company, 1983*

Contents

1. Routes to chaos

1.1. Aim

Nature is full of behaviour which we call "chaotic", the weather at Les Houches being a prime example. What we mean by this is that we cannot predict it to any significant accuracy, either because the system is complex, or maybe because some of the governing factors are indeterministic. We would like to believe, however, that the evolution of the universe is deterministic, so that it be in principle predictable. In fact, we hope to model it by dynamical systems, where the state of the system is represented by a point x on some manifold or Banach space X, and evolves in time t according to a differential equation:

$$dx/dt = F(x)$$

or in discrete time n by a difference equation:

$$x_{n+1} = F(x_n)$$

for some function F.

One might think that the notion of chaos would be excluded from a deterministic world, but there are many ways in which even very simple deterministic systems can behave chaotically. Not only can they have complicated aperiodic solutions, but their behaviour may appear random in the following sense. There are always practical limits to our observational power, but in many systems, this does not matter. For instance, a damped system may reach its equilibrium position even if we are unable to prepare the initial state with infinite precision. For others, however, orbits with nearby initial conditions tend to drift apart, and (although friction may be present) the slightest initial error gets exponentially amplified. Such behaviour we call *chaotic*, though there are weaker senses in which we will use the word. Equivalently, for a chaotic system, the information one can gain about its state from information about an initial state decays (until zero) roughly linearly in time. Thus, neither a random element, nor complexity is required to produce chaos.

The initial hope in the study of dynamical systems was to classify dynamical equations up to equivalence in some qualitative sense (Smale 1967). One would hope to obtain a partition of the space of all systems into open sets of equivalent systems, plus boundaries across which changes (bifurcations) would occur. One would also want to classify the typical bifurcations that one should expect in parametrised families. Although this goal can be achieved for certain restricted classes of systems, an appropriate and useful equivalence relation has not been found for which it can be achieved in general. An alternative approach is to look near the boundary of a set of systems which can be classified in the above way, and see what behaviour there can be (compare Newhouse's lectures). Thus we will study families of systems depending on some parameters, with simple behaviour for some parameter values, and look for transitions to more complicated behaviour. This might give insight, for example, into the transition from laminar to turbulent flow in fluids as the Reynold's number is increased. We are interested in transitions which are insensitive to perturbation of the system. After expanding on the above ideas, we will describe three scenarios for such transitions [see Eckmann (1981) for more details]. The remaining three sections will be devoted to the study of one in particular, the period doubling transition, starting from the context of one dimensional maps.

1.2. Simple systems can have complicated solutions

To illustrate the complex behaviour to which even simple dynamical systems can give rise, we consider periodically varying the length of a damped pendulum. Its angle ϕ is governed by the equation:

$$\phi_{tt} = -\omega^2(1+2\mu\cos t)\sin\phi - k\phi_t, \qquad \phi_t = \frac{d}{dt}\phi,$$

where ω is the natural frequency, k the damping rate, and μ the forcing amplitude. The system may be written instead as a first order autonomous system

$$\phi_t = v,$$

$$v_t = -\omega^2(1+2\mu\cos\tau)\sin\phi - kv,$$

$$\tau_t = 1,$$

where $(\phi, \upsilon, \tau) \in S^1 \times \mathbb{R}^2$. It is dissipative for $k > 0$, since the flow contracts volumes at rate k.

The state $\phi = 0$ is an equilibrium. Linear analysis yields fig. 1 for the zones of parametric instability of this equilibrium in the space of parameters (ω, μ, k). Note that instability is possibly even with dissipation provided the "pumping" μ is sufficiently large. What happens in these zones of instability can be quite complicated. We consider, for illustration, the orbits of an initial point $\phi(0) = 1.5$, $\upsilon(0) = -2$, $\tau(0) = 0$ with $\omega = \frac{1}{2}$, $k = 0.2$, for a sequence of values of μ. For small enough μ the orbit spirals down to the steady periodic orbit (equilibrium). But for $\mu = \frac{1}{4}$ it settles down to a periodic oscillation in ϕ, which is close to sinusoidal (fig. 2a). Note that the periodic is twice that of the pump. This is the way one usually works a swing at a playground. For $\mu = \frac{1}{2}$ one finds a periodic oscillation with the same period as the pump (fig. 2b), going now over the top. The oscillations need not be sinusoidal, as fig. 2c shows for $\mu = 0.7$. At $\mu = 1$ the orbit approaches an oscillation with period equal to four times that of the pump,

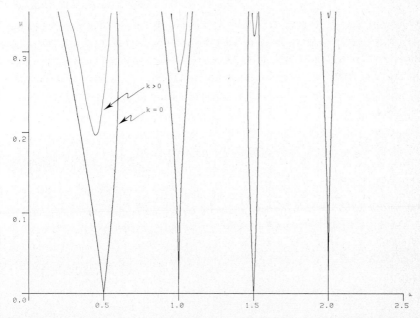

Fig. 1. Stability diagram for the equilibrium $\phi = 0$ of the parametrically forced pendulum, in the plane of the forcing amplitude μ and the natural frequency ω, for damping $k = 0$ and $k > 0$.

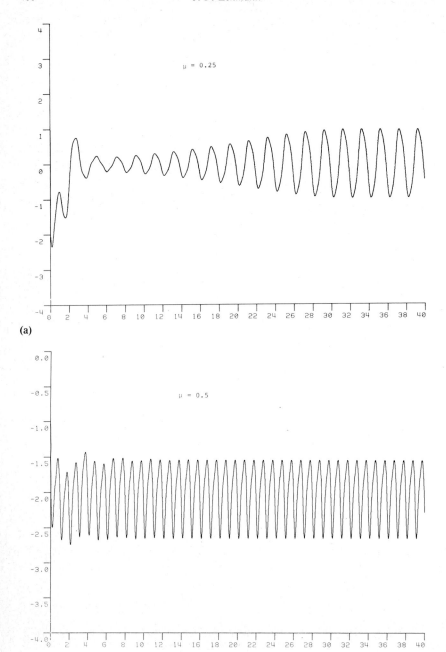

(a)

(b)   Fig. 2a,b.

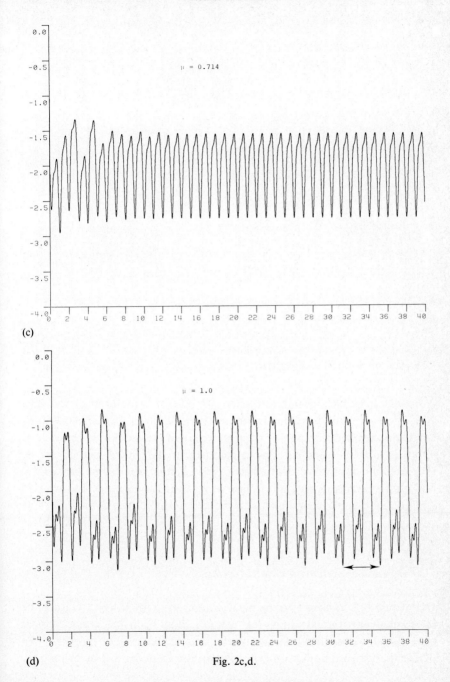

(c)

(d)   Fig. 2c,d.

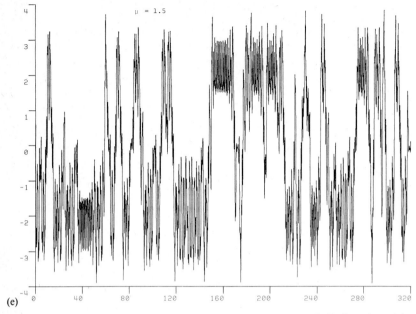

Fig. 2. Some solutions for the angular velocity $\phi_t(t)$ of the parametrically forced pendulum, for a sequence of values of forcing amplitude $\mu=$ (a) 1/4; (b) 1/2; (c) 0.714; (d) 1; (e) 1.5. Horizontal axis is t, vertical axis is ϕ_t.

although it almost repeats itself after two or even one pump period (fig. 2d). For $\mu=1.5$ the pendulum never settles down to a regular behaviour, but alternates between phases of clockwise and anticlockwise rotation at intervals for which one can see no pattern (fig. 2e). Slight changes in the initial conditions may lead to macroscopic changes after some time, since one initial condition may allow just to go over the top at some time T, while a nearby initial condition may make the pendulum fall back after T, so that some time after T the two initial conditions lead to opposite directions of rotation.

1.3. Dynamical systems are hard to classify

One would like to classify dynamical systems up to qualitative features of their behaviour, though which features one should consider is not a priori clear. Most interest has focussed on the longtime behaviour of orbits. We define the forward limit set $\omega(x)$ of the orbit of x under a flow or map f^t,

$$\omega(x) = \{y \in X: \forall \varepsilon > 0,\ T > 0,\ \exists t > T \text{ s.t. } |f^t x - y| < \varepsilon\}.$$

It is closed, invariant, and contained in the nonwandering set Ω:

$$\Omega = \{x \in X: \forall \text{ nbhds } U \text{ of } x, \ T > 0 \ \exists t > T, \ y \in U \text{ s.t. } f^t y \in U\}.$$

The hope for non-conservative systems was that, excluding possibly an exceptional set of orbits, a typical system would have only a finite or at most countable number of different forward limit sets. Thus, classification of long time behaviour would consist in classifying the possibly behaviours on ω-sets. Conservative systems (in this context, those which preserve a volume form or measure of interest), one expects to have to treat separately, for two reasons. Firstly, if they have finite volume, every point is non-wandering (see Lanford's lectures). Indeed, almost every point x is recurrent, i.e.

$$\forall \text{ nbhds } U \text{ of } x, \ T > 0 \ \exists t > T \text{ s.t. } f^t x \in U.$$

Thus $\omega(x)$ contains the whole orbit of x. Secondly, there can be no attractors if volume is preserved (apart from the whole manifold).

This sort of picture has been beautifully worked out in the context of Axiom A (see Lanford's lectures and Smale 1967). One finds a decomposition of the non-wandering set into finitely many closed invariant sets, called *basic sets*, each of which has a dense orbit so that it cannot be further decomposed. The whole manifold decomposes as the disjoint union of their stable manifold W^s. The basic sets with W^s of maximal dimension are called *Axiom A attractors*, and the corresponding W^s their *basins of attraction*. Thus most points are attracted to one of the attractors. Possible attractors include equilibrium points, periodic orbits, and more complicated objects on which the motion is strongly chaotic, with exponential separation of orbits. These may not be manifolds, in which case we call them *strange* (for example, the solenoid is locally the product of a Cantor set with a line). But there are examples too which are manifolds (e.g. the product flow of concentration onto a fixed point and a linear hyperbolic automorphism of a torus). Axiom A attractors are structurally stable (subject to a "no-cycle" condition), and have unique ergodic measures giving the asymptotic distribution of the orbit of almost every point in their basin.

The notion of attractor can be generalized to other systems than Axiom A, to be a set Λ (without loss of generality, closed), which has:

(i) a contracting neighbourhood, i.e.

$$U \supset \Lambda \text{ s.t. } fU \subset U, \qquad \Lambda = \bigcap_{n \geq 0} f^n U;$$

(ii) no wandering points, i.e. $\Lambda \subset \Omega$,

(iii) a dense orbit (topological transitivity).

The *basin* of an attractor Λ is defined to be

$$B(\Lambda) = \bigcap_{n \geq 0} f^{-n} U$$

for any contracting neighbourhood U.

This definition has problems, however. For example, the basins are open but their union is not necessarily dense. The figure of 8 in the flow of fig. 3, for instance, has a contracting neighbourhood and no wandering points (though it does have non-recurrent points). But it does not have a dense orbit, so it is not an attractor, nor can it be decomposed into smaller pieces which are attractors. In section 2.7 we will give an example of a system with a set which attracts almost every point, has every orbit dense and no wandering points, but has no contracting neighbourhood. Neither of these examples is structurally stable, but then structural stability is not generic (Smale 1966), and they at least illustrate the problems one can run into. There is no general consensus about the best notion of attractors, (see Ruelle 1981, for a new one), and much is still unknown about the possible motion on the non-wandering set. Very complicated behaviour can be quite common. For example, Newhouse showed in his lectures (and also in Newhouse, 1980) that there is an open set of maps on 2-surfaces containing a residual set on which there is a hyperbolic set with a dense orbit, with periodic attractors arbitrarily close.

A classification of dynamical systems should also cover transient motion, for example, the motion in a basin before an orbit comes close to the attractor. This is often passed by in the study of Axiom A systems as one often considers only topological conjugacy on Ω, but could be important, as a

Fig. 3. Figure of eight flow.

physical system might not reach its attractor on a practical time scale. Even simple attractors can have very complicated basins. For example, Newton's algorithm for solving $z^3 = 1$ in $\mathbb{C}$,

$$z_{n+1} = (2z_n^3 + 1)/3z_n^2$$

has three attracting fixed points, viz. the cube roots of unity. Their basins, however, can be shown to each have the same boundary, so they are highly interlaced (fig. 4)! (See e.g. Brolin 1965.)

1.4. Scenarios in 1-parameter families

In view of the difficulty of classifying general dynamical systems, we take a different approach to their study, and look near the boundary of some open set of well understood systems (c.f. Newhouse's Course). Specifically, we will look at families of systems depending smoothly on some parameters, with simple behaviour for some parameter values, and study transitions to more complicated behaviour. We hope to find results relevant, for example, to the transition to chaos as the forcing is increased on a pendulum, the transition with pressure gradient (equivalently, Reynold's number) from laminar to turbulent flow in a pipe, or that from a conducting state to a turbulent convective state in a fluid layer heated from below, depending on the heating rate.

As we never know the exact description of a system, nor whether all the relevant factors have been included, we are interested in transitions which are stable to perturbation of the system and to external noise. We will describe three scenarios for such transitions. For more details, see Eckmann (1981). Note that several scenarios may occur concurrently in phase space, or for different ranges of parameter. By no means do we claim that these exhaust all the possibilities, nor do we indicate to which physical problems (if any) they apply.

1.5. Ruelle–Takens scenario

Ruelle and Takens (1971) consider a system

$$x_t = X_\mu(x)$$

with an equilibrium at 0, which loses stability at a parameter value μ_1 by a *Hopf bifurcation*, producing a stable periodic orbit [an equilibrium 0 has

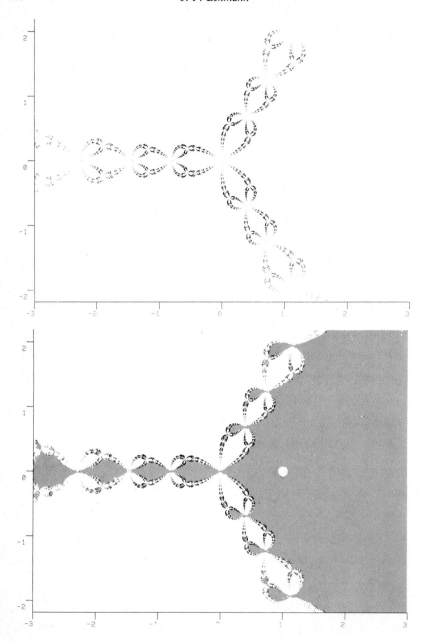

Fig. 4. Top: Boundary of basins of attraction of Newton algorithm for finding cube roots of unity. Bottom: Basin of attraction of $z = 1$.

a Hopf bifurcation when a complex conjugate pair of eigenvalues of $DX_\mu(0)$ cross the imaginary axis]. They suppose that there are some more parameter values $\mu_1 < \mu_2 < \mu_3 < \ldots$ at which further pairs of eigenvalues cross the imaginary axis, and that the corresponding degrees of freedom that are excited are weakly coupled. Thus for $\mu_k < \mu < \mu_{k+1}$ the motion is close to a contraction onto a linear flow on a k-torus, with frequencies given by the imaginary parts of the respective eigenvalues of $DX_\mu(0)$. Note that there is an attracting k-torus for $\mu_k < \mu < \mu_{k+1}$, only for sufficiently weak coupling, which requires in particular that one exclude a neighbourhood of the endpoints.

For the case of the 1- and 2-tori, there is, in fact, typically a μ_2' close to μ_2 and $\mu_3' > \mu_2'$ such that there is a stable periodic orbit (1-torus) for $\mu_1 < \mu < \mu_2'$ and an invariant attracting 2-torus for $\mu_2' < \mu < \mu_3'$. This is a Hopf bifurcation of a periodic orbit, which should be distinguished from Hopf bifurcations of equilibria. Defining the *multipliers* of a periodic orbit to be the eigenvalues of the linearization of the first return map around the orbit, a periodic orbit undergoes a Hopf bifurcation, giving birth to or absorbing an invariant 2-torus, when a complex conjugate pair of multipliers leave the unit circle (subject to some restrictions – see the courses by Chenciner and by Joseph). To get a bifurcation from a 2-torus to a 3-torus, however, requires infinitely many conditions (Chenciner and Iooss 1979), so does not typically occur. Some possibilities for producing invariant 3-tori in special cases are discussed by Iooss and Langford (1980).

If there is a k-torus, the motion on it will be close to linear. For T^2, Peixoto (Abraham and Marsden 1978) showed that there is an open dense set of flows for which the non-wandering set consists of only a finite number of fixed points and periodic orbits. Thus every neighbourhood of a linear flow contains an open dense set of phase-locked flows. Of course, the complement need not be small in the sense of measure (e.g. of parameter values $\mu_2' < \mu < \mu_3'$). In particular, there is typically a set of positive measure on which the flow has irrational winding ratio, and so is conjugate to a linear flow with that winding ratio. But nothing else much can happen.

On T^k, $k \geq 4$, however, later extended also to $k = 3$ (Newhouse et al. 1978), there is an open set of flows possessing a strange Axiom A attractor, and the linear flows are on its boundary. Thus chaotic behaviour could occur, and be stable to perturbation. This should be contrasted with the earlier Landau scenario (Landau and Lifschitz 1959). They envisaged an infinite sequence of μ_k leading to quasiperiodic flow on an infinite dimensional torus, which would appear chaotic. But such a flow is not stable to perturbation, as for any finite $k \geq 2$ a linear flow can be perturbed away. The point of the Ruelle–Takens scenario is that for $k \geq 3$, in fact, it can be perturbed into a much more chaotic flow, which is stable to perturbation.

The power spectrum $P(\omega)$ of a projection $x(t)$ in $\mathbb{R}$ of an orbit is defined by:

$$P(\omega) = |a(\omega)|^2,$$

$$a(\omega) = \lim_{T \to \infty} \frac{1}{\pi T} \int_0^T dt \, e^{-i\omega t} x(t).$$

Thus, in this scenario one would expect the power spectrum to develop first one fundamental frequency, with its harmonics, then a second, with not only harmonics but also combination frequencies, and intervals in which the frequencies lock in a rational ratio. Somewhere higher in parameter will be an interval in which the power spectrum has a third frequency, but it will be accompanied by a broadband component. Such a progression has been observed in some hydrodynamical experiments (see Gollub's Course). There is a discrepancy, however, in the form of the broadband component, as for strange Axiom A attractors the spectrum decays exponentially, whereas experimentally it falls off by a power law.

One should be aware that cases are also observed where a broadband component develops with no third frequency, directly after a two frequency phase locked state. Similar transitions to chaos, involving the breakup of an invariant 2-torus (actually a 1-torus in discrete time) have been observed numerically for two dimensional maps (Curry and Yorke 1978), and in electronic models of a forced van der Pol oscillator (Shaw 1978). This does not contradict the scenario of Ruelle and Takens, as they made no claims about what might occur when the coupling between different degrees of freedom becomes strong, nor about the behaviour in the neighbourhood of μ_3 (if it exists).

The Ruelle–Takens scenario is robust to external noise. In particular Kifer (1974) shows that the invariant measure for an Axiom A attractor, describing the asymptotic distribution of orbits, is only slightly changed when small noise is added. This might seem surprising in view of the strong hyperbolicity of Axiom A, which causes most nearby orbits to diverge exponentially. But the shadowing lemma (see the lectures by Lanford and by Katok) guarantees that any orbit perturbed by small noise actually follows some true orbit with only small error.

1.6. Transition to turbulence by intermittency

As this scenario is discussed in detail in Pomeau's lectures, we will give only a brief outline here. In experiments with fluids one often observes a transition in which a laminar flow develops turbulent bursts at irregular intervals, leading to a fully turbulent state as the bursts become more frequent. This behaviour is called *intermittency* (not to be confused with another notion of intermittency in statistically steady turbulence, meaning strong localization of vorticity).

One parameter families of one dimensional maps provide a simple model for intermittency, in the neighbourhood of a parameter value μ_c with a *saddle-node*, i.e. a periodic orbit with multiplier equal to $+1$, and the rest inside the unit circle. For μ on one side of μ_c, say below, there is a stable periodic orbit, but at μ_c it coalesces with an unstable orbit of the same period, and they annihilate each other. Above μ_c, however, orbits may still spend a considerable time in the vicinity of the old saddle-node, as indicated in fig. 5. On leaving the vicinity they may pass through some region

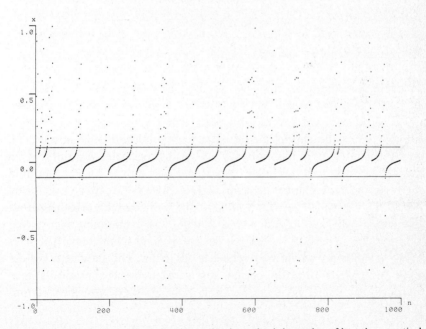

Fig. 5. Intermittency in a 1-dimensional map. Horizontal axis is number of iterations, vertical axis is position. Map is $f(x) = x + \frac{7}{4}x^2 + \varepsilon + 1 \pmod{2} - 1$, $\varepsilon > 0$ small.

where they get separated strongly, and be reinjected near the old saddle-node again. Thus one would expect "laminar" phases interrupted by "turbulent" bursts. The mean duration of the laminar phases can be estimated as $\tau \sim C \cdot |\varepsilon|^{-1/2}$, for some constant C, where $\varepsilon = \mu - \mu_c$.

For certain classes of maps on an interval, however, there will typically be intervals in parameter, maybe dense, on which a stable periodic orbit attracts almost every point, so one should not expect a monotonic transition to turbulence. Nevertheless, results of Jakobson (1981) and Collet, Eckmann (1980b) lead one to suspect that there is also typically a set of parameter values of positive measure for which truly aperiodic behaviour with sensitivity to initial conditions occurs.

One would expect external noise to shorten the laminar phases. This was investigated by Mayer-Kress and Haken (1981), and further analyzed by Eckmann et al. (1981). They found that for one dimensional maps with independent identically distributed perturbations of variance σ^2, the mean duration of the laminar phase is modified to the form

$$\tau \sim C |\varepsilon|^{-1/2} T_{\pm}(\kappa\sigma/|\varepsilon|^{3/4})$$

for some universal functions $T_{\pm}$ (for $\varepsilon > 0$, $\varepsilon < 0$), and scale κ.

1.7. Period doubling scenario

A stable periodic orbit can lose stability, as a parameter is varied, by a *flip bifurcation* (when one multiplier leaves the unit circle through -1). As it does so, it typically gives birth to, or absorbs, a periodic orbit of twice the period. In the latter case, the new orbit is unstable, and one could find intermittency as in the previous section, but in the former, the case of interest, it is initially stable. Thus one would see a periodic oscillation develop slightly different shapes in alternate cycles. This is called *period doubling*. The new periodic orbit can itself lose stability by a flip, and period double. The remarkable fact is that if a system undergoes a few successive period doublings, then it is likely to undergo an infinite sequence of them. The parameter values μ_n at which the nth flip occurs will accumulate asymptotically geometrically at some value μ_∞, with a ratio $\delta \sim 4.669...$ which is universal, i.e.

$$\mu_n - \mu_\infty \sim C \cdot \delta^{-n}$$

for some constant C. Beyond μ_∞ the flips will reverse in a similar geometric progression, but with "noisy" periodic orbits. Accompanying the universal ratio δ, there will be an asymptotic universal self-similarity in features of the orbit, expressed by a universal scaling $\lambda \sim -0.3995\dots$. This will show itself, for example, in the power spectrum (see section 3.6.2). Though some details are not yet resolved, one expects the power in successive subharmonics to decrease asymptotically by some universal factor at the accumulation point.

Such behaviour has been observed in experiments on convection (see the lectures by Libchaber, and by Gollub), and its explanation will be the goal of the rest of our lectures. We will also postpone the effect of external noise to section 3.6.4. Note that, although a significant route to aperiodicity, the period doubling transition involves only a few degrees of freedom, so one does not expect it to be relevant to fully developed turbulence. On the other hand, this makes it easier to study, and in fact we will begin our discussion of period doubling in the context of one dimensional maps.

2. Maps on an interval

2.1. Generalities

Maps on an interval to itself

$$x_{n+1} = f(x_n)$$

form one of the simplest classes of dynamical systems whose behaviour can nevertheless be non-simple (May 1976). For example, there are maps for which, excluding a set of measure zero, every orbit has an asymptotic distribution representable by a density function (absolutely continuous invariant measure), the same for almost all orbits (see Misiurewicz's lectures). Of course there are also maps with simple behaviour, such as those for which a stable periodic orbit attracts almost every point. Although there is a good classification up to topological conjugacy, the behaviour is not well understood on all the conjugacy classes. Furthermore, their arrangement in the space of all 1-dimensional maps is not fully known. Thus it is appropriate to look for transitions in 1-parameter families from stable to other behaviour, and see if there are any robust features of such transitions.

A stable periodic orbit in one dimension can lose stability in only two ways – via a saddle-node or flip (see Misiurewicz's lectures) – so there are only two candidates for how such a transition could begin. The former can give rise to intermittency, while the latter often leads to an example of the period doubling scenario, on which we shall concentrate here.

In this section, after describing the stability of periodic orbits, we will define a significant class of maps and give an introduction to their analysis by kneading theory and some important results. We conclude with some remarks on maps which exchange intervals, which occur in the period doubling transition. For a review of this material, see Collet and Eckmann (1980a).

2.2. Periodic orbits and their stability

Given a map f and an initial point x, the sequence of points $x, f(x), f^2(x), \dots$ is called its orbit. The iterates $f^n(x)$ are defined by:

$$f^{n+1}(x) = f[f^n(x)], \qquad f^0(x) = x.$$

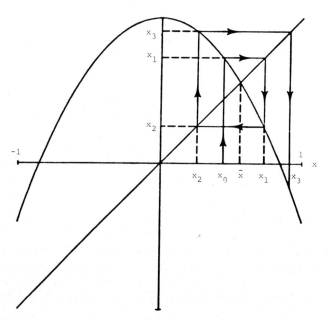

Fig. 6. Graphical iteration.

A *periodic orbit* is one which repeats itself after a finite number of steps, the smallest such being called its period. A periodic orbit of period m is also known as an *m-cycle*, and its points as *periodic points*. The orbits of a map can be visualised graphically as in fig. 6. Fixed points are given by intersections of the graph of f with the diagonal (the graph of the identity). Their stability depends on the slope of f at the fixed point. We make the following classification:

$$|f'| > 1 \quad \text{linearly unstable,}$$
$$|f'| \leq 1 \quad \text{linearly stable,}$$
$$|f'| < 1 \quad \text{strongly stable,}$$
$$|f'| = 1 \quad \text{marginally stable,}$$
$$f' = 0 \quad \text{superstable.}$$

A linearly unstable point p is always *repelling*, i.e.

$$\exists \text{ nbhd } U \text{ of } p \text{ s.t. } \forall y \in U \setminus \{p\} \; \exists n > 0 \text{ s.t. } f^n y \notin U.$$

A strongly stable point p is always *stable*, i.e.

$$\forall \text{ nbhds } U \; \exists \text{ a subnbhd } V \text{ s.t. } f^n V \subset U, \; \forall n \geq 0.$$

In fact it is *attracting*, i.e.

$$\exists \text{ nbhd } U \text{ of } p \text{ s.t. } fU \subset U, \; \bigcap_{n \geq 0} f^n U = \{p\}.$$

For the class we will define in section 2.3 one can show that marginally stable fixed points with $f' = -1$ are attracting, and with $f' = +1$ are attracting from one side, repelling on the other, unless $f'' = 0$ when they are attracting from both sides.

The above classification of fixed points can easily be extended to periodic points, by considering them as fixed points of some iterate f^m of f. Note that by the chain rule,

$$f^{m'}(x) = \prod_{j=0}^{m-1} f'(f^j x)$$

so the stability of a periodic point is given by the product of the slopes around its orbit.

2.3. Assumptions

We will consider C^1 maps f of an interval to itself, with a single maximum, which we choose to be at 0 by a coordinate shift, and no other critical points, i.e.

$$f'(0) = 0, \qquad xf'(x) < 0 \text{ for } x \neq 0. \tag{A1}$$

We suppose that the critical point maps to the right at the first step, and choose the scale so that:

$$f(0) = 1. \tag{A2}$$

We also suppose:

$$f^2(1) \geq f(1) \tag{A3}$$

so that $[f(1), 1]$ is an invariant subinterval, and we will usually restrict our attention to this interval. An assumption which will be quite significant is that of *negative Schwarzian derivative*. For $f \in \mathscr{C}^3$, the Schwarzian derivative Sf is defined whenever $f' \neq 0$ by

$$Sf = \frac{f'''}{f'} - \frac{3}{2}\left(\frac{f''}{f'}\right)^2.$$

So we assume:

$f \in \mathscr{C}^3$ except possibly at 0 where we allow
$$f(x) \sim 1 - a|x|^{1+\varepsilon}, \qquad \text{for some } \varepsilon > 0. \tag{A4}$$

$$Sf < 0 \qquad \text{for } x \neq 0. \tag{A5}$$

The class of maps satisfying (A1)–(A5) we call **C**. Their typical form is as in fig. 6, and the class includes, for example, the family

$$f_\mu(x) = 1 - \mu x^2, \qquad \mu \in [0, 2].$$

The Schwarzian derivative has several interesting properties:

$$\text{If } f, g \in \mathscr{C}^3, f', g' \neq 0, \text{ then } S(f \circ g) = g'^2 \cdot Sf \circ g + Sg \tag{S1}$$

as may be verified by direct computation (see Misiurewicz's lectures). Thus

$$Sf < 0, \ Sg < 0 \ \Rightarrow \ S(f \circ g) < 0. \tag{S2}$$

So by induction,

$$Sf < 0 \ \Rightarrow \ Sf^n < 0. \tag{S3}$$

We include negative Schwarzian derivative in our assumptions because it leads to:

Theorem 2.1. $f \in \mathbf{C}$ has at most one linearly stable periodic orbit in $[f(1), 1]$, and it attracts 0, and indeed, almost every point.

Note as a consequence that if the orbit of 0 falls on a linearly unstable periodic orbit then there can be no stable periodic orbit. An example of this phenomenon occurs for $f(x) = 1 - 2x^2$.

2.4. Kneading theory

Kneading theory is symbolic dynamics for maps of an interval, and was developed largely by Milnor and Thurston (1977), and Guckenheimer (1977). The idea, as applied to the class $\mathbf{C}$, is to associate with each point a sequence of symbols describing on which side of the critical point its successive iterates lie. In the case of the symbolic dynamics of the horseshoe, discussed in Lanford's lectures, the map from initial points to symbol sequences is bijective and in fact is a homeomorphism. For maps of an interval, there is not always such a nice correspondence, but kneading theory still gives considerable insight. For example, we will find that many 1-parameter families have an infinite sequence of parameter values μ_n for which there is a superstable 2^n-cycle, but this theory will not predict any self-similarity, since it is purely topological.

For a point x we define the *itinerary*

$$I(x) = I_0 I_1 I_2 \dots,$$

a singly infinite sequence of symbols I_n, by $I_n = L$, C, or R according as $f^n x < 0$, $f^n x = 0$, $f^n x > 0$. The map f induces a shift σ on sequences.

$$I(fx) = \sigma I(x), \qquad \text{where } \sigma I_0 I_1 I_2 \dots = I_1 I_2 \dots.$$

Defining the order $L < C < R$ on the symbols, one can extend it to an order on sequences, as follows: if A, B are two different sequences, there is a least integer m for which $A_m \neq B_m$. Call a finite sequence odd or even according as the number of Rs it contains is odd or even. Then we say

$$A < B \text{ iff } (A_m < B_m \text{ and } A_0 \dots A_{m-1} \text{ is even or}$$
$$A_m > B_m \text{ and } A_0 \dots A_{m-1} \text{ is odd}).$$

The point of this definition is that it makes the map $x \to I(x)$ weakly order preserving, i.e.

$$x < y \Rightarrow I(x) \leq I(y),$$

$$I(x) < I(y) \Rightarrow x < y.$$

Defining the *kneading sequence* K_f of f to be the itinerary $I(1)$, the above result implies that K_f is the largest itinerary that can occur for f. Also $\sigma K_f = I((1))$ is the smallest that occurs on $[f(1), 1]$. Given an itinerary $A = I(x)$, $\sigma^n A = I(F^n x)$ is also an itinerary. Thus,

$$\sigma K_f \leq \sigma^n A \leq K_f \qquad \forall n \geq 0.$$

The next theorem almost gives the converse. Note that if $A = I(x)$ contains a C, then its future is determined by the orbit of 0, and so it must have the form $A = BCK_f$, for some finite sequence B containing no C. We call an arbitrary infinite sequence A of Ls, Cs and Rs *admissible* for f if it contains no Cs or has this form.

Theorem 2.2. Every admissible sequence A for f, s.t.

$$\sigma K_f \leq \sigma^n A < \tilde{K}_f, \qquad n \geq 0,$$

occurs as the itinerary of some point in $[f(1), 1]$,

where for K containing no C, $\qquad\qquad \tilde{K} = K$
and for $K = (DC)^\infty$, D containing no C, $\quad \tilde{K} = (DL)^\infty$ if D even
$\qquad\qquad\qquad\qquad\qquad\qquad\qquad\qquad\quad \tilde{K} = (DR)^\infty$ if D odd.

Thus, apart from exceptions just below K_f, every sequence which we could expect to occur as an itinerary, does so.

Theorem 2.2 leads to two interesting consequences, both of which are discussed in detail in Misiurewicz's lecture. First we note some connections between itineraries and orbits.

Lemma 2.3. $I(x)$ is eventually periodic iff $f^n(x)$ converges to a periodic orbit.

In fact $I(x) = AB^\infty$ with B of length p, implies that x converges to a periodic orbit of period p if B is even, period p or $2p$ if B is odd. Nevertheless,

Lemma 2.4. If $f \in C$ has a periodic itinerary of period p, then it has a periodic orbit of period p (but not necessarily with the same itinerary).

The first consequence of Theorem 2.2 is Sarkovskii's theorem (1964):

Theorem 2.5. If f has a periodic orbit of period k then it has periodic orbits of all periods to the right of it in the following ordering:

$$3, 5, 7, \ldots, 2 \cdot 3, 2 \cdot 5, 2 \cdot 7, \ldots, 2^2 \cdot 3, 2^2 \cdot 5, 2^2 \cdot 7, \ldots, 2^3, 2^2, 2, 1.$$

The second is a generalization of the theorem of Li and Yorke (1975):

Theorem 2.6. If f has a periodic orbit of period which is not a power of 2, then it has topological chaos (infinitely many periodic orbits and an uncountable set of aperiodic orbits – see Misiurewicz's lectures for details).

But note that there can be topological chaos even when there is a stable periodic orbit present. The presence or absence of a stable periodic orbit is decided nicely by the following deep theorem of Guckenheimer (1979).

Theorem 2.7. $f \in C$ has a linearly stable periodic orbit iff K_f is periodic.

Note that f has a superstable periodic orbit iff K_f contains a C (as 0 is then periodic).

We have shown that $x \to I(x)$ maps $[f(1), 1]$ onto the largest set of sequences that we would expect. We would like to know to what extent it is one to one. This question is answered by Theorem 2.8, Guckenheimer (1979).

Theorem 2.8. If K_f is not periodic, then $x < y \Rightarrow I(x) < I(y)$.

Thus, in this case, $x \to I(x)$ is strictly order preserving, and in particular, one to one. Note that this says also that iterating f separates any two points at some time onto opposite sides of 0.

We would like to classify maps up to equivalence in some sense. Two maps $f: I \to I$, $g: J \to J$ are said to be *topologically conjugate* if there exists a homeomorphism $h: I \to J$ s.t.

$$f = h^{-1} g h$$

so that f and g are the same apart from a continuous coordinate change.

Theorem 2.9. If $K_f = K_g$ is not periodic, then f and g are topologically conjugate.

The conjugacy h may be constructed by requiring

$$l_f(x) = l_g(h(x)), \qquad \forall x \in I.$$

It is well defined, one to one and onto, because both maps have the same set of itineraries by Theorem 2.2, and the itineraries each occur once only for each map by Theorem 2.8. It is also monotone, and so is a homeomorphism.

Note that this theorem does not generalize to maps with linearly stable periodic orbit, i.e. when K_f is periodic. For example, for $0 < \mu < 3/4$, $f_\mu(x) = 1 - \mu x^2$ has a stable fixed point, while for $3/4 < \mu < 1$ it has a stable 2-cycle. The kneading sequence, however, is R^∞ in the whole range $0 < \mu < 1$ (see fig. 7). For periodic K_f there are in fact precisely two conjugacy classes, corresponding to these two behaviours.

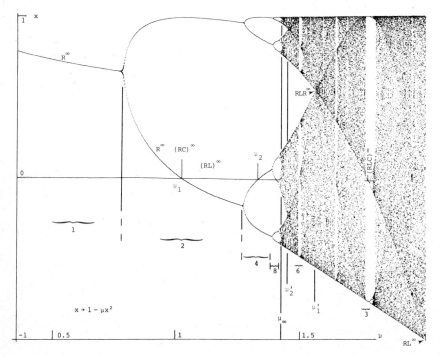

Fig. 7. Asymptotic distribution for the orbit of 0.3 under the map $x \to 1 - \mu x_2$, for $\mu \in [0, 2]$.

One would like to know what forms of chaos (stronger than topological) are possible when there is no linearly stable periodic orbit. A partial answer is given by the next theorem. We say a sequence A is *eventually periodic* if $\sigma^n A$ is periodic for some n.

Theorem 2.10. (Misiurewicz) (a) K_f eventually periodic, but not periodic $\Rightarrow 0$ is not recurrent and f has no linearly stable periodic orbit.
(b) 0 is not recurrent and f has no linearly stable periodic orbit $\Rightarrow f$ has a unique absolutely continuous invariant measure, and it is ergodic.

This situation occurs, for example, when the orbit of 0 falls on an unstable periodic orbit. Another result will be given later in Theorem 2.12, but there is not yet a complete characterization of the behaviours of maps in **C**. One would also like to know which systems are near each other. For this reason we turn to 1-parameter families.

2.5. 1-parameter families

We consider 1-parameter families f_μ such that $\mu \to f_\mu$ is a continuous map from a parameter interval into **C**, taken with the $\mathscr{C}^1$-topology ($\mathscr{C}^0$ is insufficient – see Derrida et al. 1979). One would like to know how the kneading sequence K_{f_μ} depends on μ. Note that since K_f is the largest itinerary that occurs for f, it is in particular larger than $I[f^n(1)]$. Thus,

$$\sigma^n K_f \leq K_f, \qquad \forall n \geq 0.$$

Also K_f containing a $C \Rightarrow K_f$ periodic and $= (BC)^\infty$, for some B containing no C. Sequences $\neq L^\infty$ with these two properties, we call *maximal*.

Conversely, extending a result of Metropolis et al. (1973),

Theorem 2.11. For a 1-parameter family f_μ as above, every maximal sequence K,

$$K_{f_{\mu_0}} \leq K \leq K_{f_{\mu_1}},$$

occurs as the kneading sequence for some μ between μ_0 and μ_1.

From this one can make many deductions. For example, if f_μ has a stable fixed point for some μ_0 and a stable periodic orbit of period k for some $\mu_0' > \mu_0$, then it has intermediate parameter values with stable periodic orbits (even with superstable periodic orbits) of all periods to the Sarkovskii's ordering (as in Theorem 2.5). In fact it has stable periodic orbits with all periodic itineraries between R^∞ and that of the k-cycle. Strongly stable periodic orbits are persistent, so each parameter value for which there is a superstable periodic orbit has an interval around it in which there is a stable periodic orbit of the same period. Thus, if the itinerary of the above k-cycle is large enough (e.g. k not a power of 2), there is an infinite set of intervals of μ's for which f_μ has a stable periodic orbit. Similarly one can deduce the existence of infinitely many parameter values for which the orbit of 0 falls on an unstable periodic orbit, so that f_μ has an absolutely continuous invariant measure. These features are illustrated in fig. 7. It shows the orbit of 0.3 (leaving out an initial transient phase) for the family $x \to 1 - \mu x^2$, for a range of μ.

Although the kneading sequence of the family $x \to 1 - \mu x^2$ appears to be monotone in μ, this need not be so in general. Thus the parameter values given by Theorem 2.11 are not necessarily unique. Also kneading sequences outside the range specified can occur. What the theorem says is that K_{f_μ} can not skip any maximal sequences (in the order $<$), as it varies with μ.

Particularly noticeable in fig. 7 is the infinite sequence of period doubling bifurcations from the stable fixed point. Complementary to it is an infinite sequence of "band mergings", for decreasing μ. These features follow from Theorem 2.11 and some properties of the ordering on sequences which we shall explain in the next section.

2.6. Exchange of intervals

We say a map f *exchanges intervals* if there exist disjoint intervals J_0, J_1 such that

$$0 \in J_0, \qquad fJ_0 = J_1, \qquad fJ_1 \subset J_0.$$

Figure 8 shows a map with two such intervals. Sufficient conditions for exchange of intervals are that:

$f^2(1)$ has a second pre-image $f(1)^\#$ other than $f(1)$, and
$f(1) < f(1)^\# < f^2(1) < 1$.

Equivalently, $f^2(0) < 0$ and $f^3(0) > f^4(0)$. Note that the boundary of the set of maps which exchange intervals consists of those with a superstable

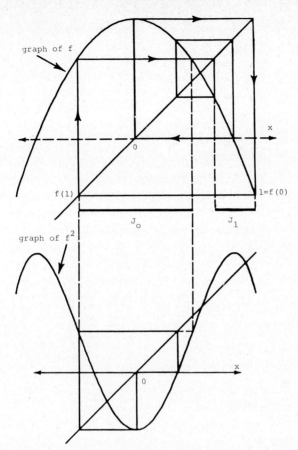

Fig. 8. A map which exchanges intervals, and its second iterate.

2-cycle ($f^2(0) = 0$) and those for which the orbit of 0 falls on an unstable fixed point [$f^3(0) = f^4(0)$].

Clearly $f \circ f|_{J_0}$ maps J_0 into itself, and one can choose J_0 to make the critical point map onto its left hand endpont. Thus, apart from a change of scale (including an inversion), $f \circ f|_{J_0}$ belongs to **C**. The operation of squaring and rescaling, we call *doubling*.

For a map which exchanges intervals, all odd iterates of 0 belong to J_1, so the kneading sequence has the form

$$K_f = RI_0 RI_1 \ldots \qquad \text{for some } I_0, I_1, \ldots,$$

while

$$K_{f \circ f} = \hat{I}_0 \hat{I}_1 \ldots \qquad \text{where } \hat{R} = L, \ \hat{L} = R, \ \hat{C} = C.$$

Equivalently,

$$K_{f \circ f} = I \implies K_f = R * I,$$

where for a finite sequence A containing no Cs, and arbitrary $I = I_0 I_1 \ldots$, the operation $*$ is defined by

$$A * I = \begin{cases} A I_0 A I_1 A I_2 \ldots & \text{if } A \text{ is even} \\ A \hat{I}_0 A \hat{I}_1 A \hat{I}_2 \ldots & \text{if } A \text{ is odd.} \end{cases}$$

Conversely, $K_f = R * I$, for some I, implies that f exchanges intervals. Thus we see that if A is the kneading sequence for f to have a certain property, then $K_f = R * A$ is the kneading sequence for f^2 to have the same property.

The operation $R *$ on sequences is order preserving and one to one onto its image. It sends maximal sequences to maximal sequences. Furthermore,

$$R * A < X < R * B, \qquad \text{with } A, B, X \text{ maximal}$$

$$\implies \exists Y \text{ s.t. } X = R * Y, \qquad \text{and } Y \text{ is maximal.}$$

So the set $M_1 = \{R * I : I \text{ maximal}\}$ forms an interval in the set M_0 of all maximal sequences (when considered with the ordering $<$). Its endpoints are

$$(RC)^\infty = R * C^\infty < R * (RL^\infty) = RLR^\infty,$$

the kneading sequences for a superstable 2-cycle, and for the orbit of 0 to fall on an unstable fixed point.

Since $f \circ f |_{J_0} \in C$ for a map which exchanges intervals, it is natural to consider whether $f \circ f |_{J_0}$ will itself exchange intervals. The condition is that $K_f = R * R * I$ for some maximal I. Thus we obtain a nested sequence of intervals

$$M_n = \{(R *)^n I : I \text{ maximal}\} \subset M_{n-1} \subset \ldots \subset M_0$$

such that when $K_f \in M_n$, $f^{2^m} |_{J_0}^{(m)}$ exchanges some intervals $J_0^{(m+1)}$, $J_1^{(m+1)}$, for each $0 \leq m < n$, where $J_0^{(0)} = [f(1), 1]$. The endpoints of M_n are $(R *)^n C^\infty$ and $(R *)^n (RL^\infty)$. Thus, for a family f_μ with kneading sequence $(RC)^\infty$ at μ_1, and RLR^∞ at $\mu_1' > \mu_1$, say, Theorem 2.11 implies that

$$\exists \mu_1 < \mu_2 < \mu_3 < \ldots < \mu_3' < \mu_2' < \mu_1'$$

s.t. f_{μ_n} has a superstable 2^n-cycle
         $f_{\mu_n'}$ has the orbit of 0 falling on an unstable 2^{n-1}-cycle.

Similarly, $\exists \bar{\mu}_n$, $\mu'_{n+1} < \bar{\mu}_n < \mu'_n$ such that $f_{\bar{\mu}_n}$ has kneading sequence $(R*)^n(RLC)^\infty$, i.e. it has a superstable $3 \cdot 2^n$-cycle. Again we emphasize that such parameter values need not be unique.

One could do a similar analysis for any operator $A*$, A finite, containing no C, for which $(AC)^\infty$ is maximal, corresponding to maps which cyclically permute more than two intervals.

Thus we see from an essentially topological theory that we expect to find infinite period doubling sequences in 1-parameter families of maps, with chaotic behaviour beyond. It gives no quantitative information, however. In the next section we will develop a quantitative theory, which will in particular explain the universal self-similarity of period doubling sequences.

2.7. The kneading sequence $(R*)^\infty$

Before we look at the theory of doubling operators, the behaviour of maps with kneading sequence $(R*)^\infty$ deserves comment (Misiurewicz 1978, Collet et al. 1980). The nonwandering set consists of one unstable periodic orbit of each period 2^n, and a Cantor set J related to the $J_0^{(n)}$ by

$$J = \bigcap_{n \geq 1} \bigcup_{m=0}^{2^n - 1} f^m(J_0^{(n)}).$$

(This can be seen directly from the kneading sequence.) The orbits of all points except the preimages of the unstable periodic orbits eventually converge to J. It is invariant and contains no periodic orbits. In fact every orbit of J is dense in J. The behaviour of f on J is not very chaotic, however. We define f to have *sensitivity to initial conditions* if

$$\exists \varepsilon > 0 \text{ s.t. } \forall x \text{ and nbhds } U \text{ of } x \; \exists y \in U, \; n \geq 0 \text{ s.t. } |f^n x - f^n y| > \varepsilon.$$

Theorem 2.12. (Guckenheimer) f has sensitivity to initial conditions iff K_f is neither periodic nor can be expressed as $A_1 * A_2 * \ldots * A_n * B$ for n arbitrarily large.

Consequently maps with kneading sequence $(R*)^\infty$ do not have sensitivity to initial conditions.

The Cantor set J provides a good example of where the intuitive notion of attractors can go wrong. It does not have a contracting neighbourhood, as every neighbourhood contains some of the unstable periodic points. Any larger set with a contracting neighbourhood, however, contains transient points. Nevertheless, J attracts a set of full measure.

3. General theory of doubling

3.1. Introduction

For maps f in $\mathbf{C}$ which exchange intervals, we noted in section 2.6 that for $f \circ f|_{J_0}$ is also in $\mathbf{C}$ apart from a change of scale and inversion. Thus we can define a *doubling operator* $\mathcal{N}$ for those f which exchange intervals, by:

$$\mathcal{N}f(x) = \frac{1}{f(1)} f \circ f[f(1)x].$$

The scale change $f(1)$ ensures that $\mathcal{N}f \in \mathbf{C}$, in particular, $\mathcal{N}f(0) = 1$. $\mathcal{N}f$ need not exchange intervals, but if it does, then we can apply $\mathcal{N}$ again. For f with kneading sequence $(R*)^\infty$, we can apply $\mathcal{N}$ infinitely often, and might expect $\mathcal{N}^n f$ to converge to some map ϕ invariant under $\mathcal{N}$. Such a ϕ would be self-similar, i.e.

$$\phi(x) = \frac{1}{\phi(1)} \phi \circ \phi[\phi(1)x],$$

so one might hope to explain the universal self-similarity in period doubling transitions in this way.

We can generalize this idea to other classes $\mathbf{F}$ of discrete time systems, acting on a space X, if they have a doubling operator $\mathcal{N}$ from some subset of $\mathbf{F}$ to $\mathbf{F}$:

$$\mathcal{N}f = \Lambda_f^{-1} \circ f \circ f \circ \Lambda_f$$

for some coordinate change Λ_f. The study of doubling operators was initiated by Feigenbaum's discovery of universality (1978) [independently rediscovered by Coullet and Tresser (1978)], and has been developed more extensively by Collet et al. (1980).

3.2. Hypotheses

The usefulness of a given doubling operator depends on the validity of hypotheses like the following three, (H1)–(H3). These have been verified in a variety of particular cases to be discussed in section 4.

(H1) $\mathcal{N}$ has a fixed point ϕ (i.e. $\mathcal{N}\phi = \phi$), and is $\mathscr{C}^2$ in a neighbourhood of ϕ.

(H2) ϕ is a hyperbolic fixed point of $\mathcal{N}$, and $\mathrm{D}\mathcal{N}_\phi$ has one simple eigenvalue $\phi > 1$, the rest of its spectrum lying strictly inside the unit disk.

Thus ϕ has spaces E^s, E^u of stable and unstable directions under $\mathcal{N}$,
which are tangent to stable and unstable manifolds W^s, W^u of codimension 1 and dimension 1, respectively.

(H3) W^u crosses (once and) transversely some piece of an interesting surface Σ_0 of codimension 1, defined by a coordinate independent property, at a point ϕ_0^*, say.

Then W^u also crosses (once and) transversely the surfaces $\Sigma_n = \mathcal{N}^{-n}\Sigma_0$, on
which f^{2^n} has the same property. The surfaces Σ_n pile up asymptotically
geometrically on W^s, at rate δ. For maps of the interval, for example, one
could consider the surface

$$\Sigma_0^s = \{f \in \mathbf{C} : f(1) = 0\},$$

for which there is a superstable 2-cycle. Then Σ_n^s is the surface for which
f has a superstable 2^{n+1}-cycle. The qualitative picture is as in fig. 9.

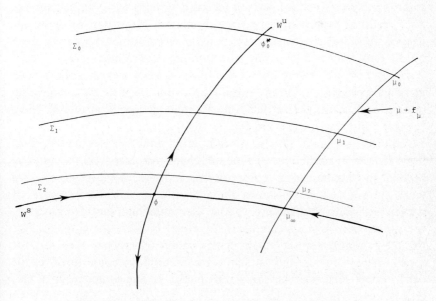

Fig. 9. The neighbourhood of the fixed point ϕ in the space $\mathbf{C}$.

3.3. Consequences

The significance of these hypotheses is that for one parameter families f_μ which cross W^s transversally at non-zero speed (at μ_∞, say), they lead to the following remarkable consequences (C1)–(C4), and further results to be given in section 3.6.

(C1) For all large enough n, f_μ will cross Σ_n transversally at parameter values μ_n, which will converge to μ_∞ asymptotically geometrically at rate δ, i.e.

$$\frac{\mu_n - \mu_{n-1}}{\mu_{n+1} - \mu_n} \to \delta \qquad \text{as } n \to \infty$$

The situation is illustrated in fig. 9.

Note that this will hold for any surface Σ_0 satisfying the hypotheses. For maps of an interval, one could take, as a second example, the primary *band merging surface*:

$$\Sigma_0^M = \{f \in \mathbf{C} : f^3(0) = f^4(0)\}$$

where the orbit of 0 falls on an unstable fixed point. Such maps possess an absolutely continuous invariant measure by Theorem 2.10. The surface Σ_0^M is called a band merging surface, because it forms part of the boundary of the set which exchanges intervals. The rest of the boundary is given by Σ_0^s (see section 2.3). Thus, for instance, the family $f_\mu(x) = 1 - \mu x^2$ possesses both an infinite period doubling sequence and an infinite band merging (perhaps "splitting" would be a more appropriate word) sequence, each converging to the same accumulation point at rate δ. These features can be seen clearly in fig. 7.

Although, speaking loosely, we call period doubling a route to chaos, there is not a continuous regime of chaos for maps of the interval beyond the accumulation point (except in a topological sense – see section 3.6.3), as for instance there is also an infinite sequence of parameter values, converging to the accumulation point from above, for which there is a superstable $3 \cdot 2^n$-cycle. In fact, for a typical one parameter family the set of parameter values (clearly an open set) for which there is a strongly stable periodic orbit is conjectured to be dense. Some such intervals are visible in fig. 7. The complement of this set, however, need not have zero measure, and has been shown to have positive measure for certain families (Jakobson 1981, Collet and Eckmann 1980). In fact, for these cases the band merging points are (one-sided) density points for the set of parameter values with absolutely continuous invariant measure.

The fixed point ϕ is self-similar:

$$\phi \circ \phi = \Lambda_\phi \circ \phi \circ \Lambda_\phi^{-1}.$$

Now for $f \in W^s$, $\mathcal{N}^n f \to \phi$ as $n \to \infty$. Thus the exact self-similarity of ϕ leads to asymptotic self-similarity of f_{μ_∞} with the same scaling Λ_ϕ. More precisely,

(C2) $(\mathcal{N}^n f_{\mu_\infty})^2 \sim \Lambda_\phi \circ \mathcal{N}^n f_{\mu_\infty} \circ \Lambda_\phi^{-1}$ for large n

and if the Λ_f commute for different f, this leads to:

$$f_{\mu_\infty}^{2^{n+1}} \sim \Lambda_\phi \circ f_{\mu_\infty}^{2^n} \circ \Lambda_\phi^{-1} \quad \text{on } \Lambda_\phi^n U \text{ for any } U \subset X, \text{ for large } n.$$

So, for example, the separation of the periodic points of period 2^n closest to 0, for a map of the interval at the accumulation of period doubling, decreases asymptotically geometrically with n, with ratio

$$\lambda = \phi(1) \sim -0.39953\ldots \quad (1/\lambda \sim -2.502908\ldots).$$

Also the derivative of the 2^n-cycles, being scale invariant, converges to some universal number $\phi'(x_0) \sim -1.601191\ldots$, where x_0 is the (unique) fixed point of ϕ on $[\phi(1), 1]$.

Furthermore, this self-similarity extends to the sequence f_{μ_n}, in the following sense:

(C3) If the Λ_f commute, then

$$f_{\mu_{n+1}}^{2^{n+1}} \sim \Lambda_\phi \circ f_{\mu_n}^{2^n} \circ \Lambda_\phi^{-1} \quad \text{on } \Lambda_\phi^n U \text{ for any } U, \text{ for large } n.$$

Thus, for example, the position $f_{\mu_n}^{2^n}(0)$, of the periodic point of the super-stable 2^{n+1}-cycle halfway round from 0, converges to 0 asymptotically geometrically with a ratio λ. Also, the derivative of the unstable 2^n cycle on which 0 lands at a band merging point converges as $n \to \infty$ to a universal number, corresponding to its value at ϕ_0^M, the intersection of Σ_0^M with W^u (Nierwetberg and Geisel 1981).

The proof of (C3) is not as simple as that of (C2). Although

$$\mathcal{N}^n f_{\mu_n} \to \phi_0^* \quad \text{as } n \to \infty$$

it is the scaling at ϕ rather than ϕ_0^* that appears in (C3). We can write

$$\mathcal{N}^n f_{\mu_n} = H_n^{-1} \circ f_{\mu_n}^{2^n} \circ H_n \quad \text{where } H_n = \Lambda_{f_{\mu_n}} \circ \Lambda_{\mathcal{N} f_{\mu_n}} \circ \ldots \circ \Lambda_{\mathcal{N}^{n-1} f_{\mu_n}}.$$

The product H_n gives the total coordinate change of interest. One can divide the orbits of f_{μ_n}, $f_{\mu_{n+1}}$ under $\mathcal{N}$ into three segments:

(i) an initial segment in which $\mathcal{N}^k f_{\mu_n}$, $\mathcal{N}^k f_{\mu_{n+1}}$ are close for large n, since they are initially close (to f_{μ_∞});

(ii) a final segment in which $\mathcal{N}^{k+1} f_{\mu_{n+1}}$, $\mathcal{N}^k f_{\mu_n}$ are close for large n, since they are close for $k = n$ (to ϕ_0^*);

(iii) an intermediate segment in which $\mathcal{N}^k f_{\mu_n}$, $\mathcal{N}^k f_{\mu_{n+1}}$ are close to ϕ.

The orbit $f_{\mu_{n+1}}, \ldots, \mathcal{N}^{n+1} f_{\mu_{n+1}}$ has one more point than $f_{\mu_n}, \ldots, \mathcal{N}^n f_{\mu_n}$, and it lies in the intermediate segment. Thus the coordinate changes H_n differ by a factor close to Λ_ϕ, leading to (C3).

Consequence (C3) can be extended to the whole one parameter family, not just its intersections with the surfaces Σ_n, as we shall now explain. The unstable manifold W^u can be parametrized in such a way as to make it a universal one parameter family, satisfying

$$\Lambda_{\phi_\mu}^{-1} \circ \phi_\mu \circ \phi_\mu \circ \Lambda_{\phi_\mu} = \phi_{\mu \cdot \delta}.$$

The parametrization is defined by this relation, plus the "initial conditions"

$$\phi_0 = \phi, \qquad \left.\frac{d\phi_\mu}{d\mu}\right|_0 = v$$

for v any nonzero vector in E^u. The choice of v sets the scale in parameter, and for maps on the interval we choose it in the direction towards Σ_0^M. Thus, for a one parameter family f_μ crossing W^s transversally at nonzero speed, without loss of generality at $\mu = 0$, we have

$$\mathcal{N}^n f_{\mu \cdot \delta^{-n}} \to \phi_{\mu/\mu_s} \qquad \text{as } n \to \infty,$$

where μ_s rescales the parameter appropriately. Hence, along the same lines as for (C3), we obtain:

(C4) If the Λ_f commute, then the whole one parameter family f_μ has asymptotic self-similarity:

$$f_{\mu/\delta}^{2^{n+1}} \sim \Lambda_\phi \circ f_\mu^{2^n} \circ \Lambda_\phi^{-1} \qquad \text{for } (x,\mu) \in \Lambda_\phi^n U \times \delta^{-n} I, \qquad \text{for large } n$$

where U is any subset of X, and I any subset of parameter values.

This is illustrated for the family $f_\mu(x) = 1 - \mu x^2$ in figs. 10 and 11, where the orbit of 0 is shown (leaving out an initial transient phase) for a range

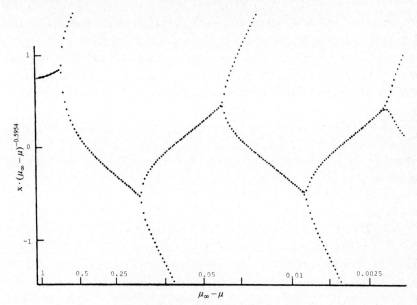

Fig. 10. Orbits of $x \to 1 - \mu x^2$ for $\mu < \mu_\infty$.

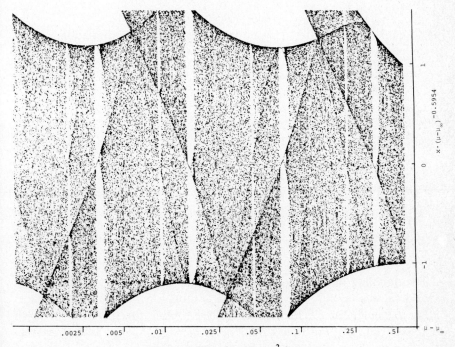

Fig. 11. Orbits of $x \mapsto 1 - \mu x^2$ for $\mu > \mu_\infty$.

of parameter values on either side of μ_∞. The μ-axis is logarithmic, and the x-axis has been expanded by a factor $|\mu - \mu_\infty|^\alpha$, where $\alpha \sim -0.595367$ is defined by $\delta^\alpha = |\lambda|$, so that the asymptotic self-similarity of f_μ in parameter and space is revealed as periodicity in $\log|\mu - \mu_\infty|$. Thus, for instance, if band mergings occur at μ_n^M and period doublings at μ_n then

$$\frac{\mu_n^M - \mu_{n+1}^M}{\mu_n - \mu_{n+1}}$$

converges to a universal limit, namely the value μ_0^M/μ_0 for the universal family ϕ_μ.

3.4. Comments

3.4.1. Renormalization

There is a close parallel between the theory of doubling operators and that of renormalization in Statistical Mechanics (Wilson and Kogut 1974). We give below a short dictionary for the correspondence between the terms used:

Doubling theory	renormalization
$\mathcal{N}$	renormalization group transformation
ϕ	critical fixed point
W^s	critical surface
E^u	relevant direction
W^u	relevant scaling field
λ	scaling
$\log \delta$	critical exponent
ϕ_0^*	scaling limit
top. entropy	order parameter

In Statistical Mechanics, however, $\mathcal{N}$ is typically a convolution operator rather than a composition, which makes the analysis a little easier.

3.4.2. Other operators

The doubling operator has no *a priori* connection with period doubling bifurcations, as W^u need not in general cross a period doubling surface. For example, one can define and analyse tripling operators, even though period tripling bifurcations cannot occur in maps of a one dimensional real space (period tripling of a periodic orbit requires a multiplier to pass through $e^{\pm 2\pi i/3}$, but for a one dimensional system a periodic orbit has only one multiplier and it is always real). One should not necessarily expect such operators to have fixed points, let alone hyperbolic fixed points with low dimensional unstable manifolds. We still lack a real understanding of why the doubling operator for maps on an interval has a fixed point with unstable manifold of dimension only one. The same appears to hold for the tripling operator for maps on an interval (Collet and Eckmann 1980a). For area preserving maps of a two dimensional surface, however the tripling operator appears to have a 2-cycle with unstable manifold of dimension only one (Greene et al. 1981). In general, regarding $\mathscr{N}$ as a dynamical system, one should be prepared for any behaviour from stable fixed points or periodic orbits to strange attractors.

3.5. *Typical difficulties in proving the hypotheses*

How to find a fixed point and check that its spectrum is as desired is outlined for some particular cases in section 4. In this section we indicate the problems involved in getting control over the stable and unstable manifolds:

(i) $\mathscr{N}$ is not invertible so the construction of the stable manifold W^s by the graph transform method (see Lanford's Course) might appear to fail. Nevertheless, Hirsch et al. (1977) show that the set theoretic inverse $\mathscr{N}^{-1}$ is sufficient to provide a differentiable manifold.

(ii) W^u must be extended far enough to check that it crosses surfaces Σ_0 of interest, and transversally.

(iii) Hartman's theorem (Hartman 1964) implies topological conjugacy of $\mathscr{N}$ in a neighbourhood of ϕ to its linearization $D\mathscr{N}_\phi$. In order to establish asymptotically geometric convergence of Σ_n to W^s, however, we require a smoother conjugacy. It suffices in fact to find differentiable coordinates (x, y) in a neighbourhood of ϕ in **F**, with $y \in \mathbb{R}$, such that

$$N(x, y) = (M(x, y), \delta \cdot y)$$

with

$$\|D_x M(x, y)\| \leq C < 1 \qquad \text{for some } C$$

and

$$M(0, y) = 0.$$

These questions have been settled in Collet et al. (1980).

3.6. Further consequences

We look at some further consequences of the existence of ϕ for the specific case of the doubling operator on maps of an interval.

3.6.1. Other period doubling sequences

Any one parameter family in **C** crossing W^s transversally at nonzero speed will give the same results, for example, f_μ^3 restricted to a neighbourhood of 0, for parameter values in a neighbourhood of a value for which f_μ has a stable 3-cycle. So one finds an infinite period doubling sequence with periods $3 \cdot 2^n$ etc. (see fig. 7). In particular there are other infinite period doubling sequences arbitrarily close to the accumulation point of any period doubling sequence.

3.6.2. Universal power spectrum

The scaling λ induces asymptotically universal self-similarity on the power spectrum of the orbit of 0 for a one parameter family f_μ. This was studied first by Feigenbaum (1979), and more recently by Nauenberg and Rudnick (1981) and Collet et al. (1981). In particular, for the fixed point ϕ, one defines the Fourier transform $a(q)$ of the orbit of 0 by

$$a(q) = \lim_{p \to \infty} \frac{1}{p} \sum_{j=0}^{p-1} e^{2\pi i j q} \phi^j(0) \qquad \text{for } q \in \mathbb{Z}/2^N.$$

This limit exists, as $\{\phi^j(0)\}$ is almost periodic. Next define the mean power A_N in the Nth subharmonics:

$$A_N = \frac{1}{2^{N-1}} \sum_{k=0}^{2^{N-1}-1} \left| a\left(\frac{2k+1}{2^N}\right) \right|^2.$$

Then one can establish the following bounds:

Theorem 3.1.

$$\left(\frac{\lambda^2}{4}\right)^M \leq \frac{A_{N+M}}{A_N} \leq C_M \left(\frac{\lambda^2}{4}\right)^M,$$

where $\lambda = \phi(1) = -0.3995$, $C_1 \leq 2.5$, $C_2 \leq 4.2$, $C_3 \leq 6.1$, $C_M \leq 1 + 14 \cdot (1.109)^M$.

Thus the mean power in successive subharmonics decreases asymptotically by between 13.45 and 14 db (i.e. a factor between 22.3 and 25.1). Figure 12 shows the power spectra for maps. Such a self-similarity in the power spectrum has been observed for real experiments on convection (see Libchaber's lectures), with a power decrease of around 10 db (fig. 13). By a more heuristic approach, Feigenbaum obtains an asymptotic mean

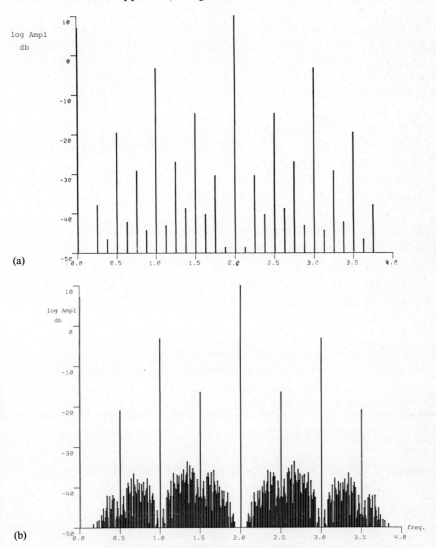

Fig. 12. Power spectra. (a) For the map ϕ; (b) for the map $1-\mu x^2$, $\mu = 1.407155 > \mu_\infty$.

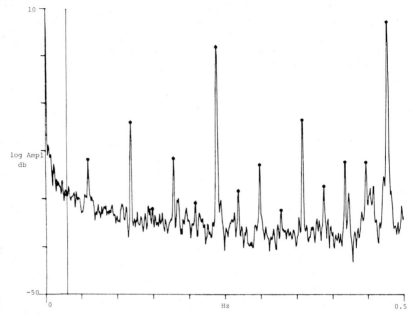

Fig. 13. Experimental power spectrum for liquid helium near the accumulation of period doubling (kindly provided by A. Libchaber). A division of the basic frequency by 16 is visible.

amplitude decrease of 8.17 db (i.e. power decrease of 16.3 db), and Nauenberg and Rudnick, a power decrease of 13.2 db, but the discrepancy remains to be resolved.

3.6.3. *Topological entropy and Liapunov exponents*

The self-similarity of ϕ_μ extends to measures of chaos, such as the topological entropy and the Liapunov exponent. For a map f on an interval, the topological entropy h_f is given by (see Misiurewicz's lectures):

$$h_f = \lim_{n \to \infty} \frac{1}{n} \log(\#\ \text{laps of}\ f^n).$$

where the laps of a map are its intervals of monotonicity. Then

$$h(f^n) = \lim_{n \to \infty} \frac{1}{n} \log(\#\ \text{laps of}\ f^{2n})$$

$$= 2 \lim_{n \to \infty} \frac{1}{2n} \log(\#\ \text{laps of}\ f^{2n}) = 2h(f).$$

This result can also be derived from kneading theory, as the topological entropy of a map depends only on its kneading sequence, and

$$h(R*K) = \tfrac{1}{2} h(K).$$

Finally,

$$K_f = R*K \;\Rightarrow\; K_{f^2} = K.$$

Thus the self-similarity of ϕ_μ implies that

$$h(\phi_{\mu/\delta}) = \tfrac{1}{2} h(\phi_\mu)$$

and so

$$h(\phi_\mu) = |\mu|^\beta U_\pm(\log |\mu|)$$

with

$$\beta = \frac{\log 2}{\log \delta} \sim 0.4498,$$

and $U_\pm$ universal functions for the "chaotic" and stable sides of the accumulation point, periodic with period $\log \delta$. The above result will hold asymptotically for μ near μ_∞ for a 1-parameter family f_μ, in the sense that:

$$h(f_\mu) \sim |\mu - \mu_\infty|^\beta U_\pm\left(\log \left|\frac{\mu - \mu_\infty}{\mu_s}\right|\right)$$

for some μ_s.

The following two hypotheses have not been checked, but are certainly true:

$$\exists \mu_1 < 0 \text{ and } n_1 \geq 0 \text{ s.t. } \mu \in [\mu_1, \delta \cdot \mu_1) \;\Rightarrow\; K_{\phi_\mu} < (R*)^{n_1} R^\infty, \qquad (h1)$$

$$\exists \mu_2 > 0 \text{ and } n_2 \geq 0 \text{ s.t. for some } \mu \in [\mu_2, \delta \cdot \mu_2)$$
$$\Rightarrow\; \phi_\mu \text{ has a } 3 \cdot 2^{n_2}\text{-cycle}. \qquad (h2)$$

In fact one suspects even more, that K_{ϕ_μ} is monotone, but in any case,

$$(h1) \;\Rightarrow\; h(\phi_\mu) = 0 \text{ on } [\mu_1, \delta \cdot \mu_1) \;\Rightarrow\; U_-(t) = 0,$$

$$(h2) \;\Rightarrow\; h(\phi_\mu) \geq 2^{-n_2} \log \tfrac{1}{2}(1 + \sqrt{5}) > 0 \text{ for some } \mu \in [\mu_2, \delta \cdot \mu_2)$$

$$\Rightarrow\; U_+(t) \geq C > 0,$$

for some constant C. The bound on the entropy is due to Misiurewicz and Szlenk (1980). Thus in a topological sense, period doubling gives a transition to chaos from none, though not necessarily in other senses, as noted in section 3.3. The entropy is shown in fig. 14, by Crutchfield, for the family $1 - \mu x^2$. Note that for the period doubling sequences of periodic orbits other than the fixed point, there is a topological chaos on both sides of the accumulation point. This does not contradict $U_-(t) = 0$, as for the restriction of the appropriate power of f_μ to the appropriate neighbourhood there is a positive entropy on one side only.

The local Liapunov exponent χ_f is defined for the orbit of 0, when the limit exists, by:

$$\chi_f = \lim_{n \to \infty} \frac{1}{n} \sum_{m=1}^{n} \log |f'[f^m(0)]| .$$

It measures how fast nearby orbits diverge or converge. A simple calculation shows that:

$$\chi_{f^2} = \lim_{n \to \infty} \frac{1}{n} \sum_{m=1}^{n} \log |f^{2\prime}[f^{2m}(0)]|$$

$$= 2 \lim_{n \to \infty} \frac{1}{2n} \sum_{m=2}^{2n+1} \log |f'[f^m(0)]| = 2\chi_f.$$

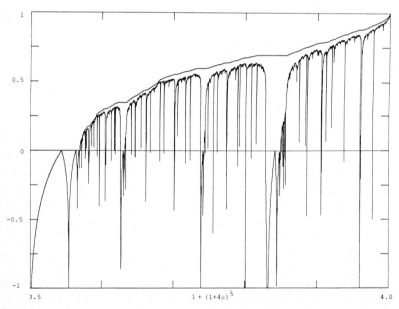

Fig. 14. Topological entropy (upper curve) and Liapunov exponent for $x \to 1 - \mu x^2$ as functions of μ (kindly provided by J. Crutchfield). Vertical scale is in units of log 2.

So we obtain

$$\chi_{\phi_\mu} = |\mu|^\beta V_\pm(\log|\mu|),$$

with $V_\pm$ universal periodic functions of period $\log\delta$. Now if the orbit of 0 falls on a periodic orbit of period p and derivative s, then

$$\chi = \frac{1}{p}\log s.$$

Thus, whenever ϕ_μ has a strongly stable periodic orbit, $\chi_{\phi_\mu} < 0$. Similarly, for marginally stable and superstable periodic orbits, $\chi = 0$, $-\infty$, respectively. Thus (h1) would imply that $V_-(t) \le 0$, oscillating between 0 and $-\infty$. For $\mu > 0$, there are parameter values for which $V_+ > 0$, for example when the orbit of 0 falls on an unstable periodic orbit, but there are also a set of intervals, believed to be dense, for which there exists a stable periodic orbit and so $V_+ \le 0$, as shown in fig. 14. Similar figures were obtained by Shaw (1978) and Feit (1978). For certain families, the set of parameter values for which the Liapunov exponent is positive has been shown to have positive measure (Jakobson 1981, Collet and Eckmann 1980b), and this is believed to be true for ϕ_μ also. It is much more delicate to decide whether the Liapunov exponent is the same for (almost) all orbits.

Note that the topological entropy is an upper bound for the Liapunov exponent.

3.6.4. *Effect of external noise*

External noise wipes out small scale structure, so one would expect to see only a finite number of period doublings. Two parallel investigations (Crutchfield et al. 1980, 1981, and Shraiman et al. 1981), however, have shown heuristically that the period doubling transition has universal behaviour with respect to noise too. This is best written as a self-similarity of the invariant distribution $v_{\mu,\varrho}(x)$, for a 1-parameter family f_μ perturbed by independent identically distributed noise with distribution ϱ:

$$|\lambda|\,v_{\mu\delta,\,\kappa\varrho\circ\kappa}(\lambda x) \sim v_{\mu,\varrho}(x),$$

where $\kappa \sim 6.619$ is a universal scaling for the noise. In other words, to see one more period doubling one must reduce the variance of the noise by a factor κ^2. Note that this is comparable to the power decrease factors suggested for the power spectrum.

4. The known examples where the doubling theory applies

4.1. Introduction

We will consider doubling operators $\mathcal{N}$ on spaces **F** of functions $f: X \to X$, defined by:

$$\mathcal{N}f = \Lambda_f^{-1} \circ f \circ f \circ \Lambda_f$$

with Λ_f a coordinate change depending on f. We want to look for $\phi \in \mathbf{F}$ such that $\mathcal{N}\phi = \phi$, $\mathcal{N}$ is $\mathscr{C}^2$ in a neighbourhood, and the spectrum of $D\mathcal{N}_\phi$ consists of a simple eigenvalue $\delta > 1$ plus a set strictly inside the unit disk. We want to show further that the unstable manifold W^u at ϕ intersects transversally some surface Σ_0 defined by a coordinate independent codimension 1 property.

This program has been carried out successfully in three cases of interest:

(a) maps of the interval of the form $x \to f(x) = h(|x|^{1+\varepsilon})$, h analytic, ε small and positive,

(b) maps of the interval with $f(x) = h(x^2)$, h analytic,

(c) analytic maps of $\mathbb{R}^n$.

Numerical work suggests that the doubling theory will also apply to:

(d) area preserving maps of a two dimensional surface.

We will sketch the results for these four function spaces.

4.2. Maps of the interval $x \to h(|x|^{1+\varepsilon})$, $\varepsilon > 0$, small

This is the first case for which rigorous results were proved. Specifically, for the doubling operator $\mathcal{N}$:

$$\mathcal{N}f(x) = \frac{1}{f(1)} f^2(f(1)x)$$

on the class $\mathbf{C}_\varepsilon$ of maps f with

$$f(x) = h(|x|^{1+\varepsilon}), \qquad h(t) \text{ analytic in } |t| < 3$$

$$h(0) = 1, \qquad h(1) > -1, \ h'(t) < 0 \text{ on } [0, 1],$$

Collet et al. (1980) proved the existence of a fixed point ϕ_ε, for all sufficiently small positive ε, given by:

$$\phi_\varepsilon(x) = 1 - (1 - \lambda_\varepsilon)|x|^{1+\varepsilon} + O(\varepsilon^2 \log \varepsilon),$$

$$\lambda_\varepsilon = \phi_\varepsilon(1) = \varepsilon \log \varepsilon + O(\varepsilon).$$

They showed that the spectrum of $D\mathcal{N}_{\phi_\varepsilon}$ lies inside a disk of radius small with ε, apart from an eigenvalue δ_ε which converges to 2 from above as $\varepsilon \to 0$. Furthermore, the unstable manifold W^u of ϕ_ε crosses transversally the surfaces Σ_0^s and Σ_0^M, for superstable 2-cycle and primary band-merging, respectively, and $\phi_\mu \in W^u$ has negative Schwarzian derivative for all μ.

One might ask why one should consider analytic functions h, rather than just $\mathscr{C}^2$, say. The reason is that composition incurs the loss of a derivative. Specifically, for $f \in \mathscr{C}^r$, the functional derivative $D\mathcal{N}_f$ belongs to $\mathscr{C}^{r-1}$, but not necessarily to $\mathscr{C}^r$. One can see this easily in the case of $r = 1$, because:

$$(f + \delta f) \circ (f + \delta f) = f \circ f + \delta f \circ f + f' \circ f \cdot \delta f + O(\delta f^2).$$

We outline briefly how the fixed point can be found. The doubling operator $\mathcal{N}$ on $f(x) = h(|x|^{1+\varepsilon})$ induces an operator $\mathscr{H}_\varepsilon$ on h:

$$\mathscr{H}_\varepsilon h(t) = \frac{1}{-a} h[|h(a^{1+\varepsilon}t)|^{1+\varepsilon}], \qquad \text{where } a = -h(1).$$

Asymptotic analysis indicates that

$$h(t) = 1 - (1 + \alpha)t,$$

is close to being a fixed point, where α is defined by:

$$\frac{-\alpha}{1 + \log \alpha} = \varepsilon.$$

Thus we write the general $h(t)$ as

$$h(t) = 1 - t + \alpha t [g(t) - 1]$$

and expect to find a solution with $g(t)$ close to zero. $\mathscr{H}_\varepsilon$ induces another operator $\mathcal{N}_\varepsilon$ on g, which can be shown to decompose into a linear part $\mathcal{N}_\varepsilon^0$, and a remainder n_ε which is small with ε for g in a neighbourhood of the zero function:

$$\mathcal{N}_\varepsilon g = \mathcal{N}_\varepsilon^0 g + n_\varepsilon g, \qquad \mathcal{N}_\varepsilon^0 g(t) = g(0) + g(1) + g'(1).$$

Note that $\mathcal{N}_\varepsilon^0$ is not only linear, but also has rank 1, as its range is the constant functions. On the space of constant functions, $\mathcal{N}_\varepsilon^0$ is multiplication by 2, as

$$g = k, \text{ constant } \Rightarrow g(0) = g(1) = k, \qquad \text{and } g'(1) = 0.$$

Thus for all small enough ε, $D\mathcal{N}_\varepsilon$ is hyperbolic with one simple eigenvalue δ near 2, and all the rest near 0.

To establish the existence of a unique fixed point g_ε of $\mathcal{N}_\varepsilon$ in some neighbourhood of 0, for small enough ε, we note that it has the same fixed points as $\mathcal{T}$:

$$\mathcal{T}g = (I - \mathcal{N}_\varepsilon^0)^{-1} n_\varepsilon g.$$

As $\mathcal{N}_\varepsilon^0$ is hyperbolic, $I - \mathcal{N}_\varepsilon^0$ is invertible (in fact it is its own inverse, since $\mathcal{N}_\varepsilon^{0^2} = 2\mathcal{N}_\varepsilon^0$), and n_ε, Dn_ε are small, the map $\mathcal{T}$ is a contraction on some neighbourhood of the zero function for ε sufficiently small, and so it has a unique fixed point there. Finally, one has to extend the unstable manifold W^u (given approximately by $g = $ constant functions) far enough to check that it intersects transversally codimension 1 surfaces of interest.

Although non-integer values of ε might appear to have only academic interest, they occur frequently, for example in Poincaré maps for families of 3-dimensional flows in the neighbourhood of parameter values for which there is a homoclinic orbit (Tresser 1981), and δ_ε has been found numerically, as a function of ε.

4.3. Analytic maps on an interval

4.3.1. Existence of a fixed point
Any analytic map with a critical point can be made even about that point by an analytic coordinate change. Thus we restrict our attention to the space of even analytic maps. A fixed point of the doubling operator $\mathcal{N}$ on this space has been found by two different methods. Lanford (1981) uses coordinates $(u, v_1, v_2, \ldots)$ defined by

$$f(x) = 1 - x^2 h(x^2)$$

$$h(t) = u/10 + \sum_{n=1}^{\infty} v_n \left(\frac{t-1}{2.5}\right)^n$$

and the norm

$$\|f\| = |u| + \sum_{n=1}^{\infty} |v_n|.$$

He finds numerically a polynomial map (of high degree) which is close to being fixed under $\mathcal{N}$, and established bounds on the derivative $D\mathcal{N}_f$ in a neighbourhood. This is sufficient to show that the map

$$f \to f - \mathcal{J}(\mathcal{N}f - f)$$

where

$$\mathcal{J}(u, v) = \left(\frac{u}{3.669}, -v\right)$$

has an invariant domain on which it is a contraction. Thus it has a fixed point which evidently satisfies $\mathcal{N}f = f$. The linear operator $\mathcal{J}$ above should be thought of as an approximation to

$$(D\mathcal{N}_f - I)^{-1}$$

which would be Newton's method.

An alternative approach has been developed by Campanino et al. (1980). Instead of using $\mathcal{N}$, they look for a scaling constant λ such that the "constant scaling" doubling operator

$$\mathcal{N}_\lambda^0 f(x) = \frac{1}{\lambda} f[f(\lambda x)]$$

has a (non-trivial) fixed point in the space of functions f with

f even and analytic,
$f(0) = 1$,
$f''(x) < 0$, for $x \in [0, 1]$.

As f is expected to exchange intervals, they rephrase the problem as follows: given f_1 on an interval $[b, 1]$, define f_2 on an interval $[-a, a]$ as the solution (guaranteed to exist, for suitable f_1) of

$$f_1 \circ f_2(\lambda x) = \lambda f_2(x).$$

Then the function f given by f_1 and f_2 on the appropriate intervals is a fixed point of $\mathcal{N}_\lambda^0$. The only problem is that they may fail to join smoothly. In the context of analytic maps this requires $f_1 = f_2$. Note that this leads to a solubility condition

$$f_1(1) = \lambda$$

so to postpone this constraint they replace the definition of f_2 by:

$$f_1 \circ f_2(\lambda x) - \lambda f_2(x) + f_1(1) - \lambda.$$

They find a contraction operator on some set giving a solution $f_1 = f_2$, depending continuously on λ, and show that $f_1(1) - \lambda$ is positive for some value of λ and negative for another, so that it is zero somewhere in between, giving a fixed point of $\mathcal{N}$.

One should note that there are believed to be other fixed points $\phi^{(m)}$ of $\mathcal{N}$, which play the same role as ϕ in the spaces $\{f(x) = h(x^{2m}),\ h\ \text{analytic}\}$, $m \geq 2$. In the space $\{f(x) = h(x^2),\ h\ \text{analytic}\}$, they would have unstable manifolds of dimension m. Campanino et al. also give an example of another fixed point with quadratic maximum.

4.3.2. *Analytic continuation of the fixed point*

The fixed point f is analytic in a disk of radius $8^{1/2}$ around 0, and real on $[-1, 1]$, and hence extends to a real analytic function on the whole real axis, by

$$\phi(x) = \frac{1}{\lambda} \phi \circ \phi(\lambda x) \tag{1}$$

and also on the imaginary axis by evenness. Figure 15 shows some of the

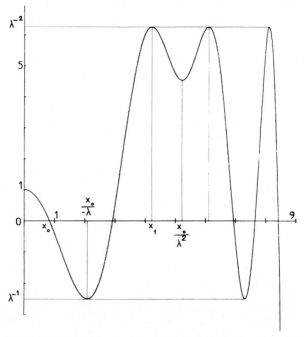

Fig. 15. The fixed point function ϕ extended outside $[-1, 1]$ (kindly provided by H. Epstein).

continuation of ϕ along $\mathbb{R}$. It cannot be continued arbitrarily into the complex plane, however. For example, Lanford (1981) shows numerically that

$$\exists c \neq \bar{c} \text{ s.t. } \phi(\lambda c) = \bar{c}.$$

Now if ϕ could be continued to c then

$$\bar{\phi}(c) = \phi(\bar{c}) = \phi \circ \phi(\lambda c) = \lambda \phi(c).$$

But $|\lambda| \neq 1$, so that $\phi(c)$ must be 0. Analyticity of ϕ in a neighbourhood of c would imply that

$$\phi(z) = (z-c)^n [a + (z-c)r(z)]$$

for some $a \neq 0$, $n \geq 1$, r analytic in a neighbourhood of c. But if ϕ satisfies eq. (1) above, then a quick calculation shows that

$$|\lambda \phi'(\lambda c)|^n = \lambda$$

and numerical evidence shows this to be false for all n. Hence ϕ cannot be continued to c.

In fact, Epstein and Lascoux (1981) show that ϕ has nonisolated singularities. They consider the analytic continuation of the inverse function u to ϕ, defined initially on an interval of monotonicity. They find that u can be analytically continued to the (open) upper and lower half-planes $\Pi_\pm$, but has square root singularities on $\mathbb{R}$ (corresponding to nondegenerate critical points of ϕ), giving rise to many branches u_j. The images $u_j(\Pi_\pm)$ are disjoint and bounded. Because of the square root singularities of u on $\mathbb{R}$, the images $u_j(\mathbb{R} \pm i0)$ consist of a doubly infinite sequence of arcs, each starting perpendicularly to the previous one, and each arc belongs to two of the $u_j(\mathbb{R} \pm i0)$. So the boundary of each $u_j(\Pi_\pm)$ consists of such a doubly infinite sequence of arcs, plus their common accumulation point $u_j(\pm i\infty)$. ϕ can be analytically continued to each $u_j(\Pi_\pm^c)$ ($\Pi_\pm^c$ are the closed half-planes), but is singular at $u_j(\pm i\infty)$ (note that c is one such point). Thus

$$D = \bigcup_{j, \pm} u_j(\Pi_\pm^c)$$

is a domain of analyticity for ϕ. But each of the singular points $u_j(\pm i\infty)$ is a limit point of others, and they are dense in the boundary ∂D is called a natural boundary (Rudin 1966). (See fig. 16.)

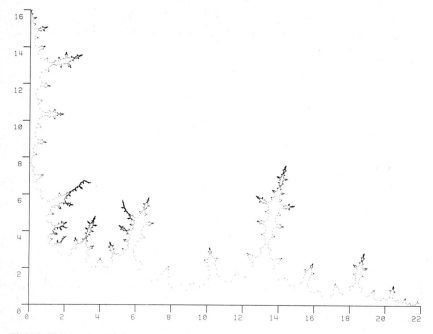

Fig. 16. 35 000 points of the natural boundary of ϕ in the first quadrant (computed according to Epstein and Lascoux 1981).

4.4. Maps of $\mathbb{R}^n$

4.4.1. Introduction

So far we have concentrated on maps of an interval. But infinite period doubling sequences with the same $\delta \sim 4.669$ have been observed numerically in higher dimensional maps and flows, and even experimentally in convection. The study of flows can often be converted to that of maps. In particular, in a neighbourhood of a periodic orbit of a flow one can take a surface of section transverse to the flow, and consider the first return map on this surface. The analysis of this map will give a lot of the essential information about the flow in the neighbourhood of the periodic orbit. Thus we consider maps on $\mathbb{R}^n$ and hope to find a doubling in $\mathbb{R}^n$ we go back to an alternative analysis of the one dimensional case.

4.4.2. Constant scaling in one dimension

The constant scaling doubling operator $\mathcal{N}^0$ (of section 4.3.1), with λ fixed at $\phi(1) \sim -0.3995$, can be applied to the space of all analytic functions. Then ϕ, although a fixed point, no longer has unstable manifold of dimension only 1. In fact,

Proposition 4.1. For $\sigma(x) = x^n$, $n \geq 0$,

$$\phi_\sigma(x) = -\sigma(\phi(x)) + \phi'(x)\,\sigma(x)$$

is an eigenvector of $D\mathcal{N}_\phi^0$ with eigenvalue λ^{n-1}.

The ϕ_σ correspond to infinitesimal coordinate changes S_t:

$$S_t: x \to x + t\sigma(x),$$

$$\phi_\sigma = \partial_t(S_t^{-1}\phi S_t)|_{t=0}.$$

Thus, in particular, $D\mathcal{N}_\phi^0$ has eigenvalues λ^{-1} and 1, which are not inside the unit disk, corresponding to shift of origin and change of scale, respectively. In fact, $\alpha^{-1}\phi(\alpha x)$ is also a fixed point, for any $\alpha \neq 0$. $D\mathcal{N}_\phi^0$ also has the eigenvalue δ, though its eigenvector $\varrho = r(x^2)$ is not necessarily the same as for $D\mathcal{N}_\phi$. These eigenvalues are simple and all others lie inside the unit disk. Note, however, that

$$\{\phi_\sigma: \sigma(x) = x^n,\ n \geq 0\} \cup \{\varrho\}$$

does not span the tangent space at ϕ. In fact, $\{\phi_\sigma: \sigma \text{ analytic}\}$ has infinite codimension. This can be seen because, being coordinate changes, they do not change the slope of any of the 2^n-cycles of ϕ.

The eigenvalues $\lambda^{-1}, 1$ can be moved inside the unit circle by including in the coordinate change Λ_f an appropriate shift of origin and change of scale. For example, our choice $\Lambda_f = f(1)$ achieves this for the eigenvalue 1, while Λ^{-1} was absent in the spectrum of $D\mathcal{N}_f$ because we restricted attention to even functions, prohibiting shifts.

4.4.3. Higher dimensions

Collet et al. (1981) define a doubling operator $\mathcal{N}^0$ on real analytic functions on a neighbourhood in $\mathbb{R}^n = \mathbb{R} \times \mathbb{R}^{n-1}$ of $[-1, 1] \times \{0\}$, by

$$\mathcal{N}^0\Psi = \Lambda^{-1} \circ \Psi \circ \Psi \circ \Lambda$$

$$\Lambda(x, y) = (\lambda x, \lambda^2 y), \qquad (x, y) \in \mathbb{R} \times \mathbb{R}^{n-1}.$$

Then

$$\Phi(x, y) = \left(\frac{1}{\alpha} g[\alpha^2 \zeta(x, y)], \mathbf{0}\right)$$

$$\zeta(x, y) = x^2 - \beta \cdot y$$

is a fixed point of $\mathcal{N}^0$ for any $\alpha, \beta \in \mathbb{R} \setminus \{0\} \times \mathbb{R}^{n-1} \setminus \{\mathbf{0}\}$, where g is defined by $\phi(x) = g(x^2)$, the 1-dimensional fixed point. The only eigenvalues of $D\mathcal{N}_\Phi^0$ not inside the unit circle are δ, λ^{-1}, λ^{-1}, 1, with spectral subspaces of dimensions 1, $n-1$, n, $n^2 - n + 1$, respectively. The eigenvalue δ has eigenvector $(r \circ \zeta, \mathbf{0})$, and the others correspond to coordinate changes Φ_σ with $\Lambda^{-1} \sigma \Lambda = \lambda^k \sigma$, $k = -2, -1, 0$. The unstable coordinate changes can be eliminated by projection as in the one dimensional case.

Lanford (private communication) has a different approach, with a coordinate change Λ_f that depends on f. It leads to a fixed point

$$\Phi(x, y) = [\phi(x), \mathbf{0}].$$

At the fixed point the scaling is

$$\Lambda_\Phi(x, y) = (\lambda x, y)$$

showing that the secondary scaling constant is somewhat arbitrary. This is not surprising as the fixed point maps the whole space onto a line, so there is no rescaling going on in any other direction. The fixed point Φ has other equivalent forms, obtainable by coordinate change. Typically coordinate changes will make the line onto which the space contracts curved and locally parabolic. Thus a scaling λ along the line will appear as a scaling λ in a direction tangent to the line, and λ^2 in all other directions. λ^2 is not an essential scaling, however, but generically it will occur.

The fixed point Φ and its unstable manifold Φ_μ in Collet, Eckmann and Koch's version, has Det $D\Phi_\mu = 0$. After one iteration it is just the 1-dimensional map ϕ_μ on a line, which is two to one in places. For invertible families of maps f_μ of $\mathbb{R}^n$, however, $f_\mu^{2^n}$ is invertible for all n, so it can never possess an invariant line on which it is two to one. The non-wandering set (when not just some periodic orbits) may be highly folded. It is not clear to what extent the 1-dimensional dynamics will carry over, however, the universal constant $\delta = 4.669...$ will certainly play an important role.

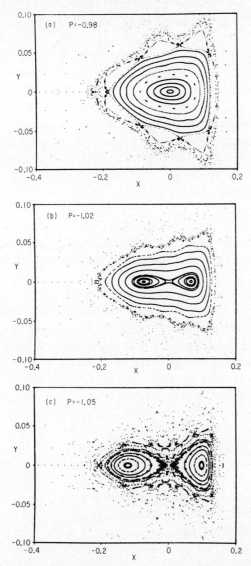

Fig. 17. Doubling of the phase space picture as a parameter *P* is varied in a family of area preserving maps. (From MacKay 1981.)

4.5. *Area preserving maps*

Finally we mention results for a simple class of conservative systems, viz.

two dimensional area preserving maps. Many physical systems, e.g. solar system, proton storage rings, charged particles in electromagnetic waves, are close to conservative. Here we use the term in its strongest sense, i.e. symplectic (Hamiltonian), though it is often used for the weaker properties of volume or measure preservation. As the fixed point Φ above for general maps of $\mathbb{R}^2$ is highly dissipative (in fact Det $D\Phi = 0$), we do not expect it to be of any relevance to conservative systems. One parameter families of area preserving maps, however, have been found to possess infinite period doubling sequences, but with a different δ ($\sim 8.72\ldots$), and scalings $1/\alpha$, $1/\beta$ in two directions, with $\alpha \sim -4.018$ and $\beta \sim 16.36$ (note that $\beta \neq \alpha^2$). The existence of a fixed point of

$$\mathcal{N}^0 f = \Lambda^{-1} f^2 \Lambda, \qquad \Lambda(x, y) = (x/\alpha, y/\beta)$$

with low dimensional unstable manifold in the space of area preserving maps, has been conjectured (Collet et al. 1980, and Greene et al. 1981) but not yet proved.

Unlike the general case, periodic orbits of conservative systems can never be attracting, by volume preservation. Elliptic periodic orbits (all multipliers on the unit circle), however, typically impose a significant degree of stability on a neighbourhood in the forms of a nested family of invariant tori on which the motion is quasiperiodic (i.e. equivalent to a linear flow). Such tori can be found arbitrarily close to the periodic orbit. In fact, for small tubes around the periodic orbit, the fraction of the volume occupied by invariant tori is arbitrarily close to 1. For non-elliptic periodic orbits, there is a neighbourhood in which there are no such tori. As elliptic points turn non-elliptic on period doubling, it is not surprising that period doubling sequences in area preserving maps lead locally to destruction of invariant tori (fig. 17). The fixed point f_0 is believed to have no invariant tori at all. Beyond the accumulation point, however, are small windows in parameter in which there is once more an elliptic periodic orbit (c.f. 1-dimensional maps) (MacKay 1981). An interesting question is whether there is similar universal behaviour in higher dimensional conservative systems.

References

Abraham, R. and J. Marsden (1978). Foundations of Mechanics (Benjamin, New York)
Brolin, H. (1965) Arkiv för Matematik 6, 103.

Campanino, M., H. Epstein and D. Ruelle (1980) On Feigenbaum's functional equation,
preprint IHES/P/80/32, to appear in Topology.

Campanino, M. and H. Epstein (1981) Commun. Math. Phys. 79, 261.

Chenciner, A. and G. Iooss (1979). Arch. Rat. Mech. Anal. 69, 109; 71, 301.

Collet, P. and J.-P. Eckmann (1980a) Iterated Maps on the Interval as Dynamical Systems
(Birkhäuser, Boston).

Collet, P. and J.-P. Eckmann (1980b) in: Nonlinear Dynamics, New York, 1979, ed., R.H.G.
Helleman; Ann. N.Y. Acad. Sci. 357, 377; Commun. Math. Phys. 73, 115.

Collet, P., J.-P. Eckmann and H. Koch (1981) On universality for area preserving maps of
the plane, Physica 3D, 457.

Collet, P., J.-P. Eckmann and H. Koch (1981) J. Stat. Phys. 25, 1.

Collet, P., J.-P. Eckmann and O.E. Lanford III (1980) Commun. Math. Phys. 76, 211.

Collet, P., J.-P. Eckmann and L. Thomas (1981) A note on the power spectrum of the iterates
of Feigenbaum's function, preprint, Geneva.

Coullet, P. and C. Tresser (1978) C.R. Acad. Sci. Paris 287, 577; J. de Phys. 39, Colloq.
C5-25.

Curry, J.H. and J.A. Yorke (1978) Lecture Notes in Mathematics, Vol. 668 (Springer, Berlin)
p. 48.

Crutchfield, J.P., J.D. Farmer and B.A. Huberman (1980) Fluctuations and simple chaotic
dynamics, reprint, UCSC.

Crutchfield, J.P., M. Nauenberg and J. Rudnick (1981) Phys. Rev. Lett. 46, 933.

Derrida, B., A. Gervois and Y. Pomeau (1979) J. Phys. A:12, 269.

Eckmann, J.-P. (1981) Roads to turbulence in dissipative dynamical systems, Rev. Mod.
Phys., in print.

Eckmann, J.-P., L. Thomas and P. Wittwer (1981) Commun. Math. Phys. 81, 261.

Epstein, H. and J. Lascoux (1981) Analyticity properties of the Feigenbaum function, preprint IHES/P/81/27.

Feigenbaum, M.J. (1978) J. Stat. Phys. 19, 25, 21 (1979) 669.

Feigenbaum, M.J. (1979) Phys. Lett. 74A, 375; Commun. Math. Phys. 77 (1980) 65.

Feit, S. (1978) Commun. Math. Phys. 61, 249.

Greene, J.M., R.S. MacKay, F. Vivaldi and M.J. Feigenbaum (1981) Universal behaviour in
families of area preserving maps, Physica 3D, 468.

Guckenheimer, J. (1977) Inv. Math. 39, 165; Bifurcations of dynamical systems, in: Dynamical Systems, Bressanone, CIME, 1978, ed., J. Guckenheimer (Birkhäuser, Boston, 1980);
Commun. Math. Phys. 70 (1979) 133.

Hartman, P. (1964) Ordinary Differential Equations (Wiley, New York).

Hirsch, M.W., C.C. Pugh, and M. Shub (1977) Lecture Notes in Math., Vol. 583 (Springer,
Berlin).

Iooss, G. and W.F. Langford (1980) Routes to turbulence, in: Nonlinear Dynamics (New
York, 1979) ed., R.H.G. Helleman, Ann. N.Y. Acad. Sci. 357, 489.

Jakobson, M.V. (1981) Commun. Math. Phys. 81, 39.

Kifer, J.I. (1974) Math. USSR Izv. 8, 1083.

Landau, L.D. and E.M. Lifschitz (1957) Fluid Mechanics (Pergamon, Oxford) III, section 103.

Lanford III, O.E. (1980) Lecture Notes in Phys., Vol. 116 (Springer, Berlin) p. 340; A computer assisted proof of the Feigenbaum conjectures, preprint IHES/P/81/17 (1981).

Lanford III, O.E. (1981) to appear.

Li, T. and J.A. Yorke (1975) Am. Math. Monthly 82, 985.

MacKay, R.S. (1981) Islets of stability beyond period doubling, preprint, Princeton PPPL-1855.

May, R.M. (1976) Nature 261, 259.

Mayer-Kress, G. and H. Haken (1981) Phys. Lett. 82A, 151; J. Stat. Phys., to appear.

Metropolis, M., M.L. Stein and P.R. Stein (1973) J. Comb. Theory 15, 25.

Milnor, J. and P. Thurston (1977) On iterated maps of the interval I, II, preprint, Princeton.

Misiurewicz, M. (1978) Structure of mappings of the interval with zero entropy preprint IHES/M/78/249.

Misiurewicz, M. and W. Szlenk (1980) Studia Math. 67, 45.

Nauenberg, M. and J. Rudnick (1981) Universality and the power spectrum at the onset of chaos, preprint, UCSC.

Newhouse, S.E. (1980) Lectures on dynamical systems, in: Dynamical Systems, Bressanone, CIME, 1978, ed., J. Guckenheimer (Birkhäuser, Boston); Publ. Math. IHES 50 (1979) 101.

Newhouse, S.E., D. Ruelle and F. Takens (1978) Commun. Math. Phys. 64, 35.

Nierwetberg, J. and T. Geisel (1981) preprint.

Rudin, W. (1966) Real and Complex Analysis (McGraw-Hill, New York).

Ruelle, D. (1981) Small random perturbations of dynamical systems and the definition of attractors, preprint IHES/P/81/23.

Ruelle, D. and F. Takens (1971) Commun. Math. Phys. 20, 167.

Sarkovskii, A.N. (1964) Ukr. Mat. Zh. 16, 61 (in Russian).

Shaw, R. (1978) Strange attractors, chaotic behaviour and information flow, Z. Naturforsch., in print.

Shraiman, B., C.E. Wayne and P.C. Martin (1981) Phys. Rev. Lett. 46, 935.

Smale, S. (1966) Am. J. Math. 88, 491.

Smale, S. (1967) Bull. AMS 73, 747.

Tresser, C. (1981) to appear.

Wilson, K. and J. Kogut (1974) Phys. Rep. 12C, 75.

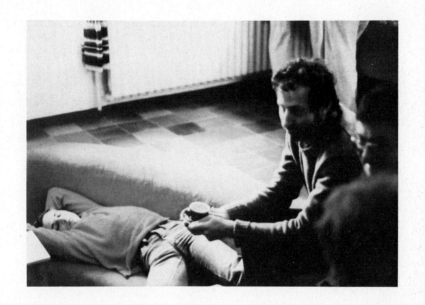

COURSE 8

NONLINEAR PROBLEMS IN ECOLOGY AND RESOURCE MANAGEMENT

Robert M. MAY

*Biology Department, Princeton University,
Princeton, NJ 08544, U.S.A.*

G. Iooss, R.H.G. Helleman and R. Stora, eds.
Les Houches, Session XXXVI, 1981 – Comportement Chaotique des Systèmes Déterministes/
Chaotic Behaviour of Deterministic Systems
© *North-Holland Publishing Company, 1983*

Contents

1. Introduction

Other Courses in this volume on "Chaotic Behaviour of Deterministic Systems" have analyzed the complicated dynamical behaviour that can arise in simple deterministic systems. Applications of this work to a variety of physical systems have also been discussed.

My lectures are focused on possible applications of this work to problems that arise in describing the dynamical behaviour of biological populations. It is appropriate that biological examples should intrude into this (previously physical) summer school, as much of the early impetus for current research on "maps of the interval" comes from biology [1, 2]. Indeed, I believe this to be the first instance when non-trivial mathematical developments have been borrowed by physics from biology; the converse process is much more common!

The lecture notes are organized as follows.

Section 2 discusses 1-dimensional difference equations, with one "hump" or critical point, which arise as models for biological populations with discrete, non-overlapping generations. The properties of such equations have been extensively reviewed in earlier Courses here, and elsewhere [3–6]; the present exposition is largely confined to presenting a simple analytic approximation to the "Feigenbaum ratio". A series of biological studies of single populations, using 1-dimensional difference equations, is then catalogued and discussed (with particular attention to estimating the commonness of cycles and chaos in natural systems).

Section 3 outlines the mathematical properties of 1-dimensional maps with 2 critical points. This work is relevant to the dynamics of "gene frequencies" when selective forces are themselves dependent on the proportions in which genotypes are present in the population ("frequency dependent selection"). Much of the material presented here has not previously been published.

Section 4 sketches the additional complications that arise in dealing with coupled systems of difference equations. This section is more a guide to further reading than a self-contained exposition.

Section 5 examines some of the first-order differential–delay equations that arise in specific contexts in population biology. The complex dynamical behaviour of these systems is revealed in numerical studies; some rough analytic understanding and some unsolved questions are also discussed.

Section 6 gives a brief indication of the further richness of dynamical behaviour that can lurk in systems of coupled first-order difference equations. As in section 4, this section serves mainly as a guide to further reading.

2. One-dimensional difference equations, with one critical point

2.1. Motivation

One of the simplest systems an ecologist can study is a seasonally breeding population in which generations do not overlap. Many natural populations, particularly among temperate zone insects (including many economically important crop and orchard pests), are of this kind. In this situation, the observational data will usually consist of information about the maximum, or the average, or the total population in each generation. The theoretician seeks to understand how the magnitude of the population in generation $t + 1$, X_{t+1}, is related to the magnitude of the population in the preceding generation t, X_t: such a relationship may be expressed in the general form

$$X_{t+1} = F(X_t). \tag{2.1}$$

The function $F(x)$ will usually be what a biologist calls "density dependent", and a mathematician calls nonlinear; eq. (2.1) is then a first-order, nonlinear difference equation.

Although I shall henceforth adopt the habit of referring to the variable X as "the population", there are countless situations outside population biology where the basic eq. (2.1) applies. There are other examples in biology, as, for example, in genetics (where the equation describes the change in gene frequence in time) or in epidemiology (with X the fraction of the population infected at time t). Examples in economics include models for the relationship between commodity quantity and price, for the theory of business cycles, and for the temporal sequences generated by various other economic quantities. The general eq. (2.1) also is germane to the social sciences, where it arises, for example, in theories of learning (where X may be the number of bits of information that can be remembered after an interval t), or in the propagation of rumours in variously structured societies (where X is the number of people to have hear the rumour after time t). References to these different kinds of examples are given in the review article by May [3].

In many of these contexts, and for biological populations in particular, there is a tendency for the variable X to increase from one generation to the next when it is small, and for it to decrease when it is large. That is, the nonlinear function $F(X)$ often has the following properties: $F(0) = 0$; $F(X)$ increases monotonically as X increases through the range $0 < X < A$ [with $F(X)$ attaining its maximum value at $X = A$]; and $F(X)$ decreases monotonically as X increases beyond $X = A$. Moreover, $F(X)$ will usually contain one or more parameters which "tune" the severity of this non-linear behaviour; parameters which tune the steepness of the hump in the $F(X)$ curve. These parameters will typically have some biological or economic or sociological significance.

A list of specific expressions for $F(X)$ that have been the subject of studies by biologists is given in table 1 [notice the functions are tabulated as $F(X)/X$]. The form labelled I has been sometimes thought of as the

Table 1
Some first-order difference equations that have been propounded in various biological contexts

Equation number	$F(X)/X$	Ref.
I	$1 + r(1 - X/K)$	Maynard Smith [7]; May [8]; Krebs [9]; Li and Yorke [1]
II	$\exp[r(1 - X/K)]$	Moran [10]; Ricker [11]; Cook [12]; May [13]
III	λX^{-b}	Varley and Gradwell [14]; Stubbs [15]
IV	$\lambda[1 + \alpha X]^{-b}$	Hassell [16]; Hassell et al. [17]
V	$\lambda[1 + (\alpha X)^b]^{-1}$	Maynard Smith and Slatkin [18]; Bellows [19]
VI	$\exp[r(1 - \{X/K\}^\theta)]$	Bellows [19]; Thomas et al. [20]
VII	$\lambda[1 + \alpha X]^{-1}$	Utida [21]; Skellam [22]; Leslie [23]
VIII	λ_+; if $X < K$ λ_-; if $X > K$	Williamson [24]
IX	$\lambda[1 + \exp(BX - A)]^{-1}$	Ullyett [25]; Pennycuik et al. [26]; Usher [27]

discrete analogue of the biologists' "logistic equation", and it can be rescaled into the canonical form

$$X_{t+1} = aX_t(1 - X_t). \tag{2.2}$$

An equivalent form is $Y_{t+1} = 1 - bY_t^2$, where $b = a(a-2)/4$. Among the other forms for $F(X)$, II has been used in the analysis of important data pertaining to insect and fish populations by Moran and Ricker, respectively, and III is the basis for "k-factor analysis" in pest control. The forms III, IV, V and VI have been used in analyses of synoptic collections of data on insect populations by Stubbs, Hassell et al., Bellows and Thomas et al., respectively; these analyses are discussed below.

2.2. Properties

2.2.1. General remarks

All the maps catalogued in table 1 are "generically quadratic" in the sense that – like eq. (2.2) – they have one critical point and negative Schwarzian derivative. Although the properties of such maps have been fully explicated in earlier lectures in this series, some brief recapitulation is necessary to lay a foundation for the later sections.

If the hump is not too steep [in eq. (2.2), $3 > a > 1$], then the fixed point of eq. (2.1) is stable. In geometrical terms [3–6, 28], stability follows as long as the slope $\lambda^{(1)}$ of the map $F(X)$ at the fixed point X^* [the point where $X^* = F(X^*)$] lies between $+1$ and -1; this is illustrated generally in fig. 1.

As the hump steepens, the fixed point becomes unstable, giving rise by successive "pitchfork" bifurcations to a *cascade* of stable cycles with periods $2, 4, 8, \ldots, 2^n$. This "period doubling" phenomenon can be understood by a simple geometrical argument, which follows. Consider the "period 2" mapping $X_{t+2} = F^{(2)}(X_t)$, which is obtained by composing $F(X)$ with itself (see the general situation depicted in fig. 1). As can easily be seen by application of the chain rule for differentiation, the slope of this period 2 mapping, $F^{(2)}$, at the fixed point X^* is the square of the slope of $F(X^*)$: $\lambda^{(2)} = [\lambda^{(1)}]^2$. Thus, as $\lambda^{(1)}$ steepens beyond -1, not only does the fixed point become unstable, but the slope of the period 2 map at X^* exceeds $+1$, and the map crosses the 45° line to give two new fixed points of period 2, corresponding to an initially stable cycle of period 2. As shown in fig. 1, this process is generic, causing a cycle of period k to cascade

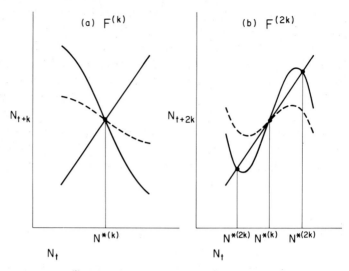

Fig. 1. (a) The map $F^{(k)}$ (relating X_{t+k} to X_t) is shown in the neighbourhood of one of its period-k fixed points, $X^{*(k)}$. For the dashed curve, the slope of $F^{(k)}$ at its intersection with the 45° line has $|\lambda^{(k)}| < 1$, and the point is stable; for the solid curve, $|\lambda^{(k)}| > 1$, and the fixed point is unstable. (b) The corresponding map $F^{(2k)}$ (relating X_{t+2k} to X_t) is shown in the neighbourhood of the same period-k fixed point, $X^{*(k)}$. As discussed in the text, as the period-k cycle becomes unstable, the map $F^{(2k)}$ undergoes bifurcation (dashed curve to solid curve) to produce two new fixed points of period $2k$, $X^{*(k)}$, on either side of $X^{*(k)}$; for details, see ref. [28]. The slope of $F^{(2k)}$ at the period-$2k$ points $X^{*(2k)}$ is denoted by $\lambda^{(2k)}$.

through a sequence of *period doublings* or *harmonics*, to give successive stable cycles of periods $2k, 4k, 8k, \ldots, 2^n k$ as $F(X)$ steepens.

Beyond the point of accumulation of the stable cycles of period 2^n [which for eq. (2.2) lie in the *window* of a-values $3.57\ldots > a > 3$], lies a regime ($4 \geq a > 3.57\ldots$) often called "chaotic". The so-called chaotic regime in eq. (2.2), and its generic relatives in table 1, comprises infinitely many small *windows* of a-values, each corresponding to a stable cycle of basic period k and its cascade of harmonics of period $2^n k$; for example, there are 93 such intervals corresponding to the 93 different stable cycles with basic period 11. The first such basic cycles to appear beyond the cascade of period doublings 2^n [in eq. (2.2), beyond $a = 3.57\ldots$] are of even period, with all even periods appearing, in descending order. Eventually (at $a = 3.67\ldots$), the first odd period cycle appears, and, as the map continues to steepen (a continues to increase), all odd periods appear, also in descending order. It follows that, when the period 3 cycle first appears (at

$a = 3.82...$), every other integer period cycle has already appeared (and, although unstable, can be attained with appropriate initial conditions); hence Li and Yorke's "period three implies chaos" [1]. All the details of the order in which the cycles appear, and of the total number of cycles of a given period, are generic and can be elucidated in various ways (as summarized in the reviews by May [3] and by Collet and Eckmann [6]). The important question of whether essentially all a-values correspond to some unique stable cycle (albeit with different cycles giving way to each other in kaleidoscopic fashion), or whether the density of a-values not corresponding to a stable cycle has finite measure, remains open and the subject of interesting conjectures [6].

2.2.2. The Feigenbaum ratio

Following earlier work by May and Oster [28], Feigenbaum [29, 30] has shown that, in such cascades of period doublings, the local structure of the attractor is reproduced at a rescaled size in successive bifurcations, with the rescaling parameter being a universal constant. These elegant and important studies by Feigenbaum have been put on a fully rigorous footing by Collet et al. [31]; a good exposition is by Collet and Eckmann [6].

Specifically, consider the first-order difference eq. (2.1),

$$X_{t+1} = F(X_t; a), \tag{2.3}$$

and let a_n denote that value of the parameter a at which bifurcation to the period-2^n cycle occurs. The asymptotic ratio δ is then defined as

$$\delta = \lim_{n \to \infty} (a_{n+1} - a_n)/(a_{n+2} - a_{n+1}). \tag{2.4}$$

Feigenbaum's [29] numerical studies establish the result $\delta = 4.6692...$. This asymptotic scaling law applies generally to the cascades of cycles with periods $2^n k$ bifurcating from basic cycles of period k in "generically quadratic" first-order maps, and in certain other circumstances [6].

Following May and Oster [32], I now present an analytic approximation to this Feigenbaum ratio. The approximation is useful, partly as a supplement to the existing numerical studies, and partly because the techniques involved in the analysis may be helpful in elucidating other aspects of period doubling and "chaotic" phenomena that are less amenable to numerical investigation.

Let $F^{(k)}(X; a)$ be the k-times-composed mapping,

$$X_{t+k} = F^{(k)}(X_t; a). \tag{2.5}$$

A period-k cycle will touch the points $X_i^{*(k)}$ (with $i = 1, 2, \ldots, k$), which are fixed points of this mapping, obtained by solving $X^{*(k)} = F^{(k)}(X^{*(k)})$. Let $\lambda^{(k)}(a)$ denote the stability-determining slope of the map $F^{(k)}$ at these fixed points, as a function of the parameter a. Any particular stable cycle of period k is born with $\lambda^{(k)} = +1$, and becomes unstable (giving birth to the period-doubled cycle with period $2k$) when $\lambda^{(k)} = -1$; see fig. 1 and ref. [32]. As $\lambda^{(k)}$ sweeps through this range from $+1$ to -1, the change in the parameter a may be denoted $\Delta a(k)$. It follows that the Feigenbaum ratio may equivalently be defined as

$$\delta = \lim_{k \to \infty} \Delta a(k)/\Delta a(2k). \tag{2.6}$$

In principle, we could now define $a_0^{(k)}$ as the a value at which the period-k cycle is born $[\lambda^{(k)}(a_0^{(k)}) = +1]$, and write subsequent a values as $a = a_0^{(k)} + \varepsilon$. A Taylor expansion then gives

$$\lambda^{(k)}(a) = 1 + \varepsilon A_0^{(k)} + O\varepsilon^2. \tag{2.7}$$

Here $A_0^{(k)}$ is defined as

$$A_0^{(k)} = d\lambda^{(k)}/da = \partial^2 F^{(k)}/\partial x \partial a, \tag{2.8}$$

evaluated at the point where the period-k cycle first appears. On this basis, δ can be calculated as the asymptotic ($k \to \infty$) ratio between $A_0^{(2k)}$ and $A_0^{(k)}$. The difficulty is that $A_0^{(k)}$ cannot in general be evaluated analytically.

An alternative method, based on the Taylor series expansion of eq. (2.7), is available if we can relate the stability-determining slope of the $2k$-times-composed map at the fixed points of period $2k$, $\lambda^{(2k)}$, to the corresponding slope of period-k points, $\lambda^{(k)}$. The stable cycle of period k is born when $\lambda^{(k)} = +1$, and becomes unstable when $\lambda^{(k)} = -1$, at which point the succeeding stable cycle of period $2k$ is born [with $\lambda^{(2k)} = +1$]; this stable cycle of period $2k$ in turn becomes unstable when $\lambda^{(k)}$ attains some (negative) value, $\lambda_c^{(k)}$ say, corresponding to $\lambda^{(2k)} = -1$. Substituting in eq. (2.7), and ignoring second-order correction terms, we thus obtain

$$\Delta a(k) = -2/A_0^{(k)}, \tag{2.9}$$

and

$$\Delta a(2K) = (\lambda_c^{(k)} + 1)/A_0^{(k)}. \tag{2.10}$$

Hence we arrive at the result

$$\delta = \lim_{k \to \infty} -2/[\lambda_c^{(k)} + 1]. \tag{2.11}$$

It remains to find an asymptotic relation between $\lambda^{(2k)}$ and $\lambda^{(k)}$, in order to determine $\lambda_c^{(k)}$. An approximate such relation has been derived by May and Oster [28], by using a cubic to approximate the way the map $F^{(2k)}$ exhibits bifurcation as the period-k cycle becomes unstable (see fig. 1). This asymptotic relation is

$$\lambda^{(2k)} = 3 - 2[\lambda^{(k)}]^2. \tag{2.12}$$

The derivation is given in detail in ref. [28]; the correction terms are of relative order δ^{-2}. It follows that

$$\lambda_c^{(k)} = -2^{1/2} + O\delta^{-2}. \tag{2.13}$$

We thus arrive at an analytic approximation of the Feigenbaum ratio,

$$\delta \simeq 2(1 + 2^{1/2}) = 4.83\ldots. \tag{2.14}$$

Comparing this with the exact numerical result above, we see the error is around 3%. An improved estimate could be obtained by going to next order in the iterative scheme outlined in ref. [28], but it hardly seems worth the effort.

(Another broadly related example of an approximate but analytic calculation of a Feigenbaum ratio is by Helleman [33]. Helleman's example is for period doubling sequences generated by a second-order difference equation arising in a problem associated with storage rings in particle accelerators. For this instance, the approximation gives $\delta \simeq 1 + 65^{1/2} = 9.06\ldots$, contrasting with the exact numerical result of $\delta = 8.72\ldots$.)

2.3. Biological examples

When maps such as those in table 1 arise as approximate descriptions of the dynamical behaviour of biological populations, the parameters [such as a in eq. (2.2)] are always subject to some degree of random variability, and exact determinacy is lost [34]. If such stochastic effects are relatively small, however, then the detailed array of dynamical possibilities outlined above gives way, in effect, to a coarser structure. As the map steepens we have, successively: a steady constant population; a population that alternates steadily between a high and a low value; a population alternating more raggedly between high and low values in successive years (roughly corresponding to period 2^n cycles with n large, or to "even period chaos"); and, eventually, populations that appear to fluctuate more-or-less randomly (roughly corresponding to "chaos" after odd periods have first appeared). This range of possibilities is illustrated in fig. 2 [which shows trajectories generated by eq. (2.2) when there is: (i) a stable point ($a = 2.7$); (ii) a stable cycle of period 2 ($a = 3.4$); (iii) "chaos" ($a = 3.99$)].

Not only are data on particular populations susceptible to detailed explanation as stable points, stable cycles, or apparent chaos, but also some laboraory populations change from stable point to cyclic behaviour as temperatures (and hence, presumably, time scales and the steepness of mapings) are altered. Several of these specific studies are reviewed elsewhere [35].

More generally, several authors have used models from the list in table

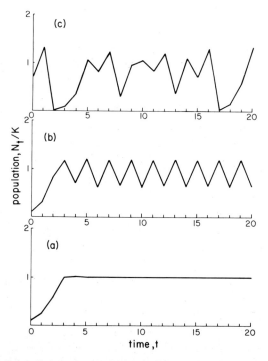

Fig. 2. The population N_t (scaled against K) is plotted as a function of time t for the simple and deterministic difference eq. (2.1) with the form I of table 1 for $F(N)$; this is equivalent to the difference equation eq. (2.2). The figure shows how this population equation can exhibit a stable equilibrium point, or stable cycles, or chaotic dynamical behaviour, depending on the severity of the nonlinear or "density dependent" effects (that is, upon the magnitude of r). Here the trajectories are for: (a) $r = 1.7$; (b) $r = 2.4$; (c) $r = 2.99$. For a more full discussion, see the text.

1 to analyze collections of data on many different populations, with a view to determining how often cyclic or chaotic dynamics seem to occur in the natural world.

The simplest models in table 1, such as I, II and VII, have essentially only one biological parameter (r or λ); the parameters K or α simply serve to set the scale of X. Biologically, this parameter r or λ is determined by the "intrinsic growth rate" of the population, corresponding to the maximum growth rate that can occur at low population densities (when density dependent limiting effects are minimized). In the 1-parameter models I, II and VII, the stability-determining slope of the map at its fixed point is thus determined by this "intrinsic growth rate" parameter. Models such as III, IV, V and VI have two parameters (in addition to the scale-setting parameter K or α); the parameter r or λ is determined from the intrinsic growth rate at very low densities, and the other parameter (b or θ) measures the severity of the nonlinear behaviour (corresponding to the steepness of the slope of the map at the fixed point).

The first such synoptic study was by Hassell et al. [17], who used the form IV for $F(X)$ to analyze data on 24 field populations and 4 laboratory studies of arthropods with discrete, non-overlapping generations. The results are shown in fig. 3, which portrays the values of the parameters b and λ for each of these 28 populations; fig. 3 also shows the boundaries between the various stability domains in this b–λ space. Subsequently Stubbs [15] used the form III in a similar, but independent, analysis of much the same compilation of population studies. Most recently, Bellows [19] has chosen a subset comprising the 14 best-documented of the population studies used by Hassell et al. and by Stubbs, and (after comparing the goodness-of-fit obtained using models III, IV, V and VI) has analyzed the data using the form V. The values of the parameters b and λ thus found for each of these 14 populations are shown in fig. 4; fig. 4 again also shows the various stability domains in b–λ space.

Bellows' results, as depicted in fig. 4, are broadly in accord with the earlier and cruder studies of Hassell et al. and of Stubbs. Essentially all the populations appear to have stable point behaviour, and *none* of the field populations have parameters lying in the chaotic domain. (Hassell et al. did find Nicholson's laboratory studies on blowflies to suggest chaotic dynamics, but this laboratory population is free from many of the density dependent mortality factors of natural populations of these blowflies.)

In an altogether different survey, Thomas et al. [20] analyzed population

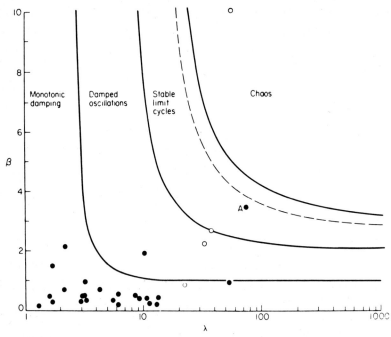

Fig. 3. This figure illustrates the regimes of dynamical behaviour of the difference equation, eq. (2.1), with the form IV of table 1 for $F(X)$. The solid curves separate the regions of monotonic and oscillatory damping to a stable point, stable limit cycles, and chaos; the broken line indicates where 2-point cycles give way to higher-order cycles. The solid circles come from the analyses of life table data on field populations, and the open circles from laboratory populations. (The point labelled A is for the Colorado potato beetle, whose contemporary role in agroecosystems is new on an evolutionary time scale.) The figure is after Hassell et al. [17], where details are given.

data for 27 species of *Drosophila* at 2 different temperatures, using the form VI for $F(X)$. Their results are shown in fig. 5, which (like figs 3 and 4) shows the parameter values for r and θ for each of these 54 populations, along with the boundaries between the different regions of dynamical behaviour in the θ–r space. Here *all* the populations appear to have a stable fixed point.

Figures 3, 4, and 5 may be taken to indicate that natural populations tend to behave stably, and rarely to exhibit dynamical behaviour corresponding to stable cycles or to "chaos". Indeed, Thomas et al. explicitly develop arguments to the effect that natural selection will tend in just such

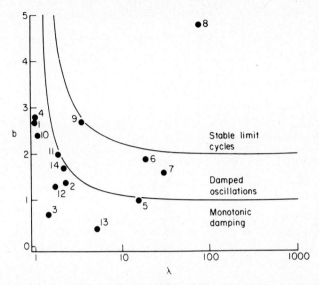

Fig. 4. This figure comes from the work of Bellows [19], and is similar to fig. 3. The main difference from fig. 3 is that the form V of table 1 has been employed for $F(X)$ in eq. (2.1): the various domains of dynamical behaviour are as indicated, and the points pertain to analyses of 14 sets of data on arthropod populations, representing a subset of those analyzed by Hassell et al. For further details, see the text.

a direction; excessively violent population fluctuations may be inconsistent with long-term survival of the population in the natural world.

Before any such generality is accepted, two severe cautions should be noted. First, there is likely to be a strong bias whereby good long-term data only exists for relatively steady populations. A population that underwent strong oscillations or chaotic fluctuations would be likely to be rare in many or most years, so that systematic year-by-year census data would simply not be available; such populations would be excluded from the data sets chosen for study by Hassell et al., Bellows, Stubbs, and Thomas et al. Second, to subsume the population's biological interactions with the world around it in passive parameters such as λ and b may do violence to the multispecies reality. Add to this the fact that it requires less severely nonlinear behaviour to take multispecies models into the chaotic regime, and figs. 3, 4, and 5 fall short of a proof that natural populations have stable point behaviour.

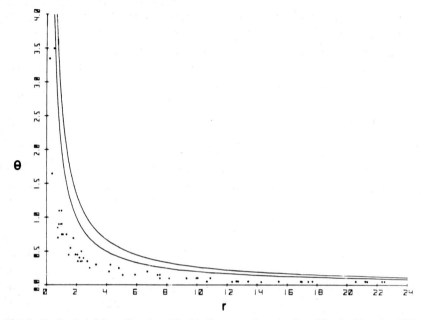

Fig. 5. Similar in spirit to figs. 3 and 4, this figure employs the form VI of table 1 for $F(X)$ in eq. (2.1), to analyze data for 27 different species of *Drosophila* (each at 2 different temperatures, whence there are 54 points in the figure). As discussed further in the text, this figure is from the work of Thomas et al. [20].

3. One-dimension difference equations, with two critical points

3.1. Motivation

In many natural situations, the interactions of organisms with their physical surroundings and with other organisms is such that, within a population, rarer genotypes have a selective advantage. Clarke [36–38], Wright [39], Kojima [40], Endler [41], Lewontin [42, 43] and others have marshalled evidence to this effect, and have argued that such ''apostatic'' or ''(gene) frequency dependent'' selection is important in promoting and preserving genetic polymorphism in natural populations.

The conventional mathematical carricature of this situation consists of one genetic locus with two alleles, A and a (with gene frequencies p_t and q_t, respectively, in the tth generation; $p + q = 1$). If the forces of natural selection are frequency dependent, it may be that the allele A enjoys a selective advantage when rare, and a concomitant disadvantage when common. As a result, the corresponding gene frequency p will tend to increase from one generation to the next when it has a low value, and to decrease when it has a high value. This general situation is depicted in fig. 6; in fig. 6 and elsewhere throughout this section, the variable p (defined on the interval $[0, 1]$) is replaced for notational convenience by the variable

$$X = 2p - 1, \tag{3.1}$$

(whence the mapping is on the interval $[-1, 1]$.

In this paradigm, it is clear that if the advantage possessed by A at low frequencies, and the disadvantage suffered at high frequencies, are both relatively modest, then the system will settle to a stable equilibrium at an intermediate gene frequency value. A stable polymorphism can thus be maintained by frequency dependent natural selection. But what happens when the map relating p_{t+1} to p_t becomes severely nonlinear? What happens to the system described by Fig. 6 as the hill at low p-values rises and the valley at high p-values deepens? To answer these questions, we begin by a general consideration of the first-order difference equation

$$X_{t+1} = F(X_t), \tag{3.2}$$

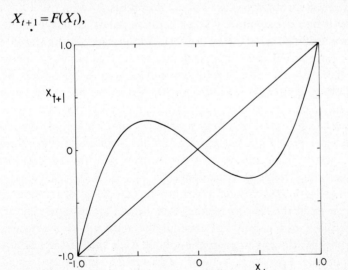

Fig. 6. The first-order difference equation, of the form of eq. (2.1), with two critical points. Here X is defined on the interval $[-1, 1]$, and the map is specifically the cubic eq. (3.3), with $a = 2$. The fixed points of period 1 (at $X = 1, 0, -1$) are the intersections of the map with the 45° line.

where the mapping $F(X)$ is defined on the interval $[-1, 1]$, and can have two critical points, a hill and a valley, as in fig. 6.

Quite apart from possible biological applications, the dynamical behaviour generated by maps with two critical points is of intrinsic mathematical interest, as the next step up from that generated by "one hump" maps.

In section 3.2, the dynamical behaviour exhibited by maps with two critical points is outlined; the presentation follows that in May [44]. Then, in section 3.3, this analysis is applied to the particular sub-class of such maps that arise in simple cases of frequency dependent selection; this work is previously unpublished.

3.2. Properties

3.2.1. General remarks

For first-order difference equations with one critical point, the maximum number of fixed points with period k is 2^k: each new iteration of the map doubles the number of possible intersections with the 45° line, so that the k-times-composed mapping can have a maximum 2^k fixed points. For a first-order difference equation with two critical points, as illustrated in fig. 6, each new folding can triple the number of intersections with the 45° line, and there can consequently be a maximum of 3^k fixed points for period k. For the map with one critical point, various techniques have been used to elucidate the generic order in which the different cycles of period k appear, and in general to classify how the totality of 2^k fixed points of period k are organized. The task of giving a generic classification of how the 3^k fixed points of period k are organized, for maps with two critical points, is more messy. A beginning is made below, but, to the best of my knowledge, much remains to be done.

For maps with one critical point, there can be at most one stable attractor; even in the chaotic regime, where there are infinitely many different periodic orbits and an uncountable number of asymptotically aperiodic orbits, there is in general a unique cycle that attracts almost all initial points. For the general first-order difference equation, Guckenheimer et al. [4] have shown that, for each parameter value, the maximum number of periodic attractors is equal to the number of critical points. Thus, for maps with two critical points, there can be domains of parameter space in which the system possesses two distinct periodic attractors. We shall see, below, how this phenomenon emerges from the bifurcation structure. Maps with two critical points consequently deserve special attention as the simplest first-order difference equations to exhibit the feature of alternative stable states.

In what follows, there is first a generic discussion of the bifurcations that occur in second, and higher, iterations of an (antisymmetric) map with two critical points, as the "hill" and "valley" steepen. Second, these processes are illustrated by the concrete example of the cubic mapping

$$X_{t+1} = aX_t^3 + 1(1-a)X_t. \tag{3.3}$$

Eq. (3.3) is the canonical exemplar of a map with two critical points, in the sense that $X_{t+1} = aX_t(1-X_t)$ is the standard example of the one-hump map.

3.2.2. Generic aspects of the bifurcation structure

Consider a first-order difference equation, as defined by eq. (3.2), with X defined on the interval $[-1, 1]$. In discussing the case where $F(X)$ has two critical points, it is convenient further to assume that $F(X)$ is antisymmetric, so that

$$F(-X) = -F(X). \tag{3.4}$$

This is the situation illustrated in fig. 6. The extent to which the subsequent conclusions are specific to these antisymmetric maps is an open question.

We further assume the map is "anchored" to (unstable) fixed points at $X = 1$ and $X = -1$: $F(1) = 1$. Interest now focuses on the fixed point at $X = 0$ [$F(0) = 0$ from eq. (3.4)]. The stability analysis follows the lines laid down for one-hump maps, and depends on the slope, $\lambda^{(1)}(0)$, of the map at the fixed point:

$$\lambda^{(1)}(0) = [dF/dX]_{X=0}. \tag{3.5}$$

So long as $|\lambda^{(1)}| < 1$, the fixed point at $X = 0$ is an attractor. However, as the hill and valley in fig. 6 steepen, $\lambda^{(1)}$ will steepen toward -1, and the fixed point at $X = 0$ will become unstable once $\lambda^{(1)}$ steepens beyond -1.

To see what happens as this fixed point of period 1 ($X = 0$) becomes unstable, we turn, as in the analysis of the one-hump map, to the map for the second iterate:

$$X_{t+2} = F^{(2)}(X_t). \tag{3.6}$$

Here $F^{(k)}$ denotes the k-times-composed mapping of F. The period-1 point at $X=0$ is obviously a degenerate period-2 point, and the slope of the $F^{(2)}$ map at this point is

$$\lambda^{(2)}(0) = [dF^{(2)}/dX]_{X=0}. \tag{3.7}$$

That is, explicitly writing $F^{(2)}(X) = F[F(X)]$,

$$\lambda^{(2)}(0) = [dF/dX]^2_{X=0} \tag{3.8}$$

$$= [\lambda^{(1)}(0)]^2. \tag{3.9}$$

The observation that $\lambda^{(2)}(0)$ is the square on $\lambda^{(1)}(0)$ is the key to the cascading bifurcations that arise in maps with one critical point: so long as the period-1 point at $X=0$ is stable, which implies $|\lambda^{(1)}|<1$, then $|\lambda^{(2)}|<1$ and the $F^{(2)}$ map intersects the 45° line only once in the neighborhood of $X=0$; as the period-1 point at $X=0$ becomes unstable, which implies $|\lambda^{(1)}|>1$, then $|\lambda^{(2)}|>1$ and the $F^{(2)}$ map steepens to make a loop which intersects the 45° line three times in the neighborhood of $X=0$. In this way, two new and initially stable fixed points of period 2 are born by a "pitchfork," bifurcation, at the same time as the period-1 fixed point becomes unstable. For the antisymmetric map under consideration here, we can label these two fixed points of period 2 as $X=\pm\Delta$:

$$\pm\Delta = F(\mp\Delta). \tag{3.10}$$

The stability of this new orbit of period 2 depends, in turn, on the slope of $F^{(2)}$ at these fixed points $X=\pm\Delta$:

$$\lambda^{(2)}(\Delta) = [dF^{(2)}/dX]_{X=\Delta}. \tag{3.11}$$

Again, writing $F^{(2)}$ explicitly,

$$\lambda^{(2)}(\Delta) = \left(\frac{dF}{dX}\right)_{X=-\Delta}\left(\frac{dF}{dX}\right)_{X=\Delta}. \tag{3.12}$$

That is,

$$\lambda^{(2)}(\pm\Delta) = \lambda^{(1)}(\Delta)\lambda^{(1)}(-\Delta). \tag{3.13}$$

Of course, the slope $\lambda^{(2)}$ is the same at each of the two points $X=\Delta$ and $X=-\Delta$.

The antisymmetry of $F(X)$ necessarily implies that $\lambda^{(1)}(\Delta) = \lambda^{(1)}(-\Delta)$, whence

$$\lambda^{(2)}(\pm\Delta) = [\lambda^{(1)}(\Delta)]^2. \tag{3.14}$$

This is a second key relationship. The stability-setting slope of the $F^{(2)}$ map at these fixed points of period 2 is *necessarily positive*. This is in complete contrast to what happens for maps with one critical point, where the period-2 cycle is born with $\lambda^{(2)} = +1$, and then as the hump continues to steepen $\lambda^{(2)}$ decreases to zero (as one of the period-2 points coincides with the critical point), beyond which $\lambda^{(2)}$ becomes negative (with the period-2 points on opposite sides of the critical point, and consequently dF/dX having opposite signs at the two points), so that eventually $\lambda^{(2)}$ steepens beyond -1 with the period-2 orbit becoming unstable and bifurcating to a stable period-4 orbit. For the antisymmetric map with two critical points, however, the stability-setting slope at $X = \pm \Delta$ is born with $\lambda^{(2)} = +1$, decreases to $\lambda^{(2)} = 0$ as the hill and valley steepen, and then increases back toward $\lambda^{(2)} = +1$. As $\lambda^{(2)}$ increases beyond $+1$, the $F^{(2)}$ map again loops to cut the 45° line three times in the neighborhood of each of the period-2 points at $X = \pm \Delta$ (in exactly the manner that initially gave rise to these two fixed points by bifurcation in the neighborhood of $X = 0$). Thus as the first cycle of period 2 becomes unstable, it bifurcates to produce two new and distinct cycles, each of period 2.

This situation is illustrated in fig. 7. The system now has two alternative stable states, each a stable cycle of period 2. Points originating in the intervals $[-1, -\Delta]$ or $[0, \Delta]$ (denoted by the hatched regions along the X-axis in fig. 7) will be attracted to one cycle; points in the intervals $[-\Delta, 0]$ or $[\Delta, 1]$ to the other cycle. This now accounts for all $3^2 = 9$ fixed points of period 2 (three of which are, of course, the period-1 points).

For each of these "second generation" cycles of period 2, the stability-setting slope $\lambda^{(2)}$ at the two fixed points will evolve in the manner made familiar by the map with one critical point: $\lambda^{(2)}$ will decrease from $+1$, through zero, toward -1, beyond which point the period-2 cycle will become unstable, giving rise to an initially stable cycle with period 4. Thus each of the two domains of attraction will exhibit the cascading bifurcation process, generating successively stable cycles of periods $2, 4, 8, 16, \ldots, 2^n$.

In summary, we see that for first-order difference equations with one critical point (as discussed in the section 2) the basic process is a bifurcating hierarchy of cycles with periods $1 \to 2 \to 4 \to 8 \cdots 2^n$: other higher-order cycles echo this theme, with stable cycles of basic period k cascading down through "harmonics" of period $k2^n$. For first-order difference equations with two antisymmetric critical points (as discussed above), the basic process is cycles with periods $1 \to 2 \to$ two distinct 2s, each with its own domain

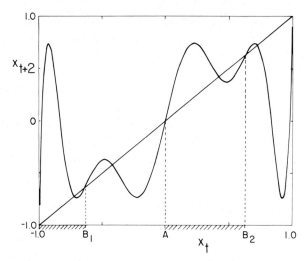

Fig. 7. This figure shows the two-times-iterated mapping $X_{t+2} = F^{(2)}(X_t)$, for the cubic mapping of eq. (3.3), with $a = 3.3$. The map intersects the 45° line at 9 points, which are the fixed points of period 2. Here the symmetric period-2 cycle between the points B_1 and B_2 is unstable, and there are two distinct periodic attractors: one attracts points originating in the hatched regions between -1 and B_1, and between A and B_2; the other attracts points originating in the two unhatched regions. For a full discussion, see the text.

of attraction, and each of which then goes $2 \rightarrow 4 \rightarrow 8 \rightarrow \cdots 2^n$.

Higher-order cycles arise by "tangent" bifurcation, and are similarly complicated. Again, in general, there will be two stable periodic attractors for each parameter value, for the reasons alluded to above [4]. As an example, consider the fixed points of period 3. For maps with one critical point, there are eight (2^3) such points: two period-1 points, and a stable and unstable pair of period-3 cycles. For maps with two critical points, there are potentially $3^3 = 27$ such points. Subtracting the three period-1 points, there remains two constellations of 12 points. The first set of 12 points originates as two pairs of 3-cycles, one initially stable and one unstable; that is, there simultaneously arise two distinct stable 3-cycles, each with its own distinct domain of attraction. The other set of 12 points behaves similarly.

The task of cataloging the way in which the 3^k fixed points of period k are organized, and the order in which the various cycles of period k originate, remains an interesting problem.

3.2.3. A specific example: the cubic map

The canonical example used in discussions of the one-hump is the quadratic difference equation $X_{t+1} = aX_t(1 - X_t)$. The above discussion of antisymmetric maps with two critical points may be similarly made concrete by considering the cubic difference eq. (3.3). This equation has nontrivial behaviour for $0 < a < 4$; for $a < 0$, the end-points at $X = \pm 1$ are attrractors, and for $a > 4$ the hill top and valley bottom lie outside $[1, -1]$.

The stability of the fixed point at $X = 0$ hinges on the slope $\lambda^{(1)}(0)$, which here is

$$\lambda^{(1)}(0) = 1 - a. \tag{3.15}$$

Thus the period-1 orbit is stable for $0 < a < 2$ (with exponential damping for $0 < a < 1$, and oscillatory damping for $1 < a < 2$).

For $a > 2$, the $F^{(2)}$ map bifurcates to give the symmetric fixed points of period 2 at $X = \pm \Delta$. These points are obtained from eq. (3.10),

$$-\Delta = a\Delta^3 + (1 - a)\Delta. \tag{3.16}$$

This gives

$$\Delta = \pm[(a - 2)/a]^{1/2}. \tag{3.17}$$

The slope of the $F^{(2)}$ map of these two points follows from eq. (3.14):

$$\lambda^{(2)}(\pm\Delta) = (2a - 5)^2. \tag{3.18}$$

Initially, at $a = 2$, this slope is $\lambda^{(2)} = +1$. It decreases to $\lambda^{(2)} = 0$ at $a = 2.5$, and then increases back to $\lambda^{(2)} = +1$ at $a = 3$. For $a > 3$, this period-2 cycle is unstable.

The subsequent nonsymmetric cycles of period 2 are obtained by finding all the fixed points of eq. (3.6). For the cubic $F(X)$ of eq. (3.3) this gives

$$X = X[aX^2 + 1 - a][aX^2(aX^2 + 1 - a)^2 + 1 - a]. \tag{3.19}$$

Writing $y = aX^2 + 1 - a$, this can, after some manipulation, be brought to the form

$$[y^2 - 1][y^2 + (a - 1)y + 1] = 0. \tag{3.20}$$

The solution $X = 0$ has been discarded. Similarly, the pair of solutions $y = \pm 1$ lead back to the 4 period-2 fixed points that have already been found: $X = \pm 1$ and the $X = \pm \Delta$ of eq. (3.17). The remaining quadratic in y in eq. (3.20) leads routinely to four other period-2 fixed points at

$X = \pm\alpha, \pm\beta$, with α and β defined by

$$\alpha, \beta = \left(\frac{(a-1) \pm [(a-1)(a-3)]^{1/2}}{2a}\right)^{1/2}. \tag{3.21}$$

It is easy to verify that these four fixed points correspond to two distinct period-2 cycles, with $\alpha \rightarrow -\beta \rightarrow \alpha$ and $\beta \rightarrow -\alpha \rightarrow \beta$, as illustrated in fig. 7. The stability of each of these cycles depends on the slope of the $F^{(2)}$ at these fixed points, namely,

$$\lambda^{(2)}(\pm\alpha, \pm\beta) = \lambda^{(1)}(\pm\alpha)\lambda^{(1)}(\mp\beta), \tag{3.22}$$

$$= 7 + 4a - 2a^2. \tag{3.23}$$

The cycles first appear at $a = 3$ [see eq. (3.21)], with $\lambda^{(2)} = +1$. They become unstable, with $\lambda^{(2)} < -1$, for $a > 1 + 5^{1/2}$.

These results are collected in table 2. Beyond $a = 1 + 5^{1/2}$, there is first (for each of the two domains of attraction) the bifurcating hierarchy of cycles of periods, 4, 8, and so forth. As for the one-hump maps, this then gives way to an apparently chaotic regime of dynamical behaviour, the details of which remain to be elucidated.

There are two distinct periodic attractors for all values of a in the range $3 < a < 4$.

3.3. Frequency-dependent natural selection

3.3.1. Analysis of a model

We now narrow the discussion to consider the subset of maps generated by simple models for frequency dependent natural selection. For biological clarity, we replace the variable X of eq. (3.1) by the variable p for the gene

Table 2
Dynamical behaviour of the mapping $X_{t+1} = aX_t^3 + (1 - a)X_t$

Value of a	Dynamical behaviour
$2 > a > 0$	stable point
$3 > a > 2$	stable cycle of period 2
$1 + 5^{1/2} > a > 3$	two distinct (unsymmetrical) cycles, each with period 2
$4 > a > 1 + 5^{1/2}$	two distinct periodic attractors, with various periods; usually the dynamics is apparently chaotic

frequency, so that the maps are of the form

$$p_{t+1} = F(p_t), \tag{3.24}$$

with p defined on the interval $[0, 1]$.

Specifically, we consider one locus with two alleles (A with frequency p, a with frequency $q = 1 - p$) in a diploid population where the genotypes AA, Aa and aa have frequency dependent fitnesses $w_{AA}(p)$, $w_{Aa}(p)$ and $w_{aa}(p)$, respectively. Two restrictive assumptions are now made. First, the heterozygotes are assumed to have fitness equal to the geometric mean of those of the homozygotes; referring all fitnesses to that of the heterozygotes, taken to be unity ($w_{Aa} = 1$), this implies $w_{AA}(p)w_{aa}(p) = 1$. Second, the frequency dependent selective advantage under discussion is assumed to be symmetric, in the sense that $w_{aa}(1 - p) = w_{AA}(p)$. If we write $w_{AA} \equiv f(p)$, the first of these assumptions means the relative fitnesses of the genotypes AA, Aa, aa can be written as $f(p), 1, 1/f(p)$ respectively, and the second assumption implies that the fitness function $f(p)$ obeys the relation

$$f(p) = 1/f(1 - p). \tag{3.25}$$

Such a symmetrical system of frequency dependent fitnesses for the three diploid genotypes is illustrated in fig. 8. The above assumptions are made in order to keep the problem manageable; they are discussed further below.

The change in gene frequency between one generation (t) and the next ($t + 1$) is given by the standard expression

$$p_{t+1} = \frac{w_{AA} p_t^2 + w_{Aa} p_t q_t}{w_{AA} p_t^2 + 2 w_{Aa} p_t q_t + w_{aa} q_t^2}. \tag{3.26}$$

Under the above assumptions, this reduces (after some manipulation) to

$$p_{t+1} = \frac{p_t f(p_t)}{p_t f(p_t) + q_t}. \tag{3.27}$$

This has precisely the form of eq. (3.24), with $F(p)$ given as

$$F(p) = \frac{pf(p)}{pf(p) + 1 - p}. \tag{3.28}$$

As is intuitively obvious, eq. (3.25) implies that $F(p)$ has the antisymmetry property of eq. (3.4).

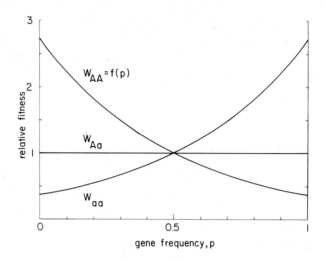

Fig. 8. This figure illustrates the assumptions made in section 3.3 about the relative fitnesses (w_{AA}, w_{Aa}, w_{aa}) of the genotypes AA, Aa, aa, as functions of the gene frequency p. Note that the fitness function $w_{AA} \equiv f(p)$ decreases monotonically. (Specifically, this illustration is for the fitness function of section 3.3.2, with $\beta = 1$.)

The slope of this map at a given point p, $\lambda^{(1)} = dF/dp$, is then

$$\lambda^{(1)}(p) = \frac{[f + p(1-p)(df/dp)]}{[pf + 1 - p]^2} . \tag{3.29}$$

In particular, the stability of the equilibrium point at $p_{\mathrm{I}}^* = \tfrac{1}{2}$ depends on the slope $\lambda^{(1)}(\tfrac{1}{2})$, which is

$$\lambda^{(1)}(\tfrac{1}{2}) = 1 + \tfrac{1}{4}(df/dp)_{p=\frac{1}{2}}. \tag{3.30}$$

If the slope of the fitness function $f(p)$ is sufficiently steep at this point (cf. figs. 6 and 8), $\lambda^{(1)}$ can steepen beyond -1, and the equilibrium point becomes unstable, bifurcating to give a stable 2-point cycle.

Our attention is then focused on this cycle, wherein the gene frequency p can settle to a stable oscillation between the values p_{II}^* and $1 - p_{\mathrm{II}}^*$. These two gene frequency values are the non-trivial solutions of the equation $(1-p) = F(p)$, with $F(p)$ defined by eq. (3.28). It follows, after some algebraic manipulation, that p_{II}^* and $1 - p_{\mathrm{II}}^*$ are the two solutions of the equation

$$(1-p)^2 = p^2 f(p). \tag{3.31}$$

The stability of this cycle depends on the slope of the period-2 map, $F^{(2)}$, at these two points, and thence [via eq. (3.14)] on the slope $\lambda^{(1)}(p_{II}^*)$. Using the relation (3.31) in eq. (3.29), this slope of the $F(p)$ map at the period-2 points can be written

$$\lambda^{(1)}(p_{II}^*) = 1 + \zeta, \tag{3.32}$$

with ζ defined as

$$\zeta \equiv p_{II}^*(1 - p_{II}^*)[d(\ln f)/dp]_{p = p_{II}^*}. \tag{3.33}$$

We now make the final assumption that $f(p)$ is a monotonic decreasing function of p. This assumption, which is as depicted in fig. 8, seems to follow from any biologically plausible mechanism producing the frequency dependent selective advantage to rare genotypes. The assumption is discussed further below. The quantity ζ is now constrained to be negative; $df/dp < 0$ implies $\zeta < 0$.

The 2-point cycle is born in the neighborhood of $p = \frac{1}{2}$, precisely when the slope at this point attains the value $\lambda^{(1)} = -1$ ($\zeta = -2$). As the map $F(p)$ steepens, the two period-2 points move away from the mid-point toward the critical points, and for any simple monotonic curve $f(p)$ the quantity ζ will increase smoothly from $\zeta = -2$ to $\zeta = -1$ (from $\lambda^{(1)} = -1$ to $\lambda^{(1)} = 0$) as the 2-cycle and the critical points coincide. As the map $F(p)$ steepens further, the two period-2 points will move beyond the critical points, and ζ will continue to increase from $\zeta = -1$ toward $\zeta = 0$ (from $\lambda^{(1)} = 0$ toward $\lambda^{(1)} = +1$). But for monotonic decreasing $f(p)$, $\zeta < 0$ for all p, and consequently $\lambda^{(1)}$ cannot attain or surpass the value $+1$. Hence the subsequent bifurcation at $\lambda^{(2)} = +1$, whereby the two new and distinct 2-point cycles are born, is not attainable.

In short, under the assumptions cataloged above, the gene frequency in diploid systems with frequency dependent natural selection can exhibit either a stable point or a stable 2-point cycle. The much richer range of dynamical behaviour that can be shown by more general maps with two critical points is not possible for the restricted class of such maps implied by eq. (3.28).

3.3.2. A particular example
As a concrete example, suppose the fitness function $f(p)$ of section 3.3.1 has the form

$$f(p) = e^{\beta(1 - 2p)}. \tag{3.34}$$

This has the properties indicated in fig. 8; it is monotonic decreasing for p on the interval $[0, 1]$, and it has the symmetry property that $f(p)f(1-p) = 1$.

Substituting eq. (3.34) into eq. (3.30), we see that the slope of the map $F(p)$ at the equilibrium point $p = \frac{1}{2}$ is here

$$\lambda^{(1)}(\tfrac{1}{2}) = 1 - (\beta/2). \tag{3.35}$$

There will therefore be a stable equilibrium gene frequency provided $\beta < 4$.

For frequency dependent selective forces so strong that $\beta > 4$, this equilibrium point gives way to a stable 2-point cycle, between two values of p determined from eq. (3.31). This eq. (3.31) now has the explicit form

$$\ln\left(\frac{1-p}{p}\right) = \beta(\tfrac{1}{2} - p). \tag{3.36}$$

(The equation has non-trivial solutions provided $\beta > 4$.) The stability of the 2-point cycle depends on the slope $\lambda^{(2)}(p_{\text{II}}^*) = [\lambda^{(1)}(p_{\text{II}}^*)]^2$, where $\lambda^{(1)}(p_{\text{II}}^*)$ is given by eqs. (3.32) and (3.33) as

$$\lambda^{(1)}(p_{\text{II}}^*) = 1 - 2\beta p_{\text{II}}^*(1 - p_{\text{II}}^*), \tag{3.37}$$

with p_{II}^* from eq. (3.36). It is easy to see that as β increases from 4 toward ∞, $\lambda^{(1)}(p_{\text{II}}^*)$ increases from -1 to approach $+1$. For $\beta \gg 1$, we have

$$p_{\text{II}}^* \simeq e^{-\beta/2}, \tag{3.38}$$

and

$$\lambda^{(1)}(p_{\text{II}}^*) \simeq 1 - 2\beta \, e^{-\beta/2}. \tag{3.39}$$

The value $\lambda^{(1)} = +1$ is, however, never attained, and the 2-cycle never becomes unstable.

Figure 9 bears out the above remarks, by showing the dynamical behaviour of the gene frequency p_t for various values of β.

Notice that in this example there is a stable equilibrium at $p = \frac{1}{2}$ unless $\beta > 4$. That is, there is a stable point unless the genotypes AA when very rare have a selective advantage of a factor $e^4 \simeq 55$ over the heterozygotes Aa, and of $e^8 \simeq 3000$ over the homozygotes aa. It is not easy to imagine ecological circumstances that will produce so large an advantage to rare alleles. This fact has previously been emphasized by Haldane and Jayaker [45].

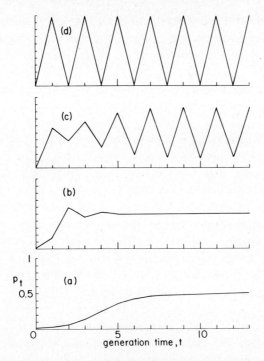

Fig. 9. The behaviour of the gene frequency p as a function of generation time, t, is shown for the specific frequency dependent fitness assumptions made in section 3.3.2. The parameter β that measures the strength of the frequency dependent effects has the values: (a) $\beta = 1$; (b) $\beta = 3$; (c) $\beta = 5$; (d) $\beta = 10$.

3.3.3. How general are these conclusions?

A key assumption in section 3.3.1 and 3.3.2 was that the fitness function $f(p)$ was monotonic decreasing (or, mutatis mutandis, monotonic increasing). Although one can imagine situations in which an allele enjoys maximum selective advantage when it is rare, but not *too* rare, such situations seem contrived. An $f(p)$ with a maximum at some intermediate value of p could perhaps be realized as the projection of some multilocus system with competing selection pressures. Once this circumstance pertains, then the constraint $\zeta < 0$ is lost, and the much wider range of dynamical behaviour indicated in section 2 is possible for the system.

For example, the frequency dependent fitness function

$$f(p) = \frac{1 + 2\alpha(1 - 2p)(1 - p)}{1 - 2\alpha(1 - 2p)p} \,, \tag{3.40}$$

(with $4 > \alpha > 0$) leads to the cubic map of eq. (3.3). But this $f(p)$ has the biologically strange feature that maximum selective advantage accrues to the genotype AA at intermediate rarity, rather than in the limit $p = 0$, once $\alpha > 1$.

On the whole the assumption that $f(p)$ decreases monotonically seems very reasonable. Other forms are likely to be the exception rather than the rule.

Less affirmative things have to be said about the other two assumptions in section 3.3.1.

The assumption that the heterozygous genotypes have fitness geometrically intermediate between the homozygotes is of a kind that is common in population genetics. It is at least as reasonable as any other choice for the heterozygotes' fitness. Here its consequence is effectively to reduce the diploid genetic system to a haploid one; for a haploid population with alleles A and a (with frequencies p and q, respectively), and with the genotype A having fitness $f(p)$ relative to the genotype a, eq. (3.27) is obtained immediately (cf. Haldane and Jayakar [45]).

If we give up this assumption about the fitness of the heterozygotes, the relative fitnesses of the genotypes may be written as $w_{AA} : w_{Aa} : w_{aa} = f(p) : 1 : g(p)$. Here the assumption that frequency dependent selection affects the rarer allele in a symmetrical way is retained by requiring that $g(p) = f(1 - p)$. What is lost is the further requirement that $f(p)f(1 - p) = 1$. Some numerical studies of specific examples of this kind have shown only a stable point, or a stable 2-point cycle. However, I have not succeeded in constructing a general proof (along the lines of section 3.3.1) that this must be so for all monotonic $f(p)$.

The final assumption is that the selective advantages possessed by the allele A when it is rare are precisely the same as those pertaining to the allele a when it is rare. This is the assumption that gives the map $F(p)$ its antisymmetry properties. While it is true that the gene frequency p can, without loss of generality, be rescaled to bring the equilibrium point to the mid-point at $p = \frac{1}{2}$, it is not in general true that the overall antisymmetry of $F(p)$ can be so preserved. Once the fitness functions for the genotypes AA and aa [$f(p)$ and $g(p)$, respectively] are unrelated, it is hard to make any general statement. Moreover, these symmetry assumptions are unlikely to be fulfilled in natural systems.

3.3.4. Conclusion

If frequency dependent natural selection is caricatured by a one locus/two allele model, the gene frequency will obey a first-order difference equation that can have two critical points. Such systems with two critical points can, in general, have very complicated dynamical behaviour, with inter alia two distinct domains of attraction. But if we assume frequency dependent selection in a diploid population where the heterozygotes are of geometrically intermediate fitness, where the selective advantage accruing to the rarer allele is symmetric, and where these selective advantages decrease monotonically as the allele becomes commoner, the consequent mappings can only exhibit either a stable point or a stable 2-point cycle.

The assumption of monotonicity in the selection effects is reasonable. The details of the assumption about the heterozygotes' intermediate fitness probably do not affect the conclusion. Whether chaotic behavior can be produced by appropriate unsymmetrical choices for the fitness functions is not clear (although I have been unable to construct such an example).

Several further comments and caveats are in order.

On the one hand, not only are the models dynamically tame in possessing only stable points or stable 2-point cycles, but even these cycles require that the rarer allele possess implausibly large selective advantages. This may be taken as an argument that frequency dependent natural selection typically produces a balanced polymorphism.

On the other hand, studies of systems of difference equations in various contexts in population biology suggest that chaotic behaviour arises more easily (requiring weaker nonlinearities) as the dimension of the system (the number of interacting species) increases. This suggests that chaotic behaviour, possibly with several alternative domains of attraction, may arise with frequency dependent selection in multi-locus systems.

4. More-than-one-dimensional difference equations

4.1. Single populations with discrete but overlapping age classes

Many populations of plants and animals are composed of several discrete, but overlapping, age classes. Such populations will be described by a set of coupled first-order difference equations, rather than by a simple 1-dimensional map.

As one among many possible examples, it may be noted that the models used by the International Whaling Commission (IWC) to set whaling quotas are of the general form

$$N_{t+1} = (1 - \mu)N_t + R(N_{t-T}). \tag{4.1}$$

Here N_t is the population of (sexually mature) adult whales in year t. The corresponding population one year later, N_{t+1}, consists of the surviving fraction, $(1 - \mu)N_t$, plus those newly recruited into the adult population from births T years ago, $R(N_{t-T})$; T is the time taken to attain sexual maturity, and R is a nonlinear recruitment function.

In more biological detail, the IWC model for the recruitment term has the form

$$R(N) = \tfrac{1}{2}(1 - \mu)^T N[P + Q\{1 - (N/K)^z\}]. \tag{4.2}$$

Here K is the pristine, unharvested equilibrium density of the whale population; P is the per capita fecundity of females at this pristine equilibrium point $N = K$; Q is the maximum increase in per capita fecundity of which the whales are capable as population densities fall to low levels; and z measures the severity with which this density dependent response is manifested. As illustrated in fig. 10, $z = 1$ corresponds to the conventional

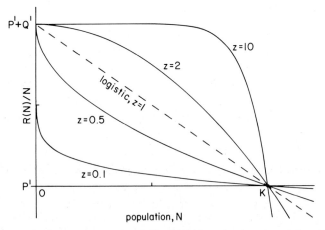

Fig. 10. The per capita recruitment, $R(N)/N$ of eq. (4.1), is plotted as a function of N, illustrating the IWC recruitment relation, eq. (4.2), for various values of the density dependence parameter z. The per capita recruitment ranges from $P' = \tfrac{1}{2}(1 - \mu)^{Tp}$ to $P' + Q' = \tfrac{1}{2}(1 - \mu)^T(P + Q)$ as N decreases from K to 0; the detailed nature of the response, however, depends on z, as illustrated. For further discussion, see the text.

logistic assumption, in which the density dependent increase in fecundity is manifested linearly, with per capita fecundity rates rising linearly from P to $P+Q$ as N falls from K to 0. For $z>1$ the density dependent response is mainly concentrated around the equilibrium point $N=K$ (and for $z\gg1$, the per capita fecundity rises from P essentially to $P+Q$ for small decreases in population density below K). Conversely, for $z<1$ the density dependent response is not manifested until the population falls to quite low levels. The factor $(1-\mu)^T$ simply measures the fraction the newly born whales that survive, with probability $1-\mu$ for each of T years, to adulthood. Finally, the factor $\frac{1}{2}$ arises because, under the assumption that the sex ratio is unity, exactly half the population is female, so that the per capita fecundity of females is to be multiplied by $\frac{1}{2}N$, the total number of females. Notice, incidentally, that the parameters μ, T and P are not independent. The equilibrium population in the unexploited system is, by assumption, $N_{t+1}=N_t=N_{t-T}=K$; substituting this in eqs. (4.1) and (4.2) gives the identity, or "balance equation",

$$\mu=\tfrac{1}{2}(1-\mu)^T P. \tag{4.3}$$

Putting eqs. (4.1), (4.2) and (4.3) together, we can write the baleen whaling equation in tidy form as

$$X_{t+1}=(1-\mu)X_t+\mu X_{t-T}[1+q\{1-X_{t-T}^z\}]. \tag{4.4}$$

Here $q\equiv Q/P$ is the maximum increase in fecundity of which the population is capable at low densities, expressed as a ratio to the natural equilibrium fecundity; X is the rescaled population variable, $X=N/K$.

Some of the simpler dynamical properties of this equation have been studied by Clark [46], and Levin and May [47] have given a general treatment of the linear stability analysis of difference–delay equations such as eq. (4.1). More recently, Levin and Goodyear [48] have discussed the highly complicated dynamical behaviour – including complex cycles and apparently chaotic orbits – that can arise in nonlinear fisheries and whaling models with many overlapping age classes.

4.2. Two interacting populations

Other, broadly similar complications arise when one considers two or more interacting populations, each of which is composed of discrete, nonoverlapping generations (with the generation times for the two or more populations in synchrony). Numerical explorations of such 2-dimensional systems have been presented for predator–prey interactions by Beddington et al. [49], and for two competitors by May [2]. An excellent account, with many analytic insights into the "deterministic chaos" that can often arise with relatively weak nonlinearities in such higher-order systems, is by Guckenheimer et al. [4].

4.3. Local versus global stability

A different kind of dynamical complexity that can arise in simple models for interacting species has to do with the distinction between local and global stability. These complications may be illustrated with a simple example (which is discussed more fully elsewhere [44]): the example is not a contrived one, but rather is a version of the model propounded by Crofton [50] to describe certain kinds of host–parasite interactions.

Consider the pair of coupled, first-order difference equations:

$$X_{t+1} = \lambda X_t [1 + Y_t]^{-k}, \tag{4.5}$$

$$Y_{t+1} = X_t Y_t [1 + Y_t]^{-(1+k)}. \tag{4.6}$$

Here X_t and Y_t are essentially the number of hosts and of parasites in generation t, respectively. The parameter λ represents the intrinsic growth factor for the host population, and the (negative binomial) parameter k measures the degree of parasite aggregation. This system has an equilibrium point which is locally stable for all λ if $k < 2$, and for $\lambda < [(k-2)/(k+2)]^k$ if $k > 2$.

Figure 11 is devoted to the specific case $\lambda = 2$, $k = 5$. The cross at $X = 2.30$, $Y = 0.15$ marks the (locally stable) equilibrium point. The figure shows the eventual dynamical fate of the system for various values of the initial populations, $X(0)$ and $Y(0)$: $X(0)$ ranges from 10^{-6} to 10^4 inclusive, and $Y(0)$ from 10^{-4} to 10^2 inclusive (both in steps of 0.25 on a logarithmic scale). The solid dots denote grid points that are attracted to the equilibrium point. All other initial grid points (except one) lead to diverging oscillations that carry Y below 10^{-99}. The initial point at $X(0) = 10^{0.75}$, $Y(0) = 10^{-3.5}$ is attracted into a stable 11-point cycle.

Figure 11 is not devoid of pattern. All points within an order or magnitude or so of the equilibrium point are attracted to it, and for large $X(0)$ there is a tendency for points lying along two "spiral arms" to converge to the equilibrium point. But, particularly for relatively small values

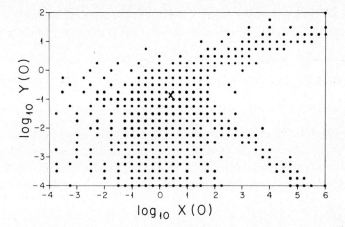

Fig. 11. The figure indicates the global stability behaviour of the pair of first-order, coupled difference equations, eqs. (4.5) and (4.6), with $\lambda = 2$ and $k = 5$. The locally stable equilibrium values of X and Y are marked by the cross, and those initial points $X(0)$, $Y(0)$ which are marked with dots on the grid do indeed give trajectories that converge to this equilibrium point. All other initial points on the grid lead to oscillations that diverge until Y is less than 10^{-99}. The point at $X(0) = 10^{0.75}$, $Y(0) = 10^{-3.5}$ is exceptional, being attracted to a stable 11-point cycle. For a more full discussion, see the text and ref. [44].

of $X(0)$, some parts of the grid look rather like a collection of go-stones thrown randomly on a go-board.

Above and beyond the questions about "deterministic chaos" that have been so much discussed in this volume, the inchoate lack of pattern determining which areas of X–Y space are attracted to the fixed point in fig. 11 raises disturbing questions. These questions of global versus local stability deserve more attention than they have typically received.

5. Differential–delay equations

5.1. Motivation

For whale populations (and for many other mammalian populations) the annual death rates and birth rates are relatively small, and maturation times are several years. It follows that the exact dynamical description of the population, in terms of many overlapping age classes and a defined breeding season, may to a good approximation be replaced by a differential equation, corresponding to continuous population growth.

Under such an approximation, the IWC equation, eq. (4.4), becomes

$$\mathrm{d}X/\mathrm{d}t = -\mu X - \mu \bar{X}[1 + q\{1 - \bar{X}^z\}], \tag{5.1}$$

with the definition

$$\bar{X} \equiv X(t - T). \tag{5.2}$$

The dynamical properties of eq. (5.1), as set out below, are very similar to those previously discovered for other differential–delay equations. For example, Mackey and Glass [51, 52] and Lasota and Wazewska [53] have studied equations of the general form

$$\mathrm{d}X/\mathrm{d}t = -\mu X + R(\bar{X}), \tag{5.3}$$

with several different "one-humped" functions for $R(\bar{X})$, as models for specific physiological processes. In the context of human demography, Frauenthal [54] and Swick [55, 56] have studied integro-differential equations that reduce to the form (5.3) if the kernels are reduced to a limiting, δ-function, form. It is remarkable that the dynamical properties of Frauenthal and Swick's integro-differential equations appear to be essentially identical with those obtained from eq. (5.1) and discussed in detail below.

Among other broadly related studies are those of Sparrow [58, 59], who has investigated the appearance of chaotic behaviour in simple feedback systems of finite dimensions. This elegant work sheds some light on the relation between the dynamical properties of first-order difference equations (as studied in section 2) and the corresponding Poincaré mappings of orbits arising from n-dimensional feedback systems. Specifically, Sparrow considers the n-dimensional system defined by the equations

$$\mathrm{d}x_1/\mathrm{d}t = n[F(x_n) - x_1], \tag{5.4}$$

$$\mathrm{d}x_i/\mathrm{d}t = n[x_{i-1} - x_i], \qquad \text{for } i = 2, 3, \ldots, n. \tag{5.5}$$

Here $F(x)$ is some nonlinear function of x. Taking the Laplace transform of these n equations gives the transformed set:

$$sL\{x_1\} = n[L\{F(x_n)\}] - L\{x_1\}], \tag{5.6}$$

$$sL\{x_i\} = n[L\{x_{i-1}\} - L\{x_i\}], \qquad \text{for } i = 2, 3, \ldots, n. \tag{5.7}$$

Here $L\{\cdot\}$ denotes the Laplace transform, with respect to the Laplace transform variable s. After some manipulation, we can write an equation for the variable x_n:

$$L\{x_n\} = (1 + s/n)^{-n} L\{F(x_n)\}. \tag{5.8}$$

In brief, this feedback system has a transfer function $G(s) = (1 + s/n)^{-n}$. For large n, $G(s)$ approximates e^{-s}, which is exactly the transfer function for a time delay of one time unit. Thus, in this limit $n \gg 1$ (as the state space becomes infinite dimensional), Sparrow's system of eqs. (5.4) and (5.5) reduce to the simple first-order difference equation of section 2:

$$x(t + 1) = F[x(t)]. \tag{5.9}$$

Sparrow's studies of the differences and similarities between his models and those of the first-order map (5.9), for moderate values of n ($3 \le n \le 50$), are discussed in section 5.2.3.

5.2. *Properties of such equations*

5.2.1. *Linear analysis*
The linearized stability analysis of differential–delay equations such as eq. (5.1) is straightforward, but the bewildering array of dynamical behaviour that can be displayed for extreme nonlinearity in the recruitment term (large z) is both fascinating and ill-understood.

For a linearized analysis of the stability of the equilibrium point at $X^* = 1$, we write $X(t) = 1 + x(t)$ in eq. (5.1), carry out a Taylor expansion (ignoring terms of order x^2 or higher), and write $x(t) = x(0) \exp(\lambda t)$ in the ensuing linear equation. This gives a transcendental equation for the eigenvalue λ:

$$\lambda = -\mu - \mu(zq - 1)e^{-\lambda T}. \tag{5.10}$$

It follows that the equilibrium point will be locally stable iff (the detailed analysis is given by Clarke [36] and by May [44]):

$$\mu T < \frac{\pi - \cos^{-1}(1/b)}{(b^2 - 1)^{1/2}}. \tag{5.11}$$

Here b is defined for notational convenience as $b = zq - 1$. This local stability criterion is illustrated in fig. 12. Notice the system always has a locally stable equilibrium point if $zq < 2$.

5.2.2. Nonlinear behaviour

What happens, however, when the equilibrium point is no longer locally stable? At first, as the stability boundary in fig. 12 is crossed, there will be a Hopf bifurcation to a stable limit cycle with period [44]

$$T_0 = \left(\frac{2\pi}{\pi - \cos^{-1}(1/b)} \right)^T . \tag{5.12}$$

As the nonlinearities become more severe (as b continues to increase), this cycle undergoes a cascade of period doublings, eventually entering a regime of apparently chaotic dynamical behavior. Finally, as b becomes very large ($zq \gg 1$), the dynamics suddenly condenses back into a single, relatively simple cycle.

This array of behaviour is shown in figs. 13–17, for eq. (5.1) with the particular parameter choices $T = 2$, $\mu = 1$, $q = 1$, and steepening nonlinearity as the density dependent exponent z takes the values $z = 3.0$, 3.5, 3.7, 3.8, and 4.0.

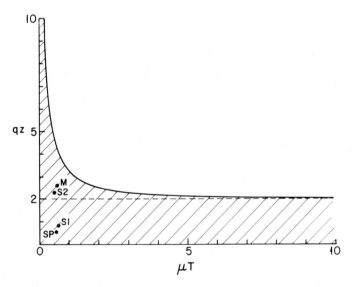

Fig. 12. The shaded area shows the domain of parameter space in which the differential-delay equation, eq. (5.1), has a locally stable equilibrium point. The IWC parameters μ, T, q and z are as defined in the text. The four points correspond to the parameter values actually used by the IWC (see table 3): the point labelled M is for minke whales; S1 is for sei whales with $z = 2.4$ (top row in table 3); S2 is for sei whales with $z = 1$ (second row in table 3); and SP is for female sperm whales. For discussion, see the text.

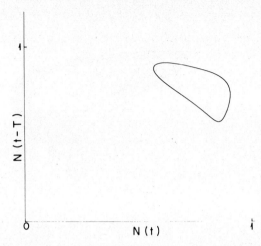

Fig. 13. This figure shows the phase-space plot of the asymptotic solution to the differential-delay eq. (5.1) for the illustrative parameter choice $\mu = 1$, $T = 2$, $q = 1$, and various z: in this figure, $z = 3.0$ (stable cycle).

The sequence of figs. 13–17 are asymptotic phase space portraits, with $N(t - T)$ plotted against $N(t)$, after initial transients have died away. The simple cycle of fig. 13 undergoes a period doubling to give fig. 14. A few further doublings, to about 4, 8 and 16 times the initial period, can be distinguished, before the detail gets lost. In figs. 15 and 16, the orbit is

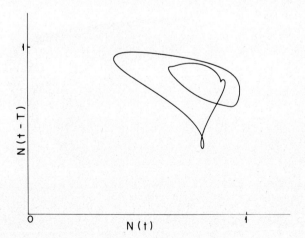

Fig. 14. As for fig. 13, but with $z = 3.5$ (stable cycle after first doubling).

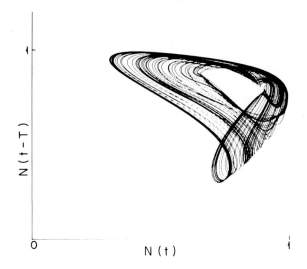

Fig. 15. As for fig. 13, but with $z = 3.7$ (apparently chaotic trajectory).

apparently chaotic. The situation up to this point is strongly reminiscent of the by now well-known properties of simple first-order difference equations with steepening nonlinearities.

The feature that is new, and perhaps surprising, for the differential-

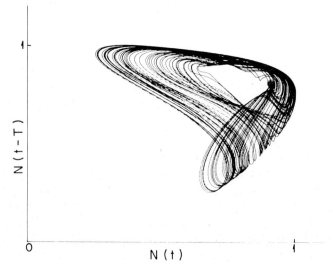

Fig. 16. As for fig. 13, but with $z = 3.8$ (apparently chaotic trajectory).

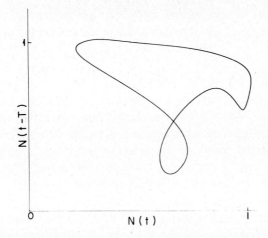

Fig. 17. As for fig. 13, but with $z = 4.0$ (stable and relatively simple cycle).

delay equation is that the orbit again becomes a simple cycle, as shown in fig. 17, for sufficiently steep nonlinearity (sufficiently large z, here $z > 3.9$). This simple orbit, which appears to persist for arbitrarily large z [corresponding to a step function $R(X)$ in eq. (5.1)], seems to condense out of the chaotic regime sharply as z increases beyond a critical value, but neither my numerical studies of this equation, nor any other studies on similar equations of which I am aware, has been sufficiently exhaustive to exclude the possibility of a very rapid "inverse cascade" leading to this simple cycle.

The general features of the linear analysis, and of the nonlinear dynamical behaviour shown in figs, 13–17, are precisely as found for other functional forms of $R(X)$ [51–53], and for related integro-differential equations [54–56].

5.2.3. *Some very crude approximations*
If other time scales tend to be shorter than the time lag T, some properties of these first-order differential–delay equations can be understood from the earlier difference equations of section 2. Since few analytic tools are available for differential-delay equations, such approximations – however crude – are worth pursuing.

If the Hopf bifurcation is to a stable limit cycle with a relatively long period, whose amplitude is (at first) relatively small, the two terms on the

RHS of eq. (5.3) will, in general, each be larger than the derivative on the LHS. Thus, some of the dynamical essentials can be deduced from the approximation

$$\mu X_{t+T} = R(X_t), \tag{5.13}$$

which gives a discrete relationship between X_{t+T} and X_t. Insofar as this difference eq. (5.13) is an accurate approximation, it suggests that there will be a bifurcation from a stable point to a stable cycle of period $2T$, and subsequent "period doubling" bifurcations to cycles of period $4T, 8T, \ldots$, followed eventually by a regime of apparent chaos. This is, indeed, qualitatively what seems to be happening in figs. 13–16.

Some numerical tests of this crude approximation are possible. if eq. (5.13) is used as an approximation to eq. (5.1), so that $R(X)$ has the form given in eq. (5.1), the fixed point in eq. (5.13) will become unstable when the slope b $[b = -(1/\mu)(dR/dX)]$ exceeds 45°; that is, when the parameter combination

$$qz > 2, \tag{5.15}$$

(recall that $b \equiv qz - 1$). This condition is illustrated by the dashed horizontal line in fig. 12, whence it can be seen that the approximate results corresponds to the exact one as μT becomes large. Moreover, for large μT [relatively long maturation times] the critical value of b (as calculated from eq. (5.11)] approaches unity, whence it follows that the exact period T_0 from eq. (5.12) approaches the value $2T$ suggested by the approximation. For the numerical example in figs. 13–17 (in which $\mu T = 2$), the Hopf bifurcation from a stable point to a stable cycle occurs exactly at $z = 2.52$ (while the approximation suggests $z = 2$), and the initial period is $2.74T$ (as opposed to the approximate $2T$).

The above approximation will fail grossly once the mapping of eq. (5.13) with the $R(X)$ of eq. (5.1) no longer maps the interval $[0, 1]$ into itself. For the $R(X)$ of eq. (5.1) this happens when

$$(1+q)z[1+z]^{-(1+z)/z} > 1. \tag{5.15}$$

Thus, for our numerical example with $q = 1$, the approximation becomes qualitatively inadequate for $z > 3.40$. At this point, the LHS of eq. (5.1) cannot be neglected; why, however, it should induce a transition from complicated to relatively simple behaviour beyond about $z \approx 3.9$ is not clear.

This rough understanding of the initial cascade of period doubling and on into an apparently chaotic regime, coupled with less of an understanding of the eventually simpler cyclic behaviour, is also found in Sparrow's numerical studies of eqs. (5.4) and (5.5). Sparrow [58, 59] uses the form II of table 1 for this nonlinear function $F(x)$:

$$F(x) = \lambda x e^{-x}. \tag{5.16}$$

For $n = 50$, he finds a cascade of period doublings, and subsequently apparent chaos, as λ increases from 8.2 to around 60. For larger λ, however, a reverse sequence of doubling bifurcations begins, and by $\lambda \simeq 200$ the system has settled to a simple periodic orbit (with period almost exactly 2), which persists for all higher values of λ. For smaller n the sequence of period doublings and the reverse bifurcations are much reduced: for $n = 30$, there is a progression from an orbit of period approximately 2, then 4, then back to remain around 2 for arbitrarily large λ; for $n = 3$, there appear to be no further bifurcations after the first stable orbit of period 2 appears.

Sparrow's results are broadly in accord with those of eq. (5.1) reported in detail above, and with the approximate analytic understanding sketched earlier in this subsection. His results suggest that an "inverse cascade" lurks between figs. 16 and 17! Clearly, further numerical and analytic work is needed here.

5.3. Biological examples

I think these results for nonlinear differential-delay equations are of great interest. Apart from their intrinsic mathematical interest, they indicate biologically that the range of dynamical behaviour exhibited by nonlinear difference equations (corresponding to populations with discrete, nonoverlapping generations) is also exhibited by differential-delay equations (corresponding to the opposite biological extreme of populations with continuously overlapping generations, provided there are recruitment delays).

Returning to the recruitment equation (5.1) for whales, in table 3 we catalog the values of the parameters P, Q, z, μ and T used by the IWC for particular species of whales. The values of the stability-determining quantities zq and μT are shown in fig. 12, for minke, sei (for two different values of the density dependence parameter z) and female sperm whales. In all cases, the dynamic lies well within the domain of stable point

Table 3
Values of the various parameters in the whaling model, eqs. (4.1) and (4.2), as used by the
IWC for some species of whales (compiled by May [57])

Species	P	Q	q	z	qz	μ	T	μT
Sei	0.19	0.18	0.95	2.4	2.3	0.06	8	0.48
Sei	0.27	0.23	0.85	1.0	0.85	0.07	9	0.63
Minke	0.34	0.37	1.09	2.4	2.6	0.095	6	0.57
Sperm								
(female)	0.20	0.05	0.25	2.4	0.60	0.055	10	0.55

behaviour. The relatively small and fast maturing sei and minke whales are
the only baleen whales still subject to significant harvesting from factory
ships; other baleen whale species (most of which now have protected
status, and therefore do not have population parameters published in
recent IWC documents) are larger, and tend to have lower q values and
longer maturation times.

Laboratory studies of various invertebrate populations have, on the
other hand, often shown patterns of regular cycles or even (arguably)
"deterministic chaos". Many of these populations can plausibly be
described by differential–delay equations (continuous population growth,
with various kinds of time lags in recruitment processes), and several such
examples are reviewed by May [35]. One much-studied example is
Nicholson's blowfly data; May's [60] original crude but successful fit to the
data using a time-delayed logistic equation has been superseded by the
more detailed and fully age-structured models of Oster and Ipaktchi [61],
Gurney et al. [62] and Brillinger et al. [63]. Another interesting laboratory
study is by Halbach [64] on the dynamical behaviour of rotifer populations
at different temperatures. Figure 18 illustrates how the populations display
increasingly oscillatory behaviour as temperature increases; these observations (which Halbach has replicated many times) can be successfully
explained by using a differential–delay equation to model the population
dynamics, and letting the intrinsic growth rate be an increasing function of
temperature.

In short, the above discussion of the nonlinear dynamical behaviour
exhibited by eqs. (5.1) and (5.3) is of mathematical interest for its own
sake, but it probably has little relevance to the dynamics of whale populations. The studies do, however, have substantial applications in explaining

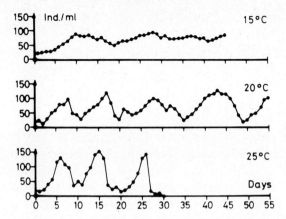

Fig. 18. The dynamical behaviour of laboratory populations of rotifers are shown as functions of time for three different temperatures (after Halbach [64]). For discussion, see the text.

the dynamical behaviour of various laboratory populations of invertebrates, and seem likely to have applications in other areas of population biology (and in physiology and elsewhere).

6. Coupled differential equations

6.1. General remarks

Up to this point, the periodic and apparently chaotic dynamics have been associated either with discrete mappings (difference equations) or with time delays in first-order differential equations. But complicated periodic and chaotic behaviour of course also arises in relatively simple systems of 3 or more coupled first-order differential equations without any time delays. (For 2-dimensional systems – 2 interacting populations – without time lags, there can only be stable points or stable limit cycles; in the plane, trajectories of first-order differential equations cannot cross, which is the basis for the Poincaré–Bendixson theorem and for the relatively simple behaviour.)

Well-known examples of chaotic behaviour in systems of 3 coupled, nonlinear, first-order differential equations come from the work of Lorenz [65] on atmospheric turbulence, and of Rossler [66] on chemical systems. Several other examples have arisen in population models.

6.2. *Three competing species*

The so-called Lotka–Volterra equations for 3 competing populations can be written in the form

$$dX_i/dt = r_i X_i \left(1 - \sum_{j=1}^{3} \alpha_{ij} X_j \right), \tag{6.1}$$

with $i = 1, 2, 3$. May and Leonard [67] and Gilpin [68] have shown that such a system of differential equations may possess any one of a number of different kinds of stable point solutions, or a special class of periodic limit cycle solutions, depending on the values of the competition coefficients α_{ij}. Most interesting, however, is a general class of solutions (corresponding, roughly, to intransitive circumstances in which in pairwise competition species 1 beats species 2, species 2 beats species 3, and species 3 beats species 1), in which the system exhibits nonperiodic population oscillations of bounded amplitude but ever-increasing cycle time. An approximate analytic treatment of this nonperiodic solution is given by May and Leonard; Gilpin's studies are numerical.

Situations of this general kind, corresponding broadly to "intransitive competition", can arise in many other circumstances in chemical reactions, behavioural interactions, physics (in interactions between beams or plasma modes), and elsewhere, so that the phenomena explored by May and Leonard are likely to find various other applications.

6.3. *Predator–prey, host–parasite and other nonlinear biological systems*

Beginning with the work of Kolmogorov in the 1930s, there is a large and growing literature on mathematical models for interacting populations of prey and predators, and on the way such models can exhibit either stable points or stable limit cycle behaviour [69].

More complicated models of this general type have recently arisen in studies of the overall population biology of the association between infectious diseases and their hosts. Such host–parasite models (with parasite defined broadly to include viruses, bacteria, fungi, protozoans, and helminths) can involve 3 or more coupled, nonlinear, first-order differential equations, depending on the complexities of the transmission process; parasites can be transmitted directly from one host to the next, or may possess free-living intermediate stages, or may be transmitted indirectly through one or more intermediate hosts. The models that have been studied in these contexts can exhibit stable points or complicated periodic cycles. As yet, none have revealed chaotic behaviour. It is, however, likely that chaotic behaviour can arise in such models for the population biology of infectious diseases; analytic and numerical studies of the nonlinear behaviour of such models are still in their infancy. (For an introduction to the contemporary literature, see the two-part review by Anderson and May [70, 71], or the detailed treatment of directly transmitted parasitic infections in invertebrate hosts by the same authors [72].

7. Conclusions

I have shown how richly complicated patterns of dynamical behaviour – including cascades of period doublings, and apparently chaotic orbits – can arise in difference equations, and in differential-delay equations, that occur in a natural way in a great variety of situations in population biology and community ecology. In some cases analytic understanding of these dynamic complexities are offered, while in other cases there are only numerical explorations and interesting unanswered questions.

An attempt has also been made to confront these dynamical studies with real data from field or laboratory populations. In general, there has been a tendency for the natural populations to exhibit dynamical behaviour that is relatively tame: for the difference equations with one critical point of section 2, most of the biological populations appear to exhibit stable point behaviour (see section 2.3); for the difference equations with two critical points arising as models of frequency dependent selection, it seems likely that only stable point or 2-point-cycle behaviour can occur (see section 3.3); for the differential-delay equations of section 5, although various artificial laboratory populations may exhibit periodic or chaotic dynamics, natural populations of whales appear to lie securely in the domain of stable point behaviour (see section 5.3). But, as repeatedly emphasized, any generalization of this kind is to be distrusted: there are biases whereby long runs of good data are likely to be mainly available only for relatively steady populations; and most simple models do not do justice to the multispecies reality (remember, higher-order systems often seem more easily to exhibit periodic or chaotic dynamics).

In short, the mathematically interesting questions are clear. But the

R.M. May

biological questions, having to do with how ubiquitous chaotic dynamics may be in nature, remain as yet unsettled.

Acknowledgements

I have benefitted from stimulating discussions with G.F. Oster, J.A. Yorke, and many other people. This work was supported, in part, by the US National Science Foundation, under grant DEB81-02783.

References

[1] Li, T.-Y and J.A. Yorke (1975) Period three implies chaos, Am. Math. Monthly, 82, 985–992.
[2] May, R.M. (1974) Biological populations with nonoverlapping generations: stable points, stable cycles, and chaos, Science 186, 645–647.
[3] May, R.M. (1976) Simple mathematical models with very complicated dynamics, Nature 261, 459–467.
[4] Guckenheimer, J., G.F. Oster, and A. Ipaktchi (1976) The dynamics of density dependent population models, J. Math. Biol. 4, 101–147.
[5] Guckenheimer, J. (1979) The bifurcation of quadratic functions, Ann. N.Y. Acad. Sci. 316, 78–85.
[6] Collett, P. and J.-P. Eckmann (1980) Iterated Maps on the Interval as Dynamical Systems (Birkhäuser, Stuttgart and Boston).
[7] Maynard Smith, J. (1968) Mathematical Ideas in Biology (Cambridge University Press, Cambridge).
[8] May, R.M. (1972) On relationships among various types of population models, Am. Natur. 107, 46–57.
[9] Krebs, C.J. (1972) Ecology: The Experimental Analysis of Distribution and Abundance (Harper and Row, New York).
[10] Moran, P.A.P. (1950) Some remarks on animal population dynamics, Biometrics, 6, 250–258.
[11] Ricker, W.E. (1954) Stock and recruitment, J. Fish. Res. Bd. Can. 11, 559–623.
[12] Cook, L.M. (1965) Oscillation in the simple logistic growth model, Nature 207, 316.
[13] May, R.M. (1974) Ecosystem patterns in randomly fluctuating environments, in: Progress in Theoretical Biology, eds., R. Rosen, F. Snell (Academic Press, New York) pp. 1–50.
[14] Varley, G.C. and G.R. Gradwell (1960) Key factors in population studies, J. Anim. Ecol. 29, 339–401.
[15] Stubbs, M. (1967) Density dependence in the life-cycles of animals and its importance in k- and r-strategies, J. Anim. Ecol. 46, 677–688.
[16] Hassell, M.P. (1974) Density dependence in single species populations, J. Anim. Ecol. 44, 283–296.

[17] Hassell, M.P., J.H. Lawton and R.M. May (1976) Patterns of dynamical behaviour in single-species populations, J. Anim. Ecol. 45, 471–486.

[18] Maynard Smith, J. and M. Slatkin (1973) The stability of predator–prey systems, Ecology 54 384–391.

[19] Bellows Jr., T.S. (1981) The descriptive properties of some models for density dependence, J. Anim. Ecol. 50, 139–156.

[20] Thomas, W.R., M.J. Pomerantz, and M.E. Gilpin (1980) Chaos, asymmetric growth and group selection for dynamical stability, Ecology 61, 1312–1320.

[21] Utida, S. (1967) Damped oscillation of population density at equilibrium, Res. Pop. Ecol. 9, 1–9.

[22] Skellam, J.G. (1951) Random dispersal in theoretical populations, Biometrika 38, 196–218.

[23] Leslie, P.H. (1957) An analysis of the data for some experiments carried out by Gause with population of the protozoa, *Paramecium aurelia* and *P. candatum*, Biometrika 44, 314–327.

[24] Williamson, M. (1974) The analysis of discrete time cycles, in: Ecological Stability, eds., M.B. Usher and M. Williamson (Chapman and Hall, London) pp. 17–23.

[25] Ullyett, G.C. (1950) Competition for food and allied phenomena in sheep and blowfly populations, Phil. Trans. Roy. Soc. B260, 77–174.

[26] Pennycuick, C.J., R.M. Compton and L. Beckingham (1968) A computer model for simulating the growth of a population, or of two interacting populations, J. Theoret. Biol. 18, 316–329.

[27] Usher, M.B. (1972) Developments in the Leslie matrix model, in: Mathematical Models in Ecology, ed., J.N.R. Jeffers (Blackwell, Oxford) pp. 29–60.

[28] May, R.M. and G.F. Oster (1976) Bifurcations and dynamic complexity in simple ecological models, Am. Natur. 110, 573–599.

[29] Feigenbaum, M.J. (1978) Qualitative universality for a class of nonlinear transformations, J. Stat. Phys. 19, 25–52.

[30] Feigenbaum, M.J. (1979) The onset spectrum of turbulence, Phys. Lett. 74A, 375–378.

[31] Collett, P., J.P. Eckmann and O.E. Lanford (1980) Universal properties of maps on an interval. Commun. Math. Phys. 76, 211–254.

[32] May, R.M. and G.F. Oster (1980) Period doubling and the onset of turbulence: an analytic estimate of the Feigenbaum ratio, Phys. Lett. 78A, 1–3.

[33] Helleman, R.H.G. (1980) Self-generated chaotic behavior in nonlinear mechanics, in: Fundamental Problems in Statistical Mechanics, Vol. 5, ed., E.G.D. Cohen (North-Holland, Amsterdam) pp. 165–233.

[34] Bunow, B. and G.H. Weiss (1979) How chaotic is chaos? Chaotic and other "noisy" dynamics in the frequency domain, Math. Biosci. 47, 221–237.

[35] May, R.M. (ed.) (1981) Theoretical Ecology: Principles and Applications, Second Edition (Blackwell, Oxford and Sinauer, Sunderland, Mass.). Chapter 2.

[36] Clarke, B.C. (1964) Frequency-dependent selection for the dominance of rare polymorphic genes, Evolution 18, 364–369.

[37] Clarke, B.C. (1969) The evidence for apostatic selection, Heredity 24, 347–352.

[38] Clarke, B.C. (1975) Frequency-dependent and density-dependent natural selection, in: The Role of Natural Selection in Human Evolution, ed., F.M. Salzano (North-Holland, Amsterdam) pp. 187–200.

[39] Wright, S. (1969) Evolution and the Genetics of Populations, Vol. 2 (Chicago University Press, Chicago).

[40] Kojima, K-I. (1971) Is there a constant fitness value for a given genotype? No! Evolution 25, 281–285.

[41] Endler, J.A. (1977) Geographic Variation, Speciation and Clines (Princeton University Press, Princeton).

[42] Lewontin, R.C. (1958) A general method for investigating the equilibrium of gene frequency in a population, Genetics 43, 419–434.

[43] Lewontin, R.C. (1974) The Genetic Basis of Evolutionary Change (Columbia University Press, New York).

[44] May, R.M. (1979) Bifurcations and dynamic complexity in ecological systems, Ann. N.Y. Acad. Sci. 316, 517–529.

[45] Haldane, J.B.S. and S.D. Jayakar (1963) Polymorphism due to selection depending on the composition of a population, J. Genet. 58, 318–323.

[46] Clark, C.W. (1976) A delayed-recruitment model of population dynamics with application to baleen whale populations, J. Math. Biol. 3, 381–391.

[47] Levin, S.A. and R.M. May (1976) A note on difference–delay equations, Theor. Pop. Biol. 9, 178–187.

[48] Levin, S.A. and C.D. Goodyear (1980) Analysis of an age-structured fishery model, J. Math. Biol. 9, 245–274.

[49] Beddington, J.R., C.A. Free and J.H. Lawton (1975) Dynamic complexity in predator–prey models framed in difference equations, Nature 255, 58–60.

[50] Crofton, H.D. (1971) A model of host–parasite relationships, Parasitology 63, 343–364.

[51] Mackey, M.C. and L. Glass (1977) Oscillation and chaos in physiological control systems, Science 197, 287–289.

[52] Glass, L. and M.C. Mackey (1979) Pathological conditions resulting from instabilities in physiological control systems, Ann. N.Y. Acad. Sci. 316, 214–235.

[53] Lasota, A. and M. Wazewska (1976) Mathematical models of the red blood cell system, Math. Stosowana 6, 25–40.

[54] Frauenthal, J.C. and K.E. Swick (1980) Limit cycle oscillations in human population dynamics (in preparation).

[55] Swick, K.E. (1980) Periodic solutions of a nonlinear age-dependent model of single species population dynamics, SIAM J. Math. Anal. 11, 901–910.

[56] Swick, K.E. (1981) A nonlinear model for human population dynamics, SIAM J. Appl. Math. 40, 266–278.

[57] May, R.M. (1980) Mathematical models in whaling and fisheries management, in: Some Mathematical Questions in Biology, Vol. 13, ed., G.F. Oster (American Mathematical Society, Providence, R.I.) pp. 1–64.

[58] Sparrow, C. (1980) Bifurcation and chaotic behavior in simple feedback systems, J. Theor. Biol. 83, 93–105.

[59] Sparrow, C. (1980) Chaos in a single loop feedback system, Chapter II, PhD thesis (Statistical Laboratory, Cambridge University).

[60] May, R.M. (1974) Stability and Complexity in Model Ecosystems, Second Ed. (Princeton University Press, Princeton).

[61] Oster, G. and A. Ipaktchi (1978) in: Theoretical Chemistry: Periodicities in Chemistry and Biology, eds., H. Eyring, D. Henderson (Academic Press, New York) pp. 111–132.

[62] Gurney, W.S.C., I.F. Blythe and R.M. Nisbet (1980) Nicholoson's blowflies revisited, Nature 287, 17-21.

[63] Brillinger, D.R., J. Guckenheimer, P. Guttorp and G.F. Oster (1980) Empirical modeling of population time series data: the case of age and density dependent vital rates, in: Some Mathematical Questions in Biology, Vol. 13, ed., G.F. Oster (American Mathematical Society, Providence, R.I.) pp. 65-90.

[64] Halbach, U. (1979) Introductory remarks: strategies in population research exemplified by rotifer population dynamics, Fortschr. Zoologie 25, 1-27.

[65] Lorenz, E.N. (1963) Deterministic nonperiodic flow, J. Atmos. Sci. 20, 130-141.

[66] Rossler, O.E. (1979) Continuous chaos - four prototype equations, Ann. N.Y. Acad. Sci. 316, 376-392.

[67] May, R.M. and W.J. Leonard (1975) Nonlinear aspects of competition between three species, SIAM J. Appl. Math. 29, 243-253.

[68] Gilpin, M.E. (1975) Limit cycles in competition communities, Am. Natur. 109, 51-60.

[69] Ref. [35], ch. 5.

[70] Anderson, R.M. and R.M. May (1979) Population biology of infectious diseases: Part I, Nature 280, 361-367.

[71] May, R.M. and R.M. Anderson (1979) Population biology of infectious diseases: Part II, Nature 280, 455-461.

[72] Anderson, R.M. and R.M. May (1981) The population dynamics of microparasites and their invertebrate hosts, Phil. Trans. Roy. Soc. B291, 451-524.

COURSE 9

MAPS OF AN INTERVAL

Michał MISIUREWICZ

Institute of Mathematics, Warsaw University, PKiN IX p.
00-901 Warsaw, Poland

G. Iooss, R.H.G. Helleman and R. Stora, eds.
Les Houches, Session XXXVI, 1981 – Comportement Chaotique des Systèmes Déterministes/
Chaotic Behaviour of Deterministic Systems
© *North-Holland Publishing Company, 1983*

Contents

567

0. Introduction

There are several reasons why people are interested in the theory of maps of an interval in itself. Some of these reasons are:

(1) they are the simplest dynamical systems which are not quite trivial; they can display various kinds of both chaotic and non-chaotic behaviour;

(2) the study of certain interesting more dimensional systems (e.g. the Lorenz attractor) can be reduced to the study of maps of interval;

(3) they may serve as models for physical, biological and other systems;

(4) they may be very easily modeled on computers.

I am able to present here only a small fragment of now well developed theory. Much more information can be found for instance in the book of Collet and Eckmann [1] and in the forthcoming article of Nitecki [2]. I have no time to describe a lot of very interesting problems and results. In particular, I will not say anything about the kneading theory which plays a crucial role in many problems connected with the content of sections 2–4.

I shall quote most theorems without proofs. Several proofs which I present, contain some basic methods and ideas used when studying similar problems.

I will always consider the interval $[0, 1]$, but of course any other closed interval interval would also do. Most of the problems are in one way or another connected with periodic points. A point $x \in [0, 1]$ is called *periodic* for a map $f: [0, 1] \circlearrowright$ if there exists $p \geq 1$ such that $f^p x = x$. The smallest such p is called the *period* of x.

For a measurable set $A \subset [0, 1]$, $|A|$ will denote the Lebesgue measure of A.

1. Topological theory of chaos

One of concepts of *chaos* is due to Li and Yorke: We say that a continuous map $f: [0, 1]] \circlearrowright$ is *chaotic*, if:

(1) there are infinitely many periodic points of f with different periods, and

(2) there is an uncountable set $S \subset J$ (containing no periodic points) which satisfies the following conditions:

A. for every $x, y \in S$, $x \neq y$,

$$\overline{\lim_{n}} \, |f^n x - f^n y| > 0,$$

$$\underline{\lim_{n}} \, |f^n x - f^n y| = 0;$$

B. for every $x \in S$ and y periodic,

$$\overline{\lim_{n}} \, |f^n x - f^n y| > 0.$$

The interpretation of a condition $\overline{\lim}_n \, |f^n x - f^n y| > 0$ is that the orbits of the points x and y have different asymptotic behaviour.

One has to notice that this is a kind of chaos which may pass unnoticed in some situations. It happens (even often) that for almost every (with respect to the Lebesgue measure) point, its trajectory converges towards a given periodic trajectory, but nevertheless the map is "chaotic" in the above meaning. So, from the "physical" point of view, this definition is not quite satisfactory. But we have to remember that in this section we are interested in *topological* theory, and we should not look especially at the Lebesgue measure.

Li and Yorke in ref. [3] prove the following fact:

Proposition (1.1). If f has a periodic point of period 3 then f is chaotic.

It is easy to see that the following holds:

Proposition (1.2). If f^k is chaotic for some $k \geqslant 1$, then f is also chaotic.

In 1963, Šarkovskiĭ [4] proved the following theorem:

Theorem (1.3). Suppose that f has a periodic point of period k. If n stands to the right of k in the following ordering: $3, 5, 7, \ldots, 2 \cdot 3, 2 \cdot 5, 2 \cdot 7, \ldots,$ $2^2 \cdot 3, 2^2 \cdot 5, 2^2 \cdot 7, \ldots, 2^3, 2^2, 2, 1$, then f has a periodic point of period n.

This gives us a sufficient condition for existence of chaos:

Theorem (1.4). If f has a periodic point of period which is not a power of 2 then f is chaotic.

Proof. Suppose that f has a periodic point of some period k which is not a power of 2. Then $k = 2^p \cdot m$, for some $p \geqslant 0$ and $m \neq 1$ odd. The integer $2^{p+1} \cdot 3$ stands to the right of k in the Šarkovskiĭ ordering. Hence, by the Šarkovskiĭ theorem, f has a periodic point of period $2^{p+1} \cdot 3$. Consequently, $f^{2^{p+1}}$ has a periodic point of period 3. By Proposition (1.1), $f^{2^{p+1}}$ is chaotic, and by Proposition (1.2), f is also chaotic.

Practically, when we are given a map, it is not easy to find a periodic point (most of the orbits are usually not periodic). To overcome this obstruction, we shall give another condition, implying chaos [5]. First, we need a definition: we say that there is *no division* for a finite sequence $(x_0, x_1, \ldots, x_n)$ of points of an interval if there is no a such that either $x_i < a$ for all i even and $x_j > a$ for all j odd, or $x_i > a$ for all i even and $x_j > a$ for all j odd.

Theorem (1.5). Assume that for some x and n we have $f^n x \leqslant x < fx$ or $f^n x \geqslant x > fx$ and that there is no division for $(x, fx, f^2 x, \ldots, f^n x)$. Then f has a periodic point of an odd ($\neq 1$) period.

Corollary. Under the assumptions of Theorem (1.5), f is chaotic.

So we see that chaos may occur as a consequence of lack of divisions.

Note that if $f^n x \leqslant x < fx$ or $f^n x \geqslant x > fx$ and n is odd, then automatically there is no divison for $(x, fx, f^2 x, \ldots, f^n x)$.

Now, we shall try to find some quantitative measure of this topological form of chaos. For this, let us introduce the following definitions: a finite set E is called (n, ε)-*separated* $(n = 1, 2, \ldots; \varepsilon > 0)$ if for every $x, y \in E, x \neq y$, there is k such that $0 \leqslant k < n$ and $|f^k x - f^k y| \geqslant \varepsilon$. Let $s(n, \varepsilon)$ be a maximal cardinality of an (n, ε)-separated set. Intuitively, this means the following: suppose that we are able to distinguish between two points only if they are at least ε apart. Then, $s(n, \varepsilon)$ measures how many different orbits of length n we are able to see. To measure how many (or rather how much) distinct asymptotic behaviours we can observe, we have to look at the asymptotic behaviour of $s(n, \varepsilon)$. For this, we define

$$s(\varepsilon) = \overline{\lim_{n}} \frac{1}{n} \log s(n, \varepsilon).$$

And now we make the last step and define the *topological entropy* of f as $h(f) = \lim_{\varepsilon \to 0} s(\varepsilon)$. It is rather clear that it measures something like an intensity of chaos. The following theorem ([6]) confirms this observation:

Theorem (1.6). A continuous map $f: [0, 1] \circlearrowleft$ has positive topological entropy if and only if f has a periodic point of period which is not a power of 2.

There are other characterizations of topological entropy in the case of maps of an interval. For example ref. [7]:

Theorem (1.7). Assume that f is piecewise monotone. Denote by m_n the number of intervals of monotonicity of f^n and by v_n the variation of f^n. Then

$$h(f) = \lim_n \frac{1}{n} \log m_n = \lim_n \frac{1}{n} \log v_n.$$

Also for every continuous map $f: [0, 1] \circlearrowleft$, $\lim_n (1/n) \log(\text{number of periodic points of period } n) \geq h(f)$ [6].

2. Smooth maps

As we mentioned in section 1, it happens that in spite of the occurrence of topological chaos, what we practically see in an experiment is a periodic motion. This is due to the existence of *attracting periodic points* (sinks). An orbit, which once comes close to such attracting periodic orbit, keeps coming closer and closer, and after some time cannot be distinguished (practically) from it.

Most of the maps which appear as models for various phenomena, are to some extent smooth (at least piecewise C^1). In this case, if we have a periodic point x of period p, then to determine if it is attracting, we have to look at the derivative of f^p at x. If $|(f^p)'(x)| < 1$ then x is attracting, if $|(f^p)'(x)| > 1$ then it is repelling. In the case $|(f^p)'(x)| = 1$ we may have various types of behaviour but this does not happen too often.

To say something more about periodic sinks, we must assume that the iterates of f are sufficiently "regular", i.e. they behave as shown in fig. 1. This means more or less that between two consecutive extrema, the absolute value of the first derivative is concave. The natural condition for

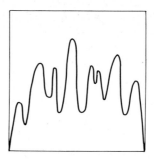

Fig. 1. "Regular" mapping.

this is (if f is of class C^3) $f'''/f' < 0$. But this condition works only for f itself, not for iterates of f. We have:

$$\frac{(f \circ g)'''}{(f \circ g)'} = \left(\frac{f'''}{f'} \circ g\right) \cdot (g')^2 + \frac{g'''}{g'} + 3\left(\frac{f''}{f'} \circ g\right) \cdot g''.$$

If $f'''/f' < 0$ and $g'''/g' < 0$ then first two summands are negative, but the third one may be positive. We want to "neutralize" this third summand. If we look carefully at the formulas for $(f \circ g)'$ and $(f \circ g)''$ we see that this can be done by subtracting $\frac{3}{2}(f''/f')^2$ from f'''/f'. Denote by Sf this difference: $Sf = (f'''/f') - \frac{3}{2}(f''/f')^2$. We see that

$$S(f \circ g) = [(Sf) \circ g] \cdot (g')^2 + Sg,$$

and consequently if $Sf < 0$ and $Sg < 0$ then $S(f \circ g) < 0$. Hence, we have:

Lemma (2.1). If $Sf < 0$ then $S(f^n) < 0$ for all $n \geqslant 1$.

When replacing f'''/f' by Sf, we have lost the convexity of $|f'|$, but the remaining "regularity" conditions are sufficient for our aims.

Lemma (2.2). If $Sf < 0$ then $|f'|$ has no positive local minima.

Proof. Suppose that f' has a non-zero local extremum at x. Then, $f''(x) = (f')'(x) = 0$, and consequently $f'''(x)/f'(x) = Sf(x) < 0$. Hence, if $f'(x) < 0$ then, $(f')''(x) = f'''(x) < 0$ and f' has a local minimum at x, and if $f'(x) > 0$ then $(f')''(x) < 0$ and f' has a local maximum at x. In both cases, $|f'|$ has a local maximum at x.

Of course, when we write $Sf < 0$, we mean that $Sf(x) < 0$ for these x at which $f'(x) \neq 0$. We assume also that the critical points of f are isolated.

Theorem (2.3). If $Sf < 0$ then for every periodic attracting orbit there is a critical point of f or an endpoint of $[0, 1]$ which is attracted by this orbit.

Proof. Notice that the critical points of f^p are the inverse images (up to the order p) of the critical points of f. Suppose $fy = y$, $|(f^p)'(y)| < 1$ (in the case of y attracting with $|(f^p)'(y)| = 1$ the arguments are similar). If y is a critical point of f^p or an endpoint of $[0, 1]$, we are done. If not, we look at the nearest critical points of f^p (to the left and right). By Lemma (2.2), on the whole interval between one of them and y we have $|(f^p)'| < 1$ (it may also happen that instead of this critical point we have an endpoint of $[0, 1]$). Under the iterates of f^p, this interval is contracted towards y.

Theorem (2.3) gives a bound for the possible number of attracting periodic orbits of f with $Sf < 0$.

The usefulness of the condition $Sf < 0$ was discovered by Singer [8]. Surprisingly, it turned out to be a well-known notion of the Schwarzian derivative, used in complex analysis since 1863. The other surprising fact was that all the most commonly considered maps of an interval into intself have a negative Schwarzian derivative. For instance: if $fx = ax(1 - x)$ then

$$Sf(x) = -\frac{3}{2}\left(\frac{2}{1 - 2x}\right)^2,$$

if $fx = \sin x$ then $Sf(x) = -1 - \frac{3}{2}\text{tg}^2 x$.

Now, we shall restrict our attention to a class $\mathscr{C}$ of maps $f : [0, 1] \circlearrowleft$ such that

(i) $f(0) = f(1) = 0$
(ii) $f'' < 0$
(iii) $Sf < 0$.

Notice that by (i) and (ii), f' vanishes at exactly one point. We shall denote this point always by c. If $f'(0) \leqslant 1$, then the whole interval $[0, 1]$ is attracted by 0, and hence, by Theorem (2.3), we get:

Theorem (2.4). If $f \in \mathscr{C}$ then f has at most one periodic attracting orbit.

Suppose that f has a periodic attracting orbit. Then, it is easy to find an open interval J such that $f^n|_J$ is a *homeomorphism* for all $n \geqslant 0$. Such a J will be called a *homterval*. An interesting problem is the existence of homtervals in absence of periodic attracting orbits. The following fact shows that a homterval may be considered as a kind of an attracting aperiodic orbit:

Proposition (2.5). Assume that all periodic points of f are repelling. Then, if J is a homterval then $|f^n J| \to 0$ as $n \to \infty$.

Proof. Notice that:
  (a) an image of a homterval is a homterval;
  (b) if $I_t, t \in T$, are homtervals and there exists $a \in [0, 1]$ such that $a \in I_t$ for every $t \in T$, then $\bigcup_{t \in T} I_t$ is also a homterval;
  (c) since all periodic points are repelling, there is no interval I such that for some n, $f^n(I) \subset I$ and $f^n|_I$ is a homeomorphism.
By (b), for every $n \geqslant 0$, $f^n(J)$ is contained in a maximal homterval (i.e. such a homterval which is not a proper subinterval of any other homterval). By (c) and (a), those maximal homtervals are pairwise disjoint. Since they are all contained in $[0, 1]$, we get $\lim_n |f^n(J)| = 0$.

In general, there are no special reasons for homtervals not to exist. Hall [9] constructed a C^∞ map of $[0, 1]$ into itself having a homterval which is not attracted by any periodic attracting orbit (perhaps the final step of his construction can be done in such a way that there is even no periodic attracting orbit). But in the case of $f \in \mathscr{C}$, this cannot happen [10]:

Theorem (2.6). If $f \in \mathscr{C}$ and all periodic orbits are repelling then f has no homtervals.

3. Bifurcations

Instead of a single map f, we often consider a one-parameter family of maps f_α, $a \leqslant \alpha \leqslant b$. Then we can observe how the properties of a map change when the parameter varies from a to b. We assume that the dependence on parameter α is either continuous or smooth, depending on the situation. The maps f_α will be smooth throughout this section.

We are especially interested in periodic orbits: how they are created or destroyed, which of them are attracting. The obvious observation is that if for some α a point x is periodic of period p and either $|(f_\alpha^p)'(x)| < 1$ or $|(f_\alpha^p)'(x)| > 1$ then for every β sufficiently close to α there will be exactly one periodic point x_p of period p close to x, and there will also be

$|(f_\beta^p)'(x_p)| < 1$ or > 1, respectively. The dependence of x_β on β is either continuous or smooth, depending on how f_α depends on α. In some sense, x_β is the same periodic point for all β close to α, it only moves as the parameter varies. therefore, the question about creation, destruction, "splitting" of periodic points (as β increases or decreases further) makes sense. These changes are called *bifurcations*. We are going to describe only basic bifurcations for maps of the interval – those which are generic (in a suitable smooth topology). The best way to understand them is to look at the graph of the corresponding iterate of f.

These are two types of such bifurcations: saddle-node and flip. The saddle-node bifurcation goes as shown in fig. 2. The graph of f^p approaches the diagonal, at the moment of bifurcation touches it (we assume that the second derivative at this point is nonzero) and then "goes through". Before bifurcation, there is no periodic point of period p in the (small) interval we are considering. At the moment of bifurcation there is one, the derivative of f^p at this point is 1, it is half-repelling, half-attracting. After the bifurcation, there are two periodic points, one of them

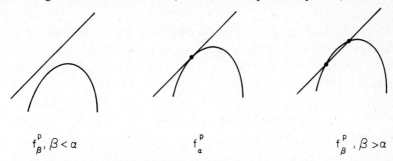

f_β^p, $\beta < \alpha$ $\qquad\qquad$ f_α^p $\qquad\qquad$ f_β^p, $\beta > \alpha$

Fig. 2. Saddle-node bifurcation. Graph of f^p.

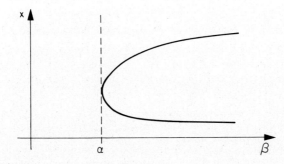

Fig. 3. Saddle-node bifurcation. Dependence of periodic points on a parameter.

is repelling and the other one attracting. The dependence of periodic points on the parameter is shown in fig. 3. I will call this bifurcation a straight one. We can also have, of course, a reverse bifurcation: as β increases, from two periodic orbits we get one and then none.

The flip (or period-doubling) bifurcation occurs when a reversing orientation period point of period p changes from attracting to repelling or vice-versa. Then, a new orbit of period $2p$ is created (or for a reverse bifurcation destroyed) as shown in figs. 4–7. The case (a) occurs if $(f_\alpha^{2p})'''(x)<0$, the case (b) if $(f_\alpha^{2p})'''(x)>0$. However, since $f_\alpha^p(x)=x$ and $(f_\alpha^p)'(x)=-1$, we have $(f_\alpha^{2p})'(x)=1$ and $(f_\alpha^{2p})''(x)=0$, and we get the following theorem:

Theorem (3.1). If $Sf_\alpha<0$ then the case (b) of a flip bifurcation cannot occur.

This explains the words "period doubling bifurcation". Since in the experiments, we do not see repelling periodic orbits, the case (a) looks like

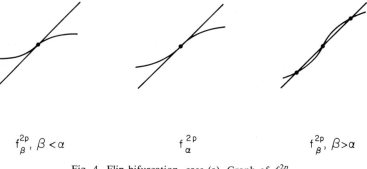

$$f_\beta^{2p},\ \beta<\alpha \qquad f_\alpha^{2p} \qquad f_\beta^{2p},\ \beta>\alpha$$

Fig. 4. Flip bifurcation, case (a). Graph of f^{2p}.

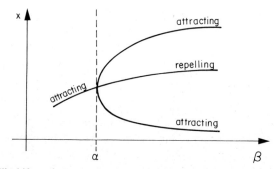

Fig. 5. Flip bifurcation, case (a). Dependence of periodic points on a parameter.

a periodic orbit of period p becoming a periodic orbit of period $2p$. Usually, when we look at the longer range of parameter values, we observe not only one bifurcation, but many of them. The typical picture is shown in fig. 8. We obtain subsequently periodic points $p, 2p, 2^2p, 2^3p, \ldots$, at some point, we get a cumulation of bifurcations. All those periodic points become repelling and they do not bifurcate any more (but of course periodic points of other periods do).

We should also say how the topological entropy depends on a parameter [7]:

Theorem (3.2). If $\alpha \mapsto f_\alpha$ is a continuous mapping from $[a, b]$ to the space of C^2 maps of $[0, 1]$ into itself with C^2 topology and every f_α is piecewise

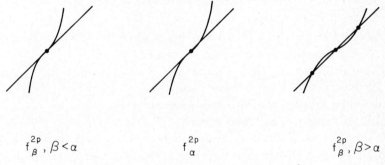

$f_\beta^{2p}, \beta < \alpha$ $\qquad\qquad$ f_α^{2p} $\qquad\qquad$ $f_\beta^{2p}, \beta > \alpha$

Fig. 6. Flip bifurcation, case (b). Graph of f^{2p}.

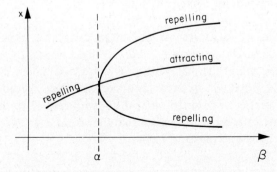

Fig. 7. Flip bifurcation, case (b). Dependence of periodic points on a parameter.

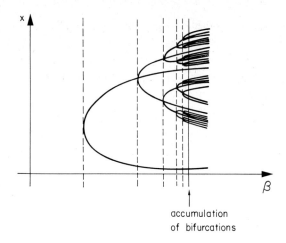

accumulation
of bifurcations

Fig. 8. Sequence of period-doubling bifurcations.

monotone with all critical points non-degenerated (i.e. $f''_\alpha \neq 0$ at the critical points) then the function $\alpha \mapsto h(f_\alpha)$ is continuous.

There are interesting open problems concerning even such simple family as $f_\alpha(x) = \alpha x(1-x)$, $1 \leqslant \alpha \leqslant 4$. For example:

(1) Is the function $\alpha \mapsto h(f_\alpha)$ non-decreasing?

(2) Can reverse bifurcations occur?

It may be shown that a negative answer to the second question would imply a positive answer to the first one. Numerical experiments suggest that indeed the answer to the second question is negative, and to the first one positive.

4. Typical behaviour of orbits

In this section, we shall restrict our attention to the elements of $\mathscr{C}$. However, some of the results are valid for larger classes of maps.

We are interested in the following problem: if we choose a point from $[0, 1]$ randomly, what will its orbit look like? In other words: we want to find some properties of orbits of *almost all* (with respect to the Lebesgue measure) points. "Almost all" means that they form a set the complement of which has Lebesgue measure zero.

Theorem (4.1). If $f \in \mathscr{C}$ and f has a periodic attracting point z then for almost every $x \in [0, 1]$ the trajectory of x is attracted towards the orbit of z.

Before we prove this theorem, we shall prove two lemmata. By c, as before, we denote the critical point of f. For a set A, by $\bar{A}$ we denote its closure.

Lemma (4.2). Let $f \in \mathscr{C}$. Let U and V be open intervals such that $\bar{V} \subset U$, $c \in V$, and all periodic orbits lying entirely outside V are repelling. Let J be a homterval. Then there exist $m \geqslant 0$ such that either $f^m(J) \subset U$ or $f^m(J) \cap V \neq \emptyset$.

Proof. Set
$$\gamma = \sup_{[0,1] \setminus V} \left| \frac{f''}{f'} \right|.$$
The function $\log |f'|$ satisfies the Lipschitz condition with the constant γ on the components of $[0, 1] \setminus V$. We show first that:
  (a) If K is an interval such that $f^k(K) \cap V = \emptyset$ for $k = 0, 1, \ldots, n-1$ then
$$\log \frac{\sup_K |(f^n)'|}{\inf_K |(f^n)'|} \leqslant \gamma \sum_{k=0}^{n-1} |f^k(K)|.$$
The ratio $\sup |(f^n)'| / \inf |(f^n)'|$ measure how f^n differs from a linear (more exactly – affine) map. Since it turns out to be bounded, this will allow us to enlarge a supposedly maximal homterval and get a contradiction.
  If $a, b \in K$ then we have:
$$\log \frac{|(f^n)'(a)|}{|(f^n)'(b)|} = \sum_{k=0}^{n-1} [\log |f'(f^k a)| - \log |f'(f^k b)|].$$
Since the set $f^k(K)$ is an interval and is contained in $[0, 1] \setminus V$, it is contained in some component of $[0, 1] \setminus V$. Hence,
$$\log |f'(f^k a)| - \log |f'(f^k b)| \leqslant \gamma |f^k a - f^k b| \leqslant \gamma |f^k(K)|,$$
and we obtain (a).
  Denote $\beta = \text{dist}(I \setminus U, \bar{V})$. Since the sets $I \setminus U$ and $\bar{V}$ are compact and disjoint, we have $\beta > 0$.
  Suppose that for every $m \geqslant 0$, $f^m(J)$ is disjoint form V and $\overline{f^n(J)}$ is not contained in U. Then:
  (b) there exists n_0 such that if $n \geqslant n_0$, H is a homterval containing

$f^n(J)$, and $f^k(H)$ is disjoint from V for all $k \geqslant 0$, then H is disjoint from U.

Suppose that (b) is false. Then there exist $n \geqslant 0$, $k > 0$ and homtervals K and L such that $f^n(J) \subset K$, $f^{n+k}(J) \subset L$, both $\bar{K}$ and $\bar{L}$ contain the same endpoint of U and a piece of U adjacent to this endpoint. Moreover, $f^k(K \cup L)$ is disjoint from V for all $k \geqslant 0$. The interval $K \cup L$ is a homterval and $f^k(K \cup L)$ intersects $K \cup L$. The same argument as in the proof of Proposition (2.5) shows that this impossible. Hence, (b) holds.

Let M be a maximal homterval containing $f^{n_0}(J)$ such that $f^k(M)$ is disjoint from V for all $k \geqslant 0$. By (b) we have:

(c) for each $n \geqslant 0$ every homterval H containing $f^n(M)$ and such that $f^k(H)$ is disjoint from V for all $k \geqslant 0$, is disjoint from U.

Now take an open interval L containing M, not equal to M such that

(d) $|L|/|M| < \beta e^{-\gamma(\beta+1)} + 1$.

We shall prove by induction that for every k:

(e) $|f^k(L)|/|f^k(M)| < \beta + 1$ and $f^k|_L$ is a homeomorphism.

For $k = 0$ this is obvious. Suppose that (e) holds for $k = 0, 1, \ldots, n-1$. We have then $|f^k(L) \setminus f^k(M)| < \beta |f^k(M)| \leqslant \beta$. From this, (c) and the definition of β it follows that $f^k(L)$ is disjoint from V for $k = 0, 1, \ldots, n-1$. Hence (since L is an interval) $f^n|_L$ is a homeomorphism. By (a) and since all $f^i(M)$ are pairwise disjoint [again we use arguments from the proof of Proposition (2.5)], we obtain:

$$\frac{|f^n(L)|}{|f^n(M)|} - 1 = \frac{|f^n(L \setminus M)|}{|f^n(M)|} < \frac{|L \setminus M|}{|M|} \exp\left(\gamma \sum_{k=0}^{n-1} |f^k(L)| \right)$$

$$\leqslant \beta \exp[-\gamma(\beta+1)] \exp\left(\gamma \sum_{k=0}^{n-1} (\beta+1) |f^k(M)| \right) \leqslant \beta.$$

This ends the proof of (e).

Since $f^n|_L$ is a homeomorphism for each $n \geqslant 0$, L is a homterval. Also $f^n(L)$ is disjoint from V for all $n \geqslant 0$. This contradicts the maximality of M.

Lemma (4.3). Let f, U, V be as in Lemma (4.2). Then there exists $m \geqslant 1$ such that if $f^j(x) \notin U$ for $j = 0, 1, \ldots, m-1$ then $|(f^m)'(x)| > 1$.

Proof. Suppose that for every $n \geqslant 1$ there exists x_n such that:

(a) $|(f^n)'(x_n)| \leqslant 1$ and

(b) $f^j(x_n) \notin U$ for $j = 0, 1, \ldots, n-1$.

Take the maximal open interval J_n containing x_n such that $f^j(J_n)$ is disjoint from V for $j=0, 1, \ldots, n-1$. The point x_n divides J_n into two subintervals. By Lemmata (2.1) and (2.2), on one of these subintervals we have $|(f^n)'| \leqslant 1$. Denote this subinterval by L_n. By the maximality of J_n, there exists $k(n) < n$ such that

(c) $f^{k(n)}(L_n)$ has a common endpoint with some component of $[0, 1] \setminus V$.

We claim that:

(d) $|L_n| \to 0$ as $n \to \infty$.

If not, then there exist: a sequence $n_1 < n_2 < n_3 < \ldots, x_0 \in [0, 1]$ and $\varepsilon > 0$ such that $x_{n_i} \to x_0$ as $i \to \infty$, $|L_{n_i}| > \varepsilon$ for every i and all intervals L_{n_i} are on the same side of x_{n_i}. For every j we have $f^j(x_{n_i}) \notin U$ if $n_i > j$, and hence $f^j(x_0) \notin U$. But one of the intervals $(x_0 - \varepsilon, x_0)$ and $(x_0, x_0 + \varepsilon)$ is a homterval. This contradicts Lemma (4.2). This proves (d).

By (a) we have $|f^n(L_n)| \leqslant |L_n|$, and therefore:

(e) $|f^{n-k(n)}[f^{k(n)}(L_n)]| \to 0$ as $n \to \infty$.

In view of (c) and the fact that $f^{k(n)}(x_n) \notin U$, $f^{k(n)}(L_n)$ contains some component of $U \setminus V$. Hence, for one of the components of $U \setminus V$ (call it K) and some sequence $n_1 < n_2 < n_3 < \cdots$ we have $f^{k(n_i)}(L_{n_i}) \supset K$ for every i. From (e) we get

$$|f^{n_i - k(n_i)}(K)| \to 0 \qquad \text{as } i \to \infty$$

and therefore

(f) $n_i - k(n_i) \to \infty$ as $i \to \infty$.

Let a be a point of condensation of the sequence $[f^{k(i)}(x_{n_i})]_{i=1}^{\infty}$. Since

$$f^j[f^{k(n_i)}(x_{n_i})] \notin U \qquad \text{for } n_i - k(n_i) > j,$$

we obtain in view of (f):

(g) $f^j(a) \notin U$ for all $j \geqslant 0$.

Let M be the minimal open interval such that $K \cup \{a\} \subset \bar{M}$. Then, by (f), M is a homterval and $f^j(M)$ is disjoint from V for every $j \geqslant 0$. This contradicts (g) and Lemma (4.2).

The above proof uses in a typical way Lemmata (2.1) and (2.2). We should mention that the proof of Theorem (2.6) goes along the same lines as the above one.

Proof of Theorem (4.1). From the proof of Theorem (2.3) it follows that for some k the whole interval joining $f^k(z)$ with c is attracted by the orbit of z. This orbit attracts also some neighbourhoods of $f^k(z)$ and c. Therefore, there exist U and V as in Lemmata (4.2) and (4.3) and such that the orbit of every point of U is attracted by the orbit of z.

Consider f^m, where m is as in the conclusion of Lemma (4.3). We want to prove that the set

$$\{x: f^n x \notin U \text{ for all } n \geqslant 0\}$$

$$= \left\{ x: (f^m)^n(x) \in \bigcap_{j=0}^{m-1} f^{-j}([0,1] \setminus U) \text{ for all } n \geqslant 0 \right\}$$

has Lebesgue measure zero.

The set $\tilde{U} = \bigcap_{j=0}^{m-1} f^{-j}([0,1] \setminus U)$ consists of a finite number of closed intervals. We slightly enlarge all of them, denote this larger set by W and redefine f^m on $W \setminus \tilde{U}$ to get a piecewise C^2 map $g: W \to [0,1]$ such that every component of W is mapped onto the whole $[0,1]$, and $|g'| > \alpha$ for some $\alpha > 1$ (see fig. 9). The set E of points at which all iterates of g are defined is the intersection of the sets E_n of points at which g^n is defined, and it contains the set $\{x: f^k x \notin U \text{ for all } k \geqslant 0\}$. Hence, it is enough to prove that $|E_n| \to 0$ as $n \to \infty$.

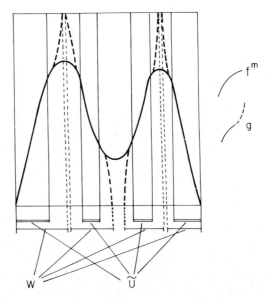

Fig. 9. Graphs of f^m and g.

By the same arguments as in the proof of Lemma (4.2), we get

$$\log \frac{\sup_K |(g^n)'|}{\inf_K |(g^n)'|} \leqslant \gamma \sum_{k=0}^{n-1} |g^k(K)|$$

for every component K of E_n (γ is a Lipschitz constant for $\log|g'|$). Since $|g'| \geqslant \alpha$, we have $1 \geqslant |g^n(K)| \geqslant \alpha^{n-k}|g^k(K)|$ and thus

$$\sum_{k=0}^{n-1} |g^k(K)| \leqslant \sum_{k=0}^{n-1} \alpha^{k-n} < \sum_{j=1}^{\infty} \alpha^{-j} = \frac{1}{\alpha-1}.$$

Hence, we get:

$$\sup_K |(g^n)'| / \inf_K |(g^n)'| \leqslant \delta,$$

where $\delta = \exp[\gamma/(\alpha-1)]$.

We have

$$g^n(K) = [0,1] \quad \text{and} \quad g^n(K \setminus E_{n+1}) = [0,1] \setminus W.$$

Therefore

$$1 \geqslant |K| \cdot \inf_K |(g^n)'| \quad \text{and} \quad |[0,1] \setminus W| \leqslant |K \setminus E_{n+1}| \sup_K |(g^n)'|.$$

We get:

$$\frac{|K|}{|K \cap E_{n+1}|} \geqslant 1 + \frac{|K \setminus E_{n+1}|}{|K|} \geqslant 1 + \frac{|[0,1] \setminus W| \cdot \inf_K |(g^n)'|}{\sup_K |(g^n)'|}$$

$$\geqslant 1 + \frac{|[0,1] \setminus W|}{\delta}.$$

Set $\eta = 1/(1 + |[0,1] \setminus W|/\delta)$. Clearly, $0 < \eta < 1$. We have $|K \cap E_{n+1}| \leqslant \eta |K|$ for every component K of E_n. Summing over components of E_n, we get $|E_{n+1}| \leqslant \eta \cdot |E_n|$. Hence, by induction, we get $|E_k| \leqslant \eta^k$ for $k \geqslant 0$. Consequently, $|E_k| \to 0$ as $k \to \infty$.

A quite different type of behaviour may be observed when there exists a probabilistic invariant measure, absolutely continuous with respect to the Lebesgue measure (p.i.a.c. measure). If this measure is ergodic then, by the ergodic theorem, almost every orbit will have a kind of "uniform" distribution, according to the density of the measure. More precisely, if ϱ is a density of a p.i.a.c. measure and A is a measurable set then

$$\lim_n \frac{\text{Card}\{k < n : f^k x \in A\}}{n} = \int_A \varrho(t) \, dt$$

for almost every x from the support of the measure. In an experiment, we see it as a "stochastic" or "chaotic" behaviour of orbits.

Let us recall several facts connected with invariant measures. let X be a non-empty compact metric space and $f : X \to X$ a continuous map. A probabilistic measure μ on X is called *invariant* if for every Borel set A, $\mu[f^{-1}(A)] = \mu(A)$. The space of all such measures will be denoted by $\mathfrak{M}(M, f)$. We can look at them from another point of view. Let $C(X)$ be the space of all continuous functions on X with the norm sup. Then the formula $\mu(\varphi) = \int \varphi \, d\mu$ gives a one-to-one correspondence between the space $\mathfrak{M}(X)$ of all probabilistic measures on X and the cone of all continuous positive functionals on $C(X)$ of norm 1. Consequently, we may endow $\mathfrak{M}(X)$ with the weak$^-$* topology: $\mu_n \to \mu$ if and only if for every $\varphi \in C(X)$, $\mu_n(\varphi) \to \mu(\varphi)$. By the Alaoglu theorem, $\mathfrak{M}(X)$ is compact in this topology.

The map $f : X \to X$ induces a map $f^* : C(X) \to C(X)$ by $f^*(\varphi) = \sigma \circ f$. Then, f^* induces $f_* : \mathfrak{M}(X) \to \mathfrak{M}(X)$ by $[f_*(\mu)](\varphi) = \mu[f^*(\varphi)]$. We have $f^*(\chi_A) = \chi_A \circ f = \chi_{f^{-1}(A)}$, and consequently

$$[f_*(\mu)](A) = [f_*(\mu)](\chi_A) = \mu(\chi_{f^{-1}(A)}) = \mu[f^{-1}(A)].$$

From this it is easy to deduce that μ is invariant if and only if $f_*(\mu) = \mu$. This gives us a natural way of looking for invariant measures. We start with an arbitrary measure $\mu \in \mathfrak{M}(X)$ and look at the sequence

$$\left(\frac{1}{n} \sum_{k=0}^{n-1} f_*^k(\mu) \right)_{n=1}^{\infty}.$$

It has a subsequence converging to some measure ν. For every $\varphi \in C(X)$ we have:

$$\left| \left[f_* \left(\frac{1}{n} \sum_{k=0}^{n-1} f_*^k(\mu) \right) \right](\varphi) - \left(\frac{1}{n} \sum_{k=0}^{n-1} f_*^k(\mu) \right)(\varphi) \right|$$

$$= \left| \frac{1}{n} [f_*^n(\mu)](\varphi) - \frac{1}{n} \mu(\varphi) \right| \leqslant \frac{2}{n} \|\varphi\| \to 0 \qquad \text{as } n \to \infty.$$

Consequently, $[f_*(\nu)](\varphi) = \nu(\varphi)$, i.e. ν is invariant (thus, we proved that $\mathfrak{M}(X, f)$ is not-empty, the so-called Krylov–Bogolubov theorem).

Since we will be interested in measures absolutely continuous with respect to a given one, we prove:

Lemma (4.4). Let $v \in \mathfrak{M}(X)$. Let measures μ_n $(n = 1, 2, \ldots)$ be absolutely continuous with respect to v with the densities $\varrho_n \geqslant 0$ respectively, and let $\varrho_n \leqslant \varrho$ for all n and for a v-integrable function ϱ. Suppose that $\mu_n \to \mu$ in the weak$^-$* topology. Then, μ is absolutely continuous with respect to v.

Proof. It is enough to show that if a set A has measure v zero then also $\mu(A) = 0$. Suppose that, on the contrary, $v(A) = 0$ but $\mu(A) > 0$. Then we can find: a compact set $B \subset A$ such that $\mu(B) > 0$, an open set $C \supset B$ such that $\int_C \varrho \, dv < \mu(B)$, and a continuous function φ such that $0 \leqslant \varphi \leqslant 1$, $\varphi = 0$ outside C, and $\varphi = 1$ on B. We have

$$\mu_n(\varphi) = \int \varphi \cdot \varrho_n \, dv \leqslant \int \varphi \cdot \varrho \, dv \leqslant \int_C \varrho \, dv$$

for all n, and consequently

$$\mu(B) \leqslant \mu(\varphi) \leqslant \int_C \varrho \, dv,$$

a contradiction.

Let us go back to the case when $X = [0, 1]$. If f is smooth then the image of a measure which is absolutely continuous with respect to the Lebesgue measure, is also absolutely continuous. Therefore, in view of Lemma (4.4) and the preceding construction, in order to prove the existence of a p.i.a.c. measure it is enough to show that all images of the Lebesgue measures have densities all bounded by the same integrable function.

It is easy to find a density of an image of a measure with a given density ϱ. One can check that it is equal to

$$P(\varrho)(x) = \sum_{y \in f^{-1}(x)} \frac{\varrho(y)}{|f'(y)|} .$$

The operator P is called the Perron–Frobenius operator. With this notation, we can restate our result as:

Proposition (4.5). If all functions $P^n(1)$ are all bounded by the same integrable function, then f has a p.i.a.c. measure.

Let me remark that one can prove Proposition (4.5) directly, but the way presented here seems to me more natural.

To illustrate how Proposition (4.5) works in the case of $f \in \mathscr{C}$, we shall prove:

Theorem (4.6). Let $f \in \mathscr{C}$ and $f(c) = 1$. Then there exists a p.i.a.c. measure.
   We start by estimating $P^n(1)$ far from the images of the critical point (i.e. in this case, far from 0 and 1):

Lemma (4.7). If $f \in \mathscr{C}$ and $f(c) = 1$ then $P^n(1)(x) \leqslant 1/\min(x, 1-x)$ for all $n \geqslant 0$ and $x \in (0, 1)$.

Proof. Let a, b be two consecutive critical points of f^n (one of them may be instead an endpoint of $[0, 1]$). Then, either $f^n a = 0$, and $f^n b = 1$, or vice versa. There is exactly one $y \in (a, b)$ such that $f^n y = x$. By Lemma (2.2), $|(f^n)'|$ is bounded by $|(f^n)'(y)|$ on one of the intervals $[a, y], [y, b]$. Therefore,

$$\min(x, 1-x) = \min\left(\int_a^y |(f^n)'|, \int_y^b |(f^n)'| \right) \leqslant |(f^n)'(y)|(b-a).$$

Hence,

$$\frac{1}{|(f^n)\in(y)|} \leqslant \frac{b-a}{\min(x, 1-x)} .$$

Summing these inequalities over all intervals of monotonicity of f^n, we get

$$\sum_{y \in f^{-1}(x)} \frac{1}{|(f^n)'(y)|} \leqslant \frac{1}{\min(x, 1-x)} .$$

But is easy to check that the left-hand side is equal to $P^n(1)(x)$.

Proof of Theorem (4.6). We want to check that the functions $P^n(1)$ are bounded by the same integrable function. Unfortunately, the function $1/\min(x, 1-x)$ is not integrable. Therefore, we have to get an additional bound in neighbourhoods of 0 and 1.
   Since f is of class C^2 and $f''(c) \neq 0$, for given $\varepsilon > 0$ there exist constants $\alpha, \omega > 0$ such that if $|x - c| < \varepsilon$ then $\alpha |x - c| \leqslant |f'(x)| \leqslant \omega |x - c|$. Consequently, $\frac{1}{2}\alpha(x-c)^2 \leqslant 1 - f(x) \leqslant \frac{1}{2}\omega(x-c)^2$. Hence, for $n \geqslant 1$, if $x \neq y$, $fx = fy = z$, $|x - c| < \varepsilon$ and $|y - c| < \varepsilon$, then

$$P^n(1)(z) = P[P^{n-1}(1)](z) = \frac{P^{n-1}(1)(x)}{|f'(x)|} + \frac{P^{n-1}(1)(y)}{|f'(y)|}$$

$$\leqslant \frac{2}{\frac{1}{2}-\varepsilon}\left(\frac{1}{|f'(x)|}+\frac{1}{|f'(y)|}\right)$$

But $|f'(x)|\geqslant\alpha|x-c|\geqslant\alpha[2\omega^{-1}(1-z)]^{1/2}$ and the same is true with y instead of x. Consequently, $P^n(1)(z)\leqslant\eta(1-z)^{-1/2}$, where $\eta=2(\frac{1}{2}-\varepsilon)^{-1}[\alpha(2/\omega)^{1/2}]^{-1}$. This holds for all z such that $1-z<\delta$, $\delta=\alpha\varepsilon^2/2$. If ε is sufficiently small then our estimate holds also for $n=0$.

Now look at a neighbourhood of 0. If $\zeta<\frac{1}{2}$ then (since 0 is repelling) there are constants β and γ such that $\gamma\geqslant\beta>1$ and $\beta\leqslant|f'(x)|\leqslant\gamma$ for all $x\in[0,\zeta]$. Also $\tilde{\beta}\leqslant|f'(x)|\leqslant\tilde{\gamma}$ (for some constants $\tilde{\gamma}\geqslant\tilde{\beta}>0$) if $x\in[1-\delta,\delta]$. If ζ is sufficiently small then we may make γ and β close to each other (but not too close to 1). In particular, we may assume that $\gamma^{1/2}<\beta$.

If $f^n y=x$ then we look at the last k such that $f^k y\notin[0,\frac{1}{2}]$. According to this, we rewrite the formula for $P^n(1)(x)$:

$$P^n(1)(x)=\sum_{k=0}^{n}\frac{P^k(1)(x_k)}{|(f^{n-k})'(x_k)|},$$

where for $k=0$, $x_0,fx_0,\ldots,f^n x_0\in[0,\zeta]$ and for $k>0$, $x_k\in(\frac{1}{2},1]$ and fx_k, $f^2 x_k,\ldots,f^{n-k}x_k\in[0,\zeta]$, and $f^{n-k}x_k=x$ for all k. If ζ is sufficiently small, then $x_k\in(1-\delta,1]$ for $k=1,2,\ldots,n$. Hence:

$$P^n(1)(x)\leqslant\frac{1}{|(f^n)'(x_0)|}+\sum_{k=1}^{n}\frac{\eta(1-x_k)^{-1/2}}{|(f^{n-k})'(x_k)|}.$$

We have $x\leqslant\gamma^n x_0$ and $x\leqslant\tilde{\gamma}\cdot\gamma^{n-k-1}(1-x_k)$ for $k\geqslant1$. Besides, $|(f^n)'(x_0)|\geqslant\beta^n$ and $|(f^{n-k})'(x_k)|\geqslant\tilde{\beta}\cdot\beta^{n-k-1}$ for $k\geqslant1$. We get:

$$P^n(1)(x)\leqslant1+\sum_{k=1}^{n}\frac{\eta}{(x/\tilde{\gamma}\cdot\gamma^{k+1-n})^{1/2}\cdot\tilde{\beta}\cdot\beta^{n-k-1}}$$

$$<1+\frac{\eta\tilde{\gamma}^{1/2}}{\tilde{\beta}}\cdot\sum_{j=0}^{\infty}\left(\frac{\gamma^{1/2}}{\beta}\right)^j\cdot\frac{1}{x^{1/2}}.$$

Since we have assumed $\gamma^{1/2}/\beta<1$, we get $P^n(1)(x)\leqslant\vartheta/x^{1/2}$ for some constant $\vartheta>0$, independent of n.

Since on neighbourhoods of 0 and 1 the functions $\vartheta/x^{1/2}$ and $\eta(1-x)^{-1/2}$ respectively, are integrable and outside these neighbourhoods the function $1/\min(x,1-x)$ is integrable, the proof is complete.

It is possible to generalize this proof to get the following theorem [11]:

Theorem (4.8). If $f\in\mathscr{C}$ and $\inf\{|c-f^n n|:n\geqslant1\}>0$ then there exists a p.i.a.c. measure.

Under the above assumption, this measure is unique, ergodic, and its support is a union of a finite number of intervals. These intervals (if there is more than one) are permuted by f. If there are p of them, then f^p restricted to one of them is exact (i.e. if a measure of A is positive then the measures of $f^n(A)$ tend to the full measure).

Ledrappier proved even more: the natural extension of this system is isomorphic to a Bernoulli shift. This means that the stochastic properties of the system are like those of the coin tossing – quite random.

The critical point is always in the interior of the support of our measure. Hence, almost every point falls after some time into the support of the measure. Therefore, the long-term behaviour of almost all points of $[0, 1]$ (not only of the support) is stochastic.

When we look instead of at a single map, at a one-parameter family of maps, we may ask what one can say about the sets of parameters for which various types of typical behaviour occur.

Assume that $f_\alpha, \alpha \in [a, b]$, is such a family, $f_\alpha \in \mathscr{C}$ for each and the dependence on α is smooth. Assume also that $(f_a)'(0) \leqslant 1$ and $f_b(c) = 1$. Then we know that:

(i) the set of these α for which f_α has an attracting periodic orbit, is open; probably, for a generic family it is dense, but this is not known for any family;

(ii) the set of these α that f_α has a p.i.a.c. measure, has positive Lebesgue measure for a generic family (Jakobson [12]) and probably even for every family;

(iii) there are α such that neither periodic attracting orbit nor a p.i.a.c. measure exists for f_α; this is proved at least for some families, but is probably true for all of them.

5. Piecewise expanding maps

We call a map $f: [0, 1] \circlearrowleft$ *piecewise expanding* if there exist points $a_0 = 0 < a_1 < a_2 < \cdots < a_m = 1$ such that f on every $[a_{i-1}, a_i]$ is C^1 and $|f'| \geqslant \alpha > 1$. We do not assume that f is continuous, it is only piecewise continuous. At the points a_i, by $f(a_i)$ and $f'(a_i)$ we mean the corresponding one-sided limits; we are not very much interested in what the actual value of f at these points is.

Since $|f'| > 1$, there cannot be any periodic attracting points. On the contrary, we have the following theorem (Rychlik [13]):

Theorem (5.1). If f is piecewise expanding and f' is piecewise Hölder continuous, then there exists a p.i.a.c. measure for f.

There is a finite number of such measures and they have similar properties as in the case of $f \in \mathscr{C}$.

There are several weaker versions of Theorem (5.1) (i.e. their assumptions are stronger) proved by various authors. We shall present the proof of the following weak version:

Theorem (5.2). If f is piecewise expanding and for each i, $f\big|_{[a_{i-1}, a_i]}$ is C^2 and onto $[0, 1]$, then there exists a p.i.a.c. measure for f.

Proof. In spite of the fact that f is not continuous, we may still use the Perron–Frobenius operator and Proposition (4.5) still holds.

We have a very similar situation as in the proof of Theorem (4.1), and in the same way, we obtain:

$$\frac{\sup_K |(f^n)'|}{\inf_K |(f^n)'|} \leqslant \delta$$

for every interval K such that $f^n\big|_K$ is continuous and monotone. Hence

$$\sup_K \frac{1}{|(f^n)'|} \leqslant \delta \inf_K \frac{1}{|(f^n)'|} .$$

Since f^n maps each such K onto the whole $[0, 1]$, we get by summing the above inequalities over all such K:

$$\sup_{[0,1]} P^n(1) \leqslant \delta \inf_{[0,1]} P^n(1).$$

It is very easy to chek that $\int P^n(1) = \int 1 = 1$. Consequently,

$$\sup_{[0,1]} P^n(1) \leqslant \delta \qquad \text{for all } n.$$

By Proposition (4.5), a p.i.a.c. measure exists.

References

[1] P. Collet and J.-P. Eckmann, Iterated Maps on the Interval as Dynamical Systems, Progr. Phys. 1 (Birkhäuser, Boston, 1980).
[2] Z. Nitecki, Topological dynamics on the interval, to appear in Erg. Theory Dyn. Syst. (Progr. Math. 21, Birkhäuser).

[3] T. Li, J. Yorke, Period three implies chaos, Am. Math. Monthly 82 (1975) 985–992.

[4] A.N. Šarkovskiĭ, Coexistence of cycles of a continuous map of a line into itself (in Russian), Ukr. Mat. Zh. 16 (1964) 61–71.

[5] T. Li, M. Misiurewicz, G. Pianigiani and J. Yorke, No division implies chaos, preprint (1981).

[6] M. Misiurewicz, Horseshoes for mappings of the interval, Bull. Acad. Pol. Sci., Ser. Sci. Math. 27 (1979) 167–169.

[7] M. Misiurewicz and W. Szlenk, Entropy of piecewise monotone mappings, Studia Math. 67 (1980) 45–63.

[8] D. Singer, Stable orbits and bifurcation of maps of the interval, SIAM J. Appl. Math. 35 (1978) 260–267.

[9] G.R. Hall, A C^{∞} Denjoy counterexample, preprint.

[10] J. Guckenheimer, Sensitive dependence on initial conditions for one dimensional maps, Commun. Math. Phys. 70 (1979) 133–160.

[11] M. Misiurewicz, Absolutely continuous measures for certain maps of an interval, to appear in Publ. Math. I.H.E.S. 53.

[12] M. Jakobson, Absolutely continuous invariant measures for one-parameter families of one-dimensional maps, to appear in Commun. Math. Phys.

[13] M. Rychlik, Invariant measures for maps of an interval of class $C^{1+\varepsilon}$, preprint.

COURSE 10

LABORATORY EXPERIMENTS ON THE TRANSITION TO CHAOS

J.P. GOLLUB

Haverford College,
Haverford, PA 19041, U.S.A.

and

A. LIBCHABER

Ecole Normale Supérieure, Laboratoire de Physique,
75231 Paris Cedex 5, France

G. Iooss, R.H.G. Helleman and R. Stora, eds.
Les Houches, Session XXXVI, 1981 – Comportement Chaotique des Systèmes Déterministes/
Chaotic Behaviour of Deterministic Systems
© *North-Holland Publishing Company, 1983*

Contents

Abstract

Recent advances in experimental methods, particularly cryogenic techniques, laser light scattering, and laboratory computer techniques, have given rise to a new generation of experiments on the transition to turbulent convection. The recent work of several groups is reviewed in this article, with particular attention to the relationship of simple nonlinear models to the experimental observations. Many experiments now indicate that nonlinear models with a few degrees of freedom are useful for describing the behaviour of fluid layers of small horizontal extent. The transition process in laterally large layers can be related to the predictions of stability theory, suitably interpreted. Related experiments on coupled electronic relaxation oscillators are also described.

1. Introduction

The literature on hydrodynamics contains hundreds of experimental articles on convection dating back at least to the experiments of Bénard [1] in 1900. While some of these were concerned with the problem of the transition to turbulence, a fundamental understanding of this phenomenon was not achieved, for a variety of reasons. Experiments have used relatively primitive measurement techniques, that do not permit turbulent and non-turbulent states to be reliably distinguished and quantitatively studied. The thermal control achieved in many of the early experiments was inadequate. Furthermore, experiments were hindered by the lack of an adequate conceptual structure for describing the transition to turbulence.

This situation has changed dramatically during the past ten years. Quantitative analysis of the secondary instabilities of convection rolls using Galerkin expansions (and visual observation of these instabilities) by Busse and co-workers [2] has certainly brought us closer to an understanding of the onset of turbulent motion. The hypothesis of Ruelle and Takens [3], and Lorenz [4] that turbulence near its onset might be described as a strange attractor in a low dimensional dynamical system has also given considerable impetus to recent experiments.

In addition to these theoretical advances, new techniques have spawned a generation of experiments that are much more quantitative than earlier ones. Major advances include: the utilization of laser Doppler techniques, which permit the measurement of local fluid velocities without perturbing the fluid significantly; cryogenic techniques, which permit sensitive measurement of the time-dependent heat flux through a convecting fluid; and computer processing of experimental data to obtain power spectra of exceedingly high resolution.

594

In this article, we review some recent experiments utilizing these new techniques. We discuss only experiments which are related to the onset of time-dependent convection, as opposed to the Rayleigh–Bénard instability itself. The period covered by this review is approximately 1974–1981, but is certainly not exhaustive. Rather, we highlight experiments that have some bearing on the relevance of nonlinear dynamics to the problem of turbulence. A more general and extensive review of hydrodynamic instabilities and the transition to turbulence is available elsewhere [5].

2. Small aspect ratio convection

The nature of the transition to turbulence is profoundly affected by the relative magnitude of the horizontal and vertical dimensions of the fluid layer, commonly known as the aspect ratio. This quantity has been defined for rectangular layers as the ratio of the largest horizontal dimension to the fluid depth, while it has been customary to use the radius to depth ratio for cylindrical layers.

Small aspect ratio layers have been found to behave like systems with a few degrees of freedom near the onset of turbulence, in the sense that they show types of time dependence that are also seen in simple dynamical systems. These phenomena include quasiperiodicity, phase locking, subharmonic bifurcations, and intermittency. Large aspect ratio layers, on the other hand, do not show such behavior. In this section, we summarize some of the experiments on small aspect ratio convection.

One important discovery [6, 8] is the observation that there are a number of stable convection patterns in cells with small aspect ratio, depending on the past history or initial conditions of the system. A given convection cell might contain either two or three rolls, for example. Each one has a distinct sequence of instabilities leading to turbulence as R is varied.

2.1. Quasiperiodicity and phase-locking

A number of groups [7–10] have detected quasiperiodic flows with several incommensurate frequencies. In some cases these incommensurate oscillations phase-lock with each other. If a quasiperiodic state is described by a trajectory lying on a torus in phase space, then the phenomenon of phase-locking produces a limit cycle on the torus. An example of this process as described by Gollub and Benson [6] follows.

The flow first becomes time-dependent by entering a strictly periodic regime at R_1. At a higher Rayleigh number R_2, a second frequency component begins to grow. The interaction of these two oscillations causes the spectrum to consist of sharp peaks at the fundamental frequencies f_1 and f_2, and all of their low order sums and differences of the form $f = m_1 f_1 + m_2 f_2$. The presence of these mixing components in the spectrum indicates that the time-dependent processes are strongly nonlinear. The ratio f_2/f_1 decreases smoothly with increasing R. Therefore, the two frequencies are generally incommensurate. At higher R, they phase-lock so that f_2/f_1 is the ratio of two small integers (typically 7/3 or 9/4). The spectrum then consists of an array of spikes at all multiples of the lowest commensurate frequency of f_1 and f_2. The phase locking persists only over a relatively narrow range in R/R_c. When f_2/f_1 begins to change again various scenarios to chaos occur, including intermittency or period doubling bifurcations, depending on the Prandtl number and other parameters. These observations are a clear example of the production of chaos in a system that has many degrees of freedom, only a few of which are excited initially. Quite similar phenomena, especially the relationship between the onset of chaos and the loss of phase locking, have been seen in systems of coupled electronic oscillators by Gollub et al. [11] (see section 3).

The fluid dynamical significance of the oscillations is not entirely clear. Gollub et al. [12] showed by Doppler mapping that the oscillations are relatively localized within the fluid layer. Dubois and Bergé [10] used an interferometric technique to show that in fluids at high Prandtl number, oscillations are associated with structures in the thermal boundary layer at the cold plate.

Quasiperiodic states characterized by three distinct frequencies were found in several cases by Gollub and Benson [6]. It is nontrivial to actually demonstrate that this is the case. The spectra typically have over 20 statistically significant peaks, but they can all be fitted to linear combinations of three basic frequencies to within about one part in 10^4. Comparable fits cannot be obtained using two independent frequencies. These facts, combined with the observation that the ratios of the three frequencies vary smoothly with R, provide strong evidence that they are incommensurate with each other. Thus this state can be represented by a 3-torus in phase space. Nonperiodicity evolves from this state at higher R.

2.2. Subharmonic bifurcations

Libchaber and Maurer [13] and Gollub et al. [12] showed that nonperiodicity was obtained after several successive subharmonic bifurcations for certain combinations of aspect ratio and Prandtl number. This behavior is at least qualitatively similar to that shown by nonlinear mappings of the unit interval of the form $x_{n+1} = f(x_n)$, where the function $f(x)$ has a single parabolic maximum in the unit interval.

A particular beautiful and reproducible set of spectra showing subharmonic bifurcations was obtained recently by Giglio et al. [14] using a laser beam deflection technique that responds to the average of the vertical temperature gradient along a horizontal line in a small aspect ratio cell containing water. Successive increments in R cause additional peaks to appear in the spectrum at subharmonics (and odd multiples) of the lowest frequency previously present in the spectrum. These phenomena are illustrated in fig. 1. The onset Rayleigh numbers are predicted to follow a geometric series after many subharmonic bifurcations:

$$(R_{n+1} - R_n)/(R_{n+2} - R_{n+1}) \rightarrow 4.6692\ldots .$$

Data for the transition points seem to be consistent with this prediction, but the comparison is not yet precise. When R exceeds the accumulation point R_∞ of the sequence, the subharmonic bifurcations are predicted to disappear successively, forming a sequence above R_∞ which is the mirror image of that below R_∞. This feature of the subharmonic route to turbulence has been observed by Libchaber and Maurer [15].

2.3. Intermittency

An interesting route to turbulence in which periodic or quasi-periodic oscillations are interrupted by turbulent bursts has been observed by Maurer and Libchaber [16] at low Prandtl number, and by Bergé et al. [17] at high Prandtl number. An example of a time record for this process is shown in fig. 2. The duration and frequency of the turbulent bursts increases with R until the motion is purely chaotic. Intermittent oscillations are known to occur in the Lorenz model, so this type of process may be another example of low dimensional chaotic behavior.

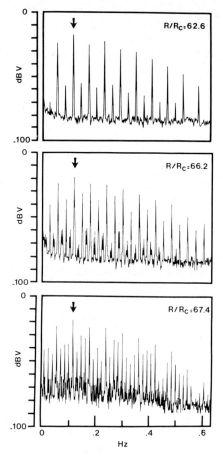

Fig. 1. Spectrum obtained by Giglio et al. [14] showing successive subharmonic bifurcations. The fundamental frequency is indicated by the arrow.

2.4. Modulation experiments

In order to examine the question of whether external noise plays any role in the onset of turbulent convection at small aspect ratio, Gollub and Steinman [18] injected spatially uniform but broadband temperature fluctuations into the lower boundary temperature. The magnitude of the applied noise was sufficient to give a root mean square fluctuation of about $0.3R_c$ in the range $20R_c < R < 50R_c$. The main observations were as follows: (1) in the periodic regime preceding the onset of turbulence, the

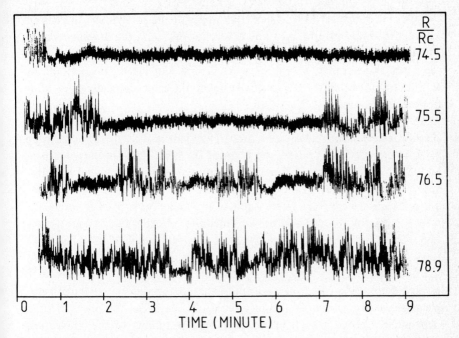

Fig. 2. Time records from the experiments of Maurer and Libchaber [16] showing intermittent turbulence. The motion is quasiperiodic between the noisy bursts.

external noise causes slow fluctuations that can be represented by a Fourier series with time-dependent amplitudes and phases. The spectral peaks are slightly broadened and a low level background is introduced. (2) In the turbulent regime ($R < 41R_c$), the external noise produces no statistically significant change in the velocity power spectra. (3) The transition Rayleigh number is not measureably depressed by the external noise.

These observations indicate that under nominally time-independent boundary conditions, stochastic driving forces can probably be ignored in constructing models of the transition to turbulence at small aspect ratio.

3. Large aspect ratio

3.1. Heat flux experiments

The aspect ratio dependence of the transition process was discovered by

Ahlers and Behringer [19] in heat flux experiments on cylindrical cells of convecting liquid helium. At small aspect ratio ($\Gamma = 2$), the time-dependent phenomena are similar to those obtained by local measurements and described in section 3. However, when Γ is 4.7 or larger, no periodic or quasiperiodic states are observed. Instead, the time dependence is non-periodic at its onset.

At $\Gamma = 57$, it is not possible to resolve any time-independent regime. Rather, the motion seems to be time-dependent and nonperiodic as soon as R_c is exceeded. The power spectrum of the heat flux is broadband even at $1.27R_c$. Thus, the fluid flow is turbulent whenever the velocity is non-zero. This data poses a serious challenge to stability theory, because convection rolls are expected to be stable over a finite interval in R above R_c.

The behavior for $\Gamma = 4.7$, an intermediate aspect ratio, is more complex. The initial experiments showed an apparent onset of (non-periodic) time dependence near $R = 2R_c$. Above this point, the power spectra $P(f)$ are broadband, flat at low frequencies, and shaped like a power law f^{-n} at high frequencies, with $n = 4.0 \pm 0.2$. These observations are surprising, because the predictions of stability theory indicate that the onset of time dependence should arise from an oscillatory instability of the convection rolls, in which transverse waves propagate along the roll axis. This instability is expected to be manisfest as a periodic oscillation in the heat flux beginning at $(3 \text{ to } 5)R_c$ rather than a much slower nonperiodic motion starting at about $2R_c$ [20].

More recent experiments on the same cell by Ahlers and Walden [21] showed even more puzzling phenomena. The flow is not in fact time-independent below $2R_c$ as had been previously believed. Rather, there is a very low level random time dependence with occasional randomly spaced major peaks in the effective thermal conductivity. As R is reduced below $2R_c$, these events become increasingly rare. Their frequency n is well represented by the equation $n = n_0 \exp(-\Delta_0/\varepsilon)$, where $\varepsilon = R/R_c - 1$. The experiments suggest that a stochastic explanation of the onset of turbulent convection at moderate and large aspect ratios may be required.

Heat flux experiments confirming the trend toward time-dependence near R_c at large aspect ratio have been performed recently by Behringer and co-workers [22]. At a radius to height ratio of 7.87, and Prandtl number 0.6, this group found broadband turbulence above $1.32R_c$.

3.2. Laser Doppler experiments

Gollub and Steinman [23] extended their laser Doppler experiments to a rectangular cell of aspect ratio 30 in order to make a comparison with the provocative heat flux experiments of Ahlers, Behringer, and Walden.

Gollub and Steinman found that the large aspect ratio system is characterized by very long transients, comparable to the horizontal thermal diffusion time from one edge of the layer to the other (about 10^5 s). In the interval $R_c < R < 5R_c$, the fluid often eventually attains a time-independent steady state, but only after a day or two. Therefore, extreme patience is required in order to make the simplest dynamical distinction: the presence or absence of intrinsic time dependence.

The observed steady state flow is not the uniform one predicted by stability theory. Rather, the rolls preferentially align with their axes perpendicular to all of the lateral walls of the cell. The resulting pattern is splayed and contained defects, where a roll ends. An example of such a pattern is shown in fig. 3, which was obtained by mapping the Doppler shift in a horizontal plane above the center of the cell. The dots mark the locations of the roll boundaries, where the velocity is entirely vertical. Stable flows were only observed below $5R_c$, and flows containing many defects show slight residual time dependence.

The structures were found to "melt" at about $5R_c$. The local velocity begins to fluctuate slowly, with statistics quite similar to those found by

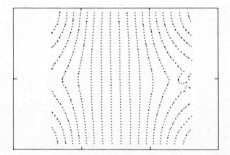

Fig. 3. Doppler map obtained by Gollub and Steinman showing the structure of the velocity field above the midplane of a large rectangular cell at $R/R_c = 4.3$. The dots mark the roll boundaries, where the flow is entirely vertical (out of or into the page). The pattern contains defects because the rolls align perpendicular to the lateral boundaries, but does not change over several days.

Ahlers and Behringer in heat flux measurements. At $5R_c$, the fluctuations are sufficiently slow that a portion of the flow field could be mapped in a time *shorter* than the characteristic time for the fluctuations. Therefore, the space and time structure could be followed in detail by repetitive laser Doppler imaging. A sequence of Doppler scans taken at 10 min intervals is shown in fig. 4. These and other observations showed that the convective rolls are unstable with respect to deformations which can create defects above about $R_1 = 5R_c$. The observed onset is consistent with the predicted onset of the "skewed varicose" instability of Busse and Clever [2]. While the linear stability theory does not predict any time dependence, it is possible that nonlinear effects would cause this instability to evolve into chaotic time dependence. Recent numerical simulations of the full Boussinesq equations by Siggia and Zippelius [24] support this view.

The transition at R_1 resembles a melting transition in two respects. First, the lattice of Rayleigh–Bénard rolls was found to be mobile only above R_1. Second, a process akin to annealing was often observed. If R is reduced (quenched) from about $15R_c$ to a value below R_1, the structure usually contains many defects. However, if the system is then held just above R_1 for a day or so, many of the defects disappear.

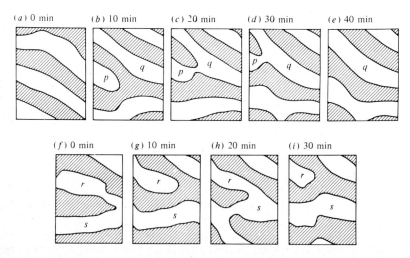

Fig. 4. Two sequences showing a portion of the velocity field in a large rectangular cell at intervals of 10 min, at $R/R_c = 5$. Shaded and unshaded regions have opposite vorticity. The rolls are clearly unstable, and the onset of this phenomenon appears to coincide with the predicted onset for the skewed varicose instability (from ref. [23]).

3.3. High Prandtl number experiments

Bergé and co-workers [25] have found many of the phenomena described in section 3.2, at much higher Prandtl numbers (of the order of 100). Provided the side walls have a thermal conductivity comparable to that of the fluid and are well-coupled to the copper plates, they also found that the axes of the rolls align preferentially in a direction perpendicular to the lateral boundaries. The onset of time dependence in flow structures containing defects is associated with a structural change known as the cross-roll instability [26]. Thus, it seems likely that the onset of obvious broadband noise in large aspect ratio convection is associated with identifiable structural instabilities at both low and high Prandtl numbers. However, the extremely slow time-dependence first observed by Ahlers and Walden closer to R_c may have a different origin.

Experiments by Wesfreid and Croquette [27] have shown that the phase variable which describes the position of the convective rolls obeys a diffusion equation, and they have measured the diffusion coefficient. The concept of phase diffusion may prove useful in understanding the time dependence close to R_c.

4. Coupled nonlinear oscillators

Many of the phenomena seen in small aspect ratio convection also occur in a much simpler dynamical system that can be easily realized in the laboratory: two coupled electronic relaxation oscillators. Gollub et al. [11] have studied such a system experimentally, and have also compared their observations with a simple numerical model. They find a variety of phase-locked and chaotic states even for this simple system.

The experimental circuit is shown in fig. 5. (For details, see ref. [11].) It consists of two tunnel diode oscillators coupled by a resistance R_c. Because of the negative resistance of part of the current/voltage characteristic of the diodes (fig. 5b), the diode voltages V_{D1} and V_{D2} and the corresponding currents I_{D1} and I_{D2} oscillate at frequencies determined by the inductances (and resistances). The voltage signals are pulse-like, while the current signals are composed of segments of exponentials. These signals can be either periodic or nonperiodic (noisy), depending on the way in which each oscillator influences the switching times of the other oscillator.

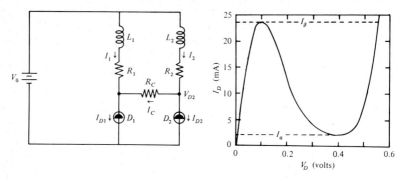

Fig. 5. (a) Tunnel diode oscillator circuit. Each oscillator consists of an inductor, a resistor, and a tunnel diode: the two oscillators are coupled by the resistor R_c. (b) Current/voltage characteristic of the tunnel diode. Arows indicate the path of oscillation.

In order to demonstrate that this circuit exhibits broadband noise, power spectra of the diode voltages were computed for different values of the natural frequency ratio F_0 of the uncoupled oscillators. In some cases the spectrum is composed of sharp peaks at multiples of a fundamental frequency. This complex but periodic behavior is due to the frequency modulation of one oscillator by the other. The modulation is such that the two oscillators lock; their frequencies are in the ratio of two integers.

Broadband spectra are obtained by varying one of the inductances and hence changing the natural frequency ratio of the two oscillators. Poincaré maps were also prepared from the data, and were used to demonstrate that nearby trajectories in phase space diverge from each other exponentially in time. Hence, these states are chaotic.

The behavior of the system was mapped out in a parameter space composed of the coupling conductance $G_c = 1/R_c$ and the uncoupled frequency ratio F_0. The goal was to compare the behavior of the experimental system and an appropriate numerical model. The experimental behavior is shown in fig. 6a. The nonperiodic (broadband) regions are labelled N, while periodic (phase-locked) regions are labelled by the frequency ratio, or by the letter P where the integers are large. The phase diagram is quite complex, with many different interlaced regions.

In order to demonstrate that noise sources are not required to reproduce this behavior, Gollub ct al. constructed a simple numerical model [11] using the standard circuit equations and a rectangular approximation to the $I–V$ characteristic of the tunnel diodes. With this approximation, the

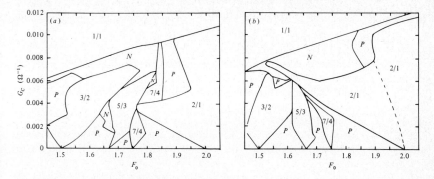

Fig. 6. Parameter space defined by the coupling conductance $G_c \equiv 1/R_c$ and the uncoupled frequency ratio F_0 of the two oscillators for (a) experiment and (b) numerical model (see text for legend).

diode voltages change only in discrete steps, and the equations become linear in the intervals between the steps. The solutions can hence be obtained by computing the switching times for 1000 or more oscillations with very little cumulative error. Spectra were then used to construct a diagram of the parameter space of the model (fig. 6b) to compare with the experimental behavior (fig. 6a). The agreement is reasonably good. Many of the periodic regions are accurately simulated, and the model correctly reproduces the large region of nonperiodic behavior near the center of the diagram. There are also some significant differences which arise from the approximations of the model.

Very recently, Tracy Allen [28] has shown that many of the complex features of fig. 6, especially the ordering of the phase-locked regions as a parameter is varied, can be understood in terms of the Farey series of number theory. The basic prediction is that adjacent phase-locked regions characterized by integer ratios p/q and p'/q' must satisfy $q'p - qp' = 1$.

5. Summary

We have described several systems of physical interest that manifest behavior characteristic of simple dynamical models: convection at small aspect ratio and coupled nonlinear oscillators. On the other hand, fluid layers of large horizontal extent require models incorporating many degrees of freedom and probably consideration of the dynamics of defects.

References

[1] H. Bénard, Rev. Gen. Sci. 12 (1900) 1261; 1309.

[2] F.H. Busse and R.M. Clever, Instabilities of convection rolls in a fluid of moderate Prandtl number, J. Fluid Mech. 91 (1979) 319.

[3] D. Ruelle and F. Takens, On the nature of turbulence, Commun. Math. Phys. 20 (1971) 167.

[4] E.N. Lorenz, Deterministic nonperiodic flow, J. Atmos. Sci. 20 (1963) 130.

[5] H.L. Swinney and J.P. Gollub, eds., Hydrodynamic Instabilities and the Transition to Turbulence, Topics in Applied Physics, Vol. 45 (Springer, Berlin, Heidelberg, New York, 1981).

[6] J.P. Gollub and S.V. Benson, Many routes to turbulent convection, J. Fluid Mech. 100 (1980) 449.

[7] G. Ahlers and R.P. Behringer, The Rayleigh–Bénard instability and the evolution of turbulence, Prog. Theor. Phys. Suppl. 64 (1978) 186.

[8] J. Maurer and A. Libchaber, Rayleigh–Bénard experiment in liquid helium: frequency locking and the onset of turbulence, J. de Phys. Lett. 40 (1979) 419.

[9] J.P. Gollub and S.V. Benson, Phase locking in the oscillations leading to turbulence, in: Pattern Formation by Dynamic Systems and Pattern Recognition, ed., H. Haken (Springer, Berlin, 1979) p. 74.

[10] M. Dubois and P. Bergé, Experimental evidence for the oscillators in a convective biperiodic regime, Phys. Lett. 76A (1980) 53.

[11] J.P. Gollub, E.J. Romer and J.E. Socolar, Trajectory divergence for coupled relaxation oscillators: measurements and models, J. Stat. Phys. 23 (1980) 321.

[12] J.P. Gollub, S.V. Benson and J.F. Steinman, A subharmonic route to turbulent convection, Ann. N.Y. Acad. Sci. 357 (1980) 22.

[13] A. Libchaber and J. Maurer, Une expérience de Rayleigh–Bénard de géométrie réduite: multiplication, accrochage et démultiplication de fréquences, J. de Phys. 41 (1980) C3-51.

[14] M. Giglio, S. Musazzi and U. Perini, Transition to turbulence via a reproducible sequence of period doubling bifurcations, Phys. Rev. Lett. 47 (1981) 243.

[15] A. Libchaber and J. Maurer, in: Nonlinear Phenomena at Phase Transitions and Instabilities, Proc. Nato Advanced Study Institute (March 1981).

[16] J. Maurer and A. Libchaber, Effect of the Prandtl number on the onset of turbulence in liquid ^{4}He, J. de Phys. Lett. 41 (1980) L515.

[17] P. Bergé, M. Dubois, P. Manneville and Y. Pomeau, Intermittency in Rayleigh–Bénard convection, J. de Phys. Lett. 41 (1980) L341.

[18] J.P. Gollub and J.F. Steinman, External noise and the onset of turbulent convection, Phys. Rev. Lett. 45 (1980) 551.

[19] G. Ahlers and R.P. Behringer, Evolution of turbulence from the Rayleigh–Bénard instability, Phys. Rev. Lett. 40 (1978) 712.

[20] A. Libchaber and J. Maurer, Local probe in a Reyleigh–Bénard experiment in liquid helium, J. de Phys. Lett. 39 (1978) L369.

[21] G. Ahlers and R.W. Walden, Turbulence near onset of convection, Phys. Rev. Lett. 44 (1980) 445

[22] R.P. Behringer, C. Agoste, J.S. Jan and J.N. Shaumeyer, Time-dependent Rayleigh–Bénard convection and instrumental attentuation, Phys. Lett. 80A (1980) 273.

[23] J.P. Gollub and J.F. Steinman, Doppler imaging of the onset of turbulent convection, Phys. Rev. Lett. 47 (1981) 505;
J.P. Gollub, A.R. McCarriar and J.F. Steinman, Convective pattern evolution and secondary instabilities, J. Fluid Mech. 125 (1982) 259.

[24] E.D. Siggia and A. Zippelius, Pattern selection in Rayleigh–Bénard convection near threshold, Phys. Rev. Lett. 47 (1981) 835.

[25] P. Bergé, Rayleigh–Bénard convection in high Prandtl number fluid, in: Synergetics, Proc. Int. Symp. on Synergetics, Schloss Elmau, 1981, ed. H. Haken (Springer, Berlin, 1981).

[26] F.H. Busse, Nonlinear properties of thermal convection, Rep. Prog. Phys. 41 (1978) 1929.

[27] J.E. Wesfreid and V. Croquette, Forced phase diffusion in Rayleigh–Bénard convection, Phys. Rev. Lett. 45 (1980) 634.

[28] T. Allen, Bifurcations of coupled relaxation oscillators isomorphic to an extended rational lattice, Dept. of Entomology, U.C.Berkeley.

COURSE 11

MODELS FOR INTERMITTENCY

Y. POMEAU

*Service de Physique Théorique, CEN-Saclay,
91191 Gif-sur-Yvette Cedex, France*

G. Iooss, R.H.G. Helleman and R. Stora, eds.
Les Houches, Session XXXVI, 1981 – Comportement Chaotique des Systèmes Déterministes/
Chaotic Behaviour of Deterministic Systems
© *North-Holland Publishing Company, 1983*

This is a brief review of dynamical models for intermittency. Emphasis is put on intermittency at transition in simple dynamical systems and in parallel flows.

Intermittency is a ubiquitous phenomenon in turbulence. In historical order below are listed some examples of its study.

(i) Intermittent transition to turbulence in parallel flows was discovered by Osborne Reynolds [1] in his famous investigation on pipe flows.

(ii) In boundary layer flows [2, 3] the fluctuations of the boundary between the potential and non-potential flow are intermittent at large distances from the solid wall.

(iii) Small scale intermittency [4, 5] in fully developed turbulence.

(iv) Intermittent transition in deterministic systems with a few degrees of freedom, as those describing the natural convection in Rayleigh–Bénard experiments with an aspect ratio of order unity [6] or in the Belousov–Zhabotinsky homogeneous chemical reaction [7].

I do not intend to give (mathematical) models for all these kinds of intermittency. Owing to the diversity of the physical situation where it appears, the existence of a unique theory or even of a unifying scheme is rather unlikely. Thus I shall be concerned with a few particular points about the theoretical aspect of these questions. The choice of these points is dictated by the general scope of this volume. I have already considered [8] intermittency in transition flows [case (i) above]. I give below some details about this. Thanks to a very nice result communicated to me by Tresser and Arneodo, the rough argument that I gave in ref. [8] is hopefully more convincing now: I shall explain below why localized perturbations with a finite amplitude appear first in transition flows. This argument is taken from the theory of dynamical systems, this is likely to be at the root of any understanding of non-trivial phenomena in turbulence.

Boundary layer fluctuations [case (ii)] seem to be a difficult problem from the point of view of *ab initio* calculations. i.e. starting from the Navier–Stokes equations and *nothing else*. A consistent theory should give the asymptotic behavior of the intermittency factor at large distances from the solid boundary.

Small scale intermittency [case (iii)] has been the subject of many theoretical studies since the work of Kolmogoroff [9, 10]. More recently some connection has been made [11] between the non-gaussian character of small scales in turbulent flows and of large scales in equilibrium second order phase transition, as the Curie point in the two dimensional Ising model for instance.

The rest of this lecture is devoted first to an informal definition of what intermittency is and then to possible explanations of this sort of behavior, based on the properties of simple deterministic dynamical systems.

Loosely speaking fluctuations of, say, one real quantity depending on time are intermittent when this quantity is approximately constant, say zero, for large periods and takes big values during relatively short intervals, then relaxes to zero up to the next burst and so on. This qualitative description defines a clear-cut concept in limit cases.

A similar concept exists in classical ergodic theory, i.e. the one of weak mixing as opposed to strong mixing [12]. In some sense weak mixing is a limit case at the boundary of strong mixing. Introducing the definition of weak mixing shall help me to give some useful ideas. Consider a map T of a measurable space Ω into itself, and let μ be a T-invariant measure on Ω. If A is a measurable subset of Ω, then

$$\mu[T^{-1}(A)] = \mu(A).$$

(Ω, T, μ) is weakly mixing iff, given two measurable subsets A and B in Ω,

$$\mu(T^{-n}A \cap B) \underset{\substack{n \to \infty \\ n \notin \mathscr{J}}}{\longrightarrow} \mu(A)\mu(B)$$

where $\mathscr{J}$ is a "small" subset of $\mathbb{N}$, such that

$$\frac{\#(\mathscr{J} \cap \{1, 2, \ldots, m\})}{m} \underset{m \to \infty}{\longrightarrow} 0$$

($\#K \equiv$ cardinality of K).

If (Ω, T, μ) is weakly but not strongly mixing,

$$\#(\mathscr{J} \cap \{1, 2, \ldots, m\}) \underset{m \to \infty}{\longrightarrow} \infty$$

and the map T "mixes" the set Ω, except that from time to time one has large deviations from this average behavior. The mean probability of occurrence of these large deviations is zero, although they surely occur at some future time, as the time given is large. In typical situations $\#(\mathcal{J} \cap \{1, 2, \ldots, m\})$ increases as $m^\alpha (0 \leq \alpha < 1)$.

An example of such subset of $\mathbb{N}$, as $\mathcal{J}$, occurs in probability theory [13] for the problem of the ruin of two head and tail players. If two players start with an equal money allocation, say zero, then most of time (and contrary to what one would expect from the idea of "equal chance") one of the two players wins. The amount of time for which the absolute difference between their fortunes is less than some uniformly fixed quantity increases with n as $n^{1/2}$, n being the number of plays.

Quite naturally these intermittent situations are a limit case for systems in which fluctuations disappear by becoming very rare when some control parameter is continuously changed. Conceptually at least* this transition can take place in qualitatively different situations [8]. For instance it may be between a steady stable and a random intrinsic** behavior. This is what occurs in the Belousov–Zhabotinsky chemical reaction [14], when it tends to become steady at large renewal times of the reagents.

This direct transition from steadiness to turbulence was also observed by Reynolds [1]. This early observation of an intermittent transition deserves more comments, as it displays a number of characteristic features.

Beyond a quite well defined value of the Reynolds number, the laminar flow in a circular pipe is interrupted from time to time by turbulent "flashes" (in the terminology of Reynolds). Each flash fills a finite portion of the pipe length. Some other experimental facts are noteworthy.

(1) The turbulent flashes are convected downstream as a whole.

(2) The front and rear velocities of the flash are different. The difference [15] increasing continuously from zero at the onset of turbulence. Henceforth the total length of the flash grows linearly downstream.

(3) The intermittency factor (=the ratio of the mean length of all turbulent flashes at a given time to the total flow length) vanishes continuously at the onset of transition.

---

* That is if one does not worry about the codimension of the transition set in the space of dynamical systems under consideration.

** I am concerned here with the so-called intrinsic stochasticity. It is well known, but not always realized, that the triggering by thermal fluctuations is completely negligible in most macroscopic non-equilibrium situations, in particular in turbulent flows. To understand this, one should always remember that the Boltzmann constant (which measures the strength of thermal fluctuations) is R/N, where N is Avogadro's number.

All these experimental facts yield a clear picture of what actually happens in transition flows*. This picture also seems to have little to do with what one would expect from a weakly nonlinear theory of stability**.

Actually, one may consider [8] the turbulent flashes as localized solutions of the fluid equations with a finite amplitude. The solutions join smoothly the laminar flow up and downstream. From rather general considerations, one may guess that these solutions exist at a well defined Reynolds number when their relative velocity is fixed. To relate these localized solutions to the general properties of the dynamical system, one has to consider "dynamics" in which the usual role of time is played by the combination $\xi = x - vt$, (x, t) being the longitudinal distance and time respectively and v the relative velocity of the localized solution. This sort of solution depends on x and t through the combination ξ only, so that it is at rest in the frame moving along x with the velocity.

I owe to Charles Tresser and Alain Arneodo the following idea. It explains that when a control parameter – the Reynolds number – is continuously changed in a parallel flow the perturbation occurring first is *localized* and of *finite amplitude* (i.e. its amplitude does *not* vanish at the onset). This gives a possible "scenario", in the sense of Jean Pierre Eckmann for this sort of transition.

Consider now a flow [here I mean by "flow" a set of first order smooth nonlinear coupled differential equations depending smoothly on a control parameter, say μ as $dx/d\tau = f_\mu(x)$, $f_\mu(x)$ being a one parameter family of vector fields. No confusion must be made with "flow" as a solution of the Navier-Stokes equations for instance] in $\mathbb{R}^d$ with a fixed point 0. This fixed point represents the simple solution of the Navier-Stokes equation corresponding to the Poiseuille flow for instance. I shall take $d = 3$, although the reasoning can be extended to higher and even infinite dimensionalities.

---

* A quite similar picture arises from studies of other transition flows, as the Blasius boundary layer [16], the Poiseuille flow [17], the flow down an inclined plane [18]. This indicates that some properties of the transition flows have a general character.

** This is in sharp contrast with, for instance, Taylor-Couette instability or Rayleigh-Bénard instability. In these cases the dominant features of the flows existing beyond the onset of instability are well described by a weakly nonlinear analysis. As shown by Sorokin [23] for the Rayleigh-Bénard instability, this is a consequence of the supercritical character of the bifurcation at the onset of instability. In parallel flows the linear instability, as predicted by the Orr-Sommerfeld equation yields a subcritical bifurcation or even no bifurcation at all.

Let us assume furthermore that the fixed point is unstable with a real index along a one dimensional direction, the *unstable* manifold W^u and that 0 is a stable node along two other dimensions, say along the *stable* manifold W^s (see fig. 1).

I shall consider a one parameter family of flows with this local structure around 0. I shall show that in this case and under suitable conditions, a closed flow line appears for a given value of the deformation parameter and bounds two parameter regions: on one side no closed trajectory exists, on the other side an infinite set of closed trajectories exists. At the critical value of the parameter the single existing closed trajectory is homoclinic: it starts from 0 at $-\infty$ and goes back to 0 at $+\infty$.

Consider the first return map in a plane π perpendicular to W^s in the vicinity of 0 (see fig. 1). Let Γ be the line of intersection of π and W^s. Our first assumption is that any point starting below Γ in π goes to infinity and never returns to this vicinity of 0.

Consider now a foliation of π by lines parallel to Γ. After one turn (more or less along W^u) starting from π the set of image points at the first return has more or less the shape of a spiral, (as shown in fig. 2), the center of the spiral being, of course, the image of Γ in the first return map, and the foliation is transformed as shown in fig. 2a.

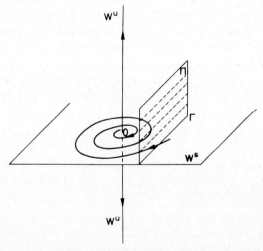

Fig. 1. This is a local picture of a flow in $\mathbb{R}^3$. 0 is a fixed point, it is stable along W^s (stable focus) and unstable (unstable node) along W^u. π is a plane perpendicular to W^s in the neighbourhood of 0.

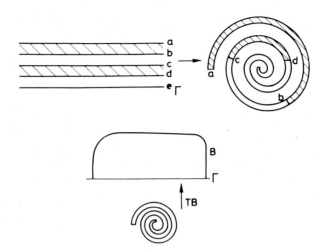

Fig. 2. (a) This is the first return map in π of a bunch (B) of points slightly above Γ. The image of the strips $(ab), (cd), \ldots$ are closer and closer to the center of an infinite spiral, the center of the spiral being the image of Γ. (b) As the parameter changes continuously, the image of B moves upward. As TB crosses B, a closed trajectory may appear. If one assumes that the image of points below Γ never return to π, and if the rate of contraction of the spiral at its center is large enough, the first closed trajectory if found at the intersection of the center of the spiral and of Γ. This is a homoclinic trajectory.

I shall consider the first return map of a bunch of points B (limited by Γ). This gives a spiral TB (see fig. 2b). To make closed trajectories appear $B \cap TB$ must be non-empty. Consider now a deformation of the flow such as the image TB starting below Γ move upward.

We shall assume that the rate of contraction toward 0 in W^s is noticeably larger than the rate of expansion along W^u. Whence a little reflection shows that, as TB moves up, the first intersection between one of the layers in B and its image in TB occurs at the center of spiral. To understand this one notices that, when the center of the spiral TB is on Γ, the image of the layer (ab) in the spiral is closer to this center than (ab) itself. The argument is still valid when one considers layers in B closer and closer to Γ.

The application of this kind of argument to a concrete situation, as in Poiseuille flow, needs some more work. Actually, the concrete situation in hydrodynamics is the one of partial differential equations, it can be considered (as explained before) as a flow with respect to a variable $\xi = x - vt$, where x is the distance along the flow lines. Perpendicular to these lines one has at least another space variable, so that the flow is actually in an infinite dimension space. Nevertheless in recent times we have been taught that partial differential equations and ordinary differential equations are not as different as commonly believed!

At the onset, the mean interval between two bursts becomes very large, when a periodic trajectory loses its stability by intermittency. This growth of the mean internal can be explained by means of the Poincaré normal form of the first return map. Let ε be a measure of the distance of the control parameter to its threshold value. When the periodic behavior loses its stability, one can guess the ε-dependence of the mean interval, say τ between two bursts. When the Floquet multiplier of the stability matrix crosses the circle at $+1$, τ is of order $\varepsilon^{-1/2}$ [19]. When this multiplier crosses the unit circle at complex values, a "naive" theory predicts $\tau \sim (-1/\ln \varepsilon)$, and computer experiments give $\tau \sim \varepsilon^{-\alpha}$ ($\alpha > 0$, but small). The failure of the naive theory is likely to be due to the strongly non-gaussian character of the fluctuations of τ at threshold.

If one considers diffeomorphisms of the circle [20] (as given by the first return map of a flow on the torus T^2) intermittency occurs when a stable period disappears, that is at the edge of stability of a period in the parameter space. In this last case, the intermittent transition is of course not between stable and random motion, and the power spectra is always made of sharp lines. At the intermittency threshold a bunch of spectral lines accumulates near the zero frequency. These small frequencies describe the locking/unlocking of the (small) frequency of occurrence of the bursts with respect to the frequency of the basic oscillation. This sort of behavior has been observed in the Belousov–Zhabotinsky chemical reaction [7].

A quite similar phenomenon occurs in the much studied one dimensional map of an interval of $\mathbb{R}$ with a single maximum. Consider for instance the first occurence of period three* in the quadratic map of $]-1,1[$ given by $x \rightarrow 1 - \mu x^2$ at $\mu = 7/4$. For μ a slightly smaller than $7/4$, most of time iteration starting at random in $]-1,1[$ will stay close to a period 3. From time to time however a turbulent "burst" will interrupt this seemingly periodic behavior. The reason is very similar to the one valid for the intermittency observed in the Lorenz model [21].

As for diffeomorphisms of the circle, the region of intermittent behavior

---

* A period exists when three real numbers (x_1, x_2, x_3), all different, satisfy: $x_1 = 1 - \mu x_3^2$; $x_2 = 1 - \mu x_1^2$; $x_3 = 1 - \mu x_2^2$.

in the parameter (μ) space (all the nice properties of maps of the interval are explained in ref. [22]) is an extremely complicated mixture of open sets wherein a stable (but in general very large) period exists although a nonperiodic behavior exists for almost all initial conditions on a Cantor set of values of μ.

As a conclusion, the intermittent transition shows a kind of continuity for many properties, despite the fact that fluctuations take large values. Due to this continuity, it is reasonable to hope that some properties, such as the critical behavior of the intermittency factor, should be accessible to *ab initio* analysis.

Acknowledgment

I take this opportunity to thank Paul Manneville for our fruitful collaboration.

References

[1] O. Reynolds, Phil. Trans. 174 (1883) 935.
[2] S. Corrsin, Natl. Adv. Comm. Aeron. Conf. report 3 (1943) L23;
    S. Corrsin and A.L. Kistler, Natl. Adv. Comm. Aeron. Report 1244 (1955).
[3] Brian L. Cantwell, Ann. Rev. of Fluid Mech. 13 (1981) 457.
[4] G.K. Batchelor and A.A. Townsend, Proc. Roy. Soc. london, Ser. A194 (1968) 527.
[5] A.S. Monin and A.M. Yaglom, Statistical Fluid Mechanics (MIT Press).
[6] P. Bergé, M. Dubois, P. Manneville and Y. Pomeau, J. de Phys. Lett. 41 (1980) L341.
[7] Y. Pomeau, J.C. Roux, A. Rossi, S. Bachelart and C. Vidal, J. de Phys. Lett. 42 (1981) L271.
[8] Y. Pomeau, Approche de la Turbulence, Proc. 1981 S.F.P. Conference at Clermont Ferrand (Editions de Physique, Orsay-Courtaboeuf).
[9] A.N. Kolmogoroff, C.R.Acad.Sci. URSS 30 (1941) 301.
[10] A.N. Kolmogoroff, Sov. Phys. Usp. 10 (1963) 734.
[11] H.A. Rose and P.L. Sulem, J. de Phys. 39 (1978) 441 and refs. therein.
[12] J.R. Brown, Ergodic theory and topological dynamics (Academic Press, New York, 1976).
[13] W. Feller, An introduction to probability theory and its applications (Wiley, New York, 1958).
[14] C. Vidal, Lectures at the Elmau Conference on synergetics (Springer, Berlin, 1981).
[15] P.J. Tritton, Physical Fluid dynamics (Van Nostrand, New York).
[16] I.J. Wygnansky, M. Sokolov and D. Friedman, J. Fluid Mech. 78 (1976) 785.
[17] I.J. Wygnansky and F.H. Champagne, J. Fluid Mech. 59 (1973) 281.

[18] H.W. Emmons, J. Aero. Sci. 18 (1951) 490.
    H.W. Emmons and A.E. Bryson, Proc. ASME (1951) 859.
[19] P. Manneville and Y. Pomeau, Phys. Lett. 75A (1980) 296.
[20] See e.g. M.R. Hermann, I.H.E.S. Pub. Math. 49 (1979) 5
[21] P. Manneville and Y. Pomeau, Commun. Math. Phys. 74 (1980) 189.
[22] P. Collet and J.P. Eckmann, Iterated maps of the interval as dynamical systems, Progress in Physics, Vol 1 (Birkhaüser, 1980).
[23] V.S. Sorokin, Priklad. Mat. i Mek. 17 (1953) 39.

COURSE 12

HYDRODYNAMIC STABILITY OF SHEAR FLOWS

Steven A. ORSZAG*

Department of Mathematics, Massachusetts Institute of Technology, Cambridge, MA 02139, U.S.A.

and

Anthony T. PATERA*

Department of Mechanical Engineering, Massachusetts Intitute of Technology, Cambridge, MA 02139, U.S.A.

* Work supported by the Office of Naval Research under Contracts N00014-77-C-0138 and N00014-79-C-0478 and NASA Langley Research Center under Contract NAS1-16722.

G. Iooss, R.H.G. Helleman and R. Stora, eds.
Les Houches, Session XXXVI, 1981 – Comportement Chaotique des Systèmes Déterministes/
Chaotic Behaviour of Deterministic Systems
© North-Holland Publishing Company, 1983

Contents

1. Introduction

In a given geometry and under the influence of given "gross" external parameters, a fluid flow can be either *laminar* or *turbulent*. Laminar flows are largely insensitive to initial conditions, and are correlated over large distances and times. Turbulent flows are highly *sensitive* to initial and external conditions, although statistically averaged properties of these flows are not. In terms of readily observed quantities, laminar flows are deterministic while turbulent flows appear chaotic.

The process by which a flow changes from a laminar state to a turbulent one is known as a *transition*. Transition to turbulence obviously requires an *instability* of the laminar flow, and thus the theory of hydrodynamic stability is the cornerstone for transition research. In essence, stability theory addresses the issue of when a laminar flow becomes sensitive to initial conditions.

The phenomenon of transition was first studied in detail experimentally in 1883 by Osborne Reynolds [1]. He found that laminar pipe flow becomes *unstable* when a non-dimensional quantity, now called the Reynolds number, attains a certain critical value, and that turbulence quickly ensues. About the same time, Lord Rayleigh made the first significant advances in understanding the stability of parallel channel flows from an analytical point of view [2]. Rayleigh studied the stability characteristics of inviscid plane-parallel channel flows to *infinitesimal* perturbations. The resulting linear theory is attractive both because it is mathematically tractable and because it predicts instability (and, perhaps, the subsequent transition to turbulence) *independently* of the initial conditions (i.e. the magnitude or the form of the perturbation).

The results of Rayleigh, although enlightening, cannot explain experimentally observed transitions. The turn of the century saw the proper formulation of the viscous parallel-flow stability problem [3, 4] and subsequent work [5–7] indicated the possibility of a resistively-driven (i.e. *viscous*) instability. Such a linear, viscous instability is indeed found in

624

plane Poiseuille flow and flat-plate boundary layer flow. The resulting disturbance is known as a Tollmein–Schlicting wave. Subsequent experimental verification of the predicted existence of such disturbances [8] gave linear stability theory physical credibility, and the concurrent work of Lin [9] gave it mathematical substance.

However, despite these advances, there were still large discrepancies between the predictions of linear theory and experimental observations. More extensive parameter searches would prove to be of no avail. For instance, linear theory predicts pipe Poiseuille flow to be stable to all disturbances at all Reynolds numbers, R, while in fact transition occurs in this flow at quite moderate Reynolds numbers. Attempts to extend linear theory to include first order nonlinear effects (beginning with the work of Meksyn and Stuart [10]) have, in general, not been very successful. The amplitude expansion techniques [11] that have proven invaluable in the study of Bénard convection [12] and Taylor–Couette flow [13] do not readily extend to parallel shear flows, the former enjoying properties (e.g. exchange of stability) which are not shared by the latter.

In the 1970s the first attempts were made at applying sophisticated numerical techniques and large modern computers to the problem of transition in the classical parallel shear flows (i.e. plane Poiseuille, plane Couette, pipe Poiseuille). Work by Zahn et al. [14] and Herbert [15] isolated significant *two-dimensional* nonlinear instabilities in plane Poiseuille flow. However, transition cannot be explained on the basis of these results. It is only in the last several years that an instability has been found that seems to obtain in all the classical shear flows [16, 17], and whose structure and parametric dependence agree quantitatively and qualitatively with that observed in experimental transitions.

In this series of lectures we will trace the evolution of the transition problem in shear flows from its initial formulation in the 1880s to its current status one hundred years later. We shall summarize the important results of linear inviscid and viscous theory as well as indicate how, formally, the linear framework can be extended to include nonlinear effects. We will then examine in detail the transition problem in the context of spectral numerical methods, starting again with linear theory but now proceeding through to the solution of the fully nonlinear three-dimensional time-dependent incompressible Navier–Stokes equation. As we incorporate more physical effects at each level of complexity a fairly comprehensive picture of transition begins to emerge.

2. Overview

In order to put the exposition that follows in perspective we summarize the main results here. The general stability problem is to investigate the evolution of a perturbation to a solution of the Navier–Stokes equation, i.e.

$$v = U(x, t) + v_p(x, t),$$ (2.1)

where U is a known solution and v_p is a perturbation. The velocity must satisfy the incompressible Navier–Stokes equations

$$\frac{\partial v}{\partial t} + (v \cdot \nabla)v = -\nabla p + \frac{1}{R}\nabla^2 v,$$ (2.2)

$$\nabla \cdot v = 0.$$ (2.3)

and no-slip boundary conditions at rigid wall. These equations have been non-dimensionalized with respect to a length h, and a velocity U_0, after which the only parameter remaining in the problem is the Reynolds number, $R = U_0 h/v$, where v is the kinematic viscosity of the fluid. The stability of the flow depends on R and the nature of v_p (U being assumed given).

Table 1 shows the extent to which analytical and numerical work has been able to explain experimentally-observed transitions in some shear flows. We characterize transition here by three parameters; the critical Reynolds number, R_c, the time scale of the instability, τ, and the (spatial) dimensionality, d, of the perturbation v_p. Here τ can eitther be convective, h/U_0, or viscous, Rh/U_0, or, in our non-dimensional variables, 1 and R, respectively.

Table 1

Flow	R_c (linear)	R_c (non-linear)	R_c (typically observed)	R_c (maximum observed)	τ (linear theory)	τ (observed)	d (linear)	d (observed)
Plane Poiseuille	5772	2900	1000	8000	Rh/U_0	h/U_0	2	3
Plane Couette	∞	$\ll 1000$	1000	?	∞	h/U_0	–	3
Pipe Poiseuille	∞	$\ll 2000$	2000	100 000+	∞	h/U_0	–	3

The three flows we consider here are plane Poiseuille flow, plane Couette flow, and pipe Poiseuille flow. Plane Poiseuille flow is defined as the flow between rigid infinite parallel plates (separated by a distance 2 in our non-dimensionalisation) driven by a constant pressure gradient in the x-direction. The corresponding laminar flow is $U(x, t) = (1 - z^2)\hat{x}$. Plane Couette flow is in the same geometry except that, instead of a pressure gradient, the top (bottom) wall moves with a speed 1 (-1) in the x-direction. The laminar flow here is given by $u(x, t) = z\hat{x}$. Pipe Poiseuille flow (also known as Hagen–Poiseuille flow) is similar to plane Poiseuille flow except that it is flow in a circular pipe of radius 1. The corresponding laminar flow is $U(x, t) = (1 - r^2)\hat{x}$.

Inviscid linear theory means that v_p is infinitesimal and $R = \infty$. *Linear theory* implies v_p infinitesimal but all R are considered. *Nonlinear 2-D theory* implies v_p depends only on x (the streamwise direction) and z (the cross-stream direction) and is of finite-amplitude. Finally, *3-D theory* here means v_p is a two-dimensional finite-amplitude disturbance together with an infinitesimal three-dimensional perturbation, and *3-D simulation* refers to a direct numerical solution of the full three-dimensional, nonlinear, time-dependent, incompressible Navier–Stokes equations.

It is seen that the three-dimensional mechanism appears to explain universally the transitions of shear flows. Indeed it appears that this instability extends to a much wider class of flows, including jets and boundary layers. However, we begin here with linear theory, for, although not relevant directly, it is important in understanding finite-amplitude behaviour.

3. Linear theory of parallel channel flows

We here take $U(x)$ to be of the form

$$U(x) = \bar{U}(z)\hat{x} \tag{3.1}$$

and v_p to be of the form

$$v_p = \varepsilon v'(x, y, z, t) \qquad \varepsilon \ll 1. \tag{3.2}$$

The linearized equations are obtained by inserting eqs. (3.1) and (3.2) into eqs. (2.2) and (2.3) and linearizing with respect to ε. The resulting equations are separable in x, y, and t, so that v' can be further specified as

$$v' = v'(z)\, e^{i(\alpha x + \beta y - \omega t)}. \tag{3.3}$$

These equations can then be reduced to a single equation for the cross-stream velocity, w (normal to the walls), $w = v' \cdot z$,

$$(D^2 - k^2)^2 w = iR[(\alpha \bar{U} - \omega)(D^2 - k^2)w - \alpha(D^2 \bar{U})w], \tag{3.4}$$

$$w = Dw = 0, \qquad z = \pm 1, \tag{3.5}$$

where

$$D = d/dz \tag{3.6}$$

and

$$k^2 = \alpha^2 + \beta^2. \tag{3.7}$$

(Note the characteristic length h has been taken as the half-channel width here so that the walls are at $z = \pm 1$.) The boundary conditions (3.5) follow from zero normal velocity and zero slip-velocity at the walls.

Equation (3.4) is the Orr–Sommerfeld equation. The homogeneous boundary conditions (3.5) lead to an eigenvalue problem and the characteristic equation

$$F(\alpha, \beta, \omega; R) = 0 \tag{3.8}$$

associated with a nontrivial eigenfunction $w \neq 0$.

There are two ways to interpret the characteristic eq. (3.8). The first, termed *temporal stability analysis*, assumes given real α, β and finds $\omega = \omega_r + i\omega_i$. From eq. (3.3) it is easily seen that instability occurs when $\omega_i > 0$. The second, termed *spatial stability analysis*, assumes that ω, β are real and finds $\alpha = \alpha_r + i\alpha_i$. From eq. (3.3) instability occurs for $\alpha_i < 0$. For small ω_i the two analyses can be easily related by Gaster's relation [18],

$$\alpha_i = -\mathrm{Re}\,\frac{\omega_i}{v_g}, \qquad v_g = \frac{\partial \omega}{\partial \alpha}. \tag{3.9}$$

This follows from the fact that F in eq. (3.8) is analytic and therefore $\omega_r \sim \omega_r + i\omega_i + \Delta\alpha(\partial\omega/\partial\alpha)$. Upon imposing that the temporal frequency ω_r be unchanged, we obtain eq. (3.9). A more physical interpretation is that the correct frame in which to view temporal growth spatially is in that of a wave packet travelling at v_g, the group velocity of the disturbance.

Our study of the Orr–Sommerfeld equation will show that observed transitions in classical shear flows cannot be explained by linear theory alone. First, we will show that the equation predicts initial two-dimensional instability, not three-dimensional instability as observed in experiments. Next, we will take the inviscid $R \to \infty$ limit to show that no inviscid instability exists, unless the laminar flow has an inflection point, so that in the absence of such a point time scales for instability will be *viscous* (if an instability exists at all). Lastly we will investigate the structure of Orr–Sommerfeld solutions for viscous flows, and will show that the critical Reynolds numbers predicted by the theory are not in agreement with experiment. For a detailed exposition of these results see Lin [19].

The fact that linear theory predicts that the critical instability (i.e. lowest R) is two-dimensional is easily shown by a simple invariance argument. Let $F(\alpha, \beta, \omega, R) = 0$ be an eigenvalue relation for the Orr–Sommerfeld equation with $\omega_i > 0$ (an unstable mode). Under the transformation

$$\alpha \to \alpha' = (\alpha^2 + \beta^2)^{1/2}, \tag{3.10}$$

$$\beta \to \beta' = 0, \tag{3.11}$$

$$\omega \to \omega' = \frac{\omega(\alpha^2 + \beta^2)^{1/2}}{\alpha}, \tag{3.12}$$

$$R \to R' = \frac{\alpha R}{(\alpha^2 + \beta^2)^{1/2}}, \tag{3.13}$$

the Orr–Sommerfeld equation is invariant, implying $F(\alpha', \beta', \omega', R') = 0$ is also a valid eigenvalue relation. This indicates that if a three-dimensional unstable mode exists at a given R, a two-dimensional unstable modes exists at a $R' < R$. This result is known as Squires' theorem [20], and it implies that the critical instability is two-dimensional.

Note that Squires' theorem does not imply that the most dangerous mode at any given R above the critical R is necessarily two-dimensional. Also, Squires' theorem does not necessarily apply to nonparallel or nonlinear flows.

We next show that plane Poiseuille and plane Couette flows are inviscidly stable to all linear disturbances. Taking the limit $R \to \infty$ we obtain the Rayleigh equation

$$(D^2 - \alpha^2)w - \frac{(D^2 \bar{U})}{(\bar{U} - c)}\,w = 0, \tag{3.14}$$

where c is the complex wave speed ω/α. This limit $R \to \infty$ is singular for two reasons:

(i) the order of eq. (3.14) is two lower than that of eq. (3.4), the Orr–Sommerfeld equation;

(ii) when $\bar{U}(z_c) = c$ the Rayleigh equation (3.14) has a singular point which does not appear in eq. (3.4).

Due to (i) we drop the no-slip (viscous) boundary condition on the Orr–Sommerfeld equation in favor of the impermeable wall condition

$$w = 0 \qquad z = \pm 1. \qquad (3.15)$$

Due to (ii), the point $z = z_c$ is known as the critical point, and the region around it, known as the critical layer, must be analyzed locally.

Multiplying eq. (3.14) by w^* ($*$ denoting complex conjugate) and integrating over the channel gives

$$\int_{-1}^{1} w^* D^2 w \, dz - \alpha^2 \int_{-1}^{1} |w|^2 \, dw = \int_{-1}^{1} \frac{(D^2 \bar{U}) |w|^2}{(\bar{U} - c)} \, dz.$$

Integrating the first term by parts [and using eq. (3.15)] gives

$$-\int_{-1}^{1} \{|Dw|^2 + \alpha^2 |w|^2\} \, dz = \int_{-1}^{1} \frac{(D^2 \bar{U}) |w|^2}{(\bar{U} - c)} \, dz. \qquad (3.16)$$

Taking the imaginary part

$$c_i \int \frac{(D^2 \bar{U}) w^2}{|\bar{U} - c|^2} \, dz = 0 \qquad (3.17)$$

from which Rayleigh's necessary condition for instability immediately follows. To wit, if $c_i \neq 0$ then

$$D^2 \bar{U}(z_i) = 0 \qquad -1 < z_i < 1 \qquad (3.18)$$

for some z_i. In other words, a necessary condition for inviscid linear instability is that $\bar{U}$ have an inflexion point (or, equivalently, the vorticity has a vanishing derivative) interior to the channel. A further refinement of this result is Fjørtoft's theorem [21], which states that the vanishing derivative of $D\bar{U}$ must correspond to an *extremum* of the vorticity (see Lin [19]).

In figs. 1a, b, c, we present schematic plots to motivate physically Rayleigh's criterion (3.18) by showing that, if $D^2 \bar{U} \neq 0$ (i.e. the vorticity is, say, increasing), then there is a force tending to restore displaced fluid

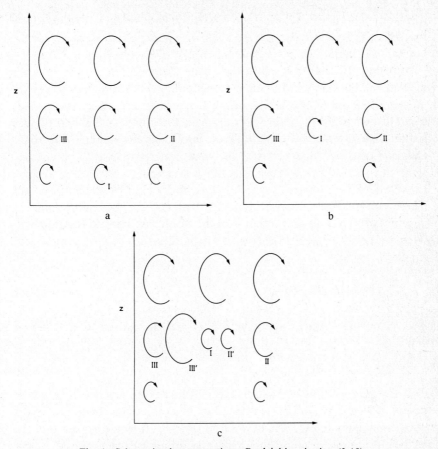

Fig. 1. Schematic plots to motivate Rayleigh's criterion (3.18).

elements. To do this, we picture the parallel shear flow as infinitely many discrete line vortices (directed in the y direction) whose magnitudes depend only on z (see fig. 1a). We recall that in a two-dimensional inviscid flow vorticity is material, i.e. carried and conserved by fluid displacements. Thus, upon displacing the vortex I up one level we obtain the situation depicted in fig. 1b.

Using the Biot–Savart law to find the velocities induced by the vortices (no longer in equilibrium), we see that III will tend to bring down vorticity (III') and II bring up low (II') (see fig. 1c). The net result of III' and II' on I is clearly to push it down, thus providing a restoring force. (Note that this argument does not preclude over-stability, which is precluded by Rayleigh's criterion if there is no inflexion point.) This argument cannot be used to obtain Fjørtoft's theorem, which is a second order effect.

Applying Rayleigh's condition to plane Poiseuille flow, $\bar{U}(z) = 1 - z^2$, indicates global stability. For plane Couette flow, inviscid theory gives only the trivial solution, $w = 0$. Thus both these flows are stable to inviscid, linear disturbances. Therefore the only chance of finding a linear instability is that viscosity be *de-stabilizing* (contrary to its usual stabilizing role). However, this will surely require the resulting growth rate to be viscous, again in contradiction to experiment. We now turn to the viscous theory.

In general we are interested in large Reynolds number instabilities, and hence a singular perturbation technique based on the Rayleigh solution is appropriate. To determine where we must modify the inviscid solution with viscous layers we balance the two dominant terms in the Orr–Sommerfeld equation, namely

$$\frac{1}{R} D^4 w \sim i\alpha(\bar{U} - c)D^2 w. \tag{3.19}$$

Away from the critical layer where $\bar{U}(z_c) = c$ the viscous layer thickness will be

$$\delta_{\text{boundary}} \sim (\alpha R)^{-1/2}, \tag{3.20}$$

upon balancing D^4 and D^2 in the layer. These viscous layers will in fact be boundary layers at $z = \pm 1$; two boundary layers can be used to satisfy the viscous boundary conditions at the wall, $Dw = 0$. It would appear that the analysis is complete, and if this were indeed the case, there would appear to be no chance for a viscous instability, as the solution in the interior is simply the stable Rayleigh solution.

However, the analysis is not complete as the Orr–Sommerfeld equation admits internal layers that derive from the second singularity in the Rayleigh limit (the critical point), just as the boundary layers result from the first. Indeed, an analysis of eq. (3.19) near $\bar{U}(z_c) = c$, noting that $\bar{U} - c \sim [DU(z_c)]Z\delta$, where $Z = (z - z_c)/\delta$ is the scaled internal layer variable, gives

$$\delta_{\text{internal}} \sim (\alpha R)^{-1/3}. \tag{3.21}$$

Note this analysis only applies where these two viscous regions are distinct, i.e. for R sufficiently large and c_i sufficiently small. A typical Orr–

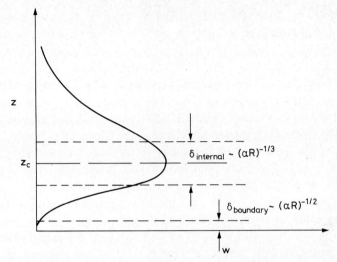

Fig. 2. Qualitative plot of the w-velocity of an Orr–Sommerfeld eigenfunction.

Sommerfeld eigenfunction is plotted qualitatively in fig. 2, where w is the (perturbation) cross-stream velocity. It is the critical layer matching (see Lin [19]) that is essential to the viscous instability.

The results of viscous linear theory for plane Poiseuille flow are plotted in fig. 3 as a neutral curve, i.e. the locus of points in α–R space for which

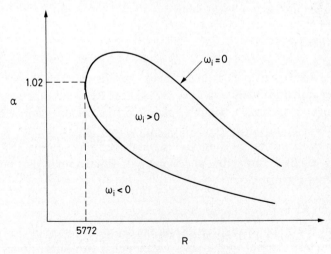

Fig. 3. Stability diagram for plane Poiseuille flow.

$\omega_i = 0$. Note that for sufficiently large R the neutral curve asymptotes to $\alpha = 0$ due to the inviscid stability of the flow. There is also a finite R at which maximum growth occurs.

The maximum linear growth rate for plane Poiseuille flow is approximately 0.0076 (in units of the convective time scale h/U_0) which occurs at $R \simeq +8000$. For a perturbation to grow by a factor of 10 requires the disturbance wave to travel approximately 50 channel widths (taking $c_r \simeq 0.3$) and a point on the centerline to travel more than 150 channel widths, which is clearly inconsistent with the explosive instabilities seen experimentally. Also, the critical Reynolds' number of 5772 is far from the observed critical Reynolds number for typical transitions of 1000.

Linear theory is even less successful in the case of plane Couette flow where, to date, no instability has been found at any α or R. The large region of parameter space searched to date (by asymptotic and numerical methods) seems to indicate global stability of this flow with respect to all linear disturbances [22].

The situation for pipe Poiseuille flow with regard to linear theory is as follows. Apart from certain results (particularly Squires' theorem) which no longer hold in cylindrical geometry, pipe flow behaves like plane Couette flow, i.e. it is *inviscidly* and *viscously* stable to all disturbances (at least in the fairly exhaustive parameter space searched to date) [23].

4. Finite-amplitude nonlinear analysis

4.1. Introduction

When v_p is infinitesimal, a linear theory results. Such a theory is very attractive in that instability depends only on the Reynolds number and not on the form or (assumed small) amplitude of the perturbation. However, as we have seen, linear theory cannot explain all the features of transition observed experimentally. Therefore, we consider a more general (but still relatively simple) form for v_p, namely a finite-amplitude, but two-dimensional perturbation:

$$v_p = v'(x, z, t), \tag{4.1}$$

where x is the streamwise direction (in the sense of the mean flow) and z is the direction normal to the walls. Although a two-dimensional theory cannot possibly explain all the details of transition (in particular its three-dimensionality), understanding two dimensional behaviour *is* important, for we shall show in the next section that the finite-amplitude secondary flows $v = \bar{U}(z)\hat{x} + v'(x, z, t)$ are explosively unstable to three-dimensional perturbations. We begin with an analysis of plane Poiseuille flow.

There are two possible "generic" behaviors for finite-amplitude, two-dimensional plane Poiseuille flow. These possibilities are shown schematically in figs. 4a and 4b as the locus of (E, R) points for which

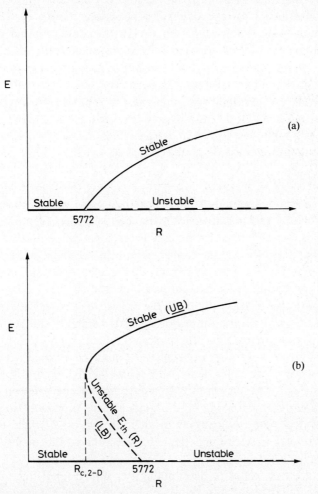

Fig. 4. (a) Supercritical stability diagram for plane Poiseuille flow. (b) Subcritical stability diagram for plane Poiseuille flow.

finite-amplitude equilibria (travelling waves with constant energy) exist (at a given wave number α). Here E is the energy of the perturbation. In both cases the $E = 0$ the laminar solution is unstable for $R > 5772$ due to the results of linear theory. In fig. 4a these unstable states evolve to an equilibrium value for $R > 5772$, and no new instabilities appear due to nonlinear interaction. (Here it is implicitly assumed that random motion is not possible in two dimensions for Reynolds numbers of order 10^4, and that all states saturate at these Reynolds numbers. There is ample numerical evidence to support these assumptions.) In fig. 4a we plot the stability diagram one would obtain if the nonlinear terms in the Navier–Stokes equations were *stabilizing* for all R. The second possible stability curve, plotted in fig. 4b, is similar to fig. 4a for $R > 5772$ (the unstable modes must saturate), however for $R_{c,2\text{-D}} < R < 5772$ the nonlinear terms are destabilizing for some threshold energy $E_{th}(R)$. These unstable finite-amplitude states themselves saturate at a higher energy.

These situations can also be understood heuristically by means of a one-dimensional autonomous evolution equation for E

$$\mathrm{d}E/\mathrm{d}t = f(E), \tag{4.2}$$

where the finite-amplitude stable (unstable) fixed points correspond to $f(E) = 0$ with $f'(E)$ negative (positive). The situation for both cases for $R > 5772$ is then described by the one-dimensional phase portrait shown in fig. 5a. For $R < 5772$ and the case shown in fig. 4a we have the phase portrait given in fig. 5b, whereas in the case depicted in fig. 4b, and $R_{c,2\text{-D}} < R < 5772$ we obtain the phase portrait shown in fig. 5c.

From the above physically plausible argument it is seen that the effect of the nonlinear terms on the two-dimensional stability characteristics can be investigated by searching for finite-amplitude equilibria (although the time scales over which they are achieved cannot be inferred). It should be noted that fig. 4 does not depict all possible behaviors, and that, in particular, the autonomous eq. (3.2) is suspect (and, as will be seen, wrong at finite-amplitudes). However, the arguments are physically sound (as is demonstrated mathematically by bifurcation theory [24]) and the results presented later (confirming fig. 4b) have been thoroughly checked numerically.

There are two routes to determining two-dimensional finite-amplitude equilibria. In both methods, the single Fourier mode ($e^{i\alpha x}$) found in linear theory is extended to include (at least, in principle) all of its harmonics,

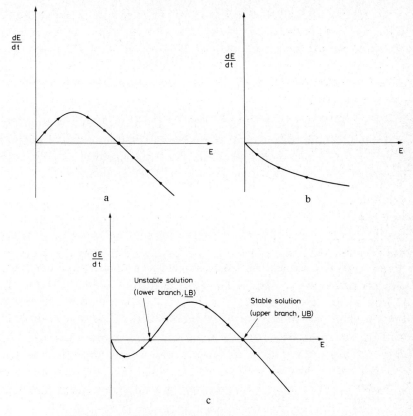

Fig. 5. Phase portrait for the autonomous evolution eq. (4.2).

$e^{i\alpha nx}$, n integer. In the first approach (termed *analytical* as it reduces the nonlinear problem to a sequence of linear ones) each Fourier mode is assumed to have an expansion in amplitude (of the primary, $e^{i\alpha x}$). It will be shown that this expansion has a small radius of convergence. The second method solves the nonlinear problem directly using numerical iteration techniques. We discuss the analytical technique first, as developed (in various forms) by Landau [11], Stuart [25] and Watson [26].

4.2. Analytical approach – amplitude equations

An expansion consistent with the Navier–Stokes (quadratic) nonlinearity is

$$\Psi = \sum_{n=-\infty}^{\infty} \sum_{m=0}^{\infty} A^{|n|+2m} \, \phi_n^{(m)}(z) \, e^{i\alpha n(x-ct)} + \bar{\Psi}(z), \qquad (4.3)$$

$$c = \sum_{m=0}^{\infty} c^{(m)} A^{2m}, \tag{4.4}$$

where Ψ is the streamfunction defined as

$$u = \partial \Psi / \partial z, \tag{4.5}$$

$$w = -\partial \Psi / \partial x, \tag{4.6}$$

and $\bar{\Psi}(z)$ is the basic flow, $z - \frac{1}{3} z^3$. Here A is an amplitude, presumably small but not infinitesimal, and $v = u\hat{x} + w\hat{z}$. To show how the method works we solve the model problem (identical in form to the two-dimensional Navier–Stokes equation for the streamfunction Ψ),

$$\frac{\partial \Psi}{\partial t} + L(\Psi) = N(\Psi, \Psi) \qquad (x \in D), \tag{4.7}$$

$$B(\Psi) = 0 \qquad (x \in \partial D), \tag{4.8}$$

where $N(a, b)$ is a bilinear operation, L and B are linear, and ∂D is the boundary of the spatial domain D. We insert eqs. (4.3) and (4.4) into eq. (4.7) and solve at each order in A for all excited Fourier modes. Proceeding in this fashion one obtains for $m = 0$ (noting that $\phi_0^{(0)} = 0$ by definition)

$$M\phi_1^{(0)} = L\phi_1^{(0)} - i\alpha c^{(0)} \phi_1^{(0)} = 0,$$

$$B\phi_1^{(0)} = 0 \qquad (x \in \partial D). \tag{4.9}$$

This is exactly the linear-theory eigenvalue problem (and M would be the fourth-order Orr–Sommerfeld operator if eq. (4.7) were exactly the Navier–Stokes equation). We require the normalization of $\phi_1^{(0)}(z^*) = 1$ (for some arbitrary z^*). The problem is then well-posed and will return the eigenfunction $\phi_1^{(0)}$ and eigenvalue $c^{(0)}$ of linear theory. There are two equations to solve at order $m = 1$

$$L\phi_0^{(1)} = N(\phi_1^{(0)}, \phi_{-1}^{(0)}) + N(\phi_{-1}^{(0)}, \phi_1^{(0)}),$$

$$B\phi_0^{(1)} = 0 \qquad (x \in \partial D), \tag{4.10}$$

$$L\phi_2^{(0)} - 2i\alpha c^{(0)} \phi_2^{(0)} = N(\phi_1^{(0)}, \phi_1^{(0)}),$$

$$B\phi_2^{(0)} = 0 \qquad (x \in \partial D), \tag{4.11}$$

giving $\phi_0^{(1)}$ and $\phi_2^{(0)}$ (the operators are non-singular and thus solution of eqs. (4.10) and (4.11) is routine). At order $m = 2$ the equation for $n = 1$ is

$$L\phi_1^{(1)} - i\alpha c^{(0)} \phi_1^{(1)} = i\alpha c^{(1)} \phi_1^{(0)} + N(\phi_0^{(1)}, \phi_1^{(0)})$$

$$+ N(\phi_1^{(0)}, \phi_0^{(1)}) + N(\phi_{-1}^{(0)}, \phi_2^{(0)}) + N(\phi_2^{(0)}, \phi_{-1}^{(0)}),$$

$$B\phi_1^{(1)} = 0 \qquad (x \in \partial D), \tag{4.12}$$

which we write symbolically as

$$M\phi_1^{(1)} = i\alpha c^{(1)} \phi_1^{(0)} + f,$$

$$B\phi_1^{(1)} = 0 \qquad (x \in \partial D). \tag{4.13}$$

The operator M is singular and thus eq. (4.13) only has a solution if

$$(\chi, [i\alpha c^{(1)} \phi_1^{(0)} + f]) = 0, \tag{4.14}$$

where χ is the eigenfunction of an adjoint operator to M

$$M^T \chi = 0, \tag{4.15}$$

defined relative to some inner product (a, b) (the precise choice of which is not important). Therefore, $c^{(1)}$ is determined as

$$c^{(1)} = i(\chi, f)/\alpha(\chi, \phi_1^{(0)}). \tag{4.16}$$

Uniqueness of the solution requires a normalization condition on $\phi_1^{(m)}$ $(m > 0)$, such as $\phi_1^{(m)}(z^*) = 0$, in order that the solution procedure continue.

To determine equilibria from eq. (4.4) we set the imaginary part of c to zero, getting

$$0 = \sum_{m=0}^{\infty} c_i^{(m)} A_{eq}^{2m}, \tag{4.17}$$

as the equation determining A_{eq}. Truncating eq. (4.17) at low order can easily yield spurious results which are not necessarily intrinsic to the method. It is also found that high-order truncations can be misleading as eq. (4.17) has a small radius of convergence (Herbert [27]). The series (4.17) converges only for A_{eq} very small and hence *very close* to the linear critical Reynolds number. The theory does correctly predict subcritical instability (see fig. 4b). However, it is of little use quantitatively even when carried to high order, especially away from the neutral curve.

Before describing the iterative approach we rewrite the imaginary part of eq. (4.4) as

$$\frac{dA}{dt} = A \sum_{m=0}^{\infty} \alpha c_i^{(m)} A^{2m}. \tag{4.18}$$

(Note this is not technically correct. The exposition given above, which is that of the method of false problems, can only be true when eq. (4.17) holds; to arrive at eq. (4.18), a modified method due to Watson should be used [27]. However in the case where we are on the neutral curve ($c_i^{(0)} = 0$) the results obtained up to order A^2 are identical to those one would get using Watson's method.)

Equation (4.18) is a classical Landau amplitude equation of which eq. (4.2) is another version. These amplitude expansions motivate the equations of the form (4.2), which we have seen are very useful in providing physical insight despite being quantitatively correct only for vanishingly small perturbation energies.

4.3. Iterative technique

The iterative technique for finding two-dimensional equilibria keeps the truncated Fourier-transform representation in x of the analytic technique, but eliminates the amplitude expansion that is the source of the convergence problems. We write the two-dimensional Navier–Stokes equations as

$$\frac{\partial}{\partial t} \nabla^2 \psi + \frac{\partial(\psi, \nabla^2 \psi)}{\partial(z, x)} = \frac{1}{R} \nabla^4 \psi, \tag{4.19}$$

where ψ is the stream function, eqs. (4.5) and (4.6), $\nabla^2 \psi$ is the y-vorticity, $(\partial u/\partial z - \partial w/\partial x)\hat{y}$, and $\partial(a, b)/\partial(x, z)$ denotes the Jacobian of a, b with respect to x and z. ψ is then expressed as a truncated Fourier series in $(x - ct)$,

$$\psi = \sum_{n=-N}^{N} \tilde{\psi}_n(z) \, e^{i\alpha n(x-ct)} + z - \frac{z^3}{3}, \tag{4.20}$$

where c is a real wave speed. Eq. (4.20) therefore represents periodic in x, steady, travelling-wave solutions to the Navier–Stokes equations.

To find the $\tilde{\psi}_n$ we use the Galerkin truncation of the Fourier-transform of eq. (4.19),

$$\frac{1}{R} (D^2 - n^2\alpha^2)^2 \tilde{\psi}_n - i\alpha n([\bar{U} - c](D^2 - n^2\alpha^2)\tilde{\psi}_n - (D^2\bar{U})\tilde{\psi}_n)$$

$$+ i\alpha \sum_{m=-N}^{N} [(n-m)\tilde{\psi}_{n-m}(D^3 - m^2\alpha^2 D)\tilde{\psi}_m$$

$$- mD\tilde{\psi}_{n-m}(D^2 - m^2\alpha^2)\tilde{\psi}_m] = 0, \tag{4.21}$$

where $\bar{U} = 1 - z^2$ as previously. This gives us a set of $N + 1$ coupled non-linear ordinary differential equations for the $\tilde{\psi}_n$, $n \geq 0$ (the negative modes are related to the positive modes by reality, $\tilde{\psi}_n^* = \tilde{\psi}_{-n}$).

These equations differ from those of the Landau approach in that they are coupled and nonlinear. The boundary conditions on the $\tilde{\psi}_n$ are

$$\tilde{\psi}_n(\pm 1) = D\tilde{\psi}_n(\pm 1) = 0 \qquad n \neq 0, \tag{4.22}$$

which corresponds to impermeability and no-slip, respectively. For $n = 0$ impermeability (no normal velocity) is automatically imposed, and so there we use the conditions

$$D\tilde{\psi}_0(\pm 1) = D^2\tilde{\psi}_0(\pm 1) = 0, \tag{4.23}$$

where the first condition here is no-slip as before, but the second is no mean stress at the wall due to the *perturbation*. We are therefore taking as our definition of plane Poiseuille flow that flow driven by a constant pressure gradient $-2/R$ (which is already completely balanced by the stress at the wall due to the laminar steady flow, $\bar{U}$).

We further impose the following symmetries on $\tilde{\psi}_n$,

$$\tilde{\psi}_n(z) = (-1)^{n+1}\tilde{\psi}_n(-z) \tag{4.24}$$

allowing us to solve for the $\tilde{\psi}_n(z)$ in only the half-channel. It is important to note that time-dependent solutions of eq. (4.19) (presented later) indicate that perturbations to eq. (4.20) breaking the symmetry (4.24) are stable, thus justifying the use of this symmetry.

Solution of eq. (4.21) proceeds as follows: the z direction is discretized by expanding the $\tilde{\psi}_n(z)$ in truncated Chebyshev polynomial expansions and projecting the differential equation on the finite space using a pseudo-spectral method (collocation) [28]; the resulting set of nonlinear algebraic equations is solved using a Newton iteration in conjuction with an arc-length continuation method [29].

A cut of the equilibrium (E, R, α) surface is plotted in (E, R) space in fig. 6, thus indicatiang that fig. 4b is the correct picture of subcritical behavior. For Reynolds numbers down to $R \simeq 2900$, *nonlinear interaction* is *destabilizing* for energies above a certain threshold.

The full (E, R, α) surface is given in fig. 7. It is seen that, not surprisingly, equilibria only exist for a finite band on streamwise wavenumbers, α. The details of the equilibrium surface will not be given here; what is important is fig. 6 and its implications. The stability characteristics implied by fig. 6

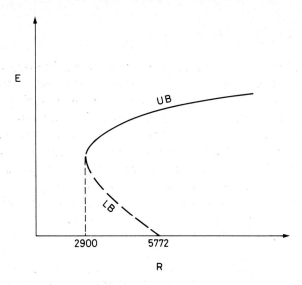

Fig. 6. E versus R for equilibrium solutions of nonlinear plane Poiseuille flow at fixed α.

have already been discussed in detail: we now address the question of how this picture fits into the general transition structure.

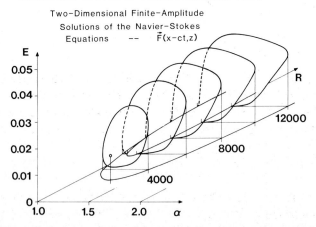

Fig. 7. The neutral surface in (E, R, α) space for plane Poiseuille flow. Finite-amplitude neutral disturbances exist down to a Reynolds number $R \approx 2900$. The critical wavenumber, α_c, is shifted from the linear value of 1.02 at $R \approx 5772$ up to 1.32 at $R \approx 2900$.

4.4. Time-dependent behavior

The mechanism for instability leading to transition to be presented shortly depends on the instability of two-dimensional secondary flows [e.g. eq. (4.20)] to three-dimensional perturbations. This instability therefore clearly requires the relative persistence of two-dimensional flows down to $R \simeq 1000$. The most simple situation would be where two-dimensional equilibria existed down to $R \simeq 1000$, however fig. 6 clearly illustrates this is not the case. It is thus necessary to investigate the time scales associated with decay of states for $R < 2900$ and compare them with the three-dimensional growth rates (which must be convective if experimental observations are to be explained). In fact, it will be shown that two-dimensional equilibration times for $R > 2900$ (and decay for $R < 2900$) involve viscous time scales.

A heuristic motivation for viscous time scales is as follows. When actual equilibria exist, eq. (4.1) in an appropriately Galilean-transformed frame is

$$\frac{\partial(\psi, \zeta)}{\partial(z, x)} = \frac{1}{R} \, \nabla^2 \zeta \qquad (\zeta = \nabla^2 \psi). \tag{4.25}$$

In the interior of the flow where the limit $R \to \infty$ is non-singular, it follows that

$$\partial(\psi, \zeta) / \partial z, x) = O(1/R)$$

so

$$\zeta = g(\psi) + O(1/R) \tag{4.26}$$

This integral of eq. (4.25) is illustrated in figs. 8a and 8b as plots of contours of ψ and ζ respectively ($R = 4000$), where it is seen that eq. (4.26) does in fact hold quite well in the interior. Equation (4.26) also holds as the flow makes final equilibrium adjustments, so as $t \to \infty$ we can write

$$\frac{\partial}{\partial t} \, \psi \sim O(1/R)\psi \tag{4.27}$$

indicating that $\tau \sim O(R)$. This fact can also be motivated from the Landau amplitude equation for the flow. For the (linearly) supercritical case, for instance,

$$\frac{\mathrm{d}A}{\mathrm{d}t} \sim a_1 A + a_2 A^3, \qquad a_1 > 0, \ a_2 < 0. \tag{4.28}$$

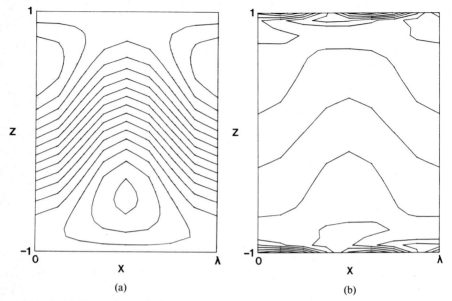

(a)   (b)

Fig. 8. (a) Streamlines of the steady (stable) finite-amplitude two-dimensional plane Poiseuille flow of the form (1) at $R = 4000$, $\alpha = 1.25$. The secondary motion appears as counter-rotating eddies. here λ ($= 2\pi/\alpha$) is the wavelength of the primary. (b) Vorticity contours of the steady (stable finite-amplitude two-dimensional plane Poiseuille flow of the form (1) at $R = 4000$, $\alpha = 1.25$. Note that in the interior of the flow (where viscosity is unimportant), the vorticity contours are very similar to the streamlines in fig. 8a. This implies that the nonlinear interaction is small away from the boundaries.

Perturbing around $A_{eq} = (-a_1/a_2)^{1/2}$, $A = A_{eq} + \varepsilon$, one obtains

$$\mathrm{d}\varepsilon/\mathrm{d}t \sim -2a_1\varepsilon. \tag{4.29}$$

Thus here the time scale is $\tau \sim 1/a_1 \sim R$ as before.

Evidence that the above arguments are valid is obtained by direct numerical simulation of the two-dimensional Navier–Stokes equations. Spectral methods are used in space, Fourier series in x and Chebyshev polynomials in z. Nonlinear terms are evaluated explicitly using collocation techniques and implicit methods are used for the pressure and viscous terms. Several time-stepping schemes (e.g. fractional time steps) and projection operators (e.g. tau, collocation) have been used for the latter, all of which give identical results [30, 31]. Results of such an experiment are given in fig. 9 at a Reynolds number of 4000 as a projection of the solution ψ on the two-dimensional phase space, $E_1^{1/2}, E_2^{1/2}$, where

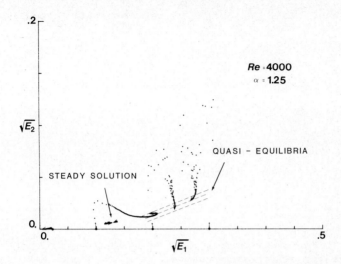

Fig. 9. A phase portrait of disturbances to laminar parallel plane Poiseuille flow in $(E_1^{1/2}, E_2^{1/2})$ space at $R = 4000$, $\alpha = 1.25$. The dots, equally spaced by 1.25 in time, indicate the trajectories of flows evolving from different initial conditions proportional to the least stable Orr–Sommerfeld mode at this (α, R). Following an initial transient, flows evolve to a state within a band of quasi-equilibria and reach the steady solution only on times of the order of R.

$$E_k = \frac{8}{15} \int_{-1}^{1} (|D\tilde{\psi}_k|^2 + k^2\alpha^2 |\psi_k|^2) \, \mathrm{d}z, \qquad (4.30)$$

i.e. it is the normalized energy of that part of the perturbation which depends on x like $\mathrm{e}^{\mathrm{i}\alpha kx}$, integrated over all z.

The arrows indicate the schematic trajectory of each orbit; the dots are the "exact" (numerical) orbit sampled at equally spaced intervals in time, clustering thus indicated slow evolution. The full triangle and the jagged line represent the equilibrium and threshold energies, respectively, obtained from the solution of eq. (4.21). As predicted, all solutions below the threshold decay and all above evolve (eventually) to the equilibrium. However, once the solutions reach a band of "quasi-equilibria" they equilibrate only very slowly, on a *viscous time* scale (as shown by the denseness of the dots). That the arguments given previously for viscous time scales hold is not surprising: what is surprising is that they hold in such a *wide* band of phase space.

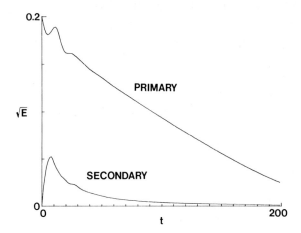

Fig. 10. The decay of a disturbance in plane Poiseuille flow at $R = 1500$, $\alpha = 1.32$. The slow decay at finite-amplitude (four times slower than the linear decay rate) reflects the existence of equilibria at higher Reynolds numbers.

The fact that quasi-equilibria exist for $R > 2900$ suggests that they also exist for $R < 2900$ where no strict equilibria exist; i.e. that the destabilizing effect of the nonlinear terms holds for all subcritical Reynolds numbers. In fig. 10 we show that this is indeed the case; at $R = 1500$ ($\sim \frac{1}{2} R_{c, 2\text{-}D}$) the finite-amplitude disturbance shown decays roughly 4 times slower than the corresponding infinitesimal (linear) perturbation. It is therefore plausible in plane Poiseuille flow that two-dimensional secondary flows will appear quasi-steady to three-dimensional rapidly growing perturbations down to Reynolds numbers as low as roughly 1000.

4.5. *Plane Couette flow and pipe flow*

The two-dimensional stability characteristics of plane Couette flow and pipe flow are markedly different from those of plane Poiseuille flow. Both of them are linearly stable to all disturbances at all R, and thus the only (non-trivial) plausible (E, R) bifurcation diagram is as shown in fig. 11.

The other (trivial) possibility is that these flows are globally stable to all finite-amplitude two-dimensional perturbations. The amplitude equation corresponding to fig. 11 is the same as that for the subcritically unstable plane Poiseuille flow case.

The amplitude expansion techniques discussed for plane Poiseuille flow

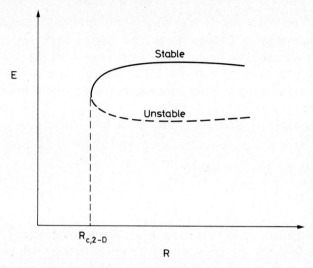

Fig. 11. Possible bifurcation diagram for plane Couette and pipe Poiseuille flows. However, there is no evidence that such finite-amplitude equilibria do exist.

can be used (in theory) for these flows as well (the Reynolds–Potter [32] false-problem method directly applies and Watson's method has been appropriately reformulated by Herbert [27] to apply to the case where no linear neutral curve exists). However, the results for plane Poiseuille flow (valid only near the neutral curve) suggest even greater problems here, where no neutral curve exists, (e.g. amplitudes will of necessity be larger and hence outside the radius of convergence). Indeed, this is found to be the case.

The amplitude expansion techniques applied (at low order) to these flows generally yield $R_{c,2\text{-D}} \ll 1000$ [33, 34]. In addition to being physically suspect (given the observed transitional Reynolds numbers), these results are, in fact, invalid. Numerical solutions of the Navier–Stokes equations starting from initial conditions with energies well above the threshold values predicted by the theory quickly decay [35, 36]. In view of Herbert's results for plane Poiseuille flow, higher order calculations using these techniques would probably be bootless.

In all numerical investigation to date, no two-dimensional equilibria have been found in either plane Couette or pipe Poiseuille flows. Furthermore, there is no evidence of quasi-equilibria, i.e. finite-amplitude flows that decay on a viscous time scale. Nonlinear effects (in two-dimensions) seem stabilizing for all (R, α). We will see in the next section that rapid two-dimensional decay does not in fact preclude three-dimensional instability, but that it does modify the picture in a way consistent with experimental observations of transition. In particular, the fact that two-dimensional equilibria exist in plane Poiseuille flow but not in pipe Poiseuille flow seems consistent with the non-intermittent transition found in the former, and the intermittent transition found in the latter [37].

5. Three-dimensional instability

5.1. Introduction

The basic mechanism to be discussed here involves the general instability of two-dimensional secondary flows to infinitesimal three-dimensional per-turbations. In general, two-dimensional equilibria do not exist, and three-dimensional growth rates must be extracted from time-dependent stability calculations. However, for the special case of plane Posieuille flow with $R > 2900$ equilibria do exist and the problem can be formulated as a classical (separable in t) stability analysis. We will thus first consider this simpler problem, and then extend the analysis to the case where equilibria do not exist.

5.2. Linear theory

We take the velocity to be of the form

$$v = (\bar{u}\hat{x} + \bar{w}\hat{z}) + \varepsilon v'(x, y, z, t) \qquad \varepsilon \ll 1, \tag{5.1}$$

where $[\bar{u}(x, z), \bar{w}(x, z)]$ here is the *combined* basic flow consisting of the parallel flow and the travelling waves found in the previous chapter, denoted $(\bar{u}_{2\text{-D}}, \bar{w}_{2\text{-D}})$. In the reference frame moving with speed c relative to the rest frame (c being the phase speed of the two-dimensional wave), $(\bar{u}, \bar{w})$ is *steady* and given by

$$(\bar{u}, \bar{w}) = [(1 - z^2) - c + \bar{u}_{2\text{-D}}(x, z), \bar{w}_{2\text{-D}}(x, z)]. \tag{5.2}$$

To obtain the perturbation equations we insert eq. (5.1) into the Navier–Stokes eqs. (2.2) and (2.3) and linearize with respect to ε [recalling that $(\bar{u}, \bar{w})$ is in fact a solution of the Navier–Stokes equations]. Upon doing so it is seen that the resulting equation is separable in t (due to the Galilean transformation) and y [as $(\bar{u}, \bar{w})$ does not depend on y], and thus we can further specify v, as

$$v' = \mathrm{Re}\left(e^{\sigma t} \sum_{n=-N}^{N} \sum_{m=-1}^{1} u_{nm}(z)\, e^{i\alpha n x'}\, e^{i\beta m y}\right), \tag{5.3}$$

where $x' = x - ct$ is the transformed coordinate and β is the spanwise (y) wavenumber. Also α is the streamwise wavenumber and is the same as that of $(\bar{u}_{\text{2-D}}, \bar{w}_{\text{2-D}})$ [see eq. (4.20)]. The number of modes kept in the x direction, N, is the same as for the travelling wave solution. In the limit as $N \to \infty$ the above representation is completely general. The perturbation equations for the $\boldsymbol{u}_{nm} = \{u_{nm}, v_{nm}, w_{nm}\}$ are

$$\left(\sigma(D^2 - k_{nm}^2) - \frac{1}{R}\, (D^2 - k_{nm}^2)^2\right) w_{nm} - im\beta D(\bar{u} * v_x + \bar{w} * v_z)_{nm}$$

$$-in\alpha D(\bar{u} * u_x + u * \bar{u}_x + \bar{w} * u_z + w * \bar{u}_z)_{nm}$$

$$-k_{nm}^2(\bar{u} * w_x + u * \bar{w}_x + \bar{w} * w_z + w * \bar{w}_z)_{nm} = 0, \tag{5.4}$$

$$\left(\sigma - \frac{1}{R}\, (D^2 - k_{nm}^2)\right) \zeta_{nm} - i\alpha n(\bar{u} * v_x + \bar{w} * v_z)_{nm}$$

$$+ im\beta(\bar{u} * u_x + u * \bar{u}_x + \bar{w} * u_z + w * \bar{u}_z)_{nm} = 0, \tag{5.5}$$

$$\zeta_{nm} = im\beta u_{nm} - in\alpha v_{nm}, \tag{5.6}$$

$$i\alpha n u_{nm} + i\beta m v_{nm} + D w_{nm} = 0, \tag{5.7}$$

where k^2 and D are as before, and $*$ denotes convolution

$$(f * g)_{nm} = \sum_{p,q} f_{p,q} g_{n-p, m-q}. \tag{5.8}$$

The above calculation is discretized in the z-direction just as in the calculation of the two-dimensional waves. If the number of modes kept in the z direction is denoted P, eqs. (5.4)–(5.7) would be an algebraic eigenvalue problem of rank $5P(2N+1)$. Minimal necessary resolution to obtain reasonable results would then require on the order of 1200 degrees of freedom (600×600 complex matrices). This requirement would be prohibitive. However, three symmetries allow us to reduce the order:

$$\{u_{n,m}, v_{n,m}, w_{n,m}\} = (\pm)\{u_{n,-m}(z), -v_{n,-m}(z), w_{n,-m}(z)\}, \tag{5.9}$$

$$\{u_{n,m}, v_{n,m}, w_{n,m}\} = (\pm)(-1)^{n+1}\{u_{nm}(-z), v_{nm}(-z), -w_{nm}(-z)\}, \tag{5.10}$$

$$\{u_{nm}, v_{nm}, w_{nm}\} = \{u_{-n-m}^{\dagger}, v_{-n-m}^{\dagger}, w_{-n-m}^{\dagger}\}, \tag{5.11}$$

where † here denotes complex conjugate. We take the upper sign in each case. The first symmetry is simply parity in y. The second is a reflection symmetry in z. The last is reality of u, which implies Im $\sigma = 0$, i.e. that the three-dimensional wave travels as the same phase speed as the two-dimensional wave. This last assumption is motivated by numerical experiments which indicate that the two-dimensional and three-dimensional components do in fact have the same frequency. Furthermore, this assumption has been verified by the solution of eqs. (5.4)-(5.7). Most importantly, direct numerical solutions which do not impose the symmetries (5.9)-(5.11) agree with the linear theory, indicating that symmetry-breaking perturbations are not much more unstable than those preserving the symmetry. this confirmation is imperative before the linear theory symmetrized results should be taked as physical.

These symmetries reduce the matrix to order $\sim 2(N+1)P$, a great aid in terms of both computation time and numerical conditioning. The linear eigenvalue problem is solved using either a global technique (QR) or a local (inverse iteration/Rayleigh quotient) technique. The results of the calculation, Re σ, are plotted as a function of β in fig. 12 at $R = 4000$, $\alpha = 1.25$.

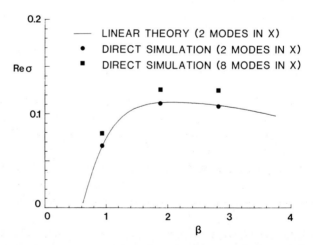

Fig. 12. A plot of the growth rate of three-dimensional perturbation, σ, as a function of β of $R = 4000$, $\alpha = 1.25$. Note the good agreement between the linear calculation and the full simulation. Increasing the number of nodes in the x-direction increases the growth rate, but the error in the 2-mode ($N = 1$) model is not large.

It is clearly seen that there is a strong three-dimensional instability.

There are two important points to be made. First, the growth rates are convective, not viscous; this suggests the instability is inviscid, and we will show much stronger evidence for this shortly. Secondly, below a certain threshold β, the instability cuts off rapidly, highlighting the three-dimensional nature of the instability. There is a weak maximum of σ at $\beta \simeq \alpha$, but above this there is little dependence of σ on β. This is consistent with experiments.

The solid line in fig. 12 is the result of linear theory. The circles are the growth rates calculated from direct numerical solutions (with no symmetries or even linearity assumed, though of course three-dimensional perturbations remain small). The very good agreement is strong evidence that the relevant instability is the linear one presented in eqs. (5.1)–(5.11).

We have found here a three-dimensional convective instability. However it remains to show that it cuts off at $R \sim 1000$, if one is to believe that this is in fact the instability responsible for transition. Although this is better done in the context of time-dependent stability analysis, we can gain some insight from the linear theory. For $R < 2900$ there are two things responsible for three-dimensional decay: two-dimensional decay, and viscous damping of the three-dimensional disturbance itself. We can isolate the latter from the former within the framework of linear theory by computing $\sigma(R)$ for $R < 2900$ using an *assumed* steady two-dimensional travelling wave which is in fact steady only at a higher R (i.e. $R > 2900$). This analysis capitalizes on the slow viscous decay rates of two-dimensional nonlinear disturbances demonstrated in section 4.

The results of such a computation are plotted in fig. 13 as $\operatorname{Re}\sigma$ versus R at $(\alpha, \beta) = (1.25, 1.25)$. There are three significant features here. First $\sigma(R)$ asymptotes as $R \to \infty$ implying that the instability is indeed *inviscid*. Secondly, the effect of dissipation cuts off growth at $R \simeq 400$ even in the absence of two-dimensional decay, thus effectively bounding the transition Reynolds number as $400 \lesssim R \lesssim 2900$. Lastly, the point at which $\sigma(R)$ abruptly changes from inviscid behavior to viscous behavior, i.e. where $\sigma(R)$ starts to change rapidly with R, is indeed in the neighborhood of $R \sim 1000$.

Better evidence for the $R \sim 1000$ cut-off can be obtained by considering the time-independent problem, i.e. solving the Navier–Stokes equations (numerically) with the initial conditions

$$v(x, 0) = (1 - z^2)\hat{x} + v_{2\text{-D}}(x, z, 0) + \varepsilon v_{3\text{-D}}(x, y, z, 0) \qquad \varepsilon \ll 1. \tag{5.12}$$

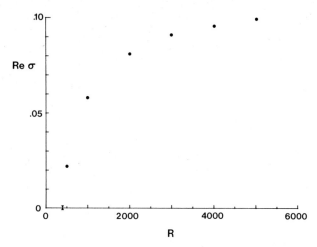

Fig. 13. A plot of the growth rate of three-dimensional disturbances, σ, as a function of R when $\alpha = \beta = 1.25$. Here, the two-dimensional background state (assumed steady in the moving frame) is an equilibrium solution of the Navier–Stokes equations at $R = 4000$, $\alpha = 1.25$. Note that viscosity eventually damps the instability even when the two-dimensional state does not decay. Note also the asymptotic independence of $\sigma(R)$ as $R \rightarrow \infty$.

The form of $v_{2\text{-}D}(x, z, 0)$ and $v_{3\text{-}D}(x, y, z, 0)$ is generally that of the most unstable Orr–Sommerfeld mode at the particular (α, β, R) in question. As a measure of two-dimensional decay and three-dimensional growth we calculate the two-dimensional and three-dimensional energies of the flow and plot their logarithm versus time. The results for several R ($500 < R < 4000$) are shown in fig. 14.

Again, note the strong exponential growth of the three-dimensional perturbations even when two-dimensional decay is present. Disturbances grow by a factor of ~10 in a time of 10. More importantly, note that a Reynolds number of the order of 1000 is singled out as being the critical R under which three-dimensional perturbations decay. Note the fact that in the absence of two-dimensional decay the cut-off Reynolds number is 400, indicating that two-dimensional decay is the more important mechanism in causing re-laminarization. For that reason we have chosen that $\alpha(\sim 1.32)$ in fig. 14 which minimizes two-dimensional decay, though it may not maximize three-dimensional growth.

The above numerical analysis and linear theory indicate that instability at transitional Reynolds numbers can be explained by three-dimensional linear instability of two-dimensional (relatively) slowly decaying secondary

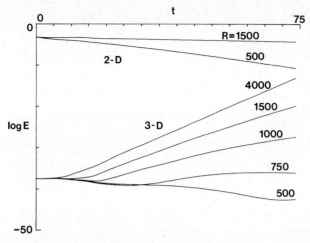

Fig. 14. A plot of the growth of three-dimensional perturbations on finite-amplitude two-dimensional states in plane Poiseuille flow at $(\alpha, \beta) = (1.32, 1.32)$. Here $E_{2\text{-D}}$ is the total energy (relative to the basic laminar flow) in wavenumbers of the form $(n\alpha, 0)$, while $E_{3\text{-D}}$ is the total energy in wave numbers $(n\alpha, \beta)$. For $R > 1000$ we obtain growth, whereas at $R = 500$ the three-dimensional perturbation ultimately decays. The growth rate of the three-dimensional disturbance amplitude at $R = 4000$ is about 0.18, and depends only weakly on R as $R \to \infty$.

flows. The fact that three-dimensional finite-amplitude perturbations need not be considered allows for a systematic analysis which otherwise could not be carried out. It also allows one to begin to analyze analytically the cause for the instability.

5.3. Vorticity dynamics

It should first be noted that the instability *cannot* be explained heuristically by inflexion point arguments, for although the secondary flow creates instantaneous (local) streamwise inflexional profiles, it is stable to all two-dimensional disturbances (see section 4 for the two-dimensional stability characteristics of plane Poiseuille flow). This is not consistent with an inflexion point picture, which would presumably predict two-dimensional as well as three-dimensional instability. Note we apply Squires theorem (non-rigorously) to the instantaneous inflexional profile, not the entire (non-parallel) flow. In fact, the vorticity dynamics of the three-dimensional instabiliy is very different than that motivated by Rayleigh's criterion (see figs. 1a–1c).

To understand these dynamics, we decompose the velocity of the flow as

$$v = v_0(x, z) + \varepsilon v_1(x, y, z, t), \tag{5.13}$$

$$\omega = \omega_0(x, z) + \varepsilon \omega_1(x, y, z, t), \tag{5.14}$$

where 0 refers to the two-dimensional flow $(\bar{u}, \bar{w})$ and 1 refers to the three-dimensional perturbation. The inviscid vorticity equation for ω_1, is then given by

$$\frac{\partial}{\partial t} \omega_1 + [(v_0 \cdot V)\omega_1 - (\omega_1 \cdot V)v_0] + [(v_1 \cdot V)\omega_0 - (\omega_0 \cdot V)v_1] = 0 \tag{5.15}$$

to order $O(\varepsilon)$.

The first bracketed term is primarily convection and stretching (in the x, z plane) of the perturbation velocity by the two-dimensional flow, and will be called the *stretching* term. The second bracket is mainly tilting of ω_0 into the (x, z) plane by ω_1, and will be denoted the *tilting* term. Stretching and tilting are the two basic mechanisms of vorticity dynamics, and thus this is a natural decomposition. (Note another mechanism is diffusion; however we consider inviscid dynamics here as fig. 13 clearly indicates that the instability is inviscid.

At present we cannot analytically determine the complete form or nature of the instability, however we can show that the mechanism requires *both* stretching and tilting. If tilting alone were present then

$$\partial\omega_1/\partial t = -(v_1 \cdot V)\omega_0 + (\omega_0 \cdot V)v_1 \tag{5.16}$$

which, by a simple vector identity can be written

$$\partial\omega_1/\partial t = V x(v_1 \times \omega_0). \tag{5.17}$$

Since $\omega_1 = V \times v_1$, it follows that

$$\partial v_1/\partial t = v_1 \times \omega_0 + V\phi, \tag{5.18}$$

where ϕ is an arbitrary potential. Forming the energy and integrating over space:

$$\frac{\partial}{\partial t} \int \tfrac{1}{2} v_1 \cdot v_1 \, dx = \int v_1 \cdot (v_1 \times \omega_0) \, dx + \int v_1 \cdot V\phi \, dx. \tag{5.19}$$

The second term is obviously zero, and the last integral, which can be rewritten

$$\int \nabla \cdot (v_1 \phi) \, \mathrm{d}x, \tag{5.20}$$

vanishes due to periodicity. Thus the energy of the perturbation cannot grow.

We next consider the stretching terms alone,

$$\frac{\partial}{\partial t} \omega_1 = -(v_0 \cdot \nabla)\omega_1 + (\omega_1 \cdot \nabla)v_0 = \nabla \times (v_0 \times \omega_1). \tag{5.21}$$

As v_0 has no y component or dependence $(\partial/\partial t) \int (\omega_1 \cdot \hat{y})^2 \, \mathrm{d}x = 0$. As we are considering eigenfunctions which must grow everywhere in space we set $\omega_1 \cdot \hat{y} = 0$, which allows us to represent ω_1 as

$$\omega_1 = \nabla \times \Psi \hat{y}. \tag{5.22}$$

Therefore

$$\frac{\partial}{\partial t} \Psi = (v_0 \times \omega_1) + \phi'(y), \tag{5.23}$$

$[\phi'(y)$ is arbitrary], which we rewrite as

$$\frac{\partial}{\partial t} \Psi + v_0 \cdot \nabla \Psi = -\frac{\partial \phi}{\partial t}(y). \tag{5.24}$$

As $v_0 \cdot \nabla \phi = 0$ this becomes

$$\frac{\partial}{\partial t} (\Psi + \phi) + v_0 \cdot \nabla (\Psi + \phi) = 0 \tag{5.25}$$

which upon integration gives

$$\frac{\partial}{\partial t} \int (\Psi + \phi)^2 \, \mathrm{d}x = 0. \tag{5.26}$$

Thus, stretching alone cannot maintain uniform exponential growth (by analogy with the two-dimensional anti-dynamo theorem of magnetohydrodynamics [38]).

Persistent exponential instability therefore requires both *stretching* and *tilting*; neither one alone will suffice. We can understand heuristically why this is so; just stretching in the (x, z) plane will line up perturbation vortex filament projections on this plane with the streaming two-dimensional flow shown in fig. 8a, after which only *algebraic* growth is possible. However, projections of ω_1 filaments in the (x, z) plane are not lined up with the streaming flow, as shown in fig. 15. This figure deserves several comments. First, the direction field, plotted here at one value of y, does not depend on y as only one Fourier mode, $e^{i\beta y}$, is present. This actually requires knowing that $\omega_1 \cdot \hat{x}$ and $\omega_1 \cdot \hat{z}$ both have the same y dependence, [see eq. (5.3)], and by noting that the projections are defined by $dz/dx = \omega_1 \cdot \hat{z}/\omega_1 \cdot \hat{x}$. Secondly, we note that although the strength of the vorticity clearly goes as $e^{\sigma t}$, the picture in fig. 15 is invariant (since we are looking at an eigenfunction in the rest frame).

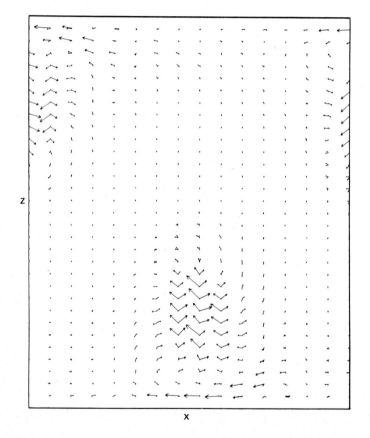

Fig. 15. Tilting and x-z perturbation vorticity fields in plane Poiseuille flow at $R = 4000$, $\alpha = \beta = 1.25$. The projection of the vorticity field is the clockwise-rotating whorl near $x = \pi/\alpha$, $z = -1$; the tilting field, T, is roughly orthogonal to the vortex filaments over the entire region of excitation.

The fact that fig. 15 does not change in appearance with time indicates that there is a mechanism which contributes to the $\omega_1 \cdot \hat{x}, \omega_1 \cdot \hat{z}$ vorticity after it has been stretched to prevent it from aligning with the streaming flow. This mechanism is precisely *tilting*. The "dynamic" term in the tilting equation (ignoring the "kinematic" convection term) is simply [from eq. (5.16)]

$$\partial \omega_1 / \partial t \sim \omega_0(x, z) \partial v_1 / \partial y, \tag{5.27}$$

and hence the *direction* of the tilting term is given by the field

$$\frac{\mathrm{d}z}{\mathrm{d}x} = \frac{\partial v_1 \cdot \hat{z}}{\partial y} \bigg/ \frac{\partial v_1 \cdot \hat{x}}{\partial y} = T(x, z), \tag{5.28}$$

which is again independent of y [see eq. (5.10)]. This direction field is also plotted in fig. 15.

We give here a very simple mechanism that models the interaction described above. Figure 16 shows two projections of ω_1 on the (x, z) plane, denoted ζ_1 and ζ_2, not aligned with the local v_0 shearing flow, here labelled by the local velocity along streamlines, u_s.

After some stretching in the (x, z) plane, ζ_1 will go to ζ_1'. The tilting term T then redirects ζ_1' so that it lines up with ζ_2 but is longer than the original projection, the final result being denoted ζ_2'. In this way, the ζ can grow continually in the same manner. Note this requires larger and larger T as well, but this follows as T and ζ are proportional. In essence, stretching provides growth and tilting rotates the filaments to allow them to continue growing. From this decomposition and fig. 16 we would therefore expect

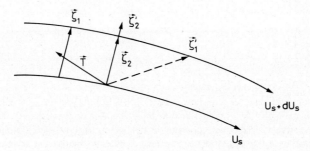

Fig. 16. A schematic plot of the combined action of vortex stretching and tilting.

the angle between T and ω_1 projections to be on the order of 90°, a hypothesis verified by fig. 15.

5.4. Plane Couette flow and pipe flow

We now consider how these results extend to plane Couette and pipe Poiseuille flows. We can no longer do a linear analysis as in the case of plane Poiseuille flow [eqs. (5.1)–(5.11)] as two-dimensional equilibria do not exist, however a time-dependent stability analysis is still possible. Taking initial conditions of the form eq. (5.12) as we did for plane Poiseuille flow for $R < 2900$, we solve the Navier–Stokes equation numerically and follow the evolution of the two-dimensional and three-dimensional components. (Note for pipe Poiseuille flow axisymmetric and non-axisymmetric is a better notation for two-dimensional and three-dimensional, respectively, however for uniformity we keep the latter.) We plot the results of this stability analysis as the logarithms of the two-dimensional and three-dimensional energies versus time for the cases of plane Couette flow and pipe Poiseuille flow in figs. 17 and 18, respectively.

In fig. 17 it is seen that, as in the case of plane Poiseuille flow (see fig. 14), plane Couette flow exhibits a three-dimensional *convective instability*,

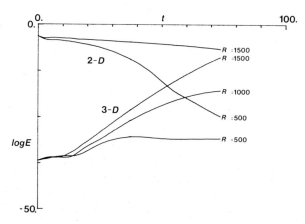

Fig. 17. A plot of the growth of three-dimensional perturbations on finite-amplitude two-dimensional states in plane Couette flow at $(\alpha, \beta) = (1.0, 1.0)$. $E_{2\text{-D}}$ and $E_{3\text{-D}}$ are defined as in fig. 14. The instability singles out a critical Reynolds number on the order of 1000 in accordance with experiment. The large decay rates of the two-dimensional states in plane Couette flow imply that larger threshold three-dimensional energies are required to force transition in this flow than in plane Poiseuille flow.

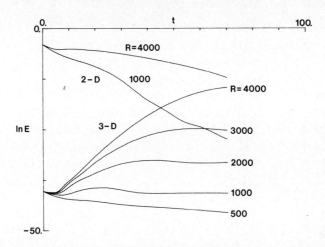

Fig. 18. A plot of the growth of non-axisymmetric states with $\alpha = 1.0$ for various R. The instability is strong for $R \gtrsim 3000$ and weak for $R \lesssim 1000$, the cut-off due primarily to increased axisymmetric decay.

the three-dimensional energy growing exponentially with $\sigma \simeq 0.2$ at $R = 4000$. Furthermore, it is between $R \simeq 500$ and $R \simeq 2000$ that the instability turns off, in good agreement with experiment. Note three-dimensional growth is maintained at $R \simeq 2000$ despite rather strong two-dimensional decay, and thus although the equilibria (and quasi-equilibria) in plane Poiseuille flow make the analysis simpler, the three-dimensional instability seems just as strong in plane Couette flow.

The case of pipe Poiseuille flow, fig. 18, is more complicated. There is no single Reynolds number at which the instability turns off rapidly (note in fig. 17 for plane Couette flow that the growth rate changes by as much between 1000 and 500 as between 4000 and 1000; this is in fact stronger evidence for $R \sim 1000$ being critical than the fact that the instability ultimately decays for $R \lesssim 1000$.) This is no doubt due to two-dimensional decay that obscures three-dimensional growth (though why this is not seen in Couette flow is not clear). To remedy this situation we force the two-dimensional solution in the following way. A time step is taken, the result of which we denote $\hat{v}(t + \Delta t)$. We then compute $v(t + \Delta t)$ (the velocity used to advance the solution to the next step) by computing the energy of $\hat{v}(t + \Delta t)$, $\hat{E}$, and then setting $v(t + \Delta T) = \hat{v}(t + \Delta t)(E_0/\hat{E})^{1/2}$, where E_0 is the desired energy. Note this forcing does not strongly affect phase relations; it is an overall energy constraint on the solution.

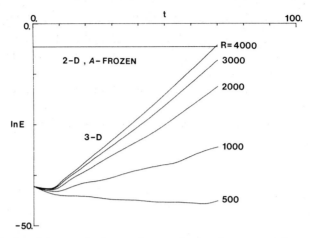

Fig. 19. A plot of the growth of non-axisymmetric perturbations on A-frozen axisymmetric states with $\alpha = 1.0$ for various R. In A-freezing the two-dimensional field, we prevent axisymmetric decay but allow non-axisymmetric growth, thus isolating the mechanism of three-dimensional instability. It is seen that at $R \simeq 2000$ there is a change from viscous behavior (strongly R-independent) to inviscid growth (approximately R-independent).

Forcing the solution in this way and repeating the time-dependent stability analysis gives the plot in fig. 19. Here there is a very strong change in character at $R \simeq 2000$; for $R \gtrsim 2000$ the growth depends only weakly on R, i.e. we have inviscid growth, whereas for $R \lesssim 2000$ σ is strongly dependent on R, i.e. we have viscosity-dominated time scales. We now justify physically the use of forcing. In pipe Poiseuille flow, transition structures seen in experiment at $R \simeq 2000$ are ultimately decaying as $t \rightarrow \infty$ though initially strongly unstable. This is exactly what we would expect if two-dimensional flows were ultimately decaying and the three-dimensional components were initially small. In order to become turbulent, general transition structures, called puffs [37], must interact. It is this interaction that we model by forcing the two-dimensional flow. Note incorrect forcing (i.e. forcing that destroys important two-dimensional phase information) such as simply freezing the flow, results in no growth [17]. This highlights the importance of the frequency-locking of the two-dimensional and three-dimensional flows.

We have presented here a nonlinear instability that obtains in all of the classical shear flows, and explains in quantitative and qualitative detail the experimental observations of transition in these flows. Numerical experiments starting from conditions predicted to be unstable by the theory do in fact exhibit the same turbulent structures (e.g. logarithmic regions in shear flows) seen in experiments [39], indicating that the instability is in fact responsible for transition. Although much is now known concerning the general structure, dimensionality, and time scale of the instability, a complete understanding of the interplay between the various elements of the instability is still lacking. Future work will address this issue as well determine the size of the class of flows to which this seemingly very general instability applies.

References

[1] O. Reynolds, An experimental investigation of the circumstances which determine whether the motion of water shall be direct or sinuous, and the law of resistance in parallel channels, Phil. Trans. Roy. Soc. London 174 (1883) 935–982.

[2] Lord Rayleigh, On the stability or instability of certain fluid motions, Proc. London Math. Soc. 11 (1880) 57–70.

[3] W.M. Orr, The stability or instability of the steady motions of a liquid, Proc. Roy. Irish Acad. A27 (1906–1907) 9–278; 69–138.

[4] A. Sommerfeld, Ein Beitrag zur hydrodynamischen Erklärung der turbulenten Flüssigkeitsbewegungen, Proc. 4th Int. Congress Mathematics Rome, Vol. III (1908) 116–124.

[5] L. Prandtl, Bemerkungen über die Entstehung der Turbulenz, Z. Angew. Math. Mech. 1 (1921) 431–436.

[6] W. Heisenberg, Uber stabilität und Turbulenz von Flüssigkeitsstromen, Ann. Phys. (4), 74 (1924) 577–627.

[7] O. Tietjens, Beiträge zur Enstehung der Turbulenz, Z. Angew. Math. Mech. 5 (1925) 200–217.

[8] G.B. Schubauer and H.K. Skramstad, Laminar boundary-layer oscillations and stability of laminar flow, J. Aeronaut. Sci. 14 (1947) 69–78.

[9] C.C. Lin, On the stability of two-dimensional parallel flows, Parts I–III, Quart. Appl. Math. 3 (1945) 117–142; 218–234; 277–301.

[10] D. Meksyn and J.T. Stuart, Stability of viscous motion between parallel planes for finite disturbances, Proc. Roy. Soc. A208 (1951) 517–526.

[11] L.D. Landau and E.M. Lifshitz, Fluid Mechanics (Pergamon Press, london, 1959).

[12] W.V.R. Malkus and G. Veronis, Finite-amplitude cellular convection, J. Fluid Mech. 4 (1958) 225–260.

[13] A. Davey, The growth of Taylor vortices in flow between rotating cylinders, J. Fluid Mech. 14 (1962) 336–368.

[14] J.-P. Zahn, J. Toomre, E.A. Spiegel and D.O. Gough, Nonlinear cellular motions in Poiseuille channel flow, J. Fluid Mech. 64 (1974) 319.

[15] T. Herbert, Finite amplitude stability of plane flows, in: Laminar–Turbulent Transition, AGARD Conf. Proc. No. 224 (1977) 3-1.

[16] S.A. Orszag and A.T. Patera, Subcritial transition to turbulence in plane channel flows, Phys. Rev. Lett. 45 (1980) 989–993.

[17] S.A. Orszag and A.T. Patera, Instability of pipe flow, in: Proc. Nonlinear Problems: Present and Future (North-Holland, Amsterdam, 1981 pp. 367–378).

[18] M.A. Gaster, A note on the relation between temporarily increasing and spatially increasing disturbances in hydrodynamic stability, J. Fluid Mech. 14 (1962) 222–224.

[19] C.C. Lin, The Theory of Hydrodynamic Stability (Cambridge University Press, Cambridge, 1955).

[20] H.B. Squires, On the stability of the three-dimensional disturbances of viscous flow between parallel walls, Proc. Roy. Soc. A142 (1933) 621–628.

[21] R. Fjørtoft, Application of integral theorems in deriving criteria of stability for laminar flows and for the baroclinic circular vortex, Geofys. Publ., Oslo 17 (1950) 1–52.

[22] A. Davey, On the stability of plane Couette flow to infinitesimal disturbances, J. Fluid Mech. 57 (1973) 369–380.

[23] H. Salwen and C.E. Grosch, The stability of Poiseuille flow in a pipe of circular cross-section, J. Fluid Mech. 54 (1972) 93–112.

[24] D.D. Joseph, Stability of fluid motions (2 vols.), Springer Tracts in Natural Philosophy Vols. 27 and 28 (Springer-Verlag, Berlin, 1976).

[25] J.T. Stuart, On the nonlinear mechanics of wave disturbances in stable and unstable parallel flows. Part 1. The basic behaviour in plane Poiseuille flow, J. Fluid Mech. 9 (1960) 353–370.

[26] J. Watson, On the nonlinear mechanics of wave disturbances in stable and unstable parallel flows. part 2. The development of a solution for plane Poiseuille flow and for plane Couette flow, J. Fluid Mech. 9 (1960) 371–389.

[27] T. Herbert, Nonlinear stability of parallel flows by high-order amplitude expansions, AIAA J. 18 (1980) 243–248.

[28] D. Gottlieb and S.A. Orszag, Numerical Analysis of Spectral Methods, Soc. Ind. Appl. Math., Philadelphia (1977).

[29] M. Lentini and H. Keller, The von Karman swirling flows, SIAM J. Appl. Math. 38 (1980) 52.

[30] S.A. Orszag and L.C. Kells, Transition to turbulence in plane Poiseuille and plane Couette flow, J. Fluid Mech. 96 (1980) 161–205.

[31] S.A Orszag and A.T. Patera, Subcritical transition to turbulence in planar shear flows, in: transition and turbulence, ed., R.E. Meyer (Academic Press, New York, 1981) pp. 127–146.

[32] W.C. Reynolds and M.C. Potter, Finite-amplitude instability of parallel shear flows, J. Fluid Mech. 27 (1967) 465–492.

[33] T. Ellingsen, B. Gjevik and E. Palm, On the non-linear stability of plane Couette flow, J. Fluid Mech. 40 (1970) 97–112.

[34] A. Davey and H.P.F. Nguyen, Finite-amplitude stability of pipe flow, J. Fluid Mech. 45 (1971) 701–720.

[35] See refs. [30] and [31].

[36] A.T. Patera and S.A. Orszag, Finite amplitude stability of axisymmetric pipe flow, J. Fluid Mech. 112 (1981) 467–474.

[37] I. Wyganski and F.H. Champagne, On transition in a pipe. Part I. The origin of puffs and the flow in a turbulent slug, J. Fluid Mech. 59 (1973) 281.

[38] H.K. Moffatt, Magnetic Field Generation in Electrically Conducting Fluids (Cambridge University Press, Cambridge, 1978).

[39] G. Comte-Bellot, University of Grenoble doctoral thesis (1963).

COURSE 13

FULLY DEVELOPED TURBULENCE AND SINGULARITIES

Uriel FRISCH

CNRS, Observatoire de Nice,
Nice, France

G. Iooss, R.H.G. Helleman and R. Stora, eds.
Les Houches, Session XXXVI, 1981 – Comportement Chaotique des Systèmes Déterministes/
Chaotic Behaviour of Deterministic Systems
© *North-Holland Publishing Company, 1983*

Contents

Abstract

It is shown that the small-scale properties of fully developed turbulence may be related to singular behavior at real or complex locations of inviscid (Euler) or viscous (Navier–Stokes) flows. Relevant mathematical and numerical results are reviewed with particular emphasis on the Taylor–Green vortex.

Aspects of the statistical theory are also examined, in particular, the breakdown of the Kolmogorov 1941 self-similarity theory and its relation to intermittency.

1. Introduction

The bulk of this Course is devoted to the onset of chaotic (or turbulent) behavior in systems (abstract, mechanical, electrical, fluids, ...) having a small number of degrees of freedom. For turbulent flows some measure of the number of degrees of freedom is provided by the Reynolds number

$$R = l_0 v_0 / v, \tag{1.1}$$

where l_0 and v_0 are a typical scale and velocity of the flow and v is the kinematic viscosity. As we shall see, the number of degrees of freedom effectively excited goes, for high R, roughly like $R^{9/4}$. In cgs units v is typically 10^{-2}, hence large values of the Reynolds number tend to be a rule in flows occurring in nature as well as in or around large vessels. The present lectures will be concerned with "fully developed turbulence", i.e. the behavior of turbulent incompressible flows at very high Reynolds numbers. Such flows display extremely chaotic motion (which can be quite spectacular; cf. the pictures of the Jovian atmosphere sent back by Voyager 1). As we shall see, experiments on fully developed turbulence indicate that the small-scale behavior is governed by a simple universal power-law for the energy spectrum and displays "high-frequency intermittency" (section 1.2).

We shall show in section 2.1 that some of the high-frequency (i.e. small-scale) behavior is possibly connected with the analytic structure of the solutions of the Euler (zero viscosity) and the Navier–Stokes equations, particularly with their singularities which may be confined to complex values of the space–time variables. Most of the remainder of sections 2.2 through 2.4 is devoted to the regularity/singularity properties of the initial value problem for inviscid and viscous flow. Mathematical and numerical results are reviewed.

Sections 3.1 through 3.4 deal with the statistical theory of turbulence. We begin by recalling in the next section a few elementary properties of the Navier–Stokes equation.

1.1. Conservation laws and invariance properties

The motion of three dimensional (3D), incompressible, viscous, electrically non-conducting fluid is governed by the Navier–Stokes equation

$$\partial_t v(t, r) + v \cdot \nabla v = -\nabla p + v \nabla^2 v + f \qquad r \in \mathcal{D}, \tag{1.2}$$

$\nabla \cdot v = 0$ + initial condition $v(t=0) = v_0$ + boundary condition $v = 0$ on $\partial \mathcal{D}$ where v is the velocity, p the pressure, f a driving force (e.g. buoyancy). When viscosity is zero one speaks of the Euler equation. Upon taking the curl of eq. (1.2) we obtain for $\omega = \nabla \times v$ the vorticity equation, which reads (for $f \equiv 0$)

$$\partial_t \omega + v \cdot \nabla \omega = \omega \cdot \nabla v + v \nabla^2 \omega. \tag{1.3}$$

The left hand side describes vorticity advection (transport), the first term on the right hand side can give rise to vortex stretching. When the flow is 2D (i.e. velocity is in the x–y plane and depends only on x and y), the vorticity $\omega = (0, 0, \xi)$ is conserved along the motion (for $v = 0$).

The energy equation is obtained from eq. (1.2) under the assumption that the velocity field is smooth, that it vanishes at the boundaries (or at infinity if $\mathcal{D} = \mathbb{R}^3$). With the notation

$$E = \tfrac{1}{2} \int_{\mathcal{D}} v^2 \, \mathrm{d}^3 r = \text{energy}, \tag{1.4}$$

$$\Omega = \tfrac{1}{2} \int_{\mathcal{D}} \omega^2 \, \mathrm{d}^3 r = \text{enstrophy}, \tag{1.5}$$

the energy equation reads

$$\frac{\mathrm{d}}{\mathrm{d}t} E = -2v\Omega + \int_{\mathcal{D}} f \cdot v \, \mathrm{d}^3 r. \tag{1.6}$$

The energy dissipation $\varepsilon = 2v\Omega$ is zero for the Euler flow. What happens to ε as $v \downarrow 0$ for fixed initial conditions depends on how well-behaved the enstrophy remains, a question related to the regularity properties of the

Euler equation. Let us assume that for $t \to \infty$ the flow achieves a (ν-dependent) statistical steady state. As $\nu \downarrow 0$, if the mean energy input $\langle \int_{\mathscr{D}} f \cdot v \, d^3 r \rangle$ approaches a non-vanishing limit, the same limit must be approached by the mean dissipation $\langle \varepsilon \rangle$. Hence, the mean enstrophy must become singular. Evidence that this is so is provided, e.g. by the experimental observation that the pressure drop in turbulent pipe flow seems to approach a finite limit as viscosity is decreased (equivalently, one can keep ν fixed, increase the Reynolds number and rescale the various quantities using the similarity property of viscous flow).

The smooth, unforced 3D Euler flow conserves, in addition to the energy, a quadratic pseudo-scalar called helicity

$$H = \tfrac{1}{2} \int_{\mathscr{D}} v \cdot \omega \, d^3 r. \tag{1.7}$$

This conservation law was discovered by Moreau (1961) and, independently, by Betchov (1961) and Moffatt (1969). The latter has shown that helicity is related to the knottedness of vortex lines (see also Arnold, 1973). A flow with a thin closed knotted vortex tube has a non-vanishing helicity (use Stokes' theorem). See fig. 1.

In the absence of boundaries and force, the Navier–stokes equation is invariant under several transformation groups (here referred to as symmetries)

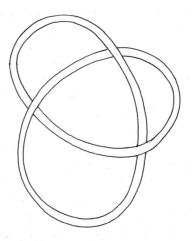

Fig. 1. A helical vortex tube.

space translations   (NC)
Time translations   (NC)
space rotations   (C)
plane symmetries   (C)
Galilean transformations   (NC)
scaling (for $\nu = 0$ or $\nu \downarrow 0$)   (NC)
Galilean transformations are defined by

$$v(t,r) \rightarrow v(t, r - Ut) + U, \qquad U = \text{constant}; \tag{1.8}$$

scaling transformations are defined by

$$r \rightarrow \lambda r, \ t \rightarrow \lambda^{1-h} t, \ v \rightarrow \lambda^h v; \qquad \lambda > 0 \tag{1.9}$$

(exponent $h \in \mathbb{R}$ arbitrary). After each group we have indicated if it is compact (C) or non-compact (NC) this being of some relevance for fully developed turbulence as we shall see in sections 3.3 and 3.4.

1.2. Some experimental results

Experimental data on fully developed turbulence can be obtained either in the laboratory (jets, wind tunnels, ...) or from geophysical flows. In a typical experiment a probe measures some velocity component [here denoted $v(t)$] as a function of time. There generally is a mean flow U with respect to the probe which is much larger than the rms velocity fluctuations v_{rms}. By performing a Galilean transformation to the frame of the mean flow it is seen that what the probe measures is, to leading order in v_{rms}/U, the spatial structure at a fixed time of the flow along a line parellel to U. Using this "Taylor hypothesis", data about the small-scale structure of fully developed turbulent flows are generally obtained by analyzing the high frequency components of the recorded signal. The highest Reynolds number experiment with reliable data on small scales is probably the tidal channel flow used by Grant et al. (1962). Measurements of the energy spectrum (the Fourier transform of the velocity correlation function) show that, over more than three decades, it follows a power-law with an exponent remarkably close to the value $-5/3$ predicted by Kolmogorov (1941). The same law is found in almost every high Reynolds number flow. So this scaling appears to be a universal asymptotic property of very high Reynolds number turbulence. When all the relevant scales are resolved, as is the case in wind-tunnel experiments, a typical energy spectrum has the shape shown

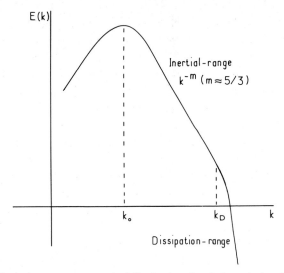

Fig. 2. Typical energy spectrum in fully developed turbulence in log–log plot.

in fig. 2. We define $E(k)$ to be the mean kinetic energy per unit wave-number and unit mass. The energy spectrum peaks at a wavenumber k_0, roughly the inverse of the scale l_0 for the generation of the turbulence (e.g. the mesh for grid-generated turbulence). The inertial-range where the spectrum follows the 5/3 law extends up to the (Kolmogorov) dissipation wave-number k_D such that $k_D/k_0 \sim R^{3/4}$, a result also predicted by Kolmogorov (1941); this implies that the number of effectively excited degrees of freedom in a volume $\sim l_0^3$ is $\sim R^{9/4}$. Beyond k_D there is a "dissipation range" with approximately exponential fall-off.

More information about the small-scale structure can be obtained by subjecting the recorded signal $\upsilon(t)$ to high-pass filtering (a lower cut-off frequency Ω corresponds to a wavenumber $K = \Omega/U$ where U is the mean flow). The high-pass filtered signal $\upsilon_\Omega(t)$ is here defined by

$$\upsilon_\Omega(t) = \int_{|\omega|>\Omega} e^{-i\omega t} \hat{\upsilon}(\omega)\, d\omega/(2\pi), \tag{1.10}$$

where $\hat{\upsilon}(\omega)$ is the Fourier transform of υ

$$\hat{\upsilon}(\omega) = \int_{-\infty}^{+\infty} e^{i\omega t} \upsilon(t)\, dt. \tag{1.11}$$

In the absence of filtering the single-time probability distribution of $\upsilon(t)$ is found to be roughly Gaussian. It becomes more and more non-Gaussian as the filter frequency is increased. For high Ω the filtered signal displays intermittent bursts. This was found for the first time in experiments performed by Batchelor and Townsend (1949) who used differentiation rather than filtering. With modern data processing techniques it is possible to raise the filter frequency up to about twice the value corresponding to k_D, well into the dissipation-range. A sample of the high-pass filtered velocity signal obtained by Gagne (1980) is shown in fig. 3. The high frequency intermittency displayed in such experiments is not to be confused with transition intermittency which requires no high-pass filtering (cf. the Courses by Libchaber and Pomeau in this volume).

Intermittency in the inertial range is not consistent with the Kolmogorov (1941) theory (see sections 3.3 and 3.4). Kolmogorov (1962) himself suggested a modification to account for intermittency. This modified theory predicts a power law $E_\varepsilon(k) \propto k^{-1+\mu}$ for the inertial-range behavior of the spectrum of dissipation fluctuations $\varepsilon(r)$, the latter being defined as $v\omega^2(r)$ minus its mean value. This is consistent with experimental measurements (Gibson et al. 1970, Monin and Yaglom 1975, ch. 25) and recent numerical simulations (see section 2.4) which give $\mu \simeq 0.5$. The exponent μ has a geometric interpretation as the codimension of a fractal set (Mandelbrot 1975, 1976; Frisch et al. 1978). Other experimentally measurable exponents can also be associated with intermittency (see section 3.4).

2. Regularity/singularity of Euler and Navier–Stokes flows

The Euler and Navier–Stokes equations governing the flow of incompres-

Fig. 3. High-pass filtered turbulent signal (Gagne 1980).

sible fluids have been known for a long time. Still, it is not yet known whether or not the corresponding initial-value problems in 3D are well-posed for all times. Other than the mathematical challenge, why should this be of interest?

Global and uniform regularity properties of Navier–Stokes flows are required to ensure that no breakdown takes place in the conditions which validate the Navier–Stokes approximation. For example, the largest hydrodynamical velocity must be small compared to the velocity of sound and the smallest hydrodynamical scale must be large compared to the molecular mean free path. So much for the regularity. The relevance of singularities (real or complex) is explained in the next subsection.

2.1. High frequency asymptotics and singularities

The experimental results reported in section 1.2 all involve looking at very high frequency (or wavenumber) behavior of the Fourier transform of some function of time (or space). Consider the Fourier representation

$$v(x) = \int_{-\infty}^{+\infty} e^{ikx} \hat{v}(k) \, dk/(2\pi),\tag{2.1}$$

(use Fourier series if $v(\cdot)$ is periodic). There are several results relating the behavior of $\hat{v}(k)$ for large k to singularities of $v(x)$ at real locations or complex locations [when $v(\cdot)$ can be continued analytically to complex values of x].

(a) Power-law behavior of $v(k)$ corresponds to lack of differentiability of $v(x)$; loosely expressed, if $v(x)$ is characterized by an exponent α near x_0, then $\hat{v}(k)$ will behave for $k \to \infty$ as $\exp(-ikx_0)k^{-\alpha-1}$. For example if $v(\cdot)$ is discontinuous ($\alpha = 0$), $\hat{v}(\cdot)$ has a k^{-1} tail. This is what happens for Burgers' model (Burgers 1974) which is governed by the partial differential equation

$$\partial_t v(t,x) + v \frac{\partial v}{\partial x} = v \frac{\partial^2 v}{\partial x^2}.\tag{2.2}$$

In the limit $v \downarrow 0$ the solutions develop shocks and, thereby a k^{-2} energy spectrum (Saffman 1968). Note that Burgers' equation is integrable (by the change of variable $v = -[1/(2v)](\partial/\partial x) \ln \psi$ it reduces to the heat equation). It therefore does not, with deterministic initial conditions, lead to any chaotic behavior (Fournier and Frisch 1983).

For Navier–Stokes homogeneous turbulence the evidence is that the inertial-range spectrum proportional to k^{-m} $(m \approx 5/3)$ extends to infinity as $v \downarrow 0$. In physical space this implies scaling for $h \to 0$ of the structure function

$$\langle |v(r+h) - v(r)|^2 \rangle \propto |h|^\varrho; \qquad \varrho \approx 2/3. \tag{2.3}$$

As noticed by Onsager (1949) this means that the velocity field is approximately $C^{1/3}$ (in a mean square sense). Hence, we are led to investigate the possible existence of singularities of the solutions of the Navier–Stokes for $v \downarrow 0$ and, thus of the Euler equation. There is however nothing compelling about this: a random function can have smooth (even analytic) samples and a singular correlation function. This is the case e.g. for the Holtzmark process $E(t)$. The random function $E(t)$ can be viewed as the electric field at a given point due to a homogeneous Poisson distribution of identical point charges moving with uniform identically, independently and isotropically distributed velocities; the correlation function is (Frisch and Brissaud 1971)

$$\langle E(t) \cdot E(t') \rangle \propto |t - t'|^{-1}. \tag{2.4}$$

(b) Exponential decrease of $\hat{v}(k)$ at high k implies that $v(x)$ can be continued analytically to complex values $x + iy$ in some strip $|y| < \delta$. Indeed, the formal analytic continuation

$$v(x + iy) = \int_{-\infty}^{+\infty} e^{ik(x+iy)} \hat{v}(k) \, dk/(2\pi) \tag{2.5}$$

is a convergent integral if we assume $|(\hat{v}(k)| \leq C e^{-\delta |k|}$. The high wavenumber (or frequency) tail of the spectrum of turbulent flow appears to be approximately exponential (for fixed Reynolds number). It is thus tempting to relate dissipation-range properties to the analytic structure of the flow. This was already observed by Von Neumann (1949) in a remarkable review paper on turbulence which remained unpublished for a long time; he pointed out that Heisenberg's (1948) predicted k^{-7} dissipation-range spectrum was difficult to reconcile with the (presumably) analytic properties of viscous flow.

(b) High-pass filtering of an analytic function with isolated singularities in the complex domain will make these conspicuous as intermittent bursts centered at the real part of the singularities. Let us show this by working in the time domain. Consider an analytic function $v(t + i\tau) = v(z)$ with singularities at complex locations $z_j = t_j + i\tau_j$ in the neighborhood of which it can be characterized by a fractional exponent

$$v(z) \approx a_j (x - z_j)^\varrho, \qquad z \to z_j. \tag{2.6}$$

To evaluate the Fourier integral

$$\hat{v}(\omega) = \int_{-\infty}^{+\infty} e^{i\omega t} v(t)\, dt \tag{2.7}$$

for $\omega \to +\infty$ we can shift the integration contour so that it wraps the singularities as shown in fig. 4. This replaces the oscillating Fourier integral by Laplace integrals which are easy to evaluate asymptotically [assuming that $v(z)$ does not grow faster than exponentially for $z \to \infty$]. To leading order one obtains (Carrier et al. 1966)

$$\hat{v}(\omega) \underset{\omega \to +\infty}{\approx} -\frac{2\sin(\pi\varrho)\Gamma(\varrho+1)}{\omega^{\varrho+1}} e^{-i\pi\varrho/2} \sum_{j}^{UP} \varepsilon_j a_j e^{i\omega z_j}, \tag{2.8}$$

where Σ^{UP} is a sum over the half plane Im $z > 0$ and the ε_j's are determination factors. If $v(t)$ is real for real times (as we shall assume) the behavior of $\hat{v}(\omega)$ for $\omega \to -\infty$ is obtained from its hermitean symmetry. Using the asymptotic expansion (2.8) it is easy to find the high-pass filtered signal $v_\Omega(t)$ defined as

$$v_\Omega(t) = \int_{|\omega|>\Omega} e^{-i\omega t} \hat{v}(\omega)\, d\omega/(2\pi); \tag{2.9}$$

for $\Omega \to +\infty$, to leading order we obtain (Frisch and Morf 1981)

$$v_\Omega(t) \underset{\Omega \to \infty}{\approx} \frac{\Gamma(\varrho+1)}{\Omega^{\varrho+1}} \sum_{j}^{UP} \varepsilon_j a_j e^{-\Omega\tau_j} \mathrm{Re}\left(\frac{\exp[-i\Omega(t-t_j) - i\pi\varrho/2]}{t - z_j} \right). \tag{2.10}$$

It is easily checked that $v_\Omega(t)$ is a sum of burst-like structures; for each singularity at $t_j + i\tau_j$ there is a burst centered at t_j with an overall amplitude factor proportional to $\exp(-\Omega\tau_j)$ which very effectively singles out the singularities close to the real axis. The intermittent bursts observed for

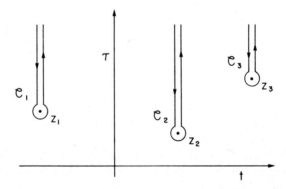

Fig. 4. Contour for the Fourier integral in the complex t-plane.

example in Gagne's (1980) experiment (cf. section 1.2) provide some evidence that the solution of the Navier–Stokes equation have complex time (and also complex space) singularities. This question is discussed further in Frisch and Morf (1981) in connection with several dynamical models substantially simpler than the Navier–Stokes equation.

Finally, we mention that the nature of singularities at complex locations may give hints as to the integrability of the equations of motion. For a number of dynamical systems a striking correlation has been found between integrability and the Painlevé property, i.e. the property that the only movable singularities are poles (Ablowitz et al. 1980, Bountis 1982, Weiss et al. 1982).

2.2. *Regularity results for Euler flows*

The Euler equation

$$\partial_t v + v \cdot \nabla v = -\nabla p; \quad \nabla \cdot v = 0; \quad v \cdot n = 0 \quad \text{on } \partial \mathscr{D} \tag{2.11}$$

has a remarkable property which goes back to the work of Lagrange but has been mainly stressed by Arnold (1966, 1974). The motion of incompressible inviscid fluid in a 2D or 3D domain corresponds to a geodesic flow on the infinite dimensional manifold $S\,\text{Diff}(\mathscr{D})$ of the volume-preserving diffeomorphisms of $\mathscr{D}$ of the form $G: a \to r(a)$ with

$$\det(\partial r_i / \partial a_j) = 1. \tag{2.12}$$

An arbitrary (possibly non-physical) motion of the fluid may be described by a curve

$$t \to g_t; \quad g_0 = \text{identity}.$$

The velocity is $v(t, a) = \partial r(t, a)/\partial t$ (in Lagrangian coordinates). The kinetic energy

$$E(t) = \int \tfrac{1}{2} v^2 \, d^2 a \tag{2.13}$$

defines a Reimannian structure on $S\,\text{Diff}(\mathscr{D})$. It is a simple exercise in the calculations of variation to obtain the Euler equation by the prescription:

variation of $[\int_0^T E(t)\, dt] = 0$ subject to the constraints of incompressibility [eq. (2.12)] and to $r(0, a) = a$, $r(T, a) = \text{given}$.

For various 2D special cases Arnold (1965, 1974) has calculated the

Riemannian curvature of sections of S Diff($\mathscr{D}$) and found it to be negative. Neighboring geodesics on a manifold of negative curvature tend to diverge exponentially. According to Arnold (1972, 1974) exponential dilatation may account for the impossibility to predict large-scale atmospheric motion for extended periods of time. A similar result has been obtained by Lorenz (1969) and Leith and Kraichnan (1972) using statistical closure techniques of a somewhat *ad hoc* nature which, to our knowledge, have no geometric content; however, the advantage of the latter approach is that it also predicts how uncertainties, which may initially be confined to very small scales, will eventually migrate to larger scales.

The infinite dimensional geometric interpretation of the Euler equation has been used by Ebin (1983) to derive certain "weak integrability" results for smooth 2D Euler flows. Such results are not strong enough to rule out that the flow can become chaotic. In fact, there is good evidence that chaos may be present in 2D when the vorticity is concentrated in at least four discrete vortice (Novikov and Sedov 1978, Aref and Pomphrey 1980). For distributed smooth vorticity fields there remains the possibility that the 2D Euler flow develops increasingly fine-scale laminar structures rather than chaos; some indication that this may be the case is given by recent high resolution numerical simulations with simple initial conditions (Brachet 1983).

The fact that the Euler equation has an infinite dimensional geometric interpretation has also been used by Ebin and Marsden (1970) to derive regularity results for the initial value problem some of which were previously obtained by methods of functional analysis. The geometric approach may be potentially more powerful (particularly because it has incompressibility built-in as curvature of the space, somewhat like the gravitational force in General Relativity). However, for what has been proved so far, more classical methods, of which we shall now give the reader a flavor, are much simpler [for more detailed but still elementary presentations see Rose and Sulem (1978) and Sulem and Sulem (1983)].

We begin with the 2D Euler flow.

Theorem (1) (Wolibner 1933, Hölder 1933, Kato 1972). Consider the 2D Euler flow in a bounded (or periodic) domain $\mathscr{D}$ of diameter L. If the initial velocity field $v_0(r)$ is $C^{1+\varepsilon}$ ($\varepsilon > 0$) it remains so for all times.

Sketch of the proof. In 2D vorticity $\omega = (0,0,\xi)$ is conserved along the motion; hence $|\xi|_{L^\infty} = \sup |\xi(r)|$ is a constant. The velocity is expressible as $\nabla \times (0, 0, \psi)$ where the stream function ψ satisfies a Poisson equation

$$\nabla^2 \psi = -\xi \qquad \psi = 0 \text{ on } \partial \mathcal{D}. \tag{2.14}$$

By solving the Poisson equation the velocity may be expressed in terms of the vorticity through an integral operator

$$v(r) = \int_{\mathcal{D}} G(r, r') \xi(r') \, d^2 r'. \tag{2.15}$$

In the absence of boundaries $G(\cdot, \cdot)$ is given by the Biot and Savart formula

$$G(r, r') = \frac{1}{2\pi} \frac{e_3 \times (r - r')}{|r - r'|^2}, \tag{2.16}$$

where e_3 is the unit vector in the x_3 direction. In the presence of boundaries the short-distance behaviour remains essentially the same, as a consequence of which it may be shown that

$$|G(r, r')| \leq C_1 / |r - r'|, \qquad C_1 > 0. \tag{2.17}$$

From this one easily derives an estimate for the velocity difference across a distance ϱ, namely

$$|v(r + \varrho) - v(r)| \leq C_2 |\xi|_{L^\infty} \cdot \varrho \ln \frac{L}{\varrho}, \qquad C_2 > 0. \tag{2.18}$$

(Hint: estimate separately the integrals from the inside and outside of a circle centered at $r + \varrho/2$ of radius ϱ).

Now, consider two passive particles following the motion of the fluid and initially separated by a distance ϱ_0.

Using vorticity conservation along particle paths it is easily shown that global regularity of the flow (in the $C^{1+\varepsilon}$ space) obtains if such particles can never collide. To prove the latter one notices that eq. (2.18) implies a differential inequality for $\varrho(t)$

$$\left| \frac{d\varrho}{dt} \right| \leq C_2 |\xi|_{L^\infty} \varrho \ln \frac{L}{\varrho}. \tag{2.19}$$

Integration of eq. (2.19) gives both upper and lower bounds for the particle separation $(t > 0)$

$$\left(\frac{\varrho_0}{eL} \right)^{\exp(tC_2 |\xi|_{L^\infty})} \leq \frac{\varrho(t)}{eL} \leq \left(\frac{\varrho_0}{eL} \right)^{\exp(-tC_2 |\xi|_{L^\infty})}; \tag{2.20}$$

where e = exp 1. Uniqueness is also easily obtained as consequence of regularity.

Next we give an analyticity result.

Theorem (2) (Bardos et al. 1976). Consider the same problem as in Theorem (1) but assume furthermore that $v_0(r + i\varrho)$ is analytic in a strip $\varrho < \delta$; then, for any real t the solution is analytic in some strip $\varrho < \delta(t)$; analyticity also holds in the time variable.

The crucial estimate used to obtain this result is

$$|\text{Im } v(r + i\varrho)| \leq C_2 |\xi|_{L^\infty} \varrho \ln \frac{L}{\varrho}. \tag{2.21}$$

The derivation is analogous to that of the real-space estimate (2.18). From eq. (2.21) it follows that the width of the analyticity strip does not shrink faster than given by the estimate (2.20) for particle separation in the real domain. Thus, the analyticity strip does not go to zero faster than an exponential of an exponential. Numerical simulations of 2D Euler flows suggest that $\delta(t)$ actually decreases only exponentially (Sulem et al. 1983, Brachet 1983).

An interesting case of 2D Euler flow for which the vorticity is never smooth is the Kelvin–Helmholtz vortex sheet in which there is a velocity discontinuity across an interface Σ_t. It has been conjectured by Birkhoff (1962) and proven recently (Babenko and Petrovich 1979, Sulem et al. 1981) that, if Σ_t is initially an analytic curve, it remains so for at least a finite time. An asymptotic expansion of Moore (1979) and numerical results of Meiron et al. (1980) suggest that a singularity of the interface develops at a finite time.

We now turn to 3D Euler flow. Vorticity is no longer conserved and may be stretched by velocity gradients; the latter depend linearly on vorticity (through a Poisson-type pseudo-differential operator of order zero). Suitable norms (e.g. Hölder norms) of the velocity gradient are easily bounded in terms of corresponding norms for the vorticity. With such estimates, however, catastrophic vortex stretching cannot be ruled out. Indeed, the best regularity result at the moment is

Theorem (3) (Lichtenstein 1925, Bardos and Frisch 1975, 1976). Consider a 3D Euler flow in a bounded or unbounded domain. If $v_0 \in C^{1+\varepsilon}(\varepsilon > 0)$ it remains so for $t \in] - T, +T[$ with $T \propto 1/|v_0|_{C^{1+\varepsilon}}$.

Similarly, there are finite time analyticity results in both time and space. Of particular interest for our subsequent discussion is the following

Theorem (4) (Bardos and Benachour 1977). Consider a 3D Euler flow with an initially analytic velocity, it remains so as long as the velocity remains C^1 in the space variable.

In other words the flow cannot lose analyticity without losing first-order differentiability in the space variables.

2.3. Numerical exploration of singularities for the 3D Euler flows. The Taylor–Green vortex

Existing theorems about the 3D Euler flow with smooth initial conditions fail to give better than finite time regularity. Can such flow develop a singularity in a finite time? This is a question the importance of which was stressed by Saffman (1981). That there may indeed be a singularity is suggested by the application to the Euler equation of statistical closure techniques (Kraichnan 1959, Orszag 1973, Rose and Sulem 1978). The closure equations which unequivocally predict finite-time singularity are not derived from the Euler equation but from certain statistical models having a number of structural properties in common with the Euler equation (Kraichnan 1961, Frisch, Lesieur, and Brissaud 1974).

At present, numerical techniques seem to be the best way to explore the question of singularities for the actual Euler flow. An early attempt was made by Orszag (1973) using numerical simulations of the Navier–Stokes equation for a particular spatially periodic flow, the Taylor–Green vortex [see section 2.3.3] at Reynolds numbers up to 400. Extrapolation to infinite Reynolds numbers suggested a finite positive limit for the dissipation beyond a certain time t_*. This would imply an infinite enstrophy in the limit $v \downarrow 0$ after t_* and thereby a singularity for the Euler flow (without boundaries the Navier–Stokes flow tends to the Euler flow as $v \downarrow 0$ as long as the latter is regular). More recent simulations at Reynolds numbers up to 3000 do not support this conclusion because there is a change-over in the behavior of the dissipation at Reynolds numbers around 500 (Brachet et al. 1983).

Using the Navier–Stokes equations to investigate the singularities of the Euler flow does not seem particularly advisable. New direct methods making use of analyticity properties have been recently applied to the Euler flow and other inviscid flows. The general ideas involved are presented in sections 2.3.1 and 2.3.2. Application to a particular 3D Euler flow, the Taylor–Green vortex, is discussed in section 2.3.3. We shall here be concerned only with flow defined on the $\mathbb{T}^3$ torus, i.e. which is 2π-periodic in x_1, x_2 and x_3 and can therefore be expanded in a spatial Fourier series

$$v(t,r) = \sum_{k \in Z_3} e^{ik \cdot r} \hat{v}(t,k). \tag{2.22}$$

2.3.1. The Taylor expansion method

For initially analytic 3D Euler flow we know from section 2.2 that there exists a unique solution analytic in space and time for at least a finite time. Therefore the solution can be expanded in a convergent Taylor series around $t = 0$.

$$v(t) = \sum_{n \geq 0} v^{(n)} t^n \tag{2.23}$$

(space dependence is here implicit). The v_n's can easily be obtained recursively. For this it is convenient to rewrite the Euler equation as an abstract quadratic equation. We start from the Euler equation written as

$$\partial_t v_i = -\partial_j (v_i v_j) - \partial_i p; \qquad \partial_j v_j = 0; \qquad \partial_j \equiv \partial/\partial x_j. \tag{2.24}$$

The pressure is eliminated by taking the divergence of the Euler equation and solving the resulting Poisson equation to give

$$\partial_t v_i = -\partial_j [v_i v_j - \Delta^{-2} \partial_{il}^2 (v_l v_j)], \tag{2.25}$$

where Δ^{-1} is the inverse Laplace operator (multiplication by one over the square of wavenumber in Fourier space). The right hand side of eq. (2.25) is a quadratic first order pseudo-differential operator (defined in suitable function spaces). There is a corresponding bilinear symmetric form which we denote $B(\cdot, \cdot)$. With this notation the Euler equations take the form

$$\partial_t v = B(v, v). \tag{2.26}$$

Substitution of eq. (2.23) into eq. (2.26) gives recursion relations for the temporal Taylor coefficients

$$(n+1)v^{(n+1)} = \sum_{\substack{r,s \geq 0 \\ r+s=n}} B(v^{(r)}, v^{(s)}); \qquad n \geq 1; \ v^{(0)} = v_0. \tag{2.27}$$

Exact numerical implementation is possible for a particular class of flows as a consequence of the following

Lemma (Taylor and Green 1937). If the initial condition v_0 has only a finite number of non-vanishing spatial Fourier components, the same holds for all $v^{(n)}$'s, the number of components increasing generally with n.

The proof follows from eqs. (2.24) and (2.27) and the observations that (i) the operators ∂_j, ∂_i and Δ^{-1} are diagonal in Fourier space, (ii) the product of two functions with a finite number of nonvanishing Fourier components has a finite number of nonvanishing Fourier components. Note also that all the $v^{(n)}$'s are entire functions in the space variables (no singularity at finite complex locations). The sum of the series (2.23) however is analytic (at least for short times) but need not be entire. If v_0 has integer or rational Fourier coefficients it is easily checked that all the Fourier coefficients of the $v^{(n)}$'s are rational and can in principle be calculated exactly. From the Taylor series (2.23) one can generate the Taylor coefficients of various spatial integral norms depending only on time. Particularly convenient are the squared Sobolev norms $\Omega_p(t)$ called p-enstrophies; they are given by

$$\Omega_p(t) = \sum_{k \in Z^3} |k|^{2p} |\hat{v}(t,k)|^2 = \sum_{n \geq 0} B_p^{(n)} t^n. \tag{2.28}$$

The $B_p^{(n)}$'s corresponding to the first few values of n can be calculated by hand (Taylor and Green 1937). If the initial condition is sufficiently simple, particularly if it has many symmetries, a large number of $B_p^{(n)}$'s can be calculated with great accuracy on a digital computer (Orszag 1973, Van Dyke 1975, Morf et al. 1981, Morf et al. 1980, Brachet et al. 1983 and section 2.3.3). On the resulting temporal Taylor series for the Ω_p's it is possible to search for singularities (at both real and complex times).

The mathematical problems that arise in doing an analytic continuation of Taylor series and searching for singularities are beautifully described in the book by Hadamard and Mandelbrojt (1926). The procedures for analyzing truncated Taylor series are quite similar to those used in analyzing high temperature series in statistical mechanics (Grant and Guttmann 1974, Gaunt 1981). Let us just stress a few points. The singularity (singularities) closest to the origin in the complex t-plane determines (determine) the radius of convergence of the Taylor series. In the simplest case the nearest singularity can be obtained by evaluating ratios of successive coefficients. For singularities located beyond the convergence disk methods of analytic continuation must be used, based, on Padé approximants (Baker 1975). It is then particularly important to have a "large" number of Taylor coefficients with "great" precision. How "large" and "great" is difficult to say because analytic continuation of power series by numerical techniques is at present more of an art than a science; there are numerous known pitfalls and no demonstrably safe methods (see Baker 1981 and other contributions in the same Cargèse Proceedings). The Taylor method seems therefore best suited for problems in which the singularity closest to the origin occurs at a real time. This is the case for Burgers' equation at zero viscosity (Morf et al. 1981) and for the 2D or 3D Euler flow investigated so far. Let us now describe a method which may be somewhat better suited for the latter case because it does not involve any analytic continuation in time.

2.3.2. The analyticity strip method

We have seen in section 2.2 that 3D Euler flow, initially analytic in space, remains analytic in space in a strip in $\mathbb{C}^3$ $[r + i\varrho$ s.t. $|\varrho| < \delta'(t)]$ for at least a finite time. Let $\delta(t)$ be the maximum analyticity strip width at time t. By Theorem (4) we know that the flow ceases to be (spatially) C^1 at precisely the real time t_* where $\delta(t)$ vanishes, if it ever does. The width of the analyticity strip is a very important characterization of the analytic structure of the flow; it also provides us with an objective unambiguous definition of the "smallest scale" in the flow (see section 3.1).

It turns out to be possible to quite accurately measure $\delta(t)$ without performing any analytic continuation in space. This is because $\delta(t)$ is related to the (angle-averaged) energy spectrum defined as

$$E(t,K) = \tfrac{1}{2} \sum_{K \le |k| < K+1} |\hat{v}(t,k)|^2, \qquad K = 0, 1, 2, \dots. \tag{2.29}$$

For high K the energy spectrum is essentially $\exp[-2\delta(t)K]$; more precisely, to leading order, the log of the energy spectrum is given by

$$\log E(t,K) \sim -2\delta(t)K, \qquad K \to \infty. \tag{2.30}$$

This relation is derived under the assumption that the spatial analytic continuation $v(t, r + i\varrho)$ is singular on some set $\mathscr{S} \in \mathbb{C}^3$ (mod 2π) such that $\delta(t)$ is the modulus of the imaginary part of the point $r_* + i\varrho_*$ closest to the real domain $\mathbb{R}^3$ (mod 2π). An intermediate step in the proof is to show that when the wavevector k tends to infinity with a fixed direction ($k = kn$, $k \to \infty$) the Fourier transform $\hat{v}(t,k)$ behaves exponentially with a logarithmic decrement which achieves its minimum $\delta(t)$ when n is parallel to ϱ_*. Essentially, the result is a generalization to several complex variables of what we presented in section 2.1.3. (The author is grateful to A. Douady, B. Malgrange and F. Pham for advising him on these matters).

Exponential spectra at high wavenumbers can actually be observed in high resolution numerical simulations of the 3D Euler equation and other equations having similar analytical properties. The appropriate numerical procedure is based on the pseudo-spectral method (Orszag 1971, Gottlieb and Orszag 1977). This method makes use of truncated Fourier series with a large number of modes (resolution of 512^2 modes in 2D and of 64^3 modes in 3D are accessible on existing fast vectorized computers; higher resolutions can be achieved for problems having symmetries). In the pseudo-spectral method evaluation of space derivatives and resolution of Poisson equations are done in Fourier space whereas multiplications are performed in physical space. As a consequence the accuracy with which nonlinear terms are calculated can vary exponentially with resolution (rather than algebraically as in finite difference methods). In high resolution simulations of this kind exponential spectra are easily observed provided the maximum wavenumber is larger than the inverse of the analyticity strip $\delta(t)$ (a condition which may not hold uniformly in time).

As an illustration of the procedure we show in fig. 5 in lin–log coordinates an energy spectrum obtained from numerical integration at resolu-

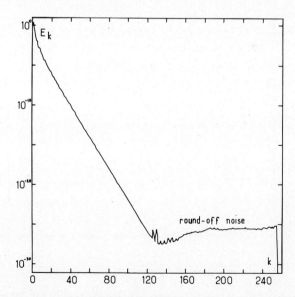

Fig. 5. Kinetic energy spectrum of 2D MHD inviscid flow in lin–log coordinates. Initial stream function and magnetic potential are $\psi_0 = 2\cos x + 2\cos y$ and $a_0 = 2\cos x + \cos 2y$; time is $t = 0.4$; spectral simulation 512^2.

tion 512^2 of an inviscid two dimensional MHD problem with analytic initial conditions (Morf et al. 1981, Frisch et al. 1983). Notice that the exponential tail is very conspicuous, the round-off noise being at a surprisingly low level (stemming probably from good cancellation properties of the FFT algorithm used). It is then possible to accurately measure the logarithmic decrement of the spectrum and, thereby, the width of the analyticity strip $\delta(t)$. The temporal evolution of $\delta(t)$ can then be followed in time. Figure 6 shows a lin–log plot of $\delta(t)$ for the 2D MHD case. The nearly perfect exponential which is obtained suggests that $\delta(t)$ becomes arbitrarily small as time increases but never vanishes, thereby providing evidence that there is no real singularity at a finite time. This result, if cor-

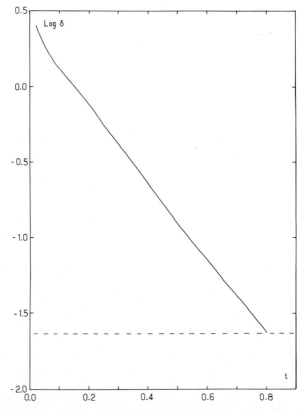

Fig. 6. Temporal variation of width of analyticity strip for the same 2D MHD problem as in fig. 5. Dashed line corresponds to $\delta = 2$ meshes; mesh is $2\pi/512$. Unreliable values of δ are obtained after times shown.

rect, is particularly interesting because it contradicts closure calculations which predict a finite-time singularity (Pouquet 1978). For an explanation of why closure calculations fail to give global regularity, see Frisch et al. (1983). Application of the analyticity strip method to a variety of one-dimensional problems may be found in Sulem et al. (1983).

Let us examine now what the Taylor expansion method and the analyticity strip method have to say about the 3D Euler equation and its singularities for a particular flow.

2.3.3. The Taylor–Green vortex

The Taylor and Green (1937) vortex is a 3D incompressible flow with initial condition

$$v_1 = \cos x_1 \sin x_2 \cos x_3, \tag{2.31}$$

$$v_2 = v_1(x_2, -x_1, x_3), \tag{2.32}$$

$$v_3 = 0. \tag{2.33}$$

Note that the v_3 component does not remain zero for $t > 0$. The Taylor–Green vortex is one of the simplest 3D flows with spatial periodicity and nontrivial dynamics. It is invariant under six discrete symmetry transformations. It is possible to use these symmetries to reduce by a factor 2^6 the amount of storage needed to describe the flow with a given resolution and thereby achieve very high resolutions (Orszag 1980, Brachet et al. 1983). The Taylor–Green vortex has been discussed several times in the literature (Taylor and Green 1937, Orszag 1973, van Dyke 1975, Chorin 1981 and 1982). Here we wish to concentrate on recent results relating to singularities. The Taylor method of section 2.3.1 has been applied by Morf et al. (1980) to generate the Taylor series (2.28) to order t^{44} for various enstrophies. This was done using CDC 7600 double precision (96 bits) and direct evaluation of the convolutions. Notice that, because of the symmetries, the coefficients of odd powers of t in the enstrophies vanish. Morf et al. (1980) find that the radius of convergence of the t^2-series for the enstrophy ($p = 1$) is determined by pure imaginary singularities located at $t^2 \approx -5$, tinuation of the series for Ω by Padé and Dlog Padé approximants suggest a real singularity at $t_* \approx 5.2$. Subsequent calculations reported in Brachet et al. (1982) have pushed the expansion up to t^{80}, using Amdahl quadruple precision (112 bits) and Fourier transform methods. The Padé approximants based on series extending beyond t_* do not support the existence of the $t_* \approx 5.2$ singularity. However, a real singularity, possibly near $t=4$, cannot be ruled out. The problem of analytically continuing the series for the Taylor–Green vortex appears to be much more ill-conditioned than for high temperature series in statistical mechanics. In the latter case, given about 10 nonvanishing coefficients, Padé approximants will typically predict the next coefficient with a 6 digit accuracy whereas for the Taylor–Green series nearly 40 coefficients are required before 6 digits are accurately predicted (Nickel 1981).

It may be that the series for the enstrophies, although perfectly deterministic, have intrinsically chaotic properties. If this is the case, then one expects to find a very complicated analytic structure with natural boundaries and not just isolated singularities (Weiss 1980, Pomeau 1980). Indeed, the simplest example of this is the series

$$\sum_{n \geq 0} a_n t^n$$

with independent random a_n's equal to ± 1, which has the unit circle $|t|=1$ as a natural boundary. In this context it is of interest to mention a result concerning the analytic structure of a Hamiltonian system with only two degrees of freedom. Chang et al. (1982) studied the Hénon–Heiles system governed by the equations

$$\ddot{x} = -x - 2Dxy, \tag{2.34}$$

$$\ddot{y} = -y - 2Dx^2 + Cy^2. \tag{2.35}$$

This system which can have intrinsic chaotic behavior in time has no real-time singularity. Chang et al. (1982) found numerically that from each complex-time singularity emanates a natural boundary with a self-similar fractal structure.

The Euler flow is an infinite dimensional Hamiltonian system (Arnold 1974). If complications similar to the above (or worse) are present in the Taylor–Green-based series then it seems preferable to avoid analytic continuation methods in searching for real singularities.

The analyticity strip method of section 2.3.2 has also been applied to the Taylor–Green vortex at zero viscosity using a 256^3 numerical simulation (Brachet et al. 1983). The results are quite similar to 2D MHD case. The width of the analyticity strip $\delta(t)$ can be accurately measured up to about $t=2.5$. Except for very short times [$\delta(0)$ is infinite because the initial condition is entire] $\delta(t)$ fits onto a decreasing exponential $\delta(t) \propto \exp(-\alpha t)$ with

$\alpha \approx 1.7$. If this behavior is extrapolated to larger times, no real singularity is ever obtained. It is however conceivable that a change-over to another more singular type of behavior takes place beyond $t = 2.5$.

Finally we mention some pecularities and possible pathologies of the Taylor–Green vortex. Of the six discrete symmetries, three can be used in the inviscid case to reduce the flow in the $(2\pi)^3$ periodicity cube to a flow in a π^3 cube with rigid walls (Brachet et al. 1983). As a consequence of the symmetries there are entire lines of vanishing vorticity, such as the x_3 axis. In contrast, a general incompressible flow will have only isolated points of vanishing vorticity. We also note that the vortex lines are closed (and unknotted) and will remain so as long as no sigularity occurs. General three-dimensional flow leaves room for considerably more complicated topologies, possibly affecting the dynamics: In the neighborhood of a closed vortex line the successive intersections of a vortex line with a transverse plane (Poincaré map) define a conservative twist map. A topologically complicated but dynamically trivial example (vorticity and velocity are parallel so that the nonlinear term in the Euler equation vanishes) has been introduced by Arnold (1965) and studied by Hénon (1966).

To conclude this section we stress that the question of the existence of real-time singularities for the 3D Euler flow remains a major challenge. We cannot rule out at present that 3D Euler flow has global regularity. It would then probably be a highly nontrivial consequence of incompressibility. Existing proofs of regularity theorems make very little use of incompressibility. In a somewhat simplified form, what may be at issue can be expressed as follows. Vorticity dynamics involves a "stretchee", the vorticity, and a "stretcher", the strain tensor (symmetrized velocity gradient). Loose identification of stretcher and stretchee makes a singularity unavoidable (the solution of $D\omega/Dt = \omega^2$ blows up in a finite time). Actually, due to incompressibility, the strain tensor has at least one positive and one negative eigenvalue. This may result in the generation of structures which have a very small scale in at least one direction and are very elongated in at least one other perpendicular direction. We then have, locally, an effective reduction in the dimensionality of the problem. For example, if quasi-one-dimensionality obtains, as is the case for two-dimensional inviscid MHD flow (section 2.3.2 and Frisch et al. 1983), then the strength of nonlinear interactions is dramatically reduced (nonlinear interactions vanish for an incompressible one-dimensional flow). Recent flow visualizations in physical space for the inviscid Taylor–Green vortex suggest that this may indeed by the case: as long as the 256^3 numerical simulation remains reliable (i.e. for $t \le 4$) all the small-scale activity seems to concentrate in sheet-like structures (Brachet et al. 1983).

2.4. Navier–Stokes flows

We now briefly discuss viscous flow. More details may be found in Rose and Sulem (1978) and Temam (1977). Viscous flow is a dissipative system which from the viewpoint of regularity should be better behaved than inviscid flow. Actually, the results proven so far do not improve much on known regularity results for the Euler flow. In 2D global regularity and analyticity are known to hold for any viscosity $v > 0$ (Leray 1934, Lions 1969). In 3D Leray (1934) has proved a global existence theorem based on the energy relation (1.6). Regularity, analyticity and uniqueness results for smooth initial data are however known only in the following cases: (i) short times and arbitrary viscosity (Kato 1972); (ii) large viscosity and arbitrary times (Leray 1934); (iii) arbitrary viscosity after sufficient time has elapsed (Leray 1934); (iv) when $v\nabla^2$ is changed into $-v'(-\nabla^2)^\alpha$ with arbitrary $v' > 0$ and a dissipativity $\alpha \ge 5/4$ (Lions 1969, Ladyzenskaya 1963). Note that very weak regularity conditions, such as boundedness of the velocity in L^∞ norm are sufficient to ensure uniqueness (Lions 1969).

Finally, it is possible to show that singularities of 3D Navier–Stokes flow, should they exist, are carried by a very small set. Indeed, the points in four dimensional space–time where the velocity is not bounded has Hausdorff dimension not greater than one (Caffarelli et al. 1982). This improves an earlier result of Scheffer (1976).

The failure to prove an all-time regularity and uniqueness theorem has led Leray (1934) to speculate that turbulent behaviour may be related to lack of uniqueness at certain discrete times. The modern theory of dynamical systems (cf. most of this volume) provides us with an alternative quite different explanation of turbulent behaviour.

Some evidence against real singularities in viscous flow comes from numerical simulations. Flows with no particular symmetries which can be simulated reliably at Reynolds number of a few hundreds, do not appear to develop any singularities. The viscous Taylor–Green vortex (cf. section 2.3.3) has been simulated by Brachet et al. (1983) at a resolution of 256^3. The Reynolds number for this flow is defined as $R = v^{-1}$ [initial scales and velocities are O(1)]. At Reynolds numbers up to about one thousand good evidence is obtained that the flow is analytic for all times. Furthermore, the width of the spatial analytical strip $\delta(t_{max})$ at the time of maximum dissipation is consistent with the Kolmogorov (1941) $R^{-3/4}$ scaling for the dissipation scale. For the highest reliably simulated Reynolds numbers ($R = 1600$ and $R = 3000$) Brachet et al. (1983) found that the energy spectrum $E(t, k)$ can be globally fitted with a function of the form $ck^{-m}\exp(-2\delta k)$. Near the time of maximum energy dissipation the exponent m fluctuates between 1.5 and 2, consistently with the Kolmogorov 5/3 value. The $R = 3000$ simulation also gives a spectrum of dissipation fluctuations which at high wavenumbers follows a power law $k^{-1+\mu}$ ($\mu = 0.5 \pm 0.2$) which is consistent with experimental data on intermittency (Gibson et al. 1970). That such a simple (and somewhat pathological) flow as the Taylor–Green vortex should behave at high Reynolds numbers so much like real world turbulence gives further evidence that there is considerable small-scale universality in fully developed turbulence. We shall now examine in a heuristic way what the basis for this universality may be.

3. The statistical theory

3.1. What is "fully developed" turbulence?

It is tempting to define fully developed turbulence as a flow with arbitrary small-scale motion. This is however a somewhat fuzzy definition. What is a scale? Taking just the inverse of a wavenumber is not appropriate since any non-trivial solution of the Navier–Stokes equation will have some excitation at arbitrary high wave numbers, whatever the value of the Reynolds number. We therefore, define the "smallest scale" as the width of the spatial analyticity strip. We then know that for wavenumbers $k \gg \delta^{-1}$ the energy spectrum decreases exponentially $E(k) \propto e^{-2\delta k}$ (cf. section 2.3.2). The study of the initial-value problem has shown that for Euler (inviscid) flow, if initially $\delta(t) > 0$ it may tend to zero as t increases and may possibly vanish after a finite time (case t_* finite) or take an infinite time to do so (case t_* infinite). For Navier–Stokes (viscous) flow the evidence is that (i) *$\delta(t)$ remains bounded away from zero unformly in time $\delta(t) \geq \Delta(v)$*, (ii) $\Delta(v)$ tends to zero with v. Hence fully developed turbulence is an idealized asymptotic regime which requires case t_* finite: $t > t_*$ and $\nu \downarrow 0$,

case t_* infinite: $t \to \infty$ first and then $\nu \downarrow 0$.

In actual experiments on fully developed turbulence the flow is usually driven and is observed for a time much longer than the time required to establish motion at the smallest scale (this is only a fraction of a second in a wind-tunnel). If, for the case t_* is infinite, we were to interchange the limits $t \to \infty$ and $\nu \downarrow 0$ the dissipation would be zero; thus if the flow is driven, energy would be accumulating indefinitely.

Finally, we stress that it is not a good idea to define fully developed turbulence merely by letting the Reynolds number tend to infinity. Indeed, even at infinite Reynolds numbers, there can be other constraints on the flow which prevent it from developing small scale motion. This is observed for example in numerical simulations of inviscid 2D MHD flow in the pre-presence of a uniform magnetic field of sufficient strength and may be caused by the dispersive action of Alfvén waves (Frisch et al. 1983).

3.2. The statistical description

There is experimental and numerical evidence that the detailed structures of turbulent flow are unpredictable for long times, particularly so for the small scales. So the goal of the theory can be, at best, to predict statistical properties; for example, how the energy spectrum changes as we change the overall characteristics of the flow (shape of boundaries, viscosity, etc.).

For the fundamental justification of the statistical description of turbulence it is possible to invoke

(i) intrinsic stochasticity: for $t \to \infty$ the flow may have one or several attractors (finite or infinite dimensional) with chaotic properties.

(ii) extrinsic stochasticity: the flow is subject to random perturbations (thermal fluctuations, roughness in the boundaries, etc.).

(iii) A mixture of (i) and (ii): weak extrinsic random perturbations can select the "correct" invariant measure on an attractor (Ruelle 1979).

There is at present no deductive statistical theory of turbulence which can handle the dynamics of inertial- and/or dissipation-ranges. Still, there is a considerable amount of qualitative understanding, based on phenomenological approaches and on the study of models related to the Navier-Stokes equations. It is not our purpose to review such material here (see Kraichnan 1959, 1961, 1965, Leslie 1973, Orszag 1973, Monin and Yaglom

1975, Rose and Sulem 1978). Let us also mention that in recent years renormalization methods borrowed from statistical mechanics and quantum field theory (Martin et al. 1973) have been applied to certain large-scale statistical fluid mechanics problems in which external random forcing is generally present (Forster et al. 1977, Fournier and Frisch 1978, DeDominicis and Martin 1979, Fournier et al. 1982, Yakhot 1981, Fournier and Frisch 1982). These problems are set up in such a way that the dynamics at wavenumber k are dominated by interactions with wavenumbers $q \gg k$, the main effect of which is to modify (renormalize) the viscosity. It has been pointed out that there is little hope that inertial- and/or dissipation-range dynamics which involve no direct forcing can be handled by such techniques (Frisch and Morf 1981, Kraichnan 1982).

3.3. The Kolmogorov 1941 self-similar picture

The lack of a systematic theory of turbulence at finite Reynolds numbers does not necessarily preclude a theory of infinite Reynolds number fully developed turbulence. Let us first make an analogy which will eventually turn out to be largely inappropriate. In the classical mechanics of N particles extremely complicated chaotic dynamics can arise with very few particles. Still, the statistical mechanics in the thermodynamic limit are reasonably well understood and appear to be strongly constrained by energy conservation. Many attempts have been made to construct a theory of turbulence along similar lines. The most successful one is the Kolmogorov (1941) theory which we present now with some emphasis on group-theoretical ideas; this will help us in the next section to bring out what may be its major shortcoming.

We first recall that the Navier–Stokes equation is invariant under various groups of transformation (section 1.1), which we refer to globally as symmetries. Turbulent Navier–Stokes flow cannot be deterministically invariant under such symmetries because the flow would be trivial (e.g. space-independent). It can however be statistically invariant. Let us use the symbol $\stackrel{\triangle}{=}$ to denote random functions which are isonomous (i.e. have identical statistical properties). We can have for example statistical time-translation-invariance (stationarity)

$$v(t+\tau,r) \stackrel{\triangle}{=} v(t,r) \qquad \forall \tau \in \mathbb{R}. \tag{3.1}$$

There are similar relations with space-translations (homogeneity), rotations (isotropy), reflections, scaling, etc.

Now, there is a difficulty: in order to maintain a non-vanishing stationary state against dissipation it is necessary to drive the flow. In actual flows the driving may be provided by convective instability, shear flows instability or some other mechanism. Typically the driving will be relevant only at large scales, the so-called energy-carrying scales $\sim l_0$. The driving, since it involves boundaries, preferred directions, etc. will usually be inconsistent with all but a few of the above symmetries. Still, we can have the following

Invariance principle (equivalent to Kolmogorov 1941). In the limit of fully developed turbulence all the symmetries of the Navier–Stokes equation, possibly broken by the driving mechanisms, are recovered asymptotically at small scales in a statistical sense.

Note that it is here implicitly assumed that the limit exists. This implies for example that, if there is a finite rate of energy input there is also a finite rate of energy dissipation. To work out some of the consequences of the invariance principle it is useful to decompose the velocity field into shell-contributions coming from various wavenumbers octave-bands. Let

$$k_{-1}=0, \; k_0=l_0^{-1} \text{ and } k_n=2^n k_0 \; (n \geq 1). \tag{3.2}$$

We define (*t*-dependence implicit)

$$v_n(r) = \int_{k_{n-1} \leq |k| < k_n} e^{ik \cdot r} \hat{v}(k) \mathrm{d}^3 k, \qquad n \geq 0, \tag{3.3}$$

where $\hat{v}(k)$ is the spatial Fourier transform of the velocity field. We denote by P_n the projection $v \mapsto v_n$ combined, if necessary, with a projection onto divergenceless functions. Applying P_n to the Navier–Stokes equation, we find that the v_n's satisfy

$$\partial_t v_n + P_n \sum_{r,s} (v_r \cdot \nabla) v_s = P_n f_n + \nu \nabla^2 v_n; \qquad n,r,s=0,1,2,\dots. \tag{3.4}$$

Note that shell variables can be used to derive rigorous bounds on inertial-range exponents (Sulem and Frisch 1975).

The invariance principle implies that for large n (i.e. at small scales) $v_n(r)$ should be statistically invariant under all the symmetries listed in section 1.1. Homogeneity allows us to define a space-independent mean energy per mass in the nth octave $E_n = \frac{1}{2} \langle v_n^2(r) \rangle$. It is related to the usual energy spectrum by

$$E_n = \int_{k_{n-1}}^{k_n} E(k) \mathrm{d}k. \tag{3.5}$$

The invariance principle also implies that in the limit $v{\downarrow}0$ the velocity field is self-similar (statistically scale-invariant) with arbitrary exponent h. Taking the spatial scaling factor to be one half, this implies

$$v_{n-1}(r) \overset{\triangle}{=} 2^{-h} v_n(r). \tag{3.6}$$

The existence of a finite energy dissipation per unit mass singles out the exponent $h = 1/3$. To prove this we first establish a detailed energy-balance equation. We take the scalar product of the Navier–Stokes equation with v_n, average and sum over n from 0 to N. This gives

$$\sum_{n=0}^{N} [-\tfrac{1}{2} \frac{d}{dt} \langle v_n^2 \gtrless + F_n] = \sum_{n=0}^{N} \left(\sum_{r|s>N} \langle v_n \cdot (v_r \cdot \nabla) v_s \rangle + v \langle |\nabla \times v_n|^2 \rangle \right). \tag{3.7}$$

Here $F_n = \langle v_n \cdot f_n \rangle$ is the energy input, confined essentially to the first few wavenumber-bands. The notation $r|s>N$ means (r and/or s) $>N$ (the contribution with n, r, s all three less than N vanishes because of energy conservation). The first term on the right hand side is called the energy flux through the Nth wavenumber-band and is denoted Π_N. In the stationary state the first term on the left hand side vanishes. The second term on the left hand side is, for N sufficiently large, independent of N and equal to $\bar{\varepsilon}$. The viscous term goes to zero with v, for fixed N. In the inertial-range, defined by first letting $v{\downarrow}0$ and then $N \to \infty$, direct input and direct dissipation are negligible, so that

$$\Pi_N = \bar{\varepsilon} \tag{3.8}$$

We shall assume that, for large N, the energy flux [first term on left hand side or eq. (3.7)] is dominated by the contributions from triads n, r, s which are all three $O(N)$. This is called (Fourier) localness and is consistent with simple phenomenology and experimental data. It is then easily checked that for eq. (3.8) to be consistent with eq. (3.6) we must have

$$h = 1/3. \tag{3.9}$$

Using eq. (3.6), we then obtain the inertial-range scaling laws of moments of the v_n's ($v{\downarrow}0, n \to \infty$)

$$\langle v_n^p \rangle = C_p \bar{\varepsilon}^{p/3} k_n^{-p/3}, \qquad p = 0, 1, 2, \dots. \tag{3.10}$$

The inclusion of the $\bar{\varepsilon}^{p/3}$ factor makes the constants C_p dimensionless. Specializing to $p = 2$, we obtain a relation which by eq. (3.5) is equivalent to the Kolmogorov (1941) energy spectrum ($v{\downarrow}0, k \to \infty$)

$$E(k) = C\bar{\varepsilon}^{2/3} k^{-5/3}. \tag{3.11}$$

It is possible to give much simpler derivations of the energy spectrum using dimensional or phenomenological arguments (Batchelor 1960, Frisch et al. 1978). Similarly, one can obtain the Kolmogorov 1941 expression for the viscous cutoff

$$k_D \sim (\varepsilon / v^3)^{1/4}. \tag{3.12}$$

Note that it is possible to set up a table of analogies between fully developed turbulence and phase transitions (Nelkin 1974, 1975, Rose and Sulem 1978). The Kolmogorov 1941 theory is then the analog of a "mean field" theory. This is probably misleading. In the language of the Renormalization-Group, the self-similar Kolmogorov 1941 picture corresponds to a fixed point (which need not be Gaussian). The breakdown of the Kolmogorov 1941 theory (to be discussed in section 3.4) corresponds to something like a chaos in the recursion relations of the Renormalization Group.

3.4. Breakdown of self-similarity, intermittency and non-compact symmetry groups

Experimental data on high Reynolds number turbulence are in good agreement with small-scale isotropy and the Kolmogorov spectrum. However, the expressions [eq. (3.10)] for moments of order $p > 3$ are not consistent with experimental data. For $p = 4$ measurements (based on a related quantity called the structure function) produce a power-law, but with an exponent 1.2 rather than 4/3; the discrepancy may be even larger for the (somewhat poorly measured) sixth order moment (Van Atta and Park 1972). As a consequence the dimensionless ratio

$$F = \langle v_n^4 \rangle / (\langle v_n^2 \rangle^2), \tag{3.13}$$

called the flatness, can take very large values at small inertial scales. This phenomenon is referred to as "inertial-range intermittency", not to be confused with dissipation-range high-frequency intermittency which was discussed in sections 1.2 and 2.1.

We are now going to argue that this (inertial-range) intermittency may be a consequence of the non-compactness of the group of scaling transformations. For this, let us return to the invariance principle of section 3.3 which we have so far not tried to justify. It is useful to think of the production of high wavenumber excitation as being due to repeated interactions among smaller wavenumbers. In the low-lying imput wavenumbers-bands most of the possible symmetries of the Navier–Stokes equations will usually be broken by the constraints imposed by the driving mechanism. As excitation cascades to higher wavenumber-bands it may lose the information relating to those constraints; particularly so if intrinsic chaos is present in the flow. For example, there can be a preferred direction in the driving. However, as we go to higher wavenumber-bands the velocity v_n can explore all possible dierctions and thereby become isotropically distributed. This is a sort of poor man's ergodic argument which has often been invoked and is based on a loose analogy with statistical mechanics of conservative dynamical systems. The point we wish to stress is that there is no basis for applying this kind of argument to non-compact groups (such as the translation group or the scaling group). To make this clear it is useful to think about the difference between Brownian motion on a sphere (compact) and Brownian motion on the plane (non-compact); in the latter case the Brownian particle just runs away and does not tend to a translation-invariant statistically stationary state.

We are thus led to state a more restricted invariance principle.

Compact invariance principle. In the limit of fully developed turbulence compact symmetries of the Navier–Stokes equations, possibly broken by the driving mechanisms, are recovered asymptotically at small scales in a statistical sense.

Particular cases of compact groups are the finite ones. Consider for example the space reflection $r \mapsto -r$ and $v \mapsto -v$. The compact invariance principle implies that the small scales of 3D turbulence should be not only isotropic but also nonhelical (i.e. statistically invariant under space-reflections). This is consistent with the phenomenological theory of helicity cascades. According to this theory, the relative helicity spectrum $k^{-1}H(k)/E(k)$ tends to zero as $k \to \infty$. $H(k)$ denotes here the spectrum of the helicity of the flow (Brissaud et al. 1973; André and Lesieur 1977).

In 2D turbulence the compact invariance principle implies small-scale isotropy and reflection symmetry. It may be shown that the former implies the latter when the velocity field is incompressible and Gaussian, but not otherwise.

A particularly interesting consequence is obtained for MHD turbulence, the equations of which are invariant under reflection of the magnetic field: unforced decay of a fully developed MHD flow in which there is initially some correlation between velocity and magnetic field leads to a growing correlation coefficient and thereby to very long-lived fossile turbulence (Pouquet et al. 1983).

Finally we turn to the scaling group which is non-compact. Since statistical self-similarity does not hold, rescaling distances by a factor two (which changes v_n into v_{n-1}) is not statistically equivalent to rescaling velocities by a constant factor. The transformation that carries v_n into v_{n+1}, if it exists, can be intrinsically chaotic; this may come, for example, from iteration of a (simple?) nonlinear transformation describing one step in the cascade. Let us model this by a relation of the form

$$v_{n+1}(r) \stackrel{\triangle}{=} a_n v_n(r), \qquad (3.14)$$

where the a_n's are identically distributed random variables with finite moments. For simplicity we shall also assume that they are independent (actually mixing would be enough). By repeated application of eq. (3.14) we obtain

$$\langle v_n^p \rangle \propto \langle a^p \rangle^n, \qquad (3.15)$$

which implies power-law behaviour as a function of wavenumber. In eq. (3.10) the exponent $-p/3$ for the pth moment of v_n becomes $p/3 + \xi_p$. The constraint (3.8), expressing independence of the energy flux on wavenumber, implies $\xi_3 = 0$ and nothing else. The spectrum need not follow the Kolmogorov (1941) law and the flatness will always increase with n unless a is sharp (nonrandom).

The first model of this kind was introduced by Kolmogorov (1962) to cope with intermittency. A whole class of probabilistic models of intermittency with a geometric interpretation in terms of fractals have been studied by Mandelbrot (1975, 1976). Following a suggestion of Kraichnan (1972), it is possible to construct a model with somewhat more dynamical content (Frisch et al. 1978). In this β-model, from one to the next wavenumber octave, the amount of space-filling by "active" eddies goes down by a factor $\beta = 2^{D-3}$, where D is the fractal dimension. Theoretical arguments involving self-avoiding random walks (Hentschel and Procaccia 1982) or distributions of Lyapounov exponents (Ruelle 1983) have been given suggesting that for the β-model $D \geq 2.5$. In the β-model the "anomalous dimension" ξ_p varies linearly with p whereas in the Kolmogorov 1962 lognormal model it varies quadratically. It may be of interest to point out that faster-than-linear variation with p implies a certain degeneracy of the single-time cumulant hierarchy: at high wavenumbers, in the rate of change of the nth order cumulant C_n, interactions involving C_2 and C_{n-1} dominate over all the $C_r - C_s$ interactions with $r + s = n + 1$ $(2 < r < n - 1)$.

The problem with all these models is that self-similarity is broken deliberately by the introduction of *ad hoc* probabilistic elements. It would be of interest to construct purely deterministic models in which the self-similarity is broken by the internal dynamics. As suggested by Siggia (1978), maybe this can be achieved with simple shell models: one starts from eqs. (3.4) for the v_n's which are then simplified by keeping only a finite number of degrees of freedom in each shell and only nearest-neighbor shell interactions. Such models are easily made scale-invariant. The simplest ones have only one degree of freedom v_n per shell and give self-similarity (Novikov and Desnyansky 1974, Bell and Nelkin 1977, 1978). Bell and Nelkin (1977) have shown that the Kolmogorov 1941 behaviour is due to the existence of a stable point in the mapping $z_n \mapsto z_{n+1}$ where $z_n = v_{n+1}/v_n$. Shell models with more degrees of freedom can have more exotic behaviour. This appears to be the case of a shell model related to Burgers's equation which has complex v_n's (Kerr and Siggia 1978) and of certain MHD shell models which behave chaotically (Gloaguen 1981).

3.5. Whither turbulence?

We wish to make a few concluding comments about the theory of fluid turbulence. Systematic scientific investigation was started about one century ago (Reynolds 1883). Very soon the emphasis was on statistical descriptions. Early 20th century turbulence research had a sizeable impact on the development of modern probability theory. More recent developments of turbulence have strongly influenced the theory of chaotic dynamical systems. The latter appears as a sort of anti-probability theory attempting to reconcile 19th century determinism and unpredictability.

We observe on the one hand that it makes little sense to oppose extrinsically and intrinsically stochastic views of fluid turbulence; in most real flows both aspects play a role and must be studied. On the other hand, some false hopes have been generated by methods based on direct (premature) averaging, that is averaging done before one understands what is happening to individual flow structures. Since the equations are nonlinear, averaging is then generally combined with ad hoc procedures which range from simple mixing length approximations (Prandtl 1925) to advanced two-points closures (Kraichnan 1959, Orszag 1973, Rose and Sulem 1978). Direct averaging methods (particularly the simplest ones) have been very useful in applications (from Astrophysics to Nuclear and Aeronautical Engineering); the more elaborate ones have led to predictions of a fundamental nature such as the existence of an inertial-range with an energy cascade from small to large scales in two-dimensional turbulence (Kraichnan 1967, Kraichnan and Montgomery 1980). Still, direct averaging methods capture neither intrinsic stochasticity nor intermittency and may be giving incorrect predictions for singularities in inviscid flows.

This is why in the present Course we have put the emphasis on deterministic aspects of high Reynolds number flows. We have seen that a number of questions which may be relevant to fully developed turbulence, (or may be academic but difficult) such as the existence of singularities, remain essentially open. We believe that particular efforts should be made to study (chaotic) deterministic systems with an infinite number of degrees of freedom either in the form of shell models or of simple partial differential equations. There are however limitations to what we can hope to extract from models of any kind. Indeed, the Navier–Stokes equation can be complicated considerably (by introduction of buoyancy, compressibility, magnetic fields, etc.); it is however unlikely that the Navier–Stokes equation can be much simplified without throwing out something essential. The reason is that there is nothing "irrelevant" in the equation (except, may be, as $\nu \downarrow 0$, the precise nature of the dissipative term). Indeed, in the inviscid limit, for unforced flow, the equation tells us that the acceleration of each fluid particle is just what is needed to keep the flow incompressible.

At the moment, high-resolution numerical simulations seem to be one of the more promising tools for the investigation of inviscid and high Reynolds number flow, particularly where we are interested in the detailed spatial structure of the flow. It is hoped that such work will eventually lead to the construction of a genuine statistical theory in which the individual objects could be coherent structures which must be captured numerically. Alternatively, it may be safer to predict that the future of turbulence research is just as unpredictable as the flow itself.

Acknowledgements

Acknowledgements are made of useful exchanges and discussions with V. Arnold, M. Brachet, J.-D. Fournier, R. Kraichnan, D. Meiron, R. Morf, B. Nickel, S. Orszag, A. Pouquet, B. Mandelbrot, P.-L. Sulem and J. Weiss. The Harvard University and the MIT are thanked for their hospitality during the completion of this manuscript. Part of the numerical work described in this paper was performed on the CRAY 1 of the National Center for Atmospheric Research which is sponsored by the National Science Foundation.

References

Ablowitz, M.J., A. Ramani and H. Segur (1980) J. Math. Phys. 21, 715; 21, 1006.
André, J.-C. and M. Lesieur (1977) J. Fluid Mech. 81, 187.
Aref, H. and N. Pomphrey (1980) Phys. lett. 78A, 297.
Arnold, V. (1965) C.R. Acad. Sci. 261, 17.
Arnold, V. (1966) Ann. Inst. H. Fourier XVI, no. 1, 319.
Arnold, V. (1972) Appl. Math. Mech. 36, 255.
Arnold, V. (1973) Akad. Nauk, Armensk. SSR (Erevan) 229.
Arnold, V. (1974) Mathematical Methods of Classical Mechanics (Russian edition, Nauka, Moscow, French edition, Mir, Moscow, 1976).
Babenko, I.I. and V.V. Petrovich (1979) Dokl. Akad. Nauk SSSR 245, 551 [Sov. Phys. Dokl. 24, 161 (1979)]; Preprint no. 68, Inst. Prikl. Mat., Moscow (1978).
Baker, G.A. Jr. (1975) Essentials of Padé Approximants (Academic Press, New York).
Baker, G.A. Jr. (1981) in: Proc. 1980 Cargèse Summer Institute on Phase Transitions, eds., M. Levy, J.C. Le Guillou, J. Zinn Justin (Plenum, New York).
Bardos, C., S. Benachour and M. Zerner (1976) C.R. Acad. Sci. 282, A 995.
Bardos, C. and S. Benachour (1977) Ann. Sc. Norm. Sup. Pisa, Serie IV 4, 648.
Bardos, C. and U. Frisch (1975) C.R. Acad. Sci. 281, A 775.
Bardos, C. and U. Frisch (1976) in: Turbulence and Navier–Stokes Equation, Orsay, 1975, ed., R. Temam, Lecture Notes in Mathematics, Vol. 565 (Springer, Berlin) p. 1.
Batchelor, G.K. and A.A. Townsend (1949) Proc. Roy. Soc. A199, 238.
Batchelor, G.K. (1960) The Theory of Homogeneous Turbulence (Cambridge University Press, Cambridge).
Bell, T.L. and M. Nelkin (1977) Phys. Fluids 20, 345.
Bell, T.L. and M. Nelkin (1978) J. Fluid Mech. 88, 369.
Betchov, R. (1961) Phys. Fluids 4, 925.
Birkhoff, G. (1962) in: Hydrodynamic Instability, Proc. Symp. Appl. Math. XIII (Am. Math. Soc., Providence) p. 55.
Bountis, T.C. (1982) Chaos in hamiltonian systems and singularities in the complex time plane, in: Proc. AIAA/AAS Astrodynamics Conf. Aug. 9–11, 1982, San Diego, CA (Am. Inst. of Aeron. Astron., New York).

Brachet, M.E. (1983) Thèse de Etat. Observatoire de Nice.

Brachet, M.E., D. Meiron, S.A. Orszag, B. Nickel, R. Morf and U. Frisch (1983) Small scale structure of the Taylor–Green vortex, J. Fluid Mech., in press.

Brissaud, A., U. Frisch, J. Léorat, M. Lesieur and A. Mazure (1973) Phys. Fluids 16, 1366.

Burgers, J.M. (1974) The Nonlinear Diffusion Equation (Reidel, Boston).

Caffarelli, L., R. Kohn and L. Nirenberg (1982) preprint, Courant Institute, New York.

Carrier, G.F., M. Krook and C.E. Pearson (1966) Functions of a Complex Variable (McGraw Hill, New York).

Chang, Y.F., M. Tabor and J. Weiss (1982) J. Math. Phys. 23, 531.

Chorin, A.J. (1981) Commun. Pure Appl. Math. 34, 853.

Chorin, A.J. (1982) Commun. Math. Phys. 83, 517.

De Dominicis, C. and P.C. Martin (1979) Phys. Rev. A19, 419.

Desnyansky, V.N. and E.A. Novikov (1974) Prikl. Mat. Mekh. 38, 507.

Ebin, D.G. (1983) Commun. Pure Appl. Math. 36, 37.

Ebin D.G. and J. Marsden (1970) Ann. Math. 92, 102.

Forster, D., D.R. Nelson and M. Stephen (1977) Phys. Rev. A16, 732.

Fournier, J.-D., P.-L. Sulem and A. Pouquet (1982) J. Phys. A:15, 1393.

Fournier, J.-D. and U. Frisch (1978) Phys. Rev. A17, 747.

Fournier, J.-D. and U. Frisch (1982) Remarks on renormalization group in statistical fluid dynamics, preprint, Observatoire de Nice.

Fournier, J.-D. and U. Frisch (1983) L'équation de Burgers déterministe et statistique, J. Méc. Théor. Appl., in press.

Frisch, U. and A. Brissaud (1971) J. Quant. Spectrosc. Radiat. Transfer 11, 1753.

Frisch, U., M. Lesieur and A. Brissaud (1974) J. Fluid Mech. 65, 145.

Frisch, U., A. Pouquet, P.-L. Sulem and M. Meneguzzi (1983) The dynamics of two dimensional ideal magnetohydrodynamics in: Two Dimensional Turbulence, J. Méc. Théor. Appl. (special issue).

Frisch, U. and R. Morf (1981) Phys. Rev. A23, 2673.

Frisch, U., P.-L. Sulem and M. Nelkin (1978) J. Fluid Mech. 87, 719.

Gagne, Y. (1980) Thesis, Institut de Mécanique de Grenoble, France (unpublished).

Gaunt, D.S. (1981) in: Proc. 1980 Cargèse Summer Institute on Phase Transitions, eds., M. Levy, J.C. Le Guillou, J. Zinn Justin (Plenum, New York).

Gaunt, D.S. and A.J. Guttmann (1974) in: Phase Transitions and Critical Phenomena, eds. C. Domb, M.S. Green (Academic Press, London), vol. 3, p. 181.

Gibson, C.H., G.R. Stegen and S. McConnell (1970) Phys. Fluids 13, 2448.

Gloaguen, C. (1981) preprint, Observatoire de Meudon.

Gottlieb, D. and S.A. Orszag (1977) Numerical analysis of spectral methods (S.I.A.M., Philadelphia).

Grant, H.L., R.W. Stewart and A. Moilliet (1962) J. Fluid Mech. 12, 241.

Hadamard, J. and S. Mandelbrojt (1926) La série de Taylor et son prolongement analytique, Scientia no. 41, vol. 1 (Gauthier-Villars, Paris).

Hentschel, H.G.E. and I. Procaccia (1982) Phys. Rev. Lett. 49, 1158.

Heisenberg, W. (1948) Z. Phys. 124, 628.

Hénon, M. (1966) C.R. Acad. Sci. 262, 312.

Hölder, E. (1933) Math. Z. 37, 327.

Kato, T. (1972) J. Funct. Anal. 9, 296.

Kerr, R.M. and E.D. Siggia (1978) J. Statist. Phys. 19, 543.

Kolmogorov, A.N. (1941) C.R. Acad. Sci. URSS 30, 301, 538.

Kolmogorov, A.N. (1962) J. Fluid Mech. 13, 82.

Kraichnan, R.H. (1959) J. Fluid Mech. 5, 497.

Kraichnan, R.H. (1961) J. Math. Phys. (N.Y.) 2, 124: 3, 205 (E) (1962).

Kraichnan, R.H. (1965) Phys. Fluids 8, 575.

Kraichnan, R.H. (1967) Phys. Fluids 10, 1417.

Kraichnan, R.H. (1972) in: Statistical Mechanics: New Concepts, New Problems, New Applications, eds., J.A. Rice, K.F. Freed, J.C. Light (University of Chicago Press, Chicago), p. 201.

Kraichnan, R.H. (1982) Phys. Rev. A25, 3281.

Kraichnan, R.H. and D. Montgomery (1980) Rep. Prog. Phys. 43, 547.

Ladyzhenskaya, O.A. (1963) The Mathematical Theory of Viscous Incompressible Flow (Gordon and Breach, New York; 2nd Ed. 1969).

Leith, C.E. and R.H. Kraichnan (1972) J. Atm. Sci. 29, 1041.

Leray, J. (1934) Acta Math. 63, 193.

Leslie, D.C. (1973) Developments in the Theory of Turbulence (Clarendon Press, Oxford).

Lichtenstein, L. (1925) Math. Z. 23.

Lions, J.L. (1969) Quelques méthodes de résolution des problèmes aux limites non-linéaires (Dunod-Gauthier-Villars, Paris).

Lorenz, E.N. (1969) Tellus 21, 289.

Mandelbrot, B. (1975) Les Objects Fractals: Forme, Hasard et Dimension (Flammarion, Paris), revised English version: Fractals: Form, Chance and Dimension (Freeman, San Francisco, 1977).

Mandelbrot, B. (1976) in: Turbulence and Navier–Stokes Equation, ed., R. Temam, Lecture Notes in Mathematics (Springer, Berlin) vol. 565, p. 121.

Martin, P.C., E.D. Siggia and H.A. Rose (1973) Phys. Rev. A8, 423.

Meiron, D.I., G.R. Baker and S.A. Orszag 1982, J. Fluid Mech. 114, 283.

Moffatt, H.K. (1969) J. Fluid Mech. 35, 117.

Monin, A.S. and A.M. Yaglom (1975) Statistical Fluid Mechanics, ed., J.L. Lumley (M.I.T. Press, Cambridge), vol. 2; updated and augmented edition of Russian original: Statisticheskaya Gidromekhanika (Nauka, Moscow, 1965).

Moore, D.W. (1979) Proc. Roy. Soc. London Ser. A365, 105.

Moreau, J.J. (1961) C.R. Acad. Sci. 252, 2810.

Morf, R.H., S.A. Orszag and U. Frisch (1980) Phys. Rev. Lett. 44, 572.

Morf, R.H. S.A. Orszag, D.I. Meiron, U. Frisch and M. Meneguzzi (1981)in: Proc. Seventh Intern., Conf. Numeric. Methods in Fluid Dynamics, Lecture Notes in Phys., vol. 141 (Springer, Berlin) p. 292.

Nelkin, M. (1974) Phys. Rev. A9, 388.

Nelkin, M. (1975) Phys. Rev. A11, 1737.

Nickel, B. (1981) private communication.

Novikov, E.A. and Yu. B. Sedov (1978) Sov. Phys. JETP 48, 440.

Onsager, L. (1949) Nuovo Cimento Suppl. 6, 279.

Orszag, S.A. (1971) Stud. Appl. Math. 50, 395.

Orszag, S.A. (1973) Statistical Theory of Turbulence in: Fluid Dynamics, les Houches Summer School 1973, eds., R. Balian, J.L. Peube (Gordon and Breach, 1977) p. 235.

Orszag, S.A. (1980) unpublished.

Pommeau, Y. (1980) private communication.

Pouquet, A. (1978) J. Fluid Mech. 88, 1.

Pouquet, A., M. Meneguzzi and U. Frisch (1983) The growth of correlations in MHD turbulence, preprint, Observatoire de Nice.

Prandtl, L. (1925) Z. Angew. Math. Mech. 5, 136.

Reynolds, O. (1883) Phil. Trans. R. Soc. London 174, 935.

Rose, H.A. and P.-L. Sulem (1978) J. de Phys. (Paris) 39, 441.

Ruelle, D. (1979) Phys. Lett. 72A, 81.

Ruelle, D. (1983) Large volume limit of distribution of characteristic exponents in turbulence, Commun. Math. Phys., in press.

Saffman, P. (1968) in: Topics in Non-Linear Physics, ed., N. Zabusky (Springer) p. 485.

Saffman, P. (1981) J. Fluid Mech. 106, 49.

Siggia, E.D. (1978) Phys. Rev. A17, 3, 1166.

Sulem, C. and P.-L. Sulem (1983) The well posedness of the two dimensional ideal flow, in: Two Dimensional Turbulence, J. Méc. Théor. Appl. (special issue).

Sulem, C., P.-L. Sulem, C. Bardos and U. Frisch (1981) Commun. Math. Phys. 80, 485.

Sulem, C., P.-L. Sulem and H. Frisch (1983) Tracing complex singularities with spectral methods, J. Comput. Phys., 50, xxx.

Sulem, P.-L. and U. Frisch (1975) J. Fluid Mech. 72, 417.

Taylor, G.I. and A.E. Green (1937) Proc. Roy. Soc. London, Ser. A158, 499.

Temam, R. (1977) Navier–Stokes Equation (North-Holland, Amsterdam).

Van Atta, C.W. and J. Park (1972) in: Statistical Models and Turbulence, eds., M. Rosenblatt, C.W. Van Atta (Springer, Berlin) p. 402.

Van Dyke, M.D. (1975) S.I.A.M. J. Appl. Math. 28, 720.

Von Neumann, J. (1949) Collected works 6 (1949–1963) p. 437.

Weiss, J. (1980) private communication.

Weiss, J., M. Tabor and G. Carnevale (1982) The Painlevé property for partial differential equations, preprint La Jolla Institute.

Wolibner, W. (1933) Math. Z. 37, 668.

Yakhot, V. (1981) Phys. Rev. A24, 642.

SEMINARS

[Co-authors of work reported on are indicated in parentheses.]

R. Abraham, Dynamics: The geometry of behaviour.

G. Benettin, Non existence of invariant manifolds in nearly integrable hamiltonian systems, and generalization of results of Poincaré and Fermi.

E. Benoît (with J.L. Callot, F. Diener, M. Diener), La chasse au canard.

M. Campanino (with H. Epstein, D. Ruelle), Existence of the universal map in one dimension.

J. Curry, Lyapunov exponents and Kolmogorov entropy.

J. Crutchfield, A short tour of the dissipative zoo.

R. Douady, Convex billiards.

F. Doveil (with D.F. Escande), Renormalization methods for the onset of stochasticity in two degrees of freedom hamiltonian systems.

U. Frisch, Statistical theories of turbulence.

D. Goroff, Report on Mather's variational principle for quasi-periodic billiard orbits.

M. Hénon, A dissipative quadratic mapping.

M. Herman, Diffeomorphisms of the circle.
Invariant circles for twist maps.

J. Jowett, Accelerators and the beam–beam interaction.

O.E. Lanford, Existence proof for universal maps.

W. Langford, Unfolding degenerate bifurcations.

P. Le Calvez, Birkhoff curves.

R. MacKay, Period doubling in area preserving maps.

G. Mayer-Kress (with H. Haken), Non local stability of attractors.

G. McCreadie (with G. Rowlands, R. Cohen), Diffusion in the standard map.

M. Misiurewicz, On the Lozi mapping.

T. Nadzieja, Non strange attractors.

S. Newhouse, Density of periodic points in area preserving mappings.

J. Nierwetberg (with T. Geisel), Fine structure in the chaotic regime of one dimensional maps.

L. Nüsse, Chaos and no chance to get lost.

A. Ozorio de Almeida, Tori and Wigner functions.

R. Ramaswamy, Irregular spectra and onset of chaos.

P. Rapp, Possible clinical significance of chaotic neural behaviour.

D. Ruelle, Attractors in the presence of noise.

Dynamics of two dimensional vortices.

H. Siegberg (with H. Peters), Chaotic behaviour of differential-delay equations.

J.M. Strelcyn, Billiards as dynamical systems with singularities.

B. Szewc, Smooth invariant measures.

The unstable manifold is dense in the Hénon attractor.

C. Tresser, Chaotic behaviour near homoclinic orbits for three dimensional flows.

S. Ushiki, Piecewise linear hamiltonian systems and chaos.

Analyticity of invariant manifolds and applications.

Bifurcations in Cremona transformations.

A. Voros, Asymptotics of one dimensional quantum problems by complex WKB methods.

C. Wayne (with B. Schraiman and P. Martin), Scaling theory for noisy period doubling transitions and chaos.

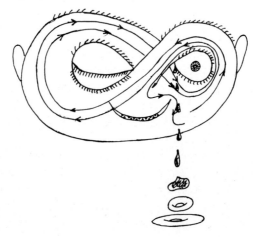

G.M.K.